"Invaluable and inspiring. This up-to-date, accessible, and comprehensive Handbook showcases the breadth of feminist contributions to philosophical studies at the nexus of science, power, social relations, and ethics."

—*Sarah S. Richardson, Harvard University, USA*

"This outstanding volume amasses a wealth of diverse voices and topics within feminist philosophy of science. It will serve as a rich resource for both those curious about how feminism intersects with philosophy of science and those familiar with the field who want to explore its most recent developments and the many fresh topics that have found a home within the field. Readers will find within its pages a cornucopia of insightful reflections on the variety of ways in which feminists have contributed to philosophy of science and harnessed it to take up specific issues of feminist concern."

—*Heidi Grasswick, Middlebury College, USA*

W0113619

THE ROUTLEDGE HANDBOOK OF FEMINIST PHILOSOPHY OF SCIENCE

The Routledge Handbook of Feminist Philosophy of Science is a comprehensive resource for feminist thinking about and in the sciences. Its 33 chapters were written exclusively for this *Handbook* by a group of leading international philosophers as well as scholars in gender studies, women's studies, psychology, economics, and political science.

The chapters of the *Handbook* are organized into four main parts:

    I. Hidden Figures and Historical Critique
    II. Theoretical Frameworks
    III. Key Concepts and Issues
    IV. Feminist Philosophy of Science in Practice.

The chapters in this extensive, fourth part examine the relevance of feminist philosophical thought for a range of scientific and professional disciplines, including biology and biomedical sciences; psychology, cognitive science, and neuroscience; the social sciences; physics; and public policy.

The *Handbook* gives a snapshot of the current state of feminist philosophy of science, allowing students and other newcomers to get up to speed quickly in the subfield and providing a handy reference for many different kinds of researchers.

Sharon Crasnow is a Distinguished Professor of Philosophy Emerita, Norco College in Norco, California, USA and Associate Researcher at the Centre for Humanities Engaging Science and Society (CHESS) at Durham University, UK. She co-edited *Out from the Shadows: Analytical Feminist Contributions to Traditional Philosophy* (2012). She currently co-edits the Lexington book series *Feminist Strategies*.

Kristen Intemann is a Professor of Philosophy and the Director for the Center for Science, Technology, Ethics & Society at Montana State University, USA. Her recent book, co-authored with Inmaculada de Melo-Martin, is *The Fight Against Doubt: How To Bridge the Gap Between Scientists and the Public* (2018).

ROUTLEDGE HANDBOOKS IN PHILOSOPHY

Routledge Handbooks in Philosophy are state-of-the-art surveys of emerging, newly refreshed, and important fields in philosophy, providing accessible yet thorough assessments of key problems, themes, thinkers, and recent developments in research.

All chapters for each volume are specially commissioned, and written by leading scholars in the field. Carefully edited and organized, *Routledge Handbooks in Philosophy* provide indispensable reference tools for students and researchers seeking a comprehensive overview of new and exciting topics in philosophy. They are also valuable teaching resources as accompaniments to textbooks, anthologies, and research-orientated publications.

Also available:

THE ROUTLEDGE HANDBOOK OF PHENOMENOLOGY OF AGENCY
Edited by Christopher Erhard and Tobias Keiling

THE ROUTLEDGE HANDBOOK OF FEMINIST PHILOSOPHY OF SCIENCE
Edited by Sharon Crasnow and Kristen Intemann

THE ROUTLEDGE HANDBOOK OF LINGUISTIC REFERENCE
Edited by Stephen Biggs and Heimir Geirsson

THE ROUTLEDGE HANDBOOK OF DEHUMANIZATION
Edited by Maria Kronfeldner

THE ROUTLEDGE HANDBOOK OF ANARCHY AND ANARCHIST THOUGHT
Edited by Gary Chartier and Chad Van Schoelandt

For more information about this series, please visit: https://www.routledge.com/Routledge-Handbooks-in-Philosophy/book-series/RHP

THE ROUTLEDGE HANDBOOK OF FEMINIST PHILOSOPHY OF SCIENCE

Edited by Sharon Crasnow and Kristen Intemann

Routledge
Taylor & Francis Group

NEW YORK AND LONDON

First published 2021
by Routledge
52 Vanderbilt Avenue, New York, NY 10017

and by Routledge
2 Park Square, Milton Park, Abingdon, Oxon, OX14 4RN

Routledge is an imprint of the Taylor & Francis Group, an informa business

© 2021 Taylor & Francis

Library of Congress Cataloging-in-Publication Data
A catalog record for this title has been requested

ISBN: 978-1-138-57985-9 (hbk)
ISBN: 978-0-429-50773-1 (ebk)

Typeset in Bembo
by codeMantra

CONTENTS

Contents

Contents

ILLUSTRATIONS

Figures

Table

CONTRIBUTORS

Drucilla K. Barker is Professor in the Department of Anthropology and the Women's and Gender Studies Program at the University of South Carolina, USA. She is a Marxist, feminist economist whose research interests are globalization, feminist political economy, feminist epistemology, and economic anthropology.

Vanessa Bentley is an Assistant Professor in the Department of Humanities and Philosophy at the University of Central Oklahoma. Her research connects philosophy of science and feminist epistemology to cognitive neuroscience to address the ethical implications of neuroscience (neuroethics), assumptions and practices in cognitive neuroscience, and issues in sex and gender research.

Robyn Bluhm is Associate Professor in the Department of Philosophy and Lyman Briggs College at Michigan State University, USA. She has recently edited *Knowing and Acting in Medicine* (2016) and co-edited *The Bloomsbury Companion to Philosophy of Psychiatry* (2019).

Kirstin Borgerson is Associate Professor in the Department of Philosophy and Gender and Women Studies Program, at Dalhousie University, Canada. Her research investigates issues arising at the intersection of epistemology and ethics in medicine.

Maria Botero is Associate Professor in the Psychology and Philosophy Department at Sam Houston State University, USA. Most of her research and publications have focused on an interdisciplinary approach to understand the development of social cognition in primates and on examining the methods used to observe the non-human primate mind.

Evelyn Brister is Professor in the Philosophy Department at the Rochester Institute of Technology, USA. She is the editor, with Robert Frodeman, of *A Guide to Field Philosophy* (2020), a collection of essays examining collaborations between philosophers and policymakers.

Sharon Crasnow is a Distinguished Professor of Philosophy Emerita, Norco College, USA, and an Associate Researcher at the Centre for Humanities Engaging Science and Society (CHESS) at Durham University, UK. She co-edited *Out from the Shadows: Analytical Feminist Contributions to Traditional Philosophy* (2012).

Sharyn Clough is Professor of Philosophy at Oregon State University, USA, where in addition to teaching courses on science and politics, she directs Phronesis Lab, designing and assessing strategies for teaching peace as phronesis (Peace Literacy) across the curriculum at all levels of education.

Catherine Clune-Taylor is Assistant Professor in the Department of Women's Studies at San Diego State University, USA. She has published articles on the management of intersex conditions in *Hypatia* and *PhaenEx* and is currently at work on a book titled *Securing Cisgendered Futures: Managing Gender in the 21st Century*.

Deepanwita Dasgupta is Assistant Professor of Philosophy at the University of Texas at El Paso, USA. Her research is in philosophy of science with an emphasis on science in non-Western contexts. Her forthcoming book analyzes a set of case studies from early 20th century India to offer a model of conceptual creativity for scientists who work from a periphery.

Inmaculada de Melo-Martín is Professor of Medical Ethics in the Division of Medical Ethics at Weill Cornell Medical College, USA. Her latest books are *Rethinking Reprogenetics* (2017) and with K. Intemann, *The Fight against Doubt* (2018).

Karen Detlefsen is Professor of Philosophy and Education at the University of Pennsylvania, USA. Her research focuses on early modern philosophy, including the history of philosophy of science, the history and philosophy of education. She has held research grants from The National Science Foundation, The American Council of Learned Societies, and the National Endowment for the Humanities.

Nicole M. Else-Quest is Associate Professor of Women's and Gender Studies at the University of North Carolina at Chapel Hill, USA. A fellow of the American Psychological Association (APA), her research focuses on examining gender equity and intersectionality in psychological research and educational interventions to increase women's participation in science.

Sara Giordano is Assistant Professor at Kennesaw State University, USA, specializing in feminist science studies. They are co-founder of HATCH: Feminist Arts and Science Shop at UC Davis. Their full-length manuscript, "Labs of Our Own: Post/Feminist Tinkerings with Science" is under review.

Maya J. Goldenberg is Associate Professor of Philosophy at the University of Guelph, Canada, and author of the forthcoming book *A Crisis of Trust: Vaccine Hesitancy, Values, and Public Debate* (2020).

Kristina Gupta is Assistant Professor in the Department of Women's, Gender, and Sexuality Studies at Wake Forest University, USA. She is the author of *Medical Entanglements: Rethinking Feminist Debates About Healthcare* (2019) and co-editor of *Queer Feminist Science Studies: A Reader* (2017).

Sandra Harding is Distinguished Research Professor Emerita, University of California, Los Angeles, USA. She co-edited *Signs: Journal of Women in Culture and Society* 2000–2005. Her most recent books are *Objectivity and Diversity: Another Logic of Scientific Research* (2015) and *Sciences from Below: Feminisms, Postcolonialities, and Modernities* (2008).

Daniel J. Hicks (they/them) is a philosopher of science and Assistant Professor of Cognitive and Information Sciences at the University of California, Merced, USA. They work primarily on science and values, but also have interest in using computational methods in philosophy.

Catherine Hundleby is Associate Professor of Philosophy at the University of Windsor, Canada, cross-appointed to the Women's and Gender Studies Program, and Director of the Interdisciplinary PhD Program in Argumentation Studies. Her latest article, co-authored with Moira Howes, is available open access: "The Epistemology of Anger in Argumentation," 2018, *Symposion* 5(2): 229–254.

Janet Shibley Hyde is Evjue-Bascom Professor and Helen Thompson Woolley Professor of Psychology and Gender and Women's Studies, and the Director of the Center for Research on Gender and Women at the University of Wisconsin—Madison, USA. A past president of APA Division 35: Society for the Psychology of Women and a fellow of APA and the American Association for the Advancement of Science, her research focuses on gender-role development and the psychology of women.

Kristen Intemann is Professor of Philosophy in the Department of History & Philosophy and serves as the Director for the Center for Science, Technology, Ethics, and Society at Montana State University, USA. Her recent book, co-authored with Inmaculada de Melo-Martin is *The Fight against Doubt: How To Bridge the Gap Between Scientists and the Public* (2018).

Ian James Kidd is Assistant Professor of Philosophy at the University of Nottingham, UK. He works mainly in epistemology, philosophy of illness and healthcare, and philosophy of science, and co-edited (with Gaile Pohlhaus, Jr. and José Medina) *The Routledge Handbook to Epistemic Injustice* (2012) and is currently co-editing (with Heather Battaly and Quassim Cassam) *Vice Epistemology* (2020).

Janet Kourany is Associate Professor of Philosophy and concurrent Associate Professor of Gender Studies at the University of Notre Dame, USA, where she is also a fellow of the John J. Reilly Center for Science, Technology, and Values. Her latest book, *Science and the Production of Ignorance: When the Quest for Knowledge Is Thwarted,* is co-edited with Martin Carrier (2020).

Edith Kuiper is Associate Professor in the Economics Department and in the Department of Women's, Gender, and Sexuality Studies at the State University of New York at New Paltz, USA. Her research is on feminist history and philosophy of economics.

Marcy P. Lascano is Professor of Philosophy focusing on early modern women philosophers at the University of Kansas, USA. She is the co-editor, with Eileen O'Neill of *Feminist History of Philosophy: The Recovery and Evaluation of Women's Philosophical Thought* (2019).

Amy G. Mazur is CO Johnson Professor in Political Science at Washington State University, USA, and an Associate Researcher at Sciences Po, Paris, France. She is currently co convening the Gender Equality Policy in Practice Network and is editor of *French Politics*.

Nancy Arden McHugh is Professor and Chair of Philosophy at Wittenberg University, USA, a Fellow of the Hagen Center for Civic and Urban Engagement, and the author of *The Limits of Knowledge: Generating Pragmatist Feminist Cases for Situated Knowing* (2015).

Zahra Meghani is a Philosophy Professor at the University of Rhode Island, USA. She edited *Women Migrant Workers: Ethical, Political and Legal Problems* (2015). She specializes in ethics (especially, health care ethics and environmental ethics), social and political philosophy, and feminist philosophy (including feminist epistemology and feminist philosophy of science).

Breny Mendoza is Professor and former Chair of the Department of Gender and Women's Studies at California State University, Northridge, USA. Her most recent publications are *Ensayos de Crítica Feminista en Nuestra América*, 2014, and "Can the Subaltern Save Us?," 2019, *Tapuya: Latin American Science, Technology and Society.*

Letitia Meynell is Associate Professor of Philosophy at Dalhousie University, USA, specializing in feminist philosophy of science. She has co-edited two collections, *Embodiment and Agency* (Penn State UP, 2009) and *Thought Experiments in Philosophy, Science and the Arts* (2012) and recently co-authored *Chimpanzee Rights: The Philosophers' Brief* (2019).

Lynn Hankinson Nelson is Professor Emerita of Philosophy at the University of Washington, USA. She is co-editor with A. Wylie of a Special Issue of *Hypatia* on Feminist Science Studies (2004); her most recent book is *Biology and Feminism: A Philosophical Introduction* (2017).

Manuela Fernández Pinto is Associate Professor in the Department of Philosophy and the Center for Applied Ethics at Universidad de Los Andes, Colombia. She is also an Affiliated Researcher in the TINT Centre for Philosophy of Social Science at the University of Helsinki, Finland.

Donna Riley is Kamyar Haghighi Head and Professor in the School of Engineering Education and at Purdue University, USA. She is the author of *Engineering and Social Justice* and *Engineering Thermodynamics and 21st Century Energy Problems,* both published by Morgan and Claypool.

Kristina Rolin is Research Fellow at Helsinki Collegium for Advanced Studies and University Lecturer in Research Ethics at Tampere University, Finland.

David A. Rubin is Associate Professor in the Department of Women's and Gender Studies at University of South Florida, USA, the author of *Intersex Matters: Biomedical Embodiment, Gender Regulation, and Transnational Activism* (2017) and co-editor of *Queer Feminist Science Studies: A Reader* (2017).

Laura Ruetsche is Louis E. Loeb Collegiate Professor of Philosophy at the University of Michigan, USA. Her *Interpreting Quantum Theories* (2011) was co-recipient of the 2013 Lakatos Award.

Dr. Shelley Lynn Tremain publishes on (feminist) philosophy of disability, Michel Foucault, ableism in philosophy, social metaphysics and epistemology, and bioethics. Tremain is the author of *Foucault and Feminist Philosophy of Disability* (2017), which received the 2016 Tobin Siebers Prize for Disability Studies in the Humanities. She was the 2016 recipient of the Tanis Doe Award for Disability Study and Culture in Canada.

Joanne Waugh is American Foundation of Greek Language and Culture Professor of Greek Culture and Associate Professor of Philosophy at the University of South Florida, USA. She is co-editor of the Book Series "Feminist Strategies" and of the book, *Philosophical Feminism and Popular Culture* (2012). She is co-author with Roger Ariew of "The Contingency of Philosophical

Problems," in *Philosophy and Its History* (2013), and "The History of Philosophy and the Philosophy of Science," in *The Routledge Companion to Philosophy of Science* (2008).

Kyle Whyte is Professor of Philosophy and Community Sustainability and Timnick Chair at Michigan State University, USA. His research addresses moral and political issues concerning climate policy and Indigenous peoples, the ethics of cooperative relationships between Indigenous peoples and science organizations, and problems of Indigenous justice in public and academic discussions of food sovereignty, environmental justice, and the anthropocene. He is an enrolled member of the Citizen Potawatomi Nation.

INTRODUCTION TO *THE ROUTLEDGE HANDBOOK OF FEMINIST PHILOSOPHY OF SCIENCE*

Sharon Crasnow and Kristen Intemann

The body of philosophical literature that might be called feminist philosophy of science has burgeoned over the past 40 years. Yet, there has not yet been a handbook devoted to this subfield. Rather, feminist philosophy of science is often represented by single chapters in both handbooks on philosophy of science, and handbooks in feminist philosophy. Both the growth in this field and the lack of a handbook that demonstrates the breadth and depth of work being done grounded the need for this project. In addition, the enthusiasm with which the idea was received by those we approached – both those who are contributors and others who were so kind as to advise us – further validated our own sense that there was a rich and distinct field of inquiry from which we would be able to provide a representative sample. As with any project like this, there are gaps, some that we are aware of and likely others of which we are not. We believe that there are also some surprises – topics not addressed elsewhere that are included here. We apologize for the lacunae that readers may find and also hope that future iterations of the handbook will be able to fill some of them and that the further suggested readings included in many of the chapters will aid those in search of more information.

In this introduction, we explain how we have conceived of *feminist philosophy of science*, offer some historical background on the subfield, as well as provide an overview of the organization and main themes of the handbook, indicating how the various chapters address the issues for each of these sections. We hope this framework will guide you in making good use of the handbook.

What Is "feminist philosophy of science?"

What constitutes "feminist philosophy of science" may well depend on how we conceive of "feminism" and "philosophy of science." Philosophy of science, itself, has often been treated in analytic philosophy as a subfield of epistemology, or the study of knowledge. Indeed, analytic epistemology often took the sciences as the paradigm instance of knowledge production and aimed to understand (and in some cases logically reconstruct) what makes scientific knowledge a form of knowledge.

Similarly, much of the literature during the 1990s discusses feminist epistemology and feminist philosophy of science as if they were equivalent. This is misleading in some ways, since it is probably more correct to acknowledge epistemology as the broader category. If an epistemology provides a theory of knowledge, then a philosophy of science might be thought of as including, but not equal to, a theory of *scientific* knowledge (since there are also, for example, questions of ontology that are addressed in philosophy of science). In addition, there are other sorts of knowledge besides scientific knowledge; thus philosophy of science might be thought of as a subgenre of epistemology, but not its whole. In many discussions of feminist philosophy of science, however, feminist philosophy of science and feminist epistemology are treated as if they were equivalent or nearly so.

Some feminist scholars have characterized feminist epistemology as a branch of social epistemology that examines the ways in which power dynamics and social relations influence the production of knowledge (Anderson 1995; Grasswick and Webb 2002). Feminist epistemologists maintain that gender (necessarily in combination with other, intersecting, social categories such as race, class, sexual orientation, and ability) is one of the axes along which power is distributed in society and has potential to influence our epistemic practices and judgments. Feminist epistemologists seek to understand not only *how* our social relations of gender have shaped our knowledge practices but also whether and how these relations *should* play a role in responsible epistemic judgments.

Scientific communities are clearly one type of knowledge producing community and one where certain groups have been historically excluded. Thus, certainly some feminist philosophy of science is a subset of feminist epistemology. That is, feminist philosophy of science involves the study of how intersecting systems of oppression influence the production of scientific knowledge and the development of normative recommendations for how scientific practices and methodologies might better serve our epistemic aims while also producing the kind of knowledge and practices that might aid in achieving social justice.

Nonetheless, feminist epistemologists have also been concerned with aspects of knowledge production outside of a scientific context, and some feminist philosophy of science has been attentive to issues in science that are not only *epistemological*. Work on various forms of epistemic injustice that arise in non-scientific contexts illustrates the latter point. The former is evidenced by the re-examination of key scientific, metaphysical, and ethical concepts that are seen throughout feminist philosophy of science and will be discussed in more detail below.

Our understanding of philosophy of science is thus that while it includes epistemological questions, science also raises a variety of metaphysical, ethical, and policy questions. For example, what are these things that we are studying, how are they being classified and might they be conceived differently? There are questions of ethics, including the ethical obligations of researchers and the ethics of the subjects researched or prioritized. There are also policy questions that have to do with the relationship between the knowledge sought and the purposes for which is used. Which questions need to be answered by science and how should those answers affect what we do in the world? And indeed there are other ways that values are relevant to thinking about science and how they might or perhaps even should shape scientific knowledge. Consequently, although feminist epistemology and philosophy of science are often conflated, there are good reasons to think of them as distinct. Thus, we interpret philosophy of science broadly to include any philosophical question that can be asked about scientific knowledge, scientific theories, scientific practice, scientifically informed policy, or the make-up of scientific communities as well as the relationship among any and all of these. Several chapters in this volume note the ways in which philosophy of science may be deeply connected with work being done in science and technology studies more broadly from a variety of disciplinary perspectives.

As noted earlier, "feminist philosophy of science" has been understood in different and contested ways not only because what constitutes "philosophy of science" is ambiguous. The term "feminist" itself has been alternatively employed to refer to a political identity, as a struggle to end sexist oppression, as a movement to achieve social justice for *all*, and as a methodological approach to analyzing phenomena. As conceptions of feminism have changed and branched out, so too has the notion of "feminist philosophy of science." Feminist theorists have developed different accounts of oppression, different analyses of the causes of oppression, and different conceptions of what the ideal society might look like, as well as what policies might bring that about. Within feminist philosophy of science, this diverse set of theoretical tools have been brought to bear to consider the ways in which scientific practices might perpetuate oppression or might be put to liberatory ends. Feminist work, in particular, has challenged traditional notions of what *counts as* "science," "knowledge," or scientific "expertise," how we ought to understand the way science works (or best works), and what sorts of methodological approaches will best serve scientific aims. In challenging and critically evaluating these sorts of assumptions, feminist scholars have identified and pursued new lines of inquiry in philosophy of science and epistemology that have often been neglected or ignored by traditional philosophers of science (and hence are often not included in other philosophy of science handbooks).

Thus, just as we have aimed to conceive of philosophy of science broadly, we have similarly interpreted what it is for a philosophy of science to be feminist inclusively. Minimally, an approach is *feminist* if it committed to equal rights and privileges for women. In addition, given that gender distinctions *do* differentially structure the lives of individuals in human societies, feminism directs us to be alert to such effects and how they might affect science. However, there is good reason to believe that these egalitarian goals should not be confined to women, for a variety of reasons. First, because there are other vectors of oppression linked to social categories that affect women differentially. Second, the category "woman" is a social gender category that is itself open to dispute. Finally, the relationships among social categories insofar as they affect freedoms are not well-understood and vary with context. Consequently, we believe that an egalitarian commitment requires the recognition that the goals of various emancipatory movements are entangled in ways that should always be open to investigation. The commitment to egalitarian goals for women means that questions on whether, in which ways, and to what extent gender matters always need to be asked – but because of the entanglement it means that other social categories and how they intersect with gender to enforce oppression should also always be among the questions we ask. This orientation is reflected in the contributions to this handbook.

A Brief History

Feminist philosophy of science, while a relatively new subdiscipline in philosophy of science, does nonetheless have a history. The influx of women into the sciences during the twentieth century is probably partly responsible for the appearance of modes of thinking that came to be called feminist philosophy of science. This change in the scientific workforce coincided with women's suffrage movements in democracies of the industrialized north and the greater participation of women in public life that had begun during the nineteenth century. This is not to claim that women were absent from scientific endeavors prior to this period. Whether they were fully acknowledged or even whether their involvement was in activities that were called science, they were present and active in the production of knowledge about the world in many ways. But it is from the mid-twentieth century onward that we see feminists raise questions about scientific knowledge – questions about who does it, what it studies, and how the knowledge produced is used to effect changes in the world.

The effort to consciously address both science and the philosophy of science from a feminist perspective gathered momentum in the early 1980s although some work appears during the prior decade particularly with feminist contributions to anthropology and biology. Ruth Hubbard (1983/2003) notes the work of Elaine Morgan (1973) and Sarah Blaffer Hrdy (1981) as examples of the latter. Hrdy's earlier work on primatology (1977) also qualifies (see Nelson, Chapter 20 of this volume for more discussion). Ruth Bleier's work is another example (1984). In 1983, Sandra Harding and Merrill B. Hintikka's edited collection, *Discovering Reality: Feminist Perspectives on Epistemology, Metaphysics, Methodology, and Philosophy of Science*, was among the first explicitly philosophical publications. *Discovering Reality* had originally been proposed as a special issue for the philosophy journal *Synthese*; however resistance from the editorial board (all men), some of whom did not believe such work qualified as philosophy, quashed that goal. Jaako Hintikka supported its publication in book form and consequently it made its way into the world.[1]

Other early work in feminist philosophy of science comes from both those engaged in biology and the social sciences and philosophers who were in conversation with them. Reflections on the inadequacies of the tools and frameworks that these scientific disciplines offered for studying the phenomena in which feminists were interested motivated both the critique that science was often shaped by unexamined androcentric assumptions and the search for alternative approaches.

Harding's *The Science Question in Feminism* (1986) influentially sketched out three options along these lines: feminist empiricism, feminist standpoint theory, and feminist postmodernism. Harding's interpretation of the sociologist Dorothy Smith's discussion of "what it would mean to construct a sociology from the 'standpoint of women'" illustrates the effort to ground philosophical accounts in the work of practicing feminist scientists (Harding 1986: 155).

Nancy Tuana divided her 1989 edited collection, *Feminism & Science*, into three parts: a brief Overview, Feminist Theories of Science, and Feminist Critiques of the Practice of Science. The second and third parts reflect the dual interests of feminist thinking about science during this period – interests that are still present in the field and that we also reflect in this handbook. Essays by Evelyn Fox Keller, trained in physics and molecular biology, and Ruth Hubbard, the first woman tenured in biology at Harvard University, are included in Tuana's collection, with those of philosophers again indicating an ongoing interdisciplinary interchange of ideas.

During the latter half of the twentieth century feminist philosophy of science took a defensive turn as the "Science Wars" broke out. Many of those who thought that they were protecting science from the threat of "radical relativism" targeted feminist philosophy of science, which they saw as particularly pernicious.[2] Detractors challenged the possibility of objectivity for a feminist science arguing that the political influence of the feminist position would distort science and subvert its aims. In response, many feminists who were focused on science during the late 1990s devoted their efforts to reconceptualizing objectivity so that it could be understood as consistent with the recognition of science as social and that political/social/cultural values may play both legitimate and illegitimate roles in knowledge production. A variety of approaches were developed during this period. Harding's notion of "strong objectivity" which appears in a number of publications in the 1990s and perhaps most thoroughly in *Whose Science? Whose Knowledge?* (1991) is one such account. Strong objectivity requires reflection on the means of producing knowledge, so that they are subject to the same scrutiny that the knowledge produced is (see Chapter 7).

Helen Longino's approach to objectivity proposes a set of requirements for the social context in which knowledge is produced (Longino 1990). Such contexts are more likely to allow for the identification of background assumptions that might otherwise go unexamined. She labels the approach "contextual empiricism" and so self-identifies as being in the empiricist tradition (see Chapter 6).

While postmodernist approaches are less represented in this handbook than we hoped for (it is one of the aforementioned lacunae), the question of objectivity has also been addressed by feminist philosophers of science who might be classified in this way. Donna Haraway's notion of situated knowledges (1988) calls for greater objectivity through acknowledging the reality of the knowers' situatedness. She argues that feminists as well as their challengers have wrongly seen their options as either objectivity in the sense of a God's eye view (unobtainable) or relativism as all viewpoints being equally as good (undesirable). She argues instead for recognizing that "[t]he alternative to relativism is partial, locatable, critical knowledges sustaining the possibility of webs of connections called solidarity in politics and shared conversations in epistemology" (Haraway 1988: 585). More recently, Karen Barad's agential realism reflects a similar rejection of the objectivity/

relativity dichotomy when she states, "We don't obtain knowledge by standing outside the world; we know because we are of the world" (Barad 2007: 185).

It is perhaps this focus on objectivity that has pushed the theoretical side of work in feminist philosophy of science toward a merging of approaches so that Harding's tripartite distinction no longer has the force that it once did. But the debate about objectivity is not all there is to feminist philosophy of science. The practice of science – how problems are conceptualized and research questions asked, how labs are organized, how results are reported – and the intersection of science policy with feminist concerns have also continued to engage feminist science scholars. It is this range of issues within philosophy of science that we have aimed at representing through the chapters of this handbook.

Organization

We have organized this handbook into four parts. Part I – Hidden Figures and Historical Critique – includes chapters by Joanne Waugh, Marcy Lascano, Karen Detlefsen, Deepanwita Dasgupta and Evelyn Brister and Dan Hicks. One of the central questions that feminist theorists have analyzed is whether it matters, from an epistemic perspective, *who* participates in science and/or philosophy of science. Certain groups, including women, have been historically underrepresented in both science and philosophy. In some cases, even when women made contributions their contributions were ignored, not viewed *as* contributions, or were appropriated by others. The chapters in this section highlight the ways in which women philosophers and scientists have made contributions throughout history in ways that were not previously recognized. In some cases they also touch on the factors that make these contributions difficult to recover or cause them to recede from the public record. Chapters by Waugh, Lascano, and Detlefsen demonstrate the ways in which women scientists and philosophers critiqued assumptions, methodologies, and experiments that dominated from the Ancient to the Modern period. Waugh's chapter considers some of the difficulties involved in discussing philosophy, science, and feminism in the context of Ancient Greece and how these difficulties are exacerbated by the way philosophy developed so that it sought to address universal and timeless questions – something that she suggests is not possible. She sketches the connection between this disregard for the material circumstances within which knowledge is produced and the way the material differences among human bodies have been dismissed as relevant for knowledge as well. Lascano and Defletsen recover the work of women whose engagement in both scientific discourse and practice was long ignored. Yet even more recently as women's participation has grown in STEM fields, their contributions have not received the attention they have perhaps warranted. Dasgupta demonstrates the vital role that women scientists and engineers in India played in the Indian Space Research Organisation and in particular the 2014 Mars Orbiter Mission. Although celebrated in India and producing positive effects for girls and women interested in pursuing science, the story is not as widely known elsewhere illustrating the multiple axes through which access to knowledge operates. The chapter by Evelyn Brister and Dan Hicks examines the influence of women's contributions to the philosophy of science over time and the potential impacts of their work and for the field. These chapters are varied and are not intended to represent a comprehensive overview of women's contributions to science throughout history; nonetheless we take them to be important examples of the kind of work that feminist philosophers of science and historians of science are doing to both reveal the contributions of women scholars and researchers and develop the normative implications that this may have for diversity in epistemic communities.

Part II – Frameworks – examines the development of various theoretical approaches to feminist philosophy of science. While Part I explores some of those who have not been acknowledged as contributors to scientific knowledge, the chapters of Part II theorize both the epistemological loss that results from the exclusion of knowers and the potential benefits of epistemic pluralism – a pluralism of both knowers and approaches. The chapters in this section include contributions that explain and develop feminist approaches to science and contributions that draw on theoretical frameworks previously neglected by feminist scholars to show how they might further advance issues in philosophy of science. We begin with chapters by Kirstin Borgerson and Catherine Hundleby that provide an overview of two dominant theories that Sandra Harding originally identified as feminist approaches to science: feminist empiricism and feminist standpoint theory. But, as these chapters show, these views have themselves underdone changes and taken new directions in recent years. Moreover, the changing theoretical terrain in both feminist theory and in philosophy of science has given rise to new theoretical approaches such as those that build on literature from decolonial philosophy (Harding and Mendoza), indigenous philosophy of science (Whyte), queer science studies (Gupta and Rubin), critical disability theory (Tremain), virtue epistemology (Kidd), and pragmatist epistemology (Clough and McHugh). While we do not intend this list to be exhaustive, it provides readers with an appreciation of the rich and diverse theoretical terrain that has emerged since Harding's work in the 1980s. Indeed, some of these approaches have developed (and continue to develop) out of a critique of feminist philosophy itself. Shelley Tremain, for example, shows the ways in which feminist philosophers (and feminist philosophers of science) have neglected disability and the ways in which their work intersected with considerations of disability. Harding and Mendoza demonstrate the innovative contributions of Latin American feminist philosophy that have been largely invisible in international contexts, in part because of the history of colonization, but that nonetheless offer important resources for decolonizing science and knowledge. Whyte shows how indigenous conceptions of knowledge and science are inherently intertwined with moral qualities – specifically the moral quality of consent. He shows how various indigenous philosophies of science have understood knowledge production as involving a respect for the self-determination of all entities and achieving consensus for the common good through mutual communication. We take the development of internal critique to be an indication of the maturity and robustness of feminist philosophy of science. All of the chapters in this section provide rich theoretical tools and new resources for the future of feminist philosophy of science.

Part III – Key Concepts and Issues – includes chapters that address some of the most important contested concepts and debates within feminist philosophy of science and the normative implications for scientific practices. Part III raises questions that cut across the different theoretical frameworks discussed in Part II and may be approached differently by them. For example, Clune-Taylor analyzes the concept of sex and the ways in which this has been historically contrasted with gender by early feminist theorists. Sex has often been construed as a biological category while gender is argued to be a socially constructed set of norms that govern a variety of behaviors and characteristics. Clune-Taylor questions this tendency (even by some feminists) to naturalize the category of sex and reveals that the biological, social, and environmental factors that play a role in sex categories are difficult to distinguish. The chapter by Intemann examines the role of values in science, a topic that feminist philosophers of science have significantly contributed to, even while it has received increasing attention by philosophers of science more generally. Yet, as Intemann shows, feminist theorists have challenged the view that science is "value-free" for distinct reasons and with an eye toward feminist goals in ways that are important to recognize. Moreover, different theoretical frameworks (e.g. feminist empiricism and standpoint theory) offer different normative recommendations and resources for managing values in science. The fact that knowledge production is embedded in a certain set of social circumstances, practices, and value judgments has also called into question the notion of scientific objectivity, and led many feminist scholars to rethink and reimagine our understanding of this key concept in a way that is consistent with a core feminist understanding of knowledge as socially situated. This is the discussion taken up in Rolin's chapter. Another key theme that has emerged from feminist epistemology and feminist philosophy of science is the question of how *ignorance* has been produced and maintained. Feminists, as well as critical race and disability theorists, have contributed to a large body of literature on this field, now referred to as agnotology. The chapter by Pinto examines the ways in which ignorance plays out in our cognitive practices related to science. She shows how scientific ignorance has been produced and maintained in relation to our lack of understanding of women's bodies, women's health issues, and cognitive abilities. Chapters by Kourany and Giordano examine the normative implications of much of the previous discussions, considering in different ways what more democratic science might look like from two very different theoretical approaches: one likely to be in line with a liberal feminist political philosophy, and one using the type of genealogical approach advocated by Foucault.

Part IV – Feminist Philosophy of Science in Practice – includes chapters that examine scientific research in particular disciplines from feminist perspectives. Part IV can be understood as implementing the sorts of frameworks and concepts from Parts II and III in the practice of particular disciplines and research programs. Yet this work should not merely be seen as a one-way "application" of theoretical frameworks, because investigation of the specific disciplines often furthers our understanding of the frameworks and concepts explored in earlier sections.

The chapters in Part IV cover a variety of different research programs in different disciplines, but it is perhaps no surprise that much feminist work has arisen in relation to the biological and social sciences. There are several potential reasons for this. First, it may be that the kinds of things investigated in the biological and social sciences are more directly related to gendered and social phenomena that feminists are concerned with. Second, as discussed, a lot of feminist work has arisen from increased numbers of feminist researchers in certain fields. Certain disciplines such as physics, chemistry, computer science, and engineering are disciplines where women are still persistently underrepresented. Of course, women are also underrepresented in philosophy of science, and even more so in philosophy of physics and chemistry. Contributions by Laura Ruetsche and Donna Reilly examine issues of underrepresentation, diversity, and inclusivity in these fields. Laura Ruetsche examines "what it's like" to be a woman in philosophy of physics and Donna Riley argues for transforming the way we think about inclusivity and diversity in engineering education. One theme that emerges here is that changing the culture of STEM and philosophy of science will require more than just a focus on increasing the number of women in these fields. It will require examining an array of variables from professional interactions to policies of the academy to the content and aims of disciplines themselves in order to achieve fields that are more inclusive and socially just.

Additional chapters in Part IV indicate various ways that feminist research has and continues to challenge the presuppositions, frameworks, and research practices of science that may hinder knowledge production or fail to serve the needs of women and other oppressed groups. They also offer insight into how the sciences have been or might in the future be changed through engaging with feminism. Contributions by Nelson, Botero, and de Melo-Martín focus on gendered assumptions implicit in the biological and biomedical sciences, including work in biology, primatology, and reproductive technologies. Chapters by Meynell, Bluhm, Bentley, and Else-Quest and Shibley Hyde focus on psychology, cognitive science, and neuroscience, one of the more recent areas of focus for feminist philosophy of science. Several of these authors identify ways in which sexism and androcentrism have shaped research done in these fields. Chapters by Bentley and Else-Quest and Shibley Hyde examine what alternative, more feminist, approaches and interventions might be utilized. Chapters by Barker and Kuiper, Crasnow, and Mazur examine how feminist methodologies might be employed in the social sciences. Barker and Kuiper look specifically at economics and the particular issues that first raised questions among feminists working in the discipline (such as the wage gap). Crasnow explores some general questions about methodology that have been debated by feminists working in the social sciences and Mazur's chapter picks up the theme of the importance of conceptualization, harking back to the importance of key concepts discussed in Part III.

Part IV concludes with contributions that critically analyze the ways in which science itself interacts with science-related public policy. Science has historically enjoyed a privileged position of authority in shaping policy debates and in developing interventions to address public health, environmental, and social problems. Yet many of the chapters in Parts II and III show that science and values are entangled, participation in science is limited, and systems of power shape who we take to be experts and the sorts of research that are prioritized or pursued. Thus, Part IV ends with an examination of the implications that feminist philosophy of science has for contemporary science-related policy issues. Drawing on feminist work related to trust, Maya Goldenberg shows us how disagreements about pediatric vaccine safety might be re-framed and understood in terms of trust. Zahra Meghani examines how social and cultural factors, including gender and socioeconomic conditions, partially constitute the nature of certain diseases and how they impact certain populations. She points out the ways in which scientists have tended to "biologize" public health problems in ways that have led to certain technological fixes (e.g. genetically engineered mosquitoes to address malaria) and can result in reinforcing, rather than addressing, social inequalities. Both of these contributions identify factors that should be taken into account in responsibly developing public policy and interventions, and ways in which scientists might more effectively engage with those most impacted by their research.

Conclusion

Feminist philosophy of science is a rich, diverse, and growing body of work. While much of this research aims to examine and evaluate science and traditional philosophy of science, it should not be understood as either "anti-science" or "anti-philosophy of science." Many of the authors here have substantial training in one or more of the sciences, and indeed some identify as "social scientists" more than as "philosophers." While feminist scholars have challenged several assumptions, methodologies, and frameworks in both science and philosophy of science, feminists also share a common commitment to the importance of achieving reliable knowledge, protecting the epistemic integrity of science, and reducing the negative consequences of bias. Through the contributions gathered together in this handbook we aim to show the power and promise of feminist critiques, theories, and methodological approaches in helping us to imagine alternative ways of doing and thinking about science that open up new lines of research and producing knowledge that also serves the aims of social justice. This work offers both a philosophical justification for such an endeavor and empirical evidence of its value.

Notes

1 This is the story as told by Sandra Harding at a dinner during the Association for Feminist Epistemologies, Methodologies, Metaphysics, and Science Studies (FEMMSS) Conference in Waterloo, Ontario, Canada in 2014 and confirmed in personal communication in April 2020. *Discovering Reality* was reissued in 2003 as a second edition marking the 20 year anniversary of its appearance.
2 An example is Gross and Levitt's attack on Harding (1994: 134–136) in which they misrepresent her views on feminist empiricism and feminist standpoint theory.

References

Anderson, E. (1995) "Feminist Epistemology: An Interpretation and a Defense," *Hypatia* 10(3), pp. 50–84.
Barad, K. (2007) *Meeting the Universe Halfway: Quantum Physics and the Entanglement of Matter and Meaning.* Durham, NC: Duke University Press.

Bleier, R. (1984) *Science and Gender: A Critique of Biology and Its Theories on Women*. New York: Pergamon Press.

Grasswick, H. and Webb, M. O (2002) "Feminist Epistemology as Social Epistemology," *Social Epistemology* 16(3), pp. 185–196.

Gross, P. R. and Levitt, N. (1994) *Higher Superstition: The Academic Left and Its Quarrel with Science*. Baltimore: The Johns Hopkins University Press.

Haraway, D. (1988) "Situated Knowledges: The Science Question in Feminism and the Privilege of Partial Perspective," *Feminist Studies* 14(3), pp. 575–599.

Harding, S. (1986) *The Science Question in Feminism*. Ithaca, NY: Cornell University Press.

Harding, S. (1991) *Whose Science? Whose Knowledge?* Ithaca, NY: Cornell University Press.

Harding, S. (1994)

Harding, S. and Hintikka, M. (eds.) (1983) *Discovering Reality: Feminist Perspectives on Epistemology, Metaphysics, Methodology, and Philosophy of Science*. Dordrecht: D. Reidel Publishing Company.

Harding, S. and Hintikka, M. (eds.) (1983/2003) *Discovering Reality: Feminist Perspectives on Epistemology, Metaphysics, Methodology, and Philosophy of Science* (Second Edition). Dordrecht: Kluwer Academic Publishers.

Hrdy, S. B. (1977) *The Langurs of Abu: Female and Male Strategies of Reproduction*. Cambridge, MA: Harvard University Press

Hrdy, S. B. (1981) *The Woman That Never Evolved*. Cambridge, MA: Harvard University Press.

Hubbard, R. (1983/2003) "Have Only Men Evolved?" in Harding, S. and Hintikka, M. (eds.) *Discovering Reality: Feminist Perspectives on Epistemology, Metaphysics, Methodology, and Philosophy of Science* (Second Edition). Dordrecht: Springer, pp. 45–69.

Longino, H. (1990) *Science as Social Knowledge*. Princeton, NJ: Princeton University Press.

Morgan, E. (1973) *The Descent of Women*. New York: Bantam Books.

Tuana, N. (ed.) (1989) *Feminism & Science*. Bloomington: Indiana University Press.

PART I

Hidden Figures and
Historical Critique

1

THE ORIGINS OF PHILOSOPHY AND SCIENCE IN ANCIENT GREECE

Material Culture and the Scarcity of Women

Joanne Waugh

Philosophy, Science, and History

Feminist historians and philosophers of science typically work on clearly identified and carefully articulated constellations of problems. Some of these problems come into focus when feminist scholars consider how scientists have conducted or practiced their studies of nature and society. Other problems arise when feminist scholars turn their attention to how and why historians and philosophers have arrived at their conceptions of science. This chapter will focus instead on the period and place in Western culture before philosophy and science are conceived of as particular forms of human inquiry, and before anyone spoke of "western culture." In this period we can see the origins of philosophy and science and how they differed from myth, the dominant cultural genre in Archaic Greece.

There is, of course, another less precise usage of the term *philosophy*. It can be used to indicate the answers that members of a society give to questions about their origins, about the coming-to-be and workings of the world in which they find themselves, or about the source and force of their laws and customs. *Philosophy* in this sense includes speculation of the sort that is found in traditional poetry, folk tales, and sagas, all of which are usually grouped under the umbrella term, *myth*. Myth, like the narratives that preserve it, is universal. It is and has been a universal human activity because speech is part of our biological–historical inheritance. We are the kind of animals for whom it is natural to have a culture, and speech is the primary but not sole means by which human societies keep alive the memory of the past, their religion and ritual, ethics and etiquette, arts, crafts, and technology, customs and folkways – all of the knowledge they need to maintain their culture and their cultural identity. Speaking in a relaxed sense, we can say that myth is the way in which the members of a society preserve their history and their philosophy.

Yet a defining characteristic of *philosophy* and *science* and *history*, as philosophers, scientists, and historians, including many feminists, use these terms, is their difference from myth. If philosophy and science are the opposite of myth, a problem arises. Despite the widespread, if not always acknowledged, assumption of philosophers and scientists that any rational person who thinks long and hard enough about human experience will eventually ask philosophical and scientific questions, philosophy and science in the sense in which they are the opposite of myth are not universal. They emerge during a particular period in human history, the sixth and fifth centuries BCE, and in a particular society at a particular place, the Greek-speaking cities on the Aegean and Mediterranean seas. If any rational person who thinks long and hard enough about human experience will ask philosophical

13

and scientific questions, why did philosophy and science emerge, when and where they did? Are we to conclude that these ancient Greek thinkers were "more rational" than their predecessors, assuming that rationality admits of degrees? What should we infer from the fact that most of these Greek thinkers, or at least he ones we know about, were male? Can we avoid the conclusion that those living in societies in which there is no evidence of what we call philosophy and science are less rational than contemporary society or ancient Greek society? Must we agree that there was a Greek miracle, as some historians, in lieu of another explanation, have concluded?

The goal of this chapter is to show that these questions can be answered or put aside once we recognize that the emergence of philosophy and science depended on material cultural resources. It is reasonable to ask why, if material culture is so important in the origins of science and philosophy and is a factor in explaining the near absence of women in these disciplines, this fact has yet to be widely recognized. The answer is that philosophers, including some feminists, have dismissed or distorted the history of philosophy and science. The dismissal of the history of philosophy can be traced to assertions of influential twentieth-century philosophers of science, made at the point in which the philosophy of science was gaining dominance in the English-speaking philosophical world, that the history of philosophy was history and not philosophy (Reichenbach 1951; Carnap 1930/1959; Schlick 1932/1967). When the study of the history of philosophy could not be eliminated for reasons that were economical and institutional as much as philosophical, it was replaced by "rational reconstructions," that is, the formulation of arguments based on statements "found" in a text from an earlier period in philosophy. These arguments provide the answer that the author of this text would have offered to some "perennial philosophical problem" had he known then what philosophers know now. In rationally reconstructed arguments, little or no distinction is drawn between present day concepts, distinctions, methods, and arguments, and those of the society in which the text was produced. These discussions proceed without much if any consideration of the forms of life or social practices represented in and by the concepts earlier philosopher shared with their audiences. This occurs even when the problem at issue concerns a just society, a just man, or a just act, or asks whether and how we know the natural world of which we are a part.

In the wake of Thomas Kuhn's *Structure of Scientific Revolutions*, itself a revolutionary if not paradigm-changing text, some philosophers of science admitted that the history of science had a place in philosophy of science. It is not clear, however, if one looks at contemporary work in the philosophy of science, including work by feminist philosophers of science, that history has a *necessary* place. Philosophers have been even more reluctant to admit the importance of the history of philosophy for their discipline. Nearly a century after the Logical Positivists first dismissed the history of philosophy as a history of errors (Schlick 1932/1967) or as logically untenable and therefore meaningless (Carnap 1930/1959), debate continues among philosophers about whether texts from philosophy's history need should be studied in their historical context, whether "appropriationist" approaches (as rational reconstructions are sometimes called) are desirable or inevitable (Laerke, Smith, and Schliesser 2013), and whether the "problems of philosophy" are perennial or contingent (Waugh and Ariew 2013). Indeed, appropriationist approaches are so widely adopted toward earlier philosophical texts that many philosophers continue to consider contextual approaches to the history of philosophy – when they are acquainted with them at all – as history and not philosophy.

That the emergence of science depended on material cultural resources and that science continues to depend on such resources cannot be denied. Still this fact has not influenced discussions in the philosophy of science as much as we might expect, despite reminders that the history and practice of science consist of experimental interventions in or with the world, made possible by various apparatus, instruments, and machines, as well as by more mundane things like pulleys, springs, glass, metal, string, wax, and wood (Hacking 1983: 149ff). The claim that philosophy and history also depended on material cultural resources was advanced by Eric Havelock a half century ago, and while it has been recognized by those working in Ancient Greek philosophy

(Detienne 1986; Cole 1991; Waugh, 1991; Robb 1994; Naddaf 2005) it has not received as much attention from historians of philosophy and historians and philosophers of science as it should have. If philosophy and science are dependent on material culture, then one cannot engage in the study of these disciplines – or their history – without placing them in their cultural and historical context. For feminist philosophy of science, history and material culture have not only a place but also a necessary and central place.

The History of Science and Philosophy

(H)istoria like *philosophia* and *muthos* were names of ancient Greek cultural genres before they became the names of putative cultural universals.[1] *(H)istoria* is the Greek word for inquiry, learning through inquiry, the knowledge one has obtained through inquiry, or an account of one's inquiries, that is, a narrative or history. It is the last sense in which we use *history*; Herodotus, described as the first historian, wrote a narrative of his inquiries, travels, and studies. *History* for us is knowledge through texts or documents (Veyne 1984). The earliest scientific and philosophical thinking comes down to us under the title *(h)istoria peri phuseōs*, an inquiry into nature, the title of works written by individual thinkers from the sixth and fifth centuries BCE. Taken as a group these thinkers are referred to as the Presocratics, a group from and about whom we possess little in the way of texts or documents. The absence or limited availability of the kinds of material cultural resources we take for granted matters a great deal here, but this is by no means the end of the story.

We possess some fragments from the Presocratics that appear to be in their authors' original wording, as well as summaries from later – sometimes much later – commentators. These summaries, called *testimonia*, or the doxographical tradition, were written by ancient commentators. We believe that some commentators had access to manuscripts containing works of the Presocratics, and others, only access to the works of earlier commentators having this access. It is likely that an oral tradition about the Presocratics is reflected in the doxographical record.

What is most striking about these earliest inquiries into nature is that they were, at one and the same time, poetic, philosophical, and scientific. They are written in poetry or "gnomic prose" – loosely connected axiomatic statements, together with expressive poetic language, featuring an accumulation of words and expressions having similar associative meanings and with a liberal use of literary devices such as alliteration and assonance (Thesleff 1966: 90). Nonetheless gnomic prose is the first step toward a scientific style of writing, that is, a style that employs strict argumentation, has a systematic structure of exposition, lacks emotionally evocative, ornamental, and superfluous language, and favors abstract nouns and expressions (Thesleff: 1966: 89). This is the style of writing that is exemplified in Aristotle and to a lesser degree in the Hippocratics, and is reproduced and reflected in Xenophon's technical writings and in some of Plato's later dialogues (ibid.).

Tradition has it that the Milesian philosophers, the first scientific philosophers, were the first to write in prose, but we possess only a partial fragment from Anaximander, and none from Thales or Anaximenes. In any case, the Presocratic authors of these texts would not have described their work as poetry, philosophy, or science, or to be more precise, they would not have used the Greek terms *poiēsis* and *philosophia* that we transliterate as *poetry* and *philosophy*. This raises the question as to how they did think of their work, a question that will take us to a cultural situation quite different from our own. At present, we should note that neither would they have described their work as science since *science* is a transliteration of the Latin *scientia*. There is little reason to believe that *poiēsis* was used in our sense of poetry before the fifth century BCE; the verses composed earlier by Homer and Hesiod, described by the Greeks as the teachers of Hellas, were described as songs. There are no uncontested uses of *philosophia* or its cognates before the fifth century BCE; for those thinkers we call the Presocratic philosophers *philosophy and philosopher* would have been new and unfamiliar terms. *Philosophia* or its cognates are not used in our sense of these terms until the fourth century BCE when both Plato and Isocrates claim *philosophia* for the activity each purports to teach. The most famous fifth-century occurrence of a cognate expression comes at 2.40.1 in Thucydides' *Peloponnesian War*, where Pericles proclaims: "We Athenians can philosophize without loss of manliness" (or, "without becoming effeminate"). [The disparity between women and men in positions of authority in professional philosophy may prompt the thought that somewhere in the profession's unconscious lurks still the anxiety that Pericles was trying to allay. On the other hand, since the early modern period, philosophers have sometimes talked as if the mind has no sex, but, as it turns out, it did (Schiebinger 1991).] The closest Greek equivalent to "science" in the sense of exact rigorous thought – *epistēmē* – is first found in Plato. If "science" is taken as shorthand for "natural science" the closest Greek equivalent is *historia peri phuseōs*, which is the name given to the works of the Presocratics, whether written in poetry or prose.

What is, perhaps, more puzzling than the fact that the earliest examples of scientific and philosophical thinking are found in poetry and gnomic prose, performed or read aloud before an audience, is that, until recently, this fact has received little more than a passing comment in studies of Presocratic philosophy. The disdain for the history of philosophy that resulted in appropriationist studies being far more common than contextualist studies partly accounts for this lack of discussion. More important, however, is the reason behind this disdain for the history of philosophy, the importance accorded formal languages, at the expense of natural languages, by the philosophers who ushered in the "linguistic turn," the name given to the period in the English-speaking philosophical world when the philosophy of science became dominant. The value of formal languages lies in their independence from context, unlike the "slag of historical languages" (Carnap, Hahn, and Neurath 1929/1973). As a consequence, many philosophers took formal languages as the model for language, ignoring the differences between speech and written language, or explaining natural languages and how they work on the model of logic. But the independence from context that is the virtue of formal language is achieved through graphic representation; formal languages depend on writing for the statement of their grammar and for their transmission. Natural languages are called natural languages because they are part of our biological-historical inheritance and they are not dependent on writing. All human societies have speech; writing systems are relatively recent inventions and their external representations of speech and thought have varying degrees of success.

As Eric Havelock once put it,

> The biological historical fact is that *homo sapiens* is a species that uses oral speech, manufactured by the mouth, to communicate. That is his definition. He is not by definition a writer or reader. His use of speech . . . has been acquired by processes of natural selection over a million years. The habit of using written symbols to represent such speech is just a useful trick that has existed over too short a span of time to have been built into our genes . . .
>
> *(1982/2019)*

Human societies that have accomplished the preservation and transmission of culture through speech alone have existed for far longer than those societies that have invented writing systems and have developed the material culture necessary for writing to be used in preserving and communicating culture. As members of a literate society we are inclined to speak of oral cultures as the exception.

> If something is 'oral', we tend to assume a conflict with the notion of 'written' . . . however, it is 'written' that has to be defined in terms of 'oral'. 'Written in not something that is not 'oral', rather, it is something *in addition* to being oral, and that something will vary from society to society. It is dangerous to universalize the phenomenon of literacy.
>
> *(Nagy 1989: 9)*

This is not to suggest that material culture does not preserve and transmit knowledge through things that are not linguistic. Nor is to suggest that speech can or does occur independently of material culture. That it has become fashionable to talk about material culture has not exactly been a cause for celebration among working in archaeology, the discipline that studies things. Indeed, Bjørnar Olsen reminds us that

> subjecting things to words and language as has always been the preferred intellectual taming device to cope with the objects' disquieting material obstinacy. Making things perform like words, enslaving them in semiotic webs of relations, is another but more serious expression of this encroachment.
>
> *(Olsen 2013: 179–180)*

Indeed, among the excesses of textualization is the failure to recognize that speech and reading and writing are bodily actions and, like other perceptual acts, involves interaction with things in a shared environment.

Speech, Writing, and Material Culture

Speech is itself a learned, embodied action. We learn to speak by watching the faces and bodies of those teaching us. We learn not – or not only – by hearing utterances, but also by registering their intonation and phrasing, the expression on the speaker's face, the stance of her body, her gestures and movements, or the lack thereof. When present to each other in a shared environment, we can repeat and rephrase and question until what is being said is understood. Speech exists, at least initially, in the actions of embodied speakers and hearers sharing an environment. That we as embodied speakers share this space makes knowledge and language possible; shared experiences are experiences "we register as communal," as experiences that "tie our words and thoughts to the world" (Davidson 2005: 161). Thus "thought is essentially social" (Davidson 2005: 159). Because language allows us to go beyond what we sense jointly and immediately, and to talk and think about what is not present, we are able to preserve and transmit the knowledge needed to ensure our survival. This does not mean that communication is solely or primarily linguistic; humans can and do communicate with and through their bodies without using speech. It is the capacity that bodies have to express meaning through movement, gesture, and dance – and by objects they create and manufacture – that enables humans to acquire speech. We are able to communicate because we are embodied and situated, not despite it.

The ability to go beyond what we sense jointly and immediately and to talk and think about what is not present can be greatly extended by graphic signs, symbols, diagrams, maps, and pictures – external representations of thought. External material representations enhance cognitive power, beyond saving memory and aiding computation. They can reduce the effort involved in drawing chains of inferences, provide shareable, identifiable, and persistent objects of thought, and permit enduring and repeated representations, including more natural representations of structure than mental representations do (Kirsh 2010: 441). External representations can also ease computation of explicitly coded information and provide for the construction of arbitrarily complex structures. The functions made possible or enhanced by external representations

> jointly . . . allow people to think more powerfully with external representations than without them. They allow us to think the previously unthinkable . . . .. There are cognitive things we can do outside our heads that we *simply cannot do inside*. On those occasions, external processes function as special cognitive artifacts that we are incapable of simulating internally.
>
> *(Kirsh 2010: 441, 452)*

Of particular interest to philosophers and scientists is the extent to which external representations make it possible to re-order or rearrange physical tokens of statements sentences, and logical formulae, and compare what was written earlier with what was written later. Bringing statements together side by side in physical space that were distant in logical space and introducing definitions or abbreviations to represent groups of statements can further increase the scope of statements that can be visually related (Kirsh 2010: 446). "The process of inferring, duplicating, substituting, reformulating, rearranging and redefining is the mechanism behind proofs, levels of abstraction, the lisp programming language, and indeed, symbolic computation more generally" (ibid.). The point is that thinking and sense-making are mostly interactive, a process in which a person alters the situation in the outside world, the altered situation changes the person, and the process continues. This applied to our interactions with texts: "reading a text silently is not an interactive process . . . though . . . extremely active. Reading and underlining the text, or reading and summarizing, even reading and moving one's lips are" (Kirsh 2010: 441).

Archaic Greece: Oral Poets and Reluctant Writers

Societies that have preserved and transmitted culture without texts – that is, most societies since there have been humans – are known through objects that archaeologists have been able to discover. For those societies in which writing eventually came to be used, their earliest known written texts provide evidence of the culture they preserved through external representations of thought, including written texts. The earliest documents and texts from Greek Antiquity present us with an extraordinary exemplar of an oral, traditional culture that for centuries preserved and transmitted the knowledge needed for the preservation of Greek culture and Greek cultural identity. Understanding not simply *that* but also *how* public performances of poetic speech work to preserve and transmit knowledge makes it possible to see why Presocratic philosophy was the beginning of revolutionary movement not just in their societies but also in the history of Western thought.

We are now able to understand how public performances work to preserve and transmit knowledge because of a series of studies, beginning in the 1920s and concluding with his untimely death in 1935, by Milman Parry. Parry demonstrated how the *Iliad* and the *Odyssey*, the oldest extant Western literature, attest to a centuries-long tradition of oral story-singing (Parry and Parry 1987). Poets were able to compose extemporaneously as they performed because Homeric epic verse is a fixed dactylic hexameter line that sets up a syntax of sound; the dactylic hexameter line had a fixed number of caesurae resulting in regular metric segments within a given line. Parry demonstrated that Homeric diction contains systems of formulaic expressions: for every major character (and important objects) in the narrative – gods, heroes, peoples, countries, ships, horses – there are formulae that meet the metrical needs of the hexameter verse in the requisite grammatical case, and there tends to be one and only one formula for each character (or object) that meets these metrical and grammatical requirements. Thus the singer composes extemporaneously not by choosing single words in the right case and meter, but by using fixed formulae and formulaic expressions that make up segments of the hexameter line. The language of Homeric poetry is itself the work of oral, traditional verse-making; it is not an earlier form of one Greek dialect, but includes Ionic, Aeolic, and Arcado-Cypriot elements (Parry and Parry 1987: 325–364).

The language and formulaic diction of the Homeric poetic tradition enable oral composition-in-performance because they are part of an edifice of rhythm that facilitate the memory of both the performer and of the audiences that assimilate the cultural education he provides. The rhythm on the levels of sound and diction supports rhythmic structures on the level of reiterated incidents, themes, and outlook (Russo 1976). Repeated incidents as the beaching of ships, offering prayers and sacrifices, and distributing booty, and repeated themes such as the proper role of gods and mortals, kings and princes, heroes and warriors, and families and communities result in the presentation of an epic outlook, a world picture that reminds audiences both in terms, both general and particular, what it is to be a good representative of their heritage.

The performance – and influence – of Homeric poetry was a pervasive Panhellenic influence long after a writing system became available, a system sufficiently simple, unambiguous, and exhaustive in representing speech to allow for the kinds of external representations of thought that allow us "to do cognitive things outside our heads," and "to think the previously unthinkable."[2] The cognitive artifacts made possible by external representations of thought depend on networks of material cultural resources: the production of materials needed not only for writing but also for widespread production, storage, and dissemination of texts, institution and practices for teaching writing and reading, changes in bureaucratic record-keeping and the establishment of archives and libraries, and changes in political and judicial practices and procedures, to name just the most obvious ones. These kinds of changes in material cultural practices occur if writing and reading are seen to have obvious and valuable uses, and the abilities to read and write are seen as "relevant" or empowering in some way (Thomas 2009: 13). In other words, "reading and writing tend to be learned and given meaning in a particular social, political, and cultural context" (Thomas 2009: 13).

The earliest examples of Greek alphabetic writing confirm both that writing was initially employed in a quite limited context and the pervasive influence of Homeric poetry. They are inscriptions found on pottery that often were haphazardly scratched after firing, composed of signs that are irregularly shaped, and sometimes written in *boustrophedon*. The objects speak – in verse. The fragments of a Euboean *skyphos* found in Methone and dated to 720 BCE bear an inscription that when translated reads: "I am [the cup of] Hakesandros" and continues with a threat to the effect that anyone else who drinks from it will lose his eyes, money, cakes, and pain (Papadopoulos 2016: 1244). The inscription on the smashed skyphos found in Pithekoussai, dated to 700 BCE and known as Nestor's Cup, is composed in hexameter verse and states: "Of Nestor, I am the Cup, a delight to drink from. He who drinks of *this* cup him straightaway Desire shall grip even of the fair garlanded Aphrodite" (Robb 1994: 46). A primary use of inscriptions was to dedicate offering to the gods. The Mantiklos Apollo (Thebes, 700–675 BCE) has an inscription in hexameter verse that is translated as saying: "Mantiklos dedicated me to the far darter [god] of the silver bow from the tithe. Do you Phoebus [Apollo] give something gracious back" (Robb 1994: 57). The inscription suggests the use of professional tools by a craftsman in metals, but it would be an understatement to say that this inscription is difficult to read. It winds around the right outer thigh across the groin to the left outer thigh and then around to the left inner thigh. If this inscription is representative, it would seem, as Robb points out, that early dedications were not even intended to be read by humans. Rather, they made the inscribed objects, surrogates for their donors allowing them "to *speak* continuously before the god when the dedicator had to depart from the temple" (Robb 1994: 59). In fact, Svenbro suggests that "writing itself could thus logically claim to be 'oral' . . . Greek writing was first and foremost a machine for producing sounds" (Svenbro 1993: 2).

Though it may be easy enough for us to see the possibilities of Greek alphabetic writing, the attitude of the early Greeks toward writing was ambivalent at best. In the earliest references to alphabetic letters, they are often seen as signs (*sēmata*). Signs include pictures, marks, and natural phenomena such as thunderclaps as well as utterances – when they occur under the right circumstances and are correctly interpreted – as when Odysseus asks Zeus for a portent (*teras*) and an utterance (*phēmē*) that he should move against the suitors. Zeus replies with a thunderbolt, which an old woman, grinding her wheat, recognizes as a portent, causing her to offer her prayer as well (Steiner 2015: 5, 11–13). She prays that this be the suitors' last day on earth. Odysseus hearing both the thunder and the old women's prayer is heartened, convinced that he would destroy the suitors (20.98–121).

To realize the possibilities – social, political, cognitive – that alphabetic writing might afford, the Greeks would have needed to overlook the association between writing, Eastern monarchs, and tyranny. Fifth-century Greek sources identify Eastern monarchs' inscribing their names obsessively on men, objects, land, columns, coins, and slaves, as a foreign practice

> outside the democratic government they celebrate as a distinctively Greek (and more particularly Athenian) achievement . . . Far from being an integral part of the identity that Greeks construct for themselves, writing is a technology employed by those who would destroy their uniquely successful way of life.
>
> *(Steiner 2015: 6–7)*

Their attitude toward reading was even less positive. A reader is often seen as giving voice to another's words, and as such, assuming a passive posture in relation to writer. On a drinking cup, dating from 500 BCE and found in Sicily, the translated inscription reads: "the writer of the inscription will bugger the reader" (Svenbro 1993: 2). Even in the Classical period, reading was very difficult. One had to stand at a special desk that allowed the papyrus roll to be moved and to learn how to pick out words from continuous script. It was an activity often left to slaves, as we see in the opening episode of Plato's *Theaetetus*.

It is perhaps not surprising, then, that as late as the fourth century in Athens, the most cosmopolitan of Greek cities with the literacy rate among its citizens approaching 15% (Harris 1991), we find Plato writing about the influence of Homeric poetry as the vehicle of education. In the *Republic*, Socrates reminds the audiences in and of the dialogue that some people say that Homer and the tragic poets know all the arts, all things human, all things pertaining to virtue and vice, and all things divine (598d9-e2) and that the celebrants of Homer will tell you that "Homer educated Hellas, and that on matters of culture and the management of human affairs he should be studied constantly as a guide by which to order one's whole life" (606e). Socrates had already pointed out that the great poets, Homer and Hesiod, as well as the others who tell the same stories about the gods and heroes, misrepresent their natures, and endanger their audience by employing *mimēsis*. *Mimēsis* refers to performances of poetry in all of their theatricality, the ever-present repetition of multiple layers of rhythm, the use of evocative and emotive language, the story of gods and larger-than-life heroes who perform larger-than-life deeds and give larger-than-life speeches, with no pauses in the performance to break the spell of the narrative and allow for reflection or analysis.

Poets and Prose Writers

The archaic poets and Presocratic thinkers can be forgiven, perhaps, for not recognizing at first the possibilities of writing that are easily recognized by us, we who live and interact with writing in such regular, routine, and habitual ways that social structures and institutions have been established as a result. In any case, the limited extent of writing and reading in the archaic and classical periods does not diminish, of course, what ancient Greek poets, natural scientists, and philosophers were still able to accomplish given these new skills. It should, in fact, do the opposite, especially when we view them in relation to the mythological tradition to which they were a response.

Hesiod, perhaps the first to use writing to record if not compose his poetry sometime in the decades just before or just after the turn of the seventh century BCE, provides an invaluable account of how myth worked and how it was viewed in archaic Greek society (Havelock 1963/2009). His *Theogony* presents a cosmogonic myth, one that explains how Zeus came to power on Olympus, and how as a result the current natural and social orders were established, and how any new natural or cultural events can and will be seen as continuing or completing these orders (Naddaf 2005: 37–38). Hesiod does this not just by invoking the Muses, the daughter of Zeus by *Mnemosunē*

(Memory), the patron of poets, but also by defining their powers, prerogatives, and functions. In stating that the Muses delight their father as they sing of what is, what was before now, and what will be, Hesiod sets the stage for an analogy between their place in divine society and his place in human society. He tells his audience that the Muses came to him as he was shepherding his flock under the holy mountain of Helicon, how they broke off a sturdy laurel shoot, how they breathed into him a divine voice to sing of what is, what was before now, and told him to sing the everlasting race of the immortals, putting themselves at the beginning and the end of their song (lines 29–34). Indeed, Hesiod goes on to state that there are mortal princes, chosen by Calliope, the Muse of epic poetry, and that when such a prince is born the Muses pour on his tongue, a honeyed voice, and words flow from him as he settles opposing causes with true judgments. The Muses tell Hesiod and Hesiod tells his audience that such a prince can put a quick and even and expert end to quarrels, and when people have gone wrong in an assembly can put them back on the right course, persuading them with gentle words (80–87). When such a prince walks through an assembly he stands out among mortals, and the people look on him as a god (91–92). In *Works and Days* Hesiod enlarges his account of the origins of mortals, explaining how women originated as part of Zeus' punishment for Prometheus' giving fire to men (42–59). Made with contributions from all of the gods, she looked beautiful but had lies and cunning and pretenses in her heart. She carried a jar and when she opened it evils and sickness were released onto men who until this time had lived free from toil and remote from ills, and away from deadly sickness. When she closed the jar, only hope was caught in the lid. She was called Pandora because all of the gods had a hand in fashioning this gift to mortal men.

The first Greek thinkers to study nature abandoned the mythological accounts in Hesiod in favor of naturalistic explanations. As noted above, they came from Miletus, an important seventh and sixth-century commercial and cultural center in Ionia, a region on the coast of Asia Minor. Although they composed in gnomic prose, they anticipated the scientific style of writing found in Aristotle who was the first to provide a summary of their teachings. While legend has it that Thales was the first of these Milesian philosophers, called the *phusilogikoi*, as is true of most legendary figures, it is hard to establish much about him. Most contemporary studies of Ionian philosophy begin with Anaximander, from whom we possess only a partial fragment, and the *testimonia* and commentary we find in the doxographical tradition. Most ancient and modern commentators agree his aim, and that of the other Presocratics, was to provide an inquiry into nature (*historia peri phuseōs*).

If scholars agree that what is distinctive about this inquiry into *phusis* is its search for natural rather than mythological origins, they do not agree about what the Presocratics would have understood by *phusis*. It is clear that whatever early Greek thinkers meant by *phusis* it was both broader than and different from what we mean by the physical world. The problem arises in translating *phusiologoi*, those who write or give talks about nature. Gregory maintains that it is

> a significant error to translate the Greek '*phusilogoi*' as 'physical philosophers' or even 'physicists'. A very few Presocratic thinkers may have been physicalists, mechanists, or materialists as we understand the term, but most were not. Some had a more organic concept of nature while many used biological analogues for the processes they saw around them.
>
> *(Gregory 2016: 12)*

Even Leucippus and Democritus, the early Atomists, are credited with using "biological analogues" when explaining the birth of *comsoi*, and "biological agricultural analogues" for effect in vortices called "like-to-like," an effect that is basic for the formation of the *cosmos* (Gregory 2016: 12–13). Nonetheless, commentators easily slip back into the notion that the Ionians, or at least, Anaximander is a mechanist, even when they are sympathetic to the claim that his *apeiron* is organic or biological. This is a critical error if it is true, as Gregory maintains, that Anaximander had a non-reductive biological conception of both the *apeiron* and the *cosmos*, that is, "it was not biological at one level which could then be explained in terms of underlying mechanisms, but one that is irreducibly biological" (Gregory 2016: 15). As important for this discussion is Naddaf's insistence that a *historia peri phuseōs* involved not just a cosmogony, but also an anthropogony, and a politogony. *Pace* present day philosophers who identify the Presocratic conception of philosophy with disinterested inquiry and speculation, "a keen interest in politics appears to have been the norm among these early philosophers. In fact, it is possible that their respective *historia* (investigations or inquiries) may have been politically motivated" (Naddaf 2005: 2). Indeed, Naddaf suggests that *historia* and/or *phusis* or – *historia peri phuseos* – may represent "newly minted phraseology to express the new rational approach to a way of life in conformity with the new political realities and the new comprehensive way of how the world, man and society originated and developed" (Naddaf 2005: 2).

In any case, what is reported about Anaximander is sufficient to see the stunning difference between his account and mythological accounts that explain the origins of the cosmos in terms of the sexual union and reproduction by warring anthropomorphic gods. Anaximander, we are told, asserted that the origin or source of existing things was the *apeiron*, a spatially and qualitatively indefinite stuff, stuff that was not born or created and is deathless or indestructible and therefore divine. Because of the endless movement of the *apeiron*, it generates a seed that is capable of producing the hot and the cold and it secretes them. With the hot and cold come the dry and wet. The germ of the hot and cold becomes a sphere of flames that is hot and dry, and this encloses the cold and the wet like a bark encircles a tree. The confrontation of hot and dry with the cold and wet produces the earth in the center, surrounded by water that generates mist or air between the earth and the fiery celestial bodies formed in the sky (Naddaf: 72–73). The underlying (or overarching) idea is that the cosmos grows like a living being, and the same principle is at work in explaining how animals and humans come to be, born or secreted from fish-like creatures (Naddaf: 73, 90–92). There is also evidence in the doxographical tradition that Anaximander studies the connection between geology and geography, and culture and history. He is credited with drawing the first map, using material models, and arguing that an Egyptian brought the alphabet into Greece (Naddaf: 106ff).

Women, Philosophy, and Science in Ancient Greece

In seeking a natural rather than mythological explanation for the origins of the world, humans, and societies, the *phusilogikoi* initiated what was to become a very long process in and through which women eventually were to become part of the community of inquirers and their experiences, objects of inquiry. In abandoning mythological explanations in favor of naturalistic explanations, early Greek scientific philosophers rejected the mythological account of women's creation in Hesiod. Abandoning the mythological account, coupled with the adoption of organic and biological conceptions of nature, resulted in the rejection of the traditional view of the temporal and logical priority of men to women (Naddaf 2005: 92). On the interpretation of Anaximander's conception of nature as irreducibly biological, the importance of women's role in the reproduction of humans can be neither denied nor devalued. If his conception of nature is seen as physical and mechanistic, as it had been until recently, there is little to be found in the first scientific writings that is relevant to women or to their experience. Nor on a deterministic view is it clear that Anaximander and the other Presocratics did not just abandon the notion of anthropomorphic divinities. In fact, they also reconceptualized both nature and the divine.

This rejection of the traditional views of divinity is important in indicating where the Presocratic philosophers stood in relation to other traditional beliefs preserved and sanctioned through myth. If the Presocratics were not disinterested and apolitical investigators, detached from the sphere of politics, or alternatively, mystical or metaphysical thinkers who withdrew from society, but instead active participants in the political and intellectual life of their cities, there is reason to believe that Anaximander's explanation of the ordering of the cosmos reflects accomplishments in material culture that made the cities of Ionia important intellectual and cultural centers in the seventh and sixth centuries. A strong case has been made that his engagement with the architects who were building the monumental temples of sixth-century Ionia, and his study of their prose manuals, influenced his account of the cosmos (Hahn 2001). It is also been demonstrated that the structures and practices that governed public discourse and political debate in the city are reflected in how Anaximander conceives of the organization of nature (Vernant 1983: esp. 125–134). If the *historia peri phuseōs* of the Presocratics involved not just a cosmogony, but also an anthropogony, and a politogony, challenges to traditional views of the origin of the cosmos should reflect, or be reflections of, challenges to traditional beliefs about the origins of society and its laws and customs.

Given our limited knowledge of the writings of the Presocratics, and the little they contain about women, any account of their importance in feminist history and philosophy of science may strike us as deflationary, nonetheless. But some feminist scholars have gone much further, arguing that the scientific and philosophical discourse initiated by the Presocratics was the beginning of a very long process in and through which women were precluded from science and philosophy, and their activities and their experiences distorted and devalued. The exclusion of women and the devaluation of their discourse are said to result from the very language and logic of philosophical and scientific discourse. Far from departing from the mythological tradition, philosophy and science continue to reflect its subordination and denigration of women.

There is no denying that until quite recently, women have not had a presence in science and philosophy, either as inquirers or as the objects of inquiry. Recent studies have suggested, however, that more women in ancient Greek Antiquity were engaged in these activities than we had originally thought. Sarah Pomeroy has devoted a book to history and writings of Pythagorean women (2013), and an examination of doxographical tradition suggests the belief that women were among the members of Plato's Academy goes back to Antiquity. Helen King has done an exhaustive study of the treatment of women by the Hippocratics, fifth- and fourth-century thinkers who are seen as the progenitors of scientific medicine (1998). Still, as King demonstrates, these writings are not free from the influence of myth and ritual, a fact that is not surprising in view of Plato's acknowledgment of the influence of Homer and Hesiod in the fifth and fourth centuries. Still King identifies conflicting traditions about the bodies of women. On one model, sexuality exists on a continuum, and both sexes contribute an active seed in reproduction; on the other, women's contribution to reproduction is passive. Both of these models presuppose one body, ranked according to its being more or less male. But King identifies yet another tradition that treats the female body as radically different from the male's, and on this view, women's bodies require their own branch of medicine, gynecology (King 1998: 9–11).

Aristotle's views of women's biology are, perhaps, the most widely known of Ancient Greek writings about women, and are often taken to be representative of how Greek scientific theories perpetuated the polarization of the sexes and the subordination of women to men. They did so because traditional cultural assumptions about women's nature led Aristotle and other ancient writers to misinterpret or to overlook data that would have contested their theories (Dean-Jones 1991, 1994). Because women are identified with the cold, the earth, darkness, and night, and men with the hot, sky, light and day, men are active and women passive, not just in reproduction, but in all aspects of life. Attempts by the Milesians to explain all things as originating from a primal stuff have been interpreted as evidence that they had a fear of diversity and difference (Saxenhouse 1995). This charge has also been brought against Parmenides (Cavarero 1995; Fischer 2014). There have been other charges brought against Parmenides as well. He has been read as denigrating natural processes, including sex and reproduction (Nye 1988; Saxenhouse 1995; Cavarero 1995; Irigaray 1999) and as embracing the principles of noncontradiction and excluded middle, and in so doing, providing the conceptual framework for theories and practices that have been used to exclude and subordinate women (Nye 1988).

There are, however, elements of Parmenides that run counter to what is said about women in the mythological tradition, in parts of the Hippocratic tradition, and in Aristotle. In B18 Parmenides suggests that both sexes make a substantial and formative contribution in reproduction, contrary to later claims that the woman is passive. Aristotle tells us at *Parts of Animals* (648a29-32) that Parmenides believes that women bodies are hotter than men's, and Theophrastus reports that Parmenides held that understanding that comes from what is hotter is purer and better (*De sensibus* 1.1–4). Rose Cherubin points out that on the Theophrastian interpretation, it follows that Parmenides believed that because both women and female animals in general are hotter, they have better and purer understanding than males (Cherubin 2019: 42). It should also be noted that Parmenides chooses to present his thinking as instruction that he received from a goddess who tells him listen to her words and carry them on. He says in B1 that as a youth he was taken to the goddess in a chariot pulled by wise mares, and escorted by the Daughter of the Sun through the Gates of the paths of Night and Day. It is significant that all of the figures that he encounters on his journey are female and divine, as are the divinities that compel the decision regarding the route one must follow in thinking and hold what exists in place (B 8). Nonetheless, even these aspects of Parmenides that seem not to devalue women have been analyzed as cases of devaluation. His linking of women with light and heat and intelligence is discounted because it occurs in his – or rather the goddess' – account of the thought of mortals; his actual position is claimed to be closer to Hesiod's (Songe-Møller 2002; Fischer 2014). The fact that Parmenides speaks only of female figures is seen as a device that ultimately is abandoned when he learns the true and necessary discourse of philosophy (Cavarero 1995: 37–47). Cherubin has provided a convincing and quite thorough defense against these charges, arguing that many of these criticisms are derived from later interpretations of Parmenides and not from his fragments, and that in fact he challenged many of the ideas and ways of thinking that have been attributed to him (Cherubin 2019: 30).

The problem with these criticisms, which is implicit in Cherubin's response to the charges against Parmenides, is that they work on an abstract conceptual level precisely because they require us to "abstract" the concepts from the forms of life and social practices in and from which they arose in Ancient Greece. Until or unless we examine these forms of life and social practices we cannot know if their concepts can be identified with our own. Historical study needs to precede conceptual analysis. There is no denying that the Greek poetic and philosophical traditions provide us with extensive evidence of the subordination of women, but there is also evidence for a tradition of dissent. There are the perceptive, wise, and powerful females, mortals and immortals – Circe, Calypso, Eurycleia, Penelope – with whom Odysseus must contend, the female characters – Antigone, Medea, and the Trojan Women – who give their names, their voices, and their power to Greek tragedy, the unique poetic voice of Sappho, and the remarkable figure of Aspasia. The question of women and their lot is too frequent a topic in Greek tragedy and comedy for women's subordination to be a settled matter. The dissenting tradition is forced to formulate its challenge in the language and style of the dominant tradition that it opposes, much as the Presocratics were forced to do when they criticized the mythological tradition, using the language and diction of Homeric poets, or prose that still showed signs of their influence.

Whether and to what extent the Presocratics dissented from the traditional subordination of women is a matter for argument, but the case should be made by looking not only at forms of life and social practices in archaic Greece but also at the material cultural resources that supported them. The Presocratics were not yet philosophers and scientists in our sense because the material culture that supports science and philosophy as we know it – its forms of life and social practices, its systems of ideas and systems of power — did not yet exist. They were committed to questioning traditional beliefs, beliefs that were accepted in their society not as the consequence of evidence or observation, but as what remained once the epic performance – the onslaught of rhythm and sight and sound and images that aroused emotion and passion, and fear and release, in their audiences – had ended. It is true that science and philosophy originated with the poet-philosopher-scientists of Ionia and their colonies in Magna Graecia when changes in their material, economic, and political resources – trade, writing, technology, travel – made it possible to move beyond the mythological tradition that had served them so well for centuries in preserving and passing on knowledge. What did not exist for them were the material cultural resources that were to establish science and philosophy as systems of ideas and systems of power in Western culture. Working within these systems of ideas and power is a function of having access to tools and technologies of material culture that make them possible. In a very real sense, what constitutes intelligence and authority in a society is an individual's or group's abilities to master the tools and technologies of their culture.

The earliest scientific and philosophical thought was not yet part of such a system, and it is not clear that the exclusion or denigration of women in archaic society was reflected in the Presocratic philosophers who were intensely critical of its systems of ideas and systems of power. It is clear that the exclusion of women from science and philosophy at this or any other period in their history was as much, if not more, the result of their access and mastery of material cultural resources as it was inherent in the language and logic of scientific and philosophical discourse. Once science and philosophy are part of the dominant discourse in Western society one must achieve levels of competence in writing and reading, access to the materials containing the body of acquired knowledge deemed critical to being educated, training in the technologies needed to contribute to this body of knowledge, and the opportunity to be educated or trained or qualified as an authority, and to be recognized by existing authorities as competent to participate in their activities.

The material cultural resources in archaic Greece were developed during a time in which there was more growth and expansion and hope (for the male population at least) and less war and death and struggle to return to home about which Homer sang, and of the evils and toils and sickness that Hesiod tells us is the fault of women. With the advantage of hindsight, living in a time of peace and prosperity, and with full knowledge of how science and philosophy will become implicated in the exclusion and devaluation of women and others, the movement from mythological to scientific explanations may not have appeared to be much of a step toward progress, if indeed a step toward progress at all. In a time of wars and social unrest and global pandemics, a time when magical thinking looks as if it might win in the war against reason, the search for natural rather than mythological explanations of nature, humans, and society looks like an important step in the right direction.

Related chapters: 2, 3, 13.

Notes

1 In speaking of a *cultural genre*, I am following Nightingale (1994) and Conte (1986). Conte describes a genre as

> a discursive form capable of constructing a coherent model of the world in its own image. It is a language…a lexicon and a style, but it is also a system of the imagination and a grammar of things. Genres are the expressive codification of a culture's models… they are the very models subjected to a process of stylization and formalization which gives them a literary voice.

> *(132)*

2 Greek-speaking peoples inhabited what is now Greek soil as far back as the second millennium BCE, and while there was a writing system in use in the Mycenaean period, Linear B, it was not until the introduction of the Greek alphabet that the Greeks had a writing system that made literate techniques of cultural transmission both possible and desirable. The Linear B syllabary had not been used for cultural purposes, and it is not clear that it could have been so used. John Chadwick, who worked with Ventris on

the decipherment of Linear B, described it as "little more than an elaborate mnemonic." In any case, the knowledge of Linear B appears to have died out with the collapse of the Mycenaean civilization at the end of the second millennium BCE. There is no evidence of writing on Greek soil in this period after the collapse of the Mycenaean civilization at the end of the twelfth century until the rise of the city-state in the eighth century. As a Mediterranean crossroads with a multi-ethnic population, including Greek-speaking peoples from as early as the second millennium BCE, Cyprus is a complex case. There is no evidence of writing on Cyprus in the tenth and possibly ninth centuries, but there was writing on Cyprus, prior to and following this period. In the second millennium, the Cypro-Minoan syllabaries were in use; they have not been deciphered. In the first millennium, the Cypriot syllabary was used to record Greek and at least one other unidentified language. As the first millennium Cypriot syllabary is assumed to be descendant of the second millennium Cypro-Minoan syllabaries, knowledge of writing must be presumed to have continued despite the absence of evidence for it. The Phoenician syllabary is present in Cyprus in the ninth century; the Greek alphabet is not attested until the sixth century (see Steele, 2013).

References

Carnap, R. (1930/1959) "The Old and the New Logic," in Ayer, A.J. (ed.) *Logical Positivism*, New York: The Free Press, 133–146.

Carnap, R., Hahn, H., and Neurath, O. (1929/1973) "The Scientific Conception of the World" [*Wissenschaftliche Weltauffassung*], in Neurath, M. and Cohen, R.S. (eds.) *Empiricism and Sociology*, Dordrecht: Reidel, 299–318.

Cavarero, A. (1995) *In Spite of Plato: A Feminist Rewriting of Ancient Philosophy*, London: Taylor & Francis.

Cherubin, R. (2019) "Sex, Gender, and Class in the Poem of Parmenides: Difference without Dualism?" *American Journal of Philology* 140(1), 29–66.

Cole, T. (1991) *The Origins of Rhetoric in Ancient Greece*, Baltimore, MD: Johns Hopkins University Press.

Conte, G.B. (1986) *Genres and Readers*, Baltimore, MD: Johns Hopkins University Press.

Davidson, D. (2005) *Truth, Language, and History*, Oxford: Oxford University Press.

Dean-Jones, L. (1991) "The Cultural Construct of the Female Body in Classical Greek Science," in Pomeroy, S. (ed.) *Women's History and Ancient History*, Chapel Hill: University of North Carolina Press, 111–137.

Dean-Jones, L. (1994) *Women's Bodies in Classical Greek Science*, Oxford: Oxford University Press.

Detienne, M. (1986) *The Creation of Mythology*, Chicago, IL: University of Chicago

Fischer, C. (2014) *Gendered Readings of Change: A Feminist-Pragmatist Approach*, New York: Palgrave Macmillan (St. Martin's).

Gregory, A. (2016) *Anaximander: A Re-assessment*, London: Bloomsbury Publishing.

Hacking, I. (1983) *Representing and Intervening: Introductory Topics in the Philosophy of Natural Science*, Cambridge: Cambridge University Press.

Hahn, R. (2001) *Anaximander and the Architects: The Contributions of Egyptian and Greek Architectural Technologies to the Origins of Greek Philosophy*, Albany: State University of New York Press.

Harris, W.V. (1991) *Ancient Literacy*, Cambridge, MA: Harvard University Press.

Havelock, E.A. (1982/2019) *The Literate Revolution in Greece and Its Cultural Consequences*, Princeton, NJ: Princeton University Press.

Havelock, E.A. (1963/2009) *Preface to Plato*, Cambridge, MA: Harvard University Press.

Irigaray, L. (1999) *The Forgetting of Air in Martin Heidegger*, Austin: The University of Texas.

King, H. (1998) *Hippocrates' Woman: Reading the Female Body in Ancient Greece*, London: Routledge.

Kirsh, D. (2010) "Thinking with External Representations," *AI & Society* 25(4), 441–454.

Laerke, M., Smith, J.E.H., and Schliesser, E. (eds.) (2013) *Philosophy and Its History: Aims and Methods in the Study of Modern Philosophy*, Oxford: Oxford University Press.

Naddaf, G. (2005) *Greek Concept of Nature*, Albany: State University of New York Press.

Nagy, G. (1989) "Early Greek Views of Poets and Poetry," in Kennedy, G. (ed.) *The Cambridge History of Literary Criticism*, vol. I: Classical Criticism, Cambridge: Cambridge University Press, 1–77.

Nightingale, A. (1994) *Genres in Dialogue: Plato and the Construct of Philosophy*, Cambridge: Cambridge University Press.

Nye, A. (1988) "Rethinking Male and Female: The Pre-Hellenic Philosophy of Mortal Opinion," *History of European Ideas* 9(3), 261–280.

Olsen, B. (2013) "Reclaiming Things: An Archaeology of Matter," in Carlile, P.R., Nicolini, D., Langley, A., and Tsoukas, H. (eds.) *How Matter Matters: Objects, Artifacts, and Materiality in Organization Studies*, Oxford: Oxford University Press, 1–15.

Papadopoulos, J.K. (2016) "The Early History of the Greek Alphabet: New Evidence from Eretria and Methone," *Antiquity* 90(353), 1238–1254.

Parry, M. and Parry, A. (eds.) (1987) *The Making of Homeric Verse: The Collected Papers of Milman Parry*, Oxford: Oxford University Press.

Pomeroy, S.B. (2013) *Pythagorean Women: Their History and Writings*, Baltimore, MD: Johns Hopkins University Press.

Reichenbach, H. (1951) *Rise of Scientific Philosophy*, Berkeley: University of California Press.

Robb, K. (1994) *Literacy and Paideia in Ancient Greece*, Oxford: Oxford University Press.

Russo, J. (1976) "How, and What, Does Homer Communicate? The Medium and Message of Homeric Verse," *The Classical Journal* 71(4), 289–299.

Saxenhouse, A.W. (1995) *Fear of Diversity: The Birth of Political Science in Ancient Greek Thought*, Chicago, IL: University of Chicago Press.

Schiebinger, L. (1991) *The Mind Has No Sex? Women in the Origins of Modern Science*, Cambridge, MA: Harvard University Press.

Schlick, M. (1967) "The Future of Philosophy," in Rorty, R. (ed.) *The Linguistic Turn*, Chicago, IL: The University of Chicago, 43–53.

Songe-Møller, V. (2002) *Philosophy without Women: The Birth of Sexism in Western Thought*, New York: Continuum.

Steele, P. M. (ed.) (2013) *Syllabic Writing on Cyprus and its Context*, Cambridge: Cambridge University Press.

Steiner, D.T. (2015) *The Tyrant's Writ: Myths and Images of Writing in Ancient Greece*, Princeton, NJ: Princeton University Press.

Svenbro, J. (1993) *Phrasikleia: An Anthropology of Reading in Ancient Greece*, Ithaca, NY: Cornell University Press.

Thesleff, H. (1966) "Scientific and Technical Style in Early Greek Prose," *Arctos: Acta Philologica Fennica* 4, 89–113.

Thomas, R. (2009) "Writing, Reading, Public and Private 'Literacies': Functional Literacy and Democratic Literacy in Greece," in Johnson, W.A. and Parker, H. N. (eds). *Ancient Literacies: The Culture of Reading in Greece and Rome*. Oxford: Oxford University Press, 13–45.

Vernant, J.P. (1983) *Myth and Thought among the Greeks*, London: Routledge.

Veyne, P. (1984) *Writing History: Essay on Epistemology*, Middletown, CT: Wesleyan University Press.

Waugh, J. (1991) "Heraclitus: The Postmodern Presocratic," *The Monist* 74(4), 605–623.

Waugh, J. and Ariew, R. (2013) "The Contingency of Philosophical Problems," in Laerke, M., Smith, J.E.H., and Schliesser, E. (eds.) *Philosophy and Its History: Aims and Methods in the Study of Modern Philosophy*, Oxford: Oxford University Press, 91–114.

2

MARGARET CAVENDISH AND THE NEW SCIENCE

"Boys that play with watery bubbles or fling dust into each other's eyes, or make a hobbyhorse of snow"

Marcy P. Lascano

Introduction

In the seventeenth century the new science was introduced through the works of Bacon, Hooke, Boyle, Power, and others. The advocates of the new science promised to divulge the inner workings of nature and to help man overcome his painful fallen state by means of controlling nature. The new sciences of mechanism and corpuscularism were to be based on objective experiments that would reveal the secret inner natures of minerals, vegetables, animals, the sun, moon, and stars. These experiments were done with new and improved telescopes and microscope with magnifications of up to 100 times.

One early critic of the new science was Margaret Cavendish. Cavendish was skeptical of the ambitious claims, methodology, instruments, and institutions of the new science. In her work, *Observations Upon Experimental Philosophy*, Cavendish argued against "experimental and dioptrical writers," provided her own account of the natural world, investigated aspects of chemistry, medicine, and the nature of heat and color, as well as many other topics in natural philosophy (OEP: 10). While many think Cavendish landed on the wrong side of history with respect to her skepticism regarding microscopes and telescopes, her criticisms of the new science were wide ranging and she was by no means the only one to question the value of such experiments and instruments. While several commentators, like Eve Keller (1997), have argued that Cavendish was against all things experimental, several recent commentators, Emma Wilkins (2014) and Deborah Boyle (2018), have tried to show the much more complicated relationship between Cavendish and the Royal Society and medical studies, respectively. In addition, some commentators, such as Lisa Sarasohn (1984, 2010) and Eve Keller (1997), have argued that Cavendish's criticisms of the new science are based on her belief that Nature, as a representation of the feminine, was under attack by the experimentalists desire to "penetrate" and "manipulate" her for their own ends. While it is certainly true that Cavendish and many of the experimentalists personified nature as a woman, and that Cavendish does portray the men as trying to make her into something she is not, I agree with Deborah Boyle (2004) that these descriptions are not the focus of her objections to experimentalism. Rather than hold that Cavendish is concerned with, as Sarasohn claims, "the sexual implications for both women and nature of the new philosophy" (2010: 147), it seems that Cavendish's objections were largely based on her philosophical commitments. However, I believe there is one aspect of the new science that Cavendish does critique from a feminist perspective, and that is what she sees as its institutional nature and its exclusion of women on the basis of sex, and to this I will turn in the last section of the chapter.[1]

My overall aim is to address Cavendish's three major critiques of the new science. The chapter is divided as follows: the first section provides a brief overview of Cavendish's views on the nature of bodies and perception. The second regards her critique of the methods and aims of the new science as represented by Bacon and Boyle. The third section examines her critique of Hooke and the instruments of experimentalism. The final section lays out her feminist critique of the institution of the new science.

Cavendish on the Nature of Matter, Individuals, Causation, Perception, and Bodies

In order to understand Cavendish's objections to the claims of the experimentalists, we must examine her positions regarding the nature of bodies and perception.

Cavendish holds that everything in the world is material and vital. Matter is composed of three "degrees" – animate rational, animate sensitive, and inanimate – which are completely blended in such a way that every portion of matter contains all three degrees. Cavendish denies that there is any smallest part of matter (no atoms) and rejects the possibility of a vacuum. This results in all of nature being one material substance that is self-moving, perceptive, rational, and sensitive. Due to the homogenous nature of completely blended matter, every part of matter is convertible into any sort of thing. That is, the parts of matter that now compose my ears after my dissolution may form into, for instance, a cat's tail. What makes some portion of matter one particular thing, rather than some other particular thing, are its *corporeal figurative motions*. So, while the very same portion of matter may at one time be a human ear and at another time a cat tail, it has different capabilities or powers at each of those times due to its motions, shape, and the organization of its parts. A portion of matter formed into a human ear, for example, has the ability to pattern sounds at a distance and relay this information to the brain. That same bit of matter formed into a cat tail, while still perceptive, sensitive, and rational, does not have the ability to pattern external objects in this same way.

For Cavendish, an individual, whether it is a plant, animal, or glass jar, is a collection of sympathetically moving parts. These individuals can change and grow and gain or lose matter, as long as the sympathetic movements are maintained. These sympathetic figurative movements determine the organization, color, capacities, exterior shape, knowledge, and perceptions of the objects in question. Cavendish holds that due to the regularity and harmony of Nature's motions, natural beings fall into various kinds and species. These natural kinds are contrasted with artificial kinds (or as she sometimes calls them "art" or "artiface"), which are man-made objects.

The way in which particular parts of matter can move and express their sensitive and rational capacities depends upon their circumscribed figurative motions. Since nature moves in regular ways, her parts conjoin into (generally) regular figures, which can be sorted into the natural kinds we recognize, such as animals, vegetables, and minerals, and the particular sorts of each category.

Cavendish's self-moving matter provides her with an account of causation and perception that is different from the mechanists. According to Cavendish, causation and perception are due to the self-moving nature of composed parts. Cavendish provides an example in which a hand moves a string or ball. She writes,

> Therefore when a man moves a string, or tosses a ball, the string or ball is no more sensible of the motion of the hand, than the hand is of the motion of the string or ball; but the hand is only an occasion that the string or ball moves thus or thus. I will not say, but that it may have some perception of the hand, according to the nature of its own figure; but it does not

move by the hand's motion, but by its own: for, there can be no motion imparted, without matter or substance.

(OEP: 140)

Since every part of matter is perceptive and self-moving, according to Cavendish, interactions between the composed parts of nature are usually instances of occasional causation. For example, when a person sees a chipmunk, the chipmunk serves as the occasion for the person's sense organs to pattern the external motions and figure of the chipmunk. The chipmunk is the occasional cause and the person's sense organs are the principal and primary cause of the perception.[2]

According to Cavendish, objects have both interior and exterior parts and motions. The distinction might seem merely to direct us to spatially inner and outer parts of bodies. But as Deborah Boyle (2015: 442) notes, Cavendish refers to interior parts as the "inherent nature" of a thing. But what is this inherent nature? It seems clear that for Cavendish, the interior parts are those parts that make a thing a particular part or creature by means of their motions and figures. Cavendish writes,

> . . . it is to be observed, that in composed figures, there are interior and exterior parts; the exterior are those which may be perceived by our exterior senses, with all their proprieties. . .*But the interior parts are the interior, natural, figurative motions, which cause it to be such or such a part or creature*: As for example, man has both his interior and exterior parts, as is evident; and each of them has not only their outward figure or shape, *but also their interior, natural, figurative motions, which did not only cause them to be such or such parts; (as for example, a leg, a head, a heart, a spleen, a liver, blood, etc.) but do also continue their being.*
>
> *(OEP: 162, emphasis mine)*

Exterior parts are those that are visible. These parts, by their figurative motions, have color, size, weight, etc., and are available to human sensory experience. When perception occurs, the exterior sensory organs of a creature pattern the exterior motions of an object. This patterning is a sort of information transfer to the interior of the creature and causes the creature to move in accordance with its interior nature in response to the exterior movements of the object. But interior motions are not subject to our exterior sensory organs.

Cavendish claims that a ball cannot know the interior motions of the hand, and thus cannot move by the hand's interior motion. The hand's interior motions cause it to be a hand and enable the type of movements a hand has. However, the ball has its own interior motions, which enable it to move in accordance with its figure. Different types of interior motions and figures constitute different things. When a ball perceives the exterior motions of the hand, it patterns these motions through its perceptive senses and moves itself in accordance with its nature in reaction to the motions of the hand. Thus, the hand is an occasional cause of the self-motion of the ball. The ball is the efficient cause of its own motion. Nevertheless, the ball would not move in the way that it does without the existence of the hand's particular exterior motions.

Cavendish's Critique of the Methodology and Aims of the New Science

With this general account of Cavendish's views on body and perception, we may now turn to her views concerning the experimentalists' methodologies, aims, and instruments. I limit myself to discussing Cavendish in relation to Bacon and Boyle in this section and Hooke in the next. However, I should note that Cavendish had a rather impressive command of the new science. She discusses chemistry and alchemy, medicine, optics, astronomy, and physics in her works. In doing so, she addresses the views of Hobbes, Descartes, Gassendi, Harvey, Glanvill, Jan Baptist van Helmont, and Power in addition to those already mentioned. Moreover, Cavendish was not criticizing microscopes from the armchair; as Emma Wilkins (2014: 3) notes, Cavendish had her own microscope, as well as access to the six her husband had, some of which were made by Eustachio Divino, an Italian experimental philosopher. While Eve Keller (1997) has characterized Cavendish as an outsider whose critiques of the Royal Society were gender based, Cavendish was not an outsider in the sense that she was barred from access to the writings of the experimentalists or lacked the support and finances to pursue her interest in natural philosophy.[3] Of course Cavendish was not a member of the Royal Society, as women were not allowed to join. However, Cavendish would be the only woman in the seventeenth century, and I believe the eighteenth, to be granted a visit to the society. There she observed Robert Hooke's and Robert Boyle's experiments. It is true, however, that very few philosophers engaged with Cavendish. She knew Hobbes, who praised her plays, through her husband and brother-in-law, but claims she never spoke with Descartes (despite his presence in her house) and she failed to engage Henry More despite repeated solicitations. Cavendish did have correspondence with several natural philosophers though, including Joseph Glanvill, Henry Power, and Christiaan Huygens (Akkerman and Corporaal 2004; Broad 2007, 2019).

One philosopher who is often mentioned in the scholarly literature on Cavendish and the new science but who is rarely discussed is Francis Bacon. Cavendish never cites Bacon, but it is clear that the Baconian conception of the scientific enterprise is in the background of much of her later works. We know that Cavendish counters his male-dominated scientific utopia, *New Atlantis*, in her own science fiction work, *Blazing World*, which was published with *Observations Upon Experimental Philosophy*.[4] Moreover, in *Observations*, Cavendish discusses many of the same examples that Bacon uses in the *New Organon* – the nature of heat and cold, and wind – although she draws different conclusions from her observations.

Bacon's suggested methodology in the *New Organon* is an inductive-deductive method that moves from observation to general principles and back to observation. In the *New Organon*, he begins with the production of lists or charts that are made by observation. He takes heat as his example. The first table of "instances meeting in the nature of heat" is a list of 28 cases of the phenomenon of heat (Bacon 2000: 110–111). After this list comes another "Closely related instances which are devoid of the nature of heat" (Bacon 2000: 112–119). After this, a third "table of the degrees or a comparison on heat" is provided (Bacon 2000: 119–126). These lists of observations are the starting point of investigation into the form, or structural nature, of a particular thing. Bacon thought that by examining the lists of presence, absence, and degrees, one could rule out accidental correlations and thereby come to only those that are essential. Bacon held that the observations and experiments which excluded various instances were key to divulging the form of a particular nature. For instance, we can exclude heat as having an "elemental nature" because it is instanced in the rays of the sun. Through this inductive method (though not an enumerative one), Bacon believed that one could arrive at axioms that would hold at various levels of phenomena. These are confirmed by deductive arguments. Although there are further steps in the methodology, involving privileged instances, relational instances, etc., the upshot is that observation and experimentation are aided by reasoning about which instances are perspicuous, what inductive generalizations hold, and what the next steps in testing should be.

As far as general methodology in science is concerned, it seems unlikely that Cavendish would have objected to Bacon's account. Her own methodology in *Observations* is quite similar. She often cites phenomena that we know through sense and reason. That is, she sees herself as providing explanations of the things we experience in the world. She writes, "the best study, is rational contemplation joined with the observations of regular sense" (OEP: 53). For instance, she discusses decay and death and change and generation. Cavendish spends quite a bit of time discussing the nature of heat, cold, fire, snow, as well as the motions of plants and animals. She recounts the changing of a chrysalis into a butterfly (OEP: 61–62). There is no doubt that her interest in observation comes from her reading of Bacon, and his leading proponent, Robert Boyle. We know that Cavendish read Boyle and observed his experiments at the Royal Society. However, as Emma Wilkins notes, Boyle preferred to leave the work of creating machines and running experiments to others (2014: 9) and focused, like Bacon, on methodology. Boyle, like Bacon, cautions against the hasty establishment of universal "principles and axioms" from too few experiments and observations and claims that "an absolute suspension of the exercise of Reasoning were exceeding troublesome, if not impossible" (Boyle 1669: 9).

Cavendish was not critical of observation, nor was she necessarily critical of all experiments. Commenting in *Philosophical Letters* on the work of Boyle, Cavendish notes that his method concentrates, like Bacon's, on studying "different parts and alterations, more than the motions, which cause the alterations in those parts," which are what Cavendish holds to be fundamental (PL: 496). She then goes on to note "for certainly experiments are very beneficial to man" (PL: 496).

Another point of agreement between Cavendish and Bacon was the use of hypotheses and a healthy distrust of human sense. Bacon writes,

> The senses are defective in two ways: they may fail us altogether or they may deceive. First, there are many things which escape the senses even when they are healthy and quite unimpeded; either because of the rarity of the whole body or by the extremely small size of its parts, or by the distance, or by its slowness or speed, or because the object is too familiar, or for other reasons. And even when the senses do grasp an object, their apprehensions of it are not always reliable. For the evidence and information given by the senses is always based on the analogy of man not of the universe; it is a very great error to assert that the senses are the measure of things.
>
> *(Bacon 2000: 17–18)*

Like Bacon Cavendish thinks that we are limited by our particular sensory abilities. These she believes are not keen enough to get to the interior natures of things (if indeed it is possible to do so at all). However, this does not mean that Cavendish completely distrusted sensitive knowledge. In her discussion of the various degrees of heat and cold, she writes

> . . .some degrees and sorts of heat and cold, are subject to the human perception of sight, some to the perception of touch, some to both, and some to none of them; there being so many various sorts and degrees both of heat and cold, as they cannot be altogether subject to our grosser exterior senses. . .for although our sensitive perceptions do often commit errors and mistakes, either through their own irregularity, or some other ways; yet, next to the rational, they are the best informers we have: for, no man can naturally go beyond his rational and sensitive perception.
>
> *(OEP: 109–110)*

Bacon claims that the senses are in need of assistance, which he claims comes not so much from instruments, but from experiments (2000: 18). While his general methodology and his commitment to reason were both likely in accord with Cavendish's views, she disagreed with the stated aims of the Baconian enterprise. According to Bacon (and Boyle) the end of science is the discovery of the forms of a nature, which Bacon defines as "nothing more than those laws and determinations of absolute actuality which govern and constitute any simple nature, as heat, light, weight, in every kind of matter and subject that is susceptible of them" (Bacon 2000: 145–146). This nature must be the same in all cases where the phenomenon actually occurs and will include ways in which the nature changes and reacts with other natures. Gaining knowledge of the forms was not an end in itself, but a necessary step in coming to control and modify nature.

The pursuit of the forms of nature is the main disagreement that Cavendish has with Bacon and Boyle. In her lengthy discussions of the various types of heat and cold, she comes to exactly the opposite conclusion than they come to. That there is a simple nature of every kind of thing that can be discerned and manipulated is simply false, according to Cavendish. There is no one cause or principle of heat, rather different corporeal figurative motions in individuals produce various kinds of heat. There is a sort of family resemblance between the things that we call "heat," but there is no one thing that is heat. While Bacon holds, like Cavendish, that the perception of heat is subjective, so that different people perceive heat in different ways, he believes that there is only one cause and one simple nature for all instances of heat. Cavendish, in contrast, holds that this subjectivity holds across all entities and so it is impossible that there be one nature of heat or one cause of heat. She writes,

> . . .one prime action or motion cannot produce all sorts of heat or cold. For, though all sorts of heat or cold, are still heat and cold (as all sorts of animals, are still animals) yet all the several sorts or kinds of them, are not one and the same kind, but different. Nor does one particular action, produce all those several sorts or kinds. For, if there were no differences in their productions, then would not only all men be exactly like, but all beasts also; that is, there would be no difference between a horse and a cow, a cow and a lion, a snake and an oyster.
>
> *(OEP: 115)*

Cavendish's claim here is that different kinds of animals are all still animals; yet in order to get different kinds, they must be produced in different ways that causes their different interior structures. That is, cows are produced differently than sheep. The same goes for different kinds of heat such as chemical heat, the heat of fire, or the burning sensation caused by ice. Given her account of causation and perception, an understanding of heat will always involve the motions and structure of both the occasional and the efficient cause. From this it is easy to see why Cavendish is critical of the notion that there are forms in nature that are over and above the resemblance class that constitutes what we call "species." As she notes, "form cannot be created without matter, nor matter without form; for form is no thing subsisting by itself without matter" (OEP: 203). So, Cavendish's critique of forms involves two parts: the denial of the existence of form without matter, and the denial of a single cause of every type of entity.

Moreover, Cavendish is extremely critical of philosophers who try to distill the complex workings of nature down to a few principles (or worse one principle). In section II.6 of *Observations* titled, "Whether There Be Any Prime or Principle Figure in Nature; and of the True Principles of Nature," Cavendish argues that those who try to distill the complex workings of nature down to one principle ("globular figures," salt, water, one of the four elements, etc.) make the mistake of taking one part of nature to be the principle of the whole. Rather, composed bodies are merely effects of nature or self-moving matter, and as effects they cannot be the cause of all of nature. Cavendish claims that self-moving matter can be called the "principles of nature," but this, of course, is just to say that all of matter is the principle of matter, and so talk of one principle within nature is meaningless (OEP: 206).

Cavendish's Critique of Experimental Philosophy

While Cavendish's discussions of Bacon and Boyle tend to the general methodological worries and aims of science, her criticisms of Hooke are more pointed. At issue here are three claims. First, Hooke claims that we, as human beings, are uniquely suited to understand nature's workings.

Second, he advances the mechanist conception of nature as artifice. And finally, he thinks that the true natures of bodies can be seen through the microscope.

Hooke published *Micrographia* in 1665. The book was meant to show the previously unseen world that is revealed by the microscope and features etchings of various insects and plants along with descriptions of their appearances under the lens. The "Preface" contains Hooke's vision for science. First, he claims that mankind is above the rest of nature and has the unique ability to alter, assist, and improve nature.

> It is the great prerogative of Mankind above other Creatures, that we are not only able to behold the works of Nature, or barely to sustain our lives by them, but we have also the power or considering, comparing, altering, assisting, and improving them to various uses.
>
> *(Hooke 1665: unpaginated preface)*

Cavendish's response to this claim is that no part of nature is able to understand the whole of nature. Human beings, according to Cavendish, are not special in nature. She claims that many non-human animals have abilities that far outstrip ours. Moreover, she holds that since nature is infinite in her motions, no finite being could possibly comprehend her works.

Next, Hooke makes clear that the model for understanding nature is a mechanistic one.

> [W]e may perhaps be inabled to discern all the secret workings of Nature, almost in the same manner as we do those that are the productions of Art, and are manag'd by Wheels, and Engines, and Springs, that were devised by humane Wit.
>
> *(Hooke 1665: unpaginated preface)*

Cavendish, of course, did not think that nature motions were mechanistic, but rather vitalistic. In addition, she did not think that the workings of machines made by human art could help us understand nature. As she writes, "The rules of art cannot be the rules of nature, nor the measures of art the measures of nature. . .for though art proceeds from nature, yet nature does not proceed from art, for the cause cannot proceed from the effect" (PPO: Preface XXVI). Art is an effect of nature in the sense that humans are only able to create with art what is possible according to nature, but this art does not work in the same way that nature works, according to Cavendish.

Her most extensive objections to Hooke are based on her views of perception and the nature of bodies in her criticism of the microscope's ability to get to the essence of beings. She claims that "could experimental philosophers find out more beneficial arts than our forefathers have done," for improving farming, housing, and trade, as well as clearing up disputes in the church with their experiments, then their pursuits would be praiseworthy. However, she goes on to note, "But, as boys that play with watery bubbles or fling dust into each other's eyes, or make a hobbyhorse of snow, are worthy of reproof rather than praise, for wasting their time with useless sports" (OEP: 51–52).[5]

Cavendish does not see how looking at flies under microscopes could possibly make human existence easier. Given the immediate needs of human beings there are more profitable ways of spending one's time than playing with test tubes or trying to make snow in a laboratory. She concludes that "those that invented microscopes, and such like dioptric glasses, at first, did, in my opinion, the world more injury than benefit" (OEP: 51).[6]

Cavendish claims microscopes can only reveal "phenomena or the exterior figures of objects" (OEP: 51). That is, she thinks that what is revealed through the lens of the microscope is not the real natures of things, but merely the outer surfaces of objects. Moreover, she thinks these exterior figures may not be accurate. Her reason for thinking so is two-fold, but it is due partly to Hooke's own descriptions of the difficulty of getting the "true form" of what was under the microscope. In *Micrographia*, he tells us that the engravings in the book are composites of drawings he made of various parts of his subjects. For it was often impossible to get a whole specimen under the lens and it was difficult to ascertain features of what he could see. He writes,

> . . .and that in making of them, I indeavoured (as far as I was able) first to discover the true appearance, and next to make a plain representation of it. This I mention the rather, because of these kind of Objects there is much more difficulty to discover the true shape, then of those visible to the naked eye, the same Object seeming quite differing, in one position to the Light, from what it really is, and may be discover'd in another. And therefore I never began to make any draught before by many examinations in several lights, and in several positions to those lights, I had discover'd the true form. For it is exceeding difficult in some Objects to distinguish between a prominency and a depression, between a shadow and a black stain, or a reflection and a whiteness in the color.
>
> (Hooke 1665: unpaginated preface)

While the microscopes that Hooke used were the best in his day, they were by no means what we would consider good. The images were dim and the lenses often quite irregular. In addition, the experimenter relied on ever-changing ambient light to illuminate his subject. This led to the belief that the process was unreliable. If an object looks different in different lighting conditions or in various positions, how does one determine what the "true form" is given that there is no independent verification for something that cannot be seen with the naked eye? Cavendish makes this point.

> Nay, artists do confess themselves, that flies, and the like, will appear of several figures or shapes, according to the several reflexions, refractions, mediums and positions of several lights; which if so, how can they tell or judge which is the truest light, position, or medium, that doth present the object naturally as it is?
>
> (OEP: 51)

The same worry holds for telescopes given that they claim to reveal things that are beyond the scope of human perception. This uncertainty gave Cavendish reason to withhold her approval of microscopes, but she was not skeptical of all kinds of lenses.

> But, mistake me not; I do not say, that no glass presents the true picture of an object: but only that magnifying and multiplying, and the like optic glasses, may, and do oftentimes present falsely the picture of an exterior object; I say, the picture, because it is not the real body of the object which the glass presents, but the glass only figures or patterns out the picture presented in and by the glass, and there mistakes may easily be committed in taking copies from copies.
>
> (OEP: 50–51)

Here, Cavendish hints at the second reason for her skepticism about microscopes – her view that every part of nature is perceptive. According to Cavendish, when we look into a mirror, we pattern the exterior image of the glass. But it is also the case that the glass of the mirror is patterning us. The image we pattern from the mirror involves a sort of double perception. The mirror patterns us and we pattern what the mirror patterns. Since we have no idea how a mirror patterns, we might be skeptical of the claim that a mirror patterns our external shape in an accurate manner. Of course, in the case of seeing our reflection in a mirror, independent verification is possible. Someone else can tell us that their perception, when they look directly at us, is similar to the one they see in the mirror. However, in the case of microscopes (and telescopes), we cannot be certain that the glass of the microscope is patterning the true image of its subject and there is no possibility of independent verification. Because we merely pattern the pattern of the microscope (copy the copy), we cannot be certain of the veracity of our perception when using such instruments.

We can see that Cavendish had a number of criticisms of the aims, claims, and instruments of experimental philosophy. However, it is clear that Cavendish was not against all scientific inquiry and experimentation. Cavendish was both involved with the new science and a critic of it. She thinks human beings are capable of ascertaining knowledge of nature and that some of this knowledge is capable of improving our lives. She writes,

> That the undoubted truth in Natural Philosophy, is, in my opinion, like the Philosopher's Stone in Chymistry, which has been sought for by many learned and ingenious Persons, and will be sought as long as the Art of Chymistry doth last; but although they cannot find the Philosophers Stone, yet by the help of this Art they have found out many rare things both for use and knowledg. The like in Natural Philosophy, although Natural Philosophers cannot find out the absolute truth of Nature, or Natures ground-works, or the hidden causes of natural effects; nevertheless they have found out many necessary and profitable Arts and Sciences, to benefit the life of man; for without Natural Philosophy we should have lived in dark ignorance, not knowing the motions of the Heavens, the cause of the Eclipses, the influences of the Stars, the use of Numbers, Measures, and Weights, the vertues and effects of Vegetables and Minerals, the Art of Architecture, Navigation, and the like.
>
> *(PL: 508)*

Cavendish is clearly in favor of natural philosophy and its use in certain domains for the benefit of human life. In building better ships and houses, in aiding navigation, and in improving trade and measures, science does make life better. In this way, as Emma Wilkins notes, Cavendish was in line with the Baconian belief that the fruit of scientific endeavor is practical application (Wilkins 2014: 10). However, it is clear that not just any manipulation of nature counts as beneficial. For Cavendish, much of the work done through the microscope either did not rise to the level of helpfulness or would not bear fruit in a reasonable enough time to be undertaken. Finally, although Cavendish's relation to the new science is complex, there is one aspect of it which she was clearly against and that is its complete exclusion of women.

Cavendish's Critique of the Institution of Natural Philosophy

As we saw in the previous sections, Cavendish's arguments against the methods, aims, and instruments of the new science are based on her views of the nature of bodies, human understanding, and causation and perception. In addition to these criticisms, Cavendish also confronted the institutional barriers that prevented women from participating in natural philosophy. In this section, I situate Cavendish's criticisms in terms of a feminist critique.

Following recent work by Eileen O'Neill (2019), we can trace three core components of feminism back to (at least) the seventeenth century. They are (1) criticism of misogyny and male supremacy, (2) the conviction that women's condition is not an immutable fact of nature and can be changed for the better, and (3) a sense of gender group identity, the conscious will to speak "on behalf of women," or "to defend the female sex," usually aiming to enlarge the sphere of action open to women.[7] While, Deborah Boyle has argued convincingly that Cavendish's natural philosophy was not feminist (2004) nor is she accurately described an advocate of women with respect to marriage or certain social reforms (2018: 166–188), I do think her critique of the exclusion of women from natural philosophy exemplifies all three of the core components of feminism. Cavendish recognizes the fact that she, as a member of the female sex, no matter how capable and knowledgeable, is not allowed to participate fully in the new science. Cavendish points to misogyny and the barring of women from universities as key factors in this exclusion.

There are several places in *Observations* where Cavendish complains that women are not allowed to participate in Natural Philosophy. The first of these occurs in the "To the Reader." She writes,

> But that I am not versed in learning, nobody, I hope, will blame me for it, since it is sufficiently known, that our sex being not suffered to be instructed in schools and universities, cannot be bred up to it. I will not say, but many of our sex may have as much wit, and be capable of learning as well as men; but since they want instructions, it is not possible they should attain to it: for learning is artificial, but wit is natural.
>
> *(OEP: 11)*

Here, Cavendish clearly claims that women are not innately inferior to men – they are capable of learning. The reason for the difference in their actual abilities is due to education. No amount of natural ability can make up for the fact that women were not educated in the same way as men. In this passage, she clearly is expressing component (2) discussed earlier.

Another criticism comes in a passage discussing her study of ancient philosophers. She notes that after reading the ancients, she found so much difference between their views and her own that she thought she might like to start her own school of philosophy.

> . . .were it allowable or usual for our sex, I might set up a sect or school for myself, without any prejudice to them [the ancient philosophers]: But I, being a woman, do fear they would soon cast me out of their schools; for, though the muses, graces and sciences are all of the female gender, yet they were more esteemed in former ages, than they are now; nay, could it be done handsomely, they would now turn them all from females into males: So great is grown the self-conceit of the masculine, and the disregard of the female sex.
>
> *(OEP: 249)*

Here, Cavendish clearly states that men have very little regard for women, even women who are capable of doing the things that they find valuable, like philosophy. This explicit recognition of the fact that she would not be allowed to teach philosophy or even associate with a school of philosophers is quite telling. The feminine might be used as a muse, grace, or representation of science, but she is not allowed to participate. This is clearly a lamentation of male misogyny, which is the first core component of feminism mentioned earlier.

Cavendish, as noted, often depicts Nature as female. But in *Observations*, Nature is a woman engaging in the methods of experimental science. She writes, "Nature being a wise and provident lady, governs her parts very wisely, methodically, and orderly" as a "good housewife does in brewing, baking, churning, spinning, sowing, etc. as also in preserving, for those that love sweetmeats; and in distilling, for those that take delight in cordials" (OEP: 105).

Not only does Cavendish describe Nature as a housewife, she goes on in this passage to make the connection between women's experimentation with recipes and alchemy/chemistry. She suggests rather coyly that women:

> would prove good experimental philosophers, and inform the world how to make artificial snow, by their creams, or possets beaten into froth: and ice, by their clear, candied, or crusted quiddities, or conserves of fruits: and frost, by their candied herbs and flowers: and hail, by

their small comfits made of water and sugar, with whites of eggs: And many other the like figures, which resemble beasts, birds, vegetables, minerals, etc.

(OEP: 105–106)

Here Cavendish notes the ways in which women have already succeeded in imitating nature in their kitchens. She suggests that women could help men by actually doing the experiments and then the men could "study the causes" of them. She writes, "I am confident, women would labour as much with fire and furnace, as men; for they'll make good cordials and spirits" (OEP: 106).

Cavendish acknowledges the fact that all women have been distilling and brewing medicines for generations. Here (and in the other passages) she speaks for all women (component 3 discussed earlier) who have quietly been engaged in medicine, chemistry, and experimentation throughout history. These practical sciences were handed down through generations of women and recorded for future use in households. In the seventeenth century, Cavendish was witness to the fact that the art of taking care of the sick would be given over to men who studied chemistry and medicine in schools where women were not allowed. There is no doubt that this exclusion was all the more insulting given that these arts (chemistry, alchemy, and medicine) had been practiced by women for years without credit. In forwarding these criticisms, Cavendish speaks on behalf of all women against the misogyny that prevents women from developing their intellects and participating in philosophy. In doing so, it is safe to say that Cavendish did put forth a feminist critique of the institutions of natural philosophy in addition to her other criticisms of the new science.

Conclusion

Margaret Cavendish had extensive knowledge of seventeenth-century natural philosophy. She was conversant in metaphysics, physics, astronomy, medicine, optics, and what we now call botany and biology. She successfully corresponded with several natural philosophers, developed her own philosophical system, and wrote philosophical texts criticizing what she saw as the overreach of some of the most famous experimentalists of her time. But to a large extent her work was unacknowledged. The fact is that no one bothered to engage with her because of her sex, and she acknowledged this injustice in her criticisms of the institutions of science.

Related chapters: 1, 3.

Notes

1 For an excellent refutation of earlier attempts to read Cavendish as presenting a feminist critique, see Boyle (2004).
2 Occasional causation, as defined by Steven Nadler (1994: 39),

> denotes the entire process whereby one thing, A, occasions or elicits another thing, B, to cause e. Even though it is B that A occasions or incites to engage in the activity of efficient causation in producing e, the relation of occasional causation links A not just to B, but also (and especially) to the effect, e, produced by B.

3 She was, of course, limited by her lack of formal education since women were not allowed to attend university. As a result, she could not read Latin or French.
4 O'Neill's critical edition (2001) does not include *Blazing World*, but see Susan James' edition of Cavendish's political works (Cavendish 2003).
5 It should be noted that Bacon and Hobbes both shared Cavendish's belief that the best sciences have practical uses.
6 Wilkins (2014: 4–5) notes that John Locke also objected that when we are looking at the inside of bodies we are not actually seeing the interior nature or essence of the being. In his early work with Thomas

Sydenham, "Anatomica," he argues that looking at the inner parts of a man does not show the inner operations of his organs (Dewhurst 1958).

7 O'Neill relied on Akkerman T. and Stuurman S. (1998: 3–4) for these components.

References

Akkerman, N. and Corporaal, M. (2004) "Mad Science Beyond Flattery: The Correspondence of Margaret Cavendish and Constantijn Huygens," *Early Modern Literary Studies*, Special Issue 14, 2: 1–21.

Akkerman, T. and Stuurman, S. (1998) "Introduction," in Tjitske Akkerman and Siep Stuurman (eds.) *Perspectives on Feminist Political Thought in European History: From the Middle Ages to the Present*, London: Routledge, 1–33.

Bacon, F. (2000) *The New Organon*, Lisa Jardine and Michael Silverthorne (eds.), Cambridge: Cambridge University Press.

Boyle, D. (2018) *The Well-Ordered Universe: The Philosophy of Margaret Cavendish*, Oxford: Oxford University Press.

Boyle, D. (2015) "Margaret Cavendish on Perception, Self-Knowledge, and Probable Opinion," *Philosophy Compass* 10(7): 438–450.

Boyle, D. (2004) "Margaret Cavendish's Nonfeminist Natural Philosophy," *Configurations* 12: 195–227.

Boyle, R. (1669) *Certain Physiological Essays and Other Tracts Written at Distant Times, and on Several Occasions by the Honourable Robert Boyle; Wherein Some of the Tracts Are Enlarged by Experiments and the Work Is Increased by the Addition of a Discourse about the Absolute Rest in Bodies*, London: Printed for Henry Herringman.

Broad, J. (2019) *Women Philosophers of Seventeenth-Century England: Selected Correspondence*, Oxford: Oxford University Press.

Broad, J. (2007) "Margaret Cavendish and Joseph Glanvill: Science, Religion, and Witchcraft," *Studies in History and Philosophy of Science* Part A 38(3): 493–505.

Cavendish, M. (2003) "The Description of the New World, Called the Blazing World," in Susan James (ed.) *Margaret Cavendish: Political Writings*, Cambridge: Cambridge University Press, 1–110.

Cavendish, M. (2001) *Observations upon Experimental Philosophy*, Eileen O'Neill (ed.), Cambridge: Cambridge University Press. Cited as OEP.

Cavendish, M. (1664) *Philosophical Letters: Or, Modest Reflections Upon Some Opinions in Natural Philosophy, Maintained by Several Famous and Learned Authors of This Age. . .*, London: n.p. Cited as PL.

Cavendish, M. (1663) *Philosophical and Physical Opinions*, London: William Wilson. Cited as PPO.

Dewhurst, K. (1958) "Locke and Sydenham on the Teaching of Anatomy," *Medical History* 2: 1–12.

Hooke, R. (1665) *Micrographia, or Some Physiological Descriptions of Minute Bodies made by Magnifying Glasses*, London: Jo. Martyn & Ja. Allestry.

Keller, E. (1997) "Producing Petty Gods: Margaret Cavendish's Critique of Experimental Science," *ELH* 64(2): 447–471.

Nadler, S. (1994) "Descartes and Occasional Causation," *British Journal of the History of Philosophy* 2: 35–54.

O'Neill, E. (2019) "Introduction," in Eileen O'Neill and Marcy Lascano (eds.) *Feminist History of Philosophy: The Recovery and Evaluation of Women's Philosophical Thought*, Dordrecht: Springer, x–xxxvi.

Sarasohn, L. (2010) *The Natural Philosophy of Margaret Cavendish*, Baltimore: Johns Hopkins University Press.

Sarasohn, L. (1984) "A Science Turned Upside Down: Feminism and the Natural Science of Margaret Cavendish," *Huntington Library Quarterly* 47: 289–307.

Wilkins, E. (2014) "Margaret Cavendish and the Royal Society," *The Royal Society Journal of the History of Science* 68: 245–260.

Further Reading

Clucas, S. (2000) "The Duchess and Viscountess: Negotiations between Mechanism and Vitalism in the Natural Philosophies of Margaret Cavendish and Anne Conway," *In-between: Essays and Studies in Literary Criticism* 9(1): 125–36. (Excellent discussion of the development of vitalism in Cavendish and her contemporary, Anne Conway, in reaction to the shortcomings of mechanism.)

Cunning, D. (2016) *Cavendish*, New York: Routledge. (The first book-length account of Cavendish's philosophical system written by a philosopher.)

Hutton, S. (2003) "Science and Satire: The Lucianic Voice of Margaret Cavendish's *Description of a New World Called the Blazing World*," in Line Cottegnies and Nancy Weitz (eds.) *Authorial Conquests: Essays on Genre in the Writings of Margaret Cavendish*, Madison: Fairleigh Dickinson University Press; London:

Associated University Press, 161–178. (Exploration of Cavendish's science fiction as a commentary on natural philosophy.)

O'Neill, E. (2001) "Introduction," in Eileen O'Neill (ed.) *Observations upon Experimental Philosophy*, Cambridge: Cambridge University Press. (Excellent overview of the issues in Cavendish's work dedicated to natural philosophy.)

Whitaker, K. (2002) *Mad Madge*, London: Chatto & Windus. (The authoritative biography.)

3

ÉMILIE DU CHÂTELET

Feminism, Epistemology and Natural Philosophy

Karen Detlefsen

Introduction

It is not hard to find examples of robust feminism in early modern philosophy, especially in philosophy that connects with ethical and social-political questions, but also in epistemology, metaphysics and natural philosophy (e.g. Broad and Detlefsen 2018; Thomas 2018). Émilie Du Châtelet (1706–1749) produced much excellent philosophical work in her lifetime, including works in natural philosophy—physics, optics and experimental work on the nature of fire—metaphysics and value theory, broadly conceived. Throughout her lifetime, and across all forms of her philosophical engagement, she displayed, in both the explicit written word and in her actions, an acute understanding of the ways in which her work was impacted by her being a woman. From her discussion of the long-term impact upon women's minds of limited early opportunities in education (Detlefsen 2017; Gardiner 1984) to her exclusion from institutions such as the Académie Royale des Sciences (Detlefsen forthcoming), Du Châtelet was alert to the obstacles as a result of her gender to her full participation in public intellectual life in France in the mid-eighteenth century, including in the sciences.

This essay examines some aspects of Du Châtelet's feminism as expressed in her natural philosophy—the closest early modern analogue to contemporary science. One broad way in which feminism is manifest in the philosophy of science is the myriad of ways in which science has been theorized by feminists, for example, by bringing feminist insights to epistemology in science. A second broad way in which feminism is manifest in science is on display when women engage in science, producing first-rate scientific work. My goal is to start with one of Du Châtelet's contributions to feminist theorizing about epistemology, including proper modes of knowing in science (an example of the first broad approach just noted), and then show how her theorizing in this way led to her top-drawer innovations in the scientific enterprise (an example of the second broad approach just noted).[1]

The chapter unfolds as follows. First, I examine Du Châtelet's early contributions to a topic in epistemology now known as the ethics of belief. Du Châtelet's entrance into this topic starts with her concerns about human bias when it comes to thinking about women, but she also extends her ideas to bias in science. In the next section, I introduce a debate that was at the forefront in various scientific communities in early modern Europe, including in Du Châtelet's milieu in mid-eighteenth-century France, namely the debate between speculative and experimental philosophy. I focus on this debate in part because it exposes the failure on behalf of many scientific practitioners to abide by the dictates of the kind of ethics of belief, which Du Châtelet espouses.

Moreover, this debate exposes a strange (to us) chapter in the history of science, a chapter in which many of those who embrace a commitment to experimental natural philosophy—experimental science—eschew the use of hypotheses in scientific practice. I then turn to a consideration of how Du Châtelet practices ethical belief formation in her own scientific practice as found in her masterwork, *Foundations of Physics* (hereafter *Foundations*). Specifically, I show how she refigures the landscape in the on-going debate between experimental and speculative natural philosophers because of her commitment to minimizing bias in her thinking. As a result, and as I argue in the final section of this chapter, Du Châtelet occupies a central role in the history of the integration of hypotheses into scientific practice, much out of step with her own time, and much in line with science as it would emerge in our contemporary world.

Du Châtelet and the Ethics of Belief in Theory: Bias, Women, Science

As a highly talented woman with especially notable abilities in mathematics and metaphysics, it is little wonder that Du Châtelet would be alert to biases against women and their intellectual capacities that might arise from a failure to examine such biases, that is, it is little wonder that she would be alert to ethical failures with respect to belief formation about the nature of women. Among her earlier writing—from the second half of the 1730s before she turned her attention full-force to metaphysics and physics—are those in value theory, including her Preface to her translation to Bernard Mandeville's *Fable of the Bees* and her posthumously published *Discourse on Happiness*. In both texts, she offers forceful discussions about the unexamined beliefs that many people hold about women, and the cost to women's lives, especially their intellectual lives.

In her Preface to *Fable of the Bees*, for example, she writes:

> Let us reflect briefly on why for so many centuries, not one good tragedy, one good poem, one esteemed history, one beautiful painting, one good book of physics, has come from the hands of women. Why do these creatures, whose understanding appears in all things equal to that of men, seem for all that, to be stopped by an invincible force on this side of a barrier; let someone give me some explanation, if there is one. . .. As for me, I confess that if I were king. . . I would allow women to share in all the rights of humanity, and most of all those of the mind. Women seem to have been born to deceive, and their soul is scarcely allowed any other exercise. . ..
>
> I am convinced that many women are either ignorant of their talents, because of flaws in their education, or bury them out of prejudice and for lack of a bold spirit. What I have experienced myself confirms me in this opinion . . ..
>
> I hold myself to be quite fortunate to have renounced in mid-course frivolous things that occupy most women all their lives, and I want to use what time remains to cultivate my soul . . ..

> *(Zinsser 2009: 48–49)*

In *Discourse on Happiness*, she acknowledges again the constraints women labor under as a result of institutional forces, especially the lack of education which deprives women of their natural ability to think well (Zinsser 2009: 357), and in the *Discourse* she expands on the meaning and power of prejudice. She writes, for example, that our very happiness relies upon our ability to free ourselves from prejudice, and she pointedly notes that responsibility for freeing ourselves from prejudice lies with the individual believer:

> [A] source of happiness is to be free from prejudices, *and the decision rests with us to rid ourselves of them*. We all have a sufficient share of intelligence to examine things that others want to

oblige us to believe.... Prejudice is an opinion that one has accepted without examination, because it would be indefensible otherwise. Error can never be a good thing, and it is surely a great evil in the things on which the conduct of life depends.

(Zinsser 2009: 352, emphasis added)

Du Châtelet's (surely overstated) claim in the first passage quoted earlier that women have not contributed to the intellectual goods of history leads her to an analysis of why this is the case. Women have been deprived of a first-rate education, they have been so deprived because they are assumed to be naturally, essentially different from men in that their souls are prone to focus on deception and frivolity, and this prejudice has conditioned women's lives both in the form of so-cial institutions that offer few options for women's intellectual activity and in the form of women internalizing these prejudices about themselves. Du Châtelet speaks from first-person knowledge of these prejudices about the nature of women, and of her own liberation from these internalized prejudices. Fundamental in this account is the unexamined, untested assumptions and prejudices about women's natures. That is, fundamental in this account is a collective, societal failure[2] to hold beliefs that have been duly examined, and that are thus ethically held. The second passage quoted shifts focus to the individual's responsibility to use their wholly adequate intellectual capacities to examine, question, and where relevant reject, prejudices which are shown through examination to be false. This is at the core of an ethics of belief: the individual's ethical responsibility to be responsible in the methods of belief formation, such that those beliefs that one does hold are not clearly erroneous beliefs.

Du Châtelet may start with unethically formed and thus unethically held beliefs about women, their natures and talents, but she extends her ethics of belief to other facets of human intellectual life as well, including science. In 1740, she published what we now understand to be her masterful contribution to the history of natural philosophy, even if we are still coming to grips with the depth and breadth of her extraordinary contributions to our history of science in that text.[3] She opens the work, in the Preface, by indicating that it is intended as a textbook for her son—even though it is far more ambitious in intent and accomplishment than that, going well beyond in philosophical and technical sophistication, as well as in originality, the standard popularizations of Newton, sometimes significantly simplified "for the ladies" (Hutton 2004a, 2004b; Zinsser 1998). Writing to her son, she begins:

> I have always thought that the most sacred duty of men was to give their children an educa-tion that prevented them at a more advanced age from regretting their youth, the only time when one can truly gain instruction. You are, my dear son, in this happy age when the mind begins to think . . ..
>
> You must early on accustom your mind to think, and to be self-sufficient. You will per-ceive at all times in your life what resources and what consolations one finds in study, and you will see that it can even furnish pleasure and delight . . .
>
> Guard yourself, my son, whatever side you take in this dispute among philosophers [New-ton and Descartes], against the inevitable obstinacy to which the spirit of [national] partisan-ship carries one; this frame of mind is dangerous on all occasions in life; but it is ridiculous in physics [and] the search for truth . . ..
>
> *(c.f. Brading et al 2018; Zinsser 2009: 116, 119–120)*

Once again, Du Châtelet underscores the importance of education, especially in youth, but there is also the importance of an ethics of belief coursing through these paragraphs. She implies that one's own mind and one's own thoughts can and ought to belong to oneself in the sense that they can and ought to be within our self-control and free from inappropriate external influence. This is implied by her urging her son to accustom his mind early on to be "self-sufficient", and her later cautioning that he avoid falling prey to believing something due to a nationalist sentiment. Du Châtelet also indicates that there are ethical duties associated with the cultivation of such minds. It is a parent's duty to develop such a mind in her child. But also implicit in this passage is the idea that individual thinkers have an ethical duty to make use of their well-educated minds both to cultivate their own happiness and more importantly, perhaps, to avoid falling prey to factors—such as nationalist sentiments—which negatively influence the pursuit of truth.

It is true that the idea of an ethics of belief entered into Western philosophical thought under that moniker only in the late nineteenth century. The idea was made famous by William K. Clifford in his paper "The Ethics of Belief" (Clifford 1879), in which he stated a stark, robust and unforgiving principle: "It is wrong always, everywhere, and for anyone, to believe anything on insufficient evidence". He adds: "It is wrong always, everywhere, and for anyone to ignore evidence that is relevant to his beliefs, or to dismiss relevant evidence in a facile way". And: "It is never lawful to stifle a doubt; for either it can be honestly answered by means of the inquiry already made, or it proves that the inquiry was not complete". Any other method of belief-formation and belief-holding is "one long sin against mankind" (Clifford 1879: 346).

But while the theory receives its name—"the ethics of belief"—only in Clifford's paper, the concept appears frequently throughout the early modern period. Perhaps most famously, both Descartes and Hume endorse an ethics of belief, and in both cases, this is tied with science. In Descartes' case, his arguments for an ethics of belief show up most notably in the Fourth Meditation, where he argues that blame for human errors in matters of mathematics and in the metaphysical foundations of physics lies squarely with humans, that is, due to our overreach in how we hold our beliefs. If only we withheld our assent in the case of beliefs that we don't hold clearly and distinctly, because we have not yet subjected them to proper examination, we would not go wrong. But we *do* assent to beliefs held without proper examination; we *do* hold beliefs on the basis of insufficient evidence. As a result, we err, and when we err in mathematics and metaphysics, we have set rotten foundations for our sciences (Descartes 1964–1976, IX: 42–50). In Hume's case, his arguments for an ethics of belief show up most notably in his chapter "On Miracles" of his *Enquiry Concerning Human Understanding* (Hume 1993), where he interrogates the human tendency to ignore countless previous examples of nature's laws when faced with a single counter example—a tendency that we especially find when religious iconography and poor early education come together. To use an example not used by Hume himself, but which captures his point, a marble statue of Jesus bleeding at the palms can lead the ill-educated to ignore all previous examples of marble that does not bleed in favor of declaring a miracle. These are troubling tendencies to be contrasted with the laudable mindset of the appropriately skeptical scientist. No scientist would declare a miracle in the face of evidence that counters a hitherto believed law of nature. Rather, the scientist, as an ethical belief seeker, would take such confounding evidence as an opportunity to gain more knowledge about the natural world. Moreover, the scientist's epistemic attitude is a model for us all.

Du Châtelet is in this early modern tradition of philosophers who connect belief formation with ethical responsibility, and she carves out her own territory on this issue. All three thinkers—Descartes, Hume and Du Châtelet—have authoritarian targets in their crosshairs, albeit different targets in each case. Descartes targets the Scholastics and their scientifically-unhelpful metaphysics. Hume targets religious authorities. Du Châtelet targets broad-based societal prejudices about women and their natures, as well as nationalistic prejudices related to scientific practices. What all three *do* have in common is a fiercely held commitment to education for all humans, as the source both of evidence and of self-ownership of the mind, amongst the surest guards against authorities who would use the gullible to propagate untruths, whether they be untruths about the natural world, or untruths about the social world, including women, their natures and social roles.

So, Du Châtelet starts with feminist concerns about people's failure to be ethical belief formers and holders, having experienced first-hand the deleterious effects of such beliefs upon her own life. She then extends these concerns more broadly to epistemology and guarding against bias in science. What does her commitment to an ethics of belief in science do for her?

Speculative and Experimental Natural Philosophy

Du Châtelet's admonishment to her son in the Preface to her *Foundations* is not just a theoretical posture for her. Indeed, she herself is admirably even-handed and skeptically minded when it comes to her scientific work and attitudes. If I were to speculate, I might wonder if her status as a woman—both as one who bore the brunt of prejudices against her intellectual capacities and as one who was kept on the outside of powerful scientific institutions—is especially key in her skeptical attitude toward the status quo. Whatever the source, she follows her own advice to her son, specifically by "guarding herself against the inevitable obstinacy to which the spirit of national partisanship" might carry her in the dispute among Newtonian and Cartesians. To show one especially powerful example of this, I turn to one way of characterizing scientific culture in early modern Europe that has gained some currency lately. This is Peter Anstey's framework (Anstey 2005), according to which those engaged in natural philosophy in the seventeenth and eighteenth centuries distinguished between two camps, namely speculative and experimental natural philosophers.

Anstey offers this meta-narrative as a replacement of the old rationalist-empiricist framework, which dominated Western research of philosophy in the early modern period for a significant portion of the nineteenth and virtually all of the twentieth century. Anstey believes that his meta-narrative more faithfully tracks the ways in which the early modern actors themselves saw their work. I must confess a wariness of any grand narrative, for—and, indeed, the following account of Du Châtelet only serves to underscore this point—they can result in an obscuring of the true diversity of thought. Still, Anstey's theory does capture something important about the *zeitgeist* of the scientific community in early modern Europe, and at the least, his theory captures something important about Du Châtelet's immediate intellectual milieu.

In one early paper on the distinction between speculative and experimental natural philosophy, Anstey's focus is on early modern England, and he argues that the distinction "is found in many different English writers in the latter half of the seventeenth century" (Anstey 2005: 215). In that paper, he aims to establish:

...five strong claims regarding this distinction:

1   This distinction is in evidence, in some form or another, from the late 1650s until the early decades of the eighteenth century.
2   This distinction provides the primary methodological framework within which natural philosophy was interpreted and practiced in the late seventeenth century.
3   This distinction is independent of disciplinary boundaries within and closely allied to natural philosophy.
4   This distinction crystallized in the 1690s when opposition to hypotheses in natural philosophical methodology intensified.
5   This distinction provides the terms of reference by which we should interpret Newton's strictures on the use of hypotheses in natural philosophy (Anstey 2005: 216).

As evidence for the distinction, Anstey provides a wide range of examples from natural philosophers in seventeenth-century England, and here are two, which underscore the fourth point regarding hypotheses in Anstey's list, the point I shall return to near the end of this section of the chapter, and throughout the next section. Isaac Newton writes:

> Experimental Philosophy reduces phenomena to general Rules & looks upon the Rules to be general when they hold generally in Phenomena . . .. Hypothetical Philosophy consists in imaginary explications of things & imaginary arguments for or against such explications . . .. The first sort of Philosophy is followed by me, the latter too much by Cartes, Leibniz and some others.
>
> *(Newton 1959–1977: 388–390)*

And William Wotton writes:

> I do not here reckon the several *Hypotheses of Des Cartes, Gassendi,* or *Hobbes*, as Acquisitions to real Knowledge, since they may only be Chimaera's and amusing Notions, fit to entertain working Heads. I only alledge [sic] such Doctrines are raisied upon faithful Experiments, and nice Observations . . .
>
> *(Wotton 1694: 244)*

Anstey expands on his conclusions reached in this early paper in subsequent publications aimed at showing the thoroughgoingness of this distinction in early modern European natural philosophy (e.g. Anstey 2014; Anstey and Vanzo 2012).

Whatever could be said in general about a pan-European trend on this question, it is certainly the case that Voltaire, in his laudatory *Elements of Newton's Philosophy*, draws a very bright line between the old (French) way of hypotheses, exhibited by Descartes, and the new (British) way of hard-headed experimentalism. Throughout his text, he contrasts Descartes and Newton along the lines roughly articulated by Anstey to Newton's benefit and Descartes' detriment. This is an interesting moment in mid-eighteenth-century France for my purposes, and for two reasons. First, while he himself does not fall in with nationalistic prejudices about his own nation, Voltaire does believe that the British are the leaders of the scientific game on the role of experiment, and that the French, bogged down in fanciful hypotheses, had better catch up. Du Châtelet is far more subtle and nuanced in her thinking on this point, and I return to this in the final section of the chapter. Second, for decades, Du Châtelet's thought was assimilated to Voltaire's thought, and she was seen as merely derivative of this man with whom she lived and worked for years (e.g. Barber 1967). This evidence-free claim has since been debunked by close examination of her own work, and her scientific methodology, including on this point about hypotheses, is a case in point. Both points connect with her following a robust ethics of belief.

On the divide between speculative and experimental natural philosophy and the supposed hostility of experimental natural philosophy to the use of hypotheses, Voltaire writes: "Note that all experience and calculation ruin almost all Descartes' ideas when this great philosopher bases these ideas only on hypotheses. These are bright and misleading perspectives the brightness of which diminishes as we approach them" (Voltaire: 337, notes to lines 50–57).[4] And when accepting Newton's theory of gravity, whilst also rejecting Descartes' theory based upon vortices, Voltaire writes:

> It is true that they [those who embrace vortex theories of gravity] have given no proof of this assertion: there is not the slightest experience, not the slightest analogy in the things we know, which can establish the slightest presumption in favor of this whirlwind of subtle matter; thus from this lack of empirical evidence alone, this system is a pure hypothesis; it must be rejected. It is, however, by this alone that he [Descartes] has been accredited. This vortex was conceived without effort, a vague explanation of things, given by pronouncing this claim

of subtle matter, and when philosophers felt the contradictions and absurdities attached to this philosophical fiction, they thought of correcting it rather than abandoning it.

(Voltaire [1738] 1992: 401)[5]

In fact, by the time Du Châtelet put pen to paper in order to tackle the natural philosophy related to gravity in her *Foundations*,[6] Descartes' star had faded in her immediate intellectual circle, eclipsed by Newton's natural philosophy. Voltaire's attitude throughout his *Elements of Newton's Philosophy* captures the attitude, which Du Châtelet has in mind when she cautions her son to resist partisanship in science and to think for himself.

Three points are worthy of note as background to an investigation of what Du Châtelet herself goes on to do with her own resistance of such partisanship, and with her own skepticism of attitudes of those in her intellectual milieu. First, as Anstey has well documented, the speculative versus experimental debate in natural philosophy was not just about scientific methodology and epistemology. Many other factors inflected this debate. For example, Du Châtelet's milieu is a case in point; nationalism and partisanship play a notable role in this debate. Second, this first point only serves to underscore the lesson I extracted from Du Châtelet's Preface, in which she addresses her son. For in her cautioning her son against such prejudices, she is urging he harness his education to make his mind self-sufficient, which implies the kind of self-ownership of the mind, which undergirds the kind of skepticism and independence of mind requisite for ethically developing a belief system. Her ethics of belief, that is, is meant to guard against such partisanship. Finally, to the modern mind, it is frankly *odd* to align the use of hypotheses with the non-empirical, non-rigorous speculative side in the debate within natural philosophy. If *anything* marks the scientific method as we know it today, it is surely the posing and testing of hypotheses. Indeed, the history of the nature and fate of hypotheses is central in the emergence of modern science, and perhaps the most crucial chapter in that history in Europe is the one that stars Du Châtelet.

Du Châtelet and the Ethics of Belief in Action: Refiguring the Scientific Landscape

The stark alignment suggested earlier between speculative natural philosophy and the embrace of hypotheses on the one hand, and experimental natural philosophy and the rejection of hypotheses on the other hand is too stark. Anstey knows this, as do many scholars of early modern science. One complicating factor is the wide range of meanings one can attach to the notion of the hypothesis. I. Bernard Cohen, for example, in his exhaustive survey of the ways in which Newton uses the term throughout his *corpus* identifies nine distinct—and often widely divergent—meanings of the term (Cohen 1969). The nebulousness of the concept of a hypothesis during this period is part of what makes the history of hypotheses so rich and interesting. A second complicating factor, perhaps emerging from this first factor, is the ambivalence shown toward hypotheses by those practicing science in the early modern period, including those who align themselves on the experimental side of the (supposed) divide. As Anstey argues, some of these thinkers are apologetic in their use of hypotheses, while some are looking for ways of reconciling experimentalism and the use of hypotheses (Anstey 2005: 224ff). Du Châtelet fits into the latter category, and her systematic, sustained and intentional effort to show just how hypotheses and experimental science ought to be reconciled is, I contend, *the* watershed moment in the history of hypotheses in European science. Under Du Châtelet's guidance—largely free from partisanship because of her following her ethics of belief—hypotheses arrive as central and respectable in modern scientific method.

To see how strikingly modern Du Châtelet's view of hypotheses is, I will here present the core elements of her theory of the nature and role of hypotheses when they are properly used in science. Detailed, and historically contextualized, examinations of her views on hypotheses have already enjoyed attention among philosophers (e.g. Detlefsen 2019; Kawashima 1993; Reichenberger 2016; Suisky 2019). This account is laid out in detail in her chapter 4: "Hypotheses" of her *Foundations*. In that chapter, she is clear that—despite how individuals might depict their own work—hypotheses are not uncommon in many scientists' practice, and that this is a good thing:

> And so good hypotheses will always be the product of the greatest men. Copernicus, Kepler, Huygens, Descartes, Leibniz, and even Newton himself, have all devised useful hypotheses to explain complicated and difficult phenomena. The example of these great men, and of their successes, should make us see that those who wish to ban hypotheses from philosophy, intend harm to the interests of science.
>
> *(IP §71)*

Among her targets who "intend harm to the interests of science" by banning hypotheses are surely those who latch excessively onto Newton's famous claim in the General Scholium to his *Principia* that he "feigns no hypotheses". Indeed, uncritically taking on board such a principle would undermine entire sciences, for example, "without hypotheses . . . there would be no astronomy now" (IP §57). She underscores the necessity to science of hypotheses, and she gives reasons for their necessity:

> Hypotheses are . . . sometimes very necessary . . . in all cases when we cannot discover the true reason for a phenomenon and the attendant circumstances, neither *a priori*, by means of truths [identified as principles of knowledge in §53] that we already know, nor *a posteriori*, with the help of experiments.
>
> *(IP §60)*

And: "[P]hilosophers frame hypotheses to explain the phenomena, the cause of which cannot be discovered either by experiment or by demonstration" (IP §56). On this point, Du Châtelet is remarkably in step with Descartes. Both thinkers believe that underlying principles that explain all phenomena in the world are underdetermined, and that the only way to understand exactly *how* these principles give rise to the precise world that we have is to posit hypotheses to further specify underlying causes and principles.

Still, she is cautious about the potential overuse of hypotheses—overuse that might invite "fables and dreams" (IP §55) into science. When *this* happens, hypotheses are the "poison of reason and the plague of philosophy" (§55, Du Châtelet, quoting Newton), and on this front, Du Châtelet's target is surely those who take Cartesianism to an extreme by postulating a range of untestable speculations about the structure of the subvisible world. Right away, we see Du Châtelet hewing to a middle path between the two extremes we saw depicted by the characterization earlier of the experimental-speculative divide in early modern natural philosophy.

But the real power of her view comes to the fore when we see the constraints she proposes in order to ensure hypotheses do not become "fables and dreams". Among these constraints are the following. First, empirical testing is key: "when a hypothesis is once posed, experiments are often done to ascertain if it is a good one, experiments which would never have been thought of without it" (IP §58). Second, she adheres to the principle of falsification, according to which one falsifying piece of evidence, experimental or otherwise, requires that we reject the hypothesis: a hypothesis "becomes false when it is found to contradict a well-established observation" (IP §67). Third, the fruitfulness of a hypothesis increases with more and more data that supports it, or with more and more observations that can be explained by it: a hypothesis should "not only [explain] the phenomenon that one had proposed to explain with it, but also that all the consequences drawn from it agree with the observations" (IP §58). Finally, the scientist, when posing and evaluating hypotheses, ought to take a stance of epistemic modesty and caution. On this point, she turns specifically to the hypothesis that explains the phenomena of gravity through an appeal to vortices. This is Descartes', and Leibniz's, hypothesis, and it was a hypothesis held in disregard in her intellectual circles because Newton's view—which seems to endorse action at a distance—was so highly esteemed instead. To this wholesale rejection of the hypotheses of vortices, she writes that a hypothesis "can be true in one of its parts and false in another" (IP §65), which results in her acknowledgment that "it cannot be legitimately concluded that a vortex, or several vortices, conceived of in a different way, cannot be the cause of these movements" (IP §65). She concludes with this overall characterization of the proper role of hypotheses in science, once again, a characterization that has a strikingly contemporary ring to it:

> So, hypotheses are only probable propositions, which have a greater or lesser degree of probability according to whether they satisfy a larger or fewer number of the circumstances that accompany the phenomena that we want to explain by means of the hypotheses. And since a very high degree of probability encourages our agreement so as to have nearly the effect upon us as certainty, hypotheses eventually become truths for us if their probability increases to such a point that this probability can morally pass for certainty . . .. In contrast, an hypothesis becomes improbable in proportion to the number of circumstances found for which the hypothesis does not give a reason. And finally, it becomes false when it is found to contradict a well-established observation.
>
> *(IP §67)*

With Du Châtelet's chapter on hypotheses in her *Foundations*, a new chapter in scientific methodology finally emerged in full-blown form.

Du Châtelet's Role in the Emergence of a Modern Conception of Hypotheses

As the brief precis of Du Châtelet's theory of hypotheses offered earlier shows, and against a context of her circles where we saw the rejection of hypotheses as counter-productive to experimental philosophy, we can see how critically minded she is with respect to at least some ways of thinking prominent at her time. As a result, she shakes up the alignment between speculation and hypotheses on the one hand, and experimentalism and the rejection of hypotheses on the other hand. Rather, in her carefully-considered view, there is no natural tension between experimental philosophy and the use of hypotheses, properly conceived and constrained. This understanding of the centrality of hypotheses to science is *de rigueur* in science today, but it was not so in Du Châtelet's time and place. That she reached this view is at least in part as a result of abiding by her commitment to an ethics of belief, and that she articulated this view fully and powerfully for the first time puts her at the center of one of the most important chapters in scientific methodology in our history.[7]

A close look at *her* close look at Descartes, Newton and their followers on the question of hypotheses shows that she was attune to—unlike many—the fact that Descartes abides by many of the strictures that Du Châtelet puts on the proper use of hypotheses (Detlefsen 2019), and the fact that many Newtonians make overuse of hypotheses in methodologically problematic ways. Much of her *Foundations* aims to show how a proper methodology, including proper use of hypotheses, can correct for the excesses and timidities of scientists across national divides (Brading 2019). Her importance in the history of scientific methodology is underscored by the fact that several entries in the extraordinarily influential text, Diderot's and d'Alembert's *Encyclopédie, ou dictionnaire raisonné des sciences, des arts et des métiers* (1751–1772), are reproductions of Du Châtelet's own ideas. The entries in the *Encyclopédie* include the entry on "Hypothesis", which just is Du Châtelet's fourth chapter of the *Foundations* replicated almost verbatim (Koffi 2008). Du Châtelet's importance in the history of science—an importance that emerges at least in part from her hard-headed, critically-minded commitment to an ethics of belief—was once known, and then egregiously erased,[8] an error that we today are beholden to correct, should we wish to learn from her own model and become as ethical as we can in our histories and our beliefs about those histories.

Related chapters: 1, 2, 12, 15.

Notes

1 Other approaches to Du Châtelet's feminist philosophy of science which scholars have explored include her strategizing in the *vis viva* debate in order to engage with the Secretary of the Royal Academy of Sciences, Jean-Jacques, d'Ortous de Mairan given her exclusion from the Academy (Detlefsen forthcoming; Kawashima 1995; Rey 2017, 2019; Terrall 1995), her discussion of human freedom (Jorati 2019) especially with respect to gender and as it interacts with determinism due to her physics (Hagengruber 2017) and her strategies regarding research and publishing surrounding her *Dissertation on Fire* (Kawashima 2005).

2 This kind of insight, namely that institutional and structural norms can have an influence on individuals and ought to be included in explanations for human behaviors, is plentiful in the early modern period, especially among the disenfranchised. While clearly recognized hundreds of years ago, the idea is perhaps most fully and *explicitly* theorized only in recent years by Sally Haslanger (2016).

3 While Du Châtelet was highly regarded for her mathematical and scientific abilities during her lifetime, shortly after her death, her revered reputation was quickly lost (Allen 1998). In English language work on her thought, early divergent views held by William Barber and Ira O. Wade over whether she is merely derivative of her close collaborator Voltaire (Barber 1967) or an intellectual force in her own right (Wade 1941) have since given way to Wade's position. Early papers that establish Du Châtelet's originality and prowess in physics (Iltis 1977; Janik 1982) have paved the way for the cascade of research on Du Châtelet's thought, including papers (e.g. Brading 2018; Detlefsen 2013; Gessell 2019; Hagengruber 2011; Janiak 2018) and several book-length studies (e.g. Brading 2019; Le Ru 2019; Reichenberger 2016).

4 Remarquez que toute expérience et tout calcul ruine presque toutes les idées de Descartes quand ce grand philosophe ne les fonde que sur des hypothèses. Ce sont des perspectives brillantes et trompeuses qui diminuent à mesure qu'on en approche. Tous les autres philosophes ont cherché des solutions de ce problème de la nature; mais l'expérience a renversé aussi leaur conjectures.

5 Il est vraiqu'ils n'ont donné aucune preuve de cette assertion: iln'y a pas la moindre expérience, pas la moindre analogie dans les choses que nous connaissons in peu, qui puisse fonder une présomption légère en faveur de ce tourbillon de matière subtile; ainsi de cela seul que ce système est une pure hypothèse, il doit être rejeté. C'est cependant par cela seul qu'il e été accrédité. On concevait ce tourbillon sans effort, on donnait une explication vague des choses en prononçant ce mot de matière subtile, et quant les philosophe sentaient les contradictions et les absurdités attachées à ce roman philosophique, ils songeaient à le corriger plutôt qu'à l'abandonner. See also Voltaire ([1738] 1992: 699–700). For praise of Newton for avoiding the use of hypotheses, see Voltaire ([1738] 1992: 729). For a direct comparison of the two to Descartes' disadvantage and Newton's advantage, see Voltaire ([1738] 1992: 733–734).

6 On Du Châtelet's innovations with respect to the science of gravity, see Brading (2018) and Janiak (2018).

7 Du Châtelet's broader bringing together of valuable ideas from across a wide range of scientific and philosophical points of view has been well-examined in secondary literature, for example by Locqueneux (1995).

8 For a masterful explanation for why women and their philosophical accomplishments were erased from the history of philosophy, see O'Neill (1998).

References

Allen, L. D. (1998) "Physics, Frivolity, and 'Madame Pompon-Newton': The Historical Reception of the Marquise du Châtelet from 1750–1966," PhD diss., University of Cincinnati.

Anstey, P. R. (2014) "Philosophy of Experiment in Early Modern England: The Case of Bacon, Boyle and Hooke," *Early Science and Medicine* 19.2: 103–132.

Anstey, P. R. (2005) "Experimental Versus Speculative Natural Philosophy," in P. R. Anstey and J. A Schuster (eds.) *The Science of Nature in the Seventeenth Century: Patterns of Change in Early Modern Natural Philosophy*, Cham, Switzerland: Springer, 215–242.

Anstey, P. R. and A. Vanzo (2012) "The Origins of Early Modern Experimental Philosophy," *Intellectual History Review* 22.4: 499–518.

Barber, W. H. (1967) "Mme du Châtelet and Leibnizianism: The Genesis of the *Institutions de physique*," in W. H. Barber, J. H. Brumfitt, R. A. Leigh, R. Shackleton, and S. S. B. Taylor (eds.) *The Age of Enlightenment: Studies Presented to Theodore Besterman*, Edinburgh: University Court of the University of St. Andrew, 1967, 200–222.

Brading, K. (2019) *Émilie Du Châtelet and the Foundations of Physical Science*, New York: Routledge.

Brading, K. (2018) "Émilie Du Châtelet and the Problem of Bodies," in E. Thomas (ed.) *Early Modern Women on Metaphysics*, Cambridge, UK: Cambridge University Press, 150–168.

Brading, K., et al. (2018) *Foundations of Physics* (trans. and eds. Emilie Du Châtelet). https://www.kbrading.org/translations

Broad, J. and K. Detlefsen (2017) *Women and Liberty, 1600–1800: Philosophical Essays*, Oxford: Oxford University Press.

Clifford, W. K. (ed.) (1879) "The Ethics of Belief," in *Lectures and Essays*, Charlottesville, VA: Macmillan Publishing, 339–363.

Cohen, I. B. (1969) "Hypotheses in Newton's Philosophy," *Boston Studies in Philosophy of Science* 5: 304–326.

Descartes, R. (1964–1976) *Oeuvres de Descartes*, 11 vols., C. Adam and P. Tannery (eds.). Paris: J. Vrin.

Detlefsen, K. (Forthcoming) "The Rise of a Public Science? Women and Natural Philosophy in the Early Modern Period," in D. Miller and J. Jalobeanu (eds.) *The Cambridge History of Philosophy of the Scientific Revolution*, Cambridge, MA: Cambridge University Press.

Detlefsen, K. (2019) "Du Châtelet and Descartes on the Role of Hypothesis and Metaphysics in Natural Philosophy," in E. O'Neill and M. P. Lascano (eds.) *Feminist History of Philosophy: The Recovery and Evaluation of Women's Philosophical Thought*, Cham, Switzerland: Springer, 97–128.

Detlefsen, K. (2018) "Cavendish on Laws and Order," in Emily Thomas (ed.) *Early Modern Women on Metaphysics*, Cambridge, MA: Cambridge University Press, 72–91.

Detlefsen, K. (2017) "Émilie Du Châtelet on Women's Minds and Education," in S. Hetherington (ed.) *What Makes a Philosopher Great?* London: Routledge, 128–147.

Detlefsen, K. (2013) "Emilie Du Châtelet," in E. Zalta (ed.). *Stanford Encyclopedia of Philosophy*, Stanford, CA: Stanford University. https://plato.stanford.edu/entries/emilie-du-chatelet/

Du Châtelet, E. (1740) *Institutions de physique*, Paris: Prault. Cited in the text as IP, followed by section (§) number.

Gardiner, L. (1984) "Women in Science," in S. I. Spencer (ed.) *French Women and the Age of Enlightenment*, Bloomington: Indiana University Press, 181–193.

Gessell, B. (2019) " 'Mon petit essai': Émilie du Châtelet's *Essai sur l'optique* and Her Early Natural Philosophy," in S. Hutton and R. Hagengruber (eds.) *Special Issue: Women Philosophers in Early Modern Philosophy*, *British Journal for the History of Philosophy* 27.4: 860–879.

Hagengruber, R. (2017) "If I Were King! Morals and Physics in Emilie Du Châtelet's Subtle Thoughts on Liberty," in J. Broad and K. Detlefsen (eds.) *Women and Liberty, 1600–1800: Philosophical Essays*, Oxford: Oxford University Press, 195–205.

Hagengruber, R. (2011) "Emilie Du Châtelet between Leibniz and Newton: The Transformation of Metaphysics," in R. Hagengruber (ed.) *Émilie Du Châtelet: Between Leibniz and Newton. Papers Presented at the Conference on Emilie du Châtelet in Honor of her 300th Birthday Held at the Research Center for European Enlightenment in the Autumn of 2006 in Potsdam*, London: Springer, 1–59.

Haslanger, S. (2016) "What Is a (Social) Structural Explanation?" *Philosophical Studies* 173.1: 113–130.

Hume, D. (1993) "Of Miracles," in Eric Steinberg (ed.) *An Enquiry Concerning Human Understanding*, second edition. Indianapolis, IN: Hackett Publishing Company, Inc, 72–90.

Hutton, S. (2004a) "Women, Science, and Newtonianism: Emilie Du Châtelet versus Francesco Algarotti," in J. E. Force and S. Hutton (eds.) *Newton and Newtonianism: New Studies*, Dordrecht, The Netherlands: Kluwer Publishing, 183–203.

Hutton, S. (2004b) "Women, Science, and Newtonianism: Emilie Du Châtelet Versus Francesco Algarotti," in J. E. Force and S. Hutton (eds.) *Newton and Newtonianism*, Dordrecht, The Netherlands: Kluwer Academic Publishing, 183–203.

Iltis, C. (1977) "Madame Du Châtelet's Metaphysics and Mechanics," *Studies in the History and Philosophy of Science* 8.1: 29–48.

Janiak, A. (2018) "Émilie Du Châtelet: Physics, Metaphysics and the Case of Gravity," in E. Thomas (ed.) *Early Modern Women on Metaphysics*, Cambridge, UK: Cambridge University Press, 2018, 49–71.

Janik, L. G. (1982) "Searching for the Metaphysics of Science: The Structure and Composition of Madame Du Châtelet's *Institutions de physique, 1737–1740," Studies on Voltaire and the Eighteenth Century* 201: 85–113.

Jorati, J. (2019) "Du Châtelet on Freedom, Self-Motion, and Moral Necessity," *Journal of the History of Philosophy* 57.2: 255–280.

Kawashima, K. (2005) "The Issue of Gender and Science: A Case Study of Madame du Châtelet's *Dissertation dur le feu," Historia Scientiarum* 15.1: 23–43.

Kawashima, K. (1995) "Madame Du Châtelet dans le journalisme," *LLULL* 18: 471–491.

Kawashima, K. (1993) "Les idées scientifiques de Madame du Châtelet dans ses *Institutions de phyisque:* Un rêve de femme de la haute société dans la culture scientifique au Siècle des Lumières. 1ère partie," *Historia Scientiarum* 3.1: 63–82.

Koffi, M. (2008) "Mme Du Châtelet, l'*Encyclopédia*, et la philosophie des sciences," in U. Kölving and O. Courcelle (eds.) *Émile Du Châtelet: Éclairages & document nouveaux*, Paris: Centre International d'Étude du XVIIIe Siècle Ferney-Voltaire, 255–266.

Le Ru, V. (2019) *Émilie Du Châtelet Philosophe*, Paris: Classiques Garnier.

Locqueneux, R. (1995) "Les *Institutions de physique* de Madame Du Châtelet ou d'un traité de paix entre Descartes, Leibniz et Newton," *Revue Du Nord* 77.312: 859–892.

Newton, I. (1959–77) "Draft of a Letter to Roger Cotes, March 1713," in H. W. Turnbull, J. F. Scott, A. R. Hall, and L. Tilling (eds.) *The Correspondence of Isaac Newton*, 7 vols., Cambridge, MA: Cambridge University Press.

O'Neill, E. (1998) "Disappearing Ink: Early Modern Women Philosophers and Their Fate in History," in J. A. Kourany (ed.) *Philosophy in a Feminist Voice: Critiques and Reconstructions*, Princeton, NJ: Princeton University Press, 17–61.

Reichenberger, A. (2016) *Émilie Du Châtelets Institutions physiques: Über die Rolle von Prinzipien und Hypothesen in der Physik*, Wiesbaden, Germany: Springer VS.

Rey, A.-L. (2019) "Der Streit um die lebendigen Kräfte in Du Châtelets Institutions de physique: Leibniz, Wolff und König," in R. Hagengruber and H. Hecht (eds.) *Emilie Du Châtelet und die deutsche Aufklärung*, Wiesbaden: Springer, 27–63.

Rey, A.-L. (2017) "Agonistic and Epistemic Pluralisms: A New Interpretation of the Dispute between Emilie du Châtelet and Dortous de Mairan," *Paragraph* 40.1: 43–60.

Suisky, D. (2019) "Émilie Du Châtelet und Leonhard Euler über die Rolle von Hypothesen," in R. Hagengruber and H. Hecht (eds.) *Emilie Du Châtelet und die deutsche Aufklärung*, Wiesbaden, Germany: Springer, 99–172.

Terrall, M. (1995) "Émilie Du Châtelet and the Gendering of Science," *History of Science* 23: 283–310.

Thomas, E. (ed.) (2018) *Early Modern Women on Metaphysics*, Cambridge, MA: Cambridge University Press.

Voltaire. ([1738] 1992) *Eléments de la philosophie de Newton*, in R. L. Walters and W.H. Barber (eds.) *The Complete Works of Voltaire*, volume 15, general editors W.H. Barber and U. Kölving, Oxford: University of Oxford Press.

Wade, I. O. (1941) *Voltaire and Madame Du Châtelet: An Essay on the Intellectual Activity at Cirey*. Princeton, NJ: Princeton University Press.

Wotton, W. (1694) *Reflections Upon Ancient and Modern Learning*. London.

Zinsser, J. P. (ed.) (2009) *Emilie Du Châtelet: Selected Philosophical and Scientific Writings* (trans. Isabel Bour), Chicago, IL and London: The University of Chicago Press.

Zinsser, J. P. (1998) "Emilie Du Châtelet: Genius, Gender, and Intellectual Authority," in H. L. Smith (ed.) *Women Writers and the Early Modern British Political Tradition*, Cambridge, UK: Cambridge University Press, 168–190.

4

THE ROCKET WOMEN OF INDIA

Eight Women Scientists and Their Roles in the 2014 Mars Orbiter Mission (MOM)

Deepanwita Dasgupta

On November 5, 2013, Indian Space Research Organisation (ISRO) launched a satellite in space using its Polar Satellite Launch Vehicle Rocket (PSLV C-25), intending to place the satellite in orbit around Mars. The Mars Orbit Insertion (MOI) was successfully carried out on September 24, 2014. It was India's first entry into interplanetary space, and it was a landmark event for several reasons. First, in this first ever entry into interplanetary domain, success was achieved at the very first attempt, a rare thing indeed. Second, it was completed on a shoestring budget – a mere 74 million – thus costing somewhat less than even an average Hollywood movie. To get this job done within the existing launch facilities at hand, a number of innovative strategies were applied, e.g. the satellite inserted itself into Mars' orbit by using a very elliptical trajectory. Finally, and perhaps most outstandingly from our point of view, the project included a substantial number of women in its planning and design – eight such women to be precise. Their participation and their expertise were essential for the success of the Mars Orbiter Mission (MOM).

Could the MOM project then be taken as an example of feminist science, or at least the demonstration of a very substantial stock of feminist expertise? While plenty of news articles have already been written on the Mars mission itself, and about its proud national implications, nobody yet has examined this episode from a feminist point of view, trying to tease out its epistemic implications. What does such an event mean for the women in science in India in particular and for the women of the country in general?

In this chapter, my goal will be to explore this complex question by considering various aspects of the mission. To develop an account that can handle this problem, my strategy will be as follows. First, I shall begin with a quick outline of the history of ISRO, the institution that conceived and executed this project, thus putting the Mars mission within the contexts of a national as well as normal science. Second, I shall take into account the journeys of the women scientists themselves into their profession – in so far as this is publicly available – and ask what kind of immediate and projected consequences might follow from their participation. Third, I shall briefly describe the design of the Mars Orbiter Vehicle and some of its most important engineering challenges. Once we have this triple-layered account in hand, we shall be ready to return some answers to the question about feminist epistemology.

A Very Short History of ISRO

Let me now begin this journey by taking a quick look at the origin of ISRO, the institution that carried out this project, and especially ISRO's beginning as a case of nationalist science. Formed by Vikram Sarabhai, the father of the Space Research in India, the Indian National Committee for Space Research (INCOSPAR) began its journey in 1962, as a modest first venture that hoped to gain at least some share in the newly founded field of space research, especially in the areas of cosmic rays and satellite communication. Sarabhai was clear that he was not really thinking about competing with the developed nations,[1] and yet he refused to give up the very ambitious idea of having an entire research facility devoted to space science in a country that had just freed itself from colonial rule, and was thus still struggling to feed its population (Singh, 2017). Focusing on delivering future agricultural and climate services as well as services in the area of satellite communication, Sarabhai was thus clearly stepping into the role of a visionary.

In 1969, ISRO emerged out this earlier organization, with a somewhat expanded and more ambitious aims.

The first series of satellites launched by ISRO with its own launch vehicle[2] was called *Rohini* – of which the first was a complete failure and the third a partial success in reaching the polar orbit but the fourth one was quite successfully placed on earth's orbit on April 17, 1983. Its camera continued to take pictures until April 1990. Four decades of work involving frequent collaborations with both the United States and the former Soviet Union yielded a substantial number of gains. Among them, the most notable one was the inclusion of Captain Rakesh Sharma in the Soyuz T-11 team (1984), who thus became the first Indian astronaut to travel in space. By the end of the millennium, India had acquired enough expertise in the launching and the placing of satellites on earth's orbit. Then in 2008 came the venture called *Chandrayaan*, literally the moon vehicle, a satellite sent to orbit around the moon. This was ISRO's first mission beyond the earth's orbit. With its track record thus established, one can say that ISRO had finally stepped outside Sarabhai's modest initial philosophy.

Planning the Mission

After the success of the lunar *Chandrayaan* mission, the *Mangalyaan* or the Mars Orbiter project emerged as a very natural next step for ISRO for several reasons. First, numerous international missions had already been sent to Mars, and given the fact that strong national rivalries had always been a dominant theme in space science, it is natural that the Indian scientific community too would like its fair share of the same kind of epistemic progress. Second, the red planet had always been the object of much human curiosity and imagination. The similarity between the two planets is indeed encouraging for more future missions– either for directly landing on the planet or for founding future colonies – Mars has a day/night cycle much like Earth, and its atmosphere is the result of interactions among its various kinds of spheres, e.g. the atmosphere, the exosphere, and so on. Finally, the prospect of being part of an elite space club, the possibility of staking claims to the discovery of some new data, e.g. regarding the exploration of life on Mars, made the whole project look like a race for discovery, stimulating both imagination and action.

The team that developed the MOM consisted of several hundred people, involving both men and women. As it happens, the timeline for developing this project was extremely short – a mere 18 months from its conception to execution – since the project began to be put together from April 2012. And yet within that short time, several innovative features had to be designed and developed in spite of the fact that the MOM team did not have any previous experience on such matters. While the teams' homework on various international Mars missions as well as its own lunar mission (2008) had given it a good grasp of the necessary hardware and all the constraints of the situation, the design of an onboard autonomy system remained still a great challenge, and yet the mission could not possibly function without such a system, given the 40-minute time lag of signals between Mars and the earth. The Mars-bound satellite cannot thus be controlled from the earth itself in real-time. Moreover, the satellite must "learn" how to detect its own problems as well as how to manage them, both during its long voyage and during the Mars orbit insertion and thereafter. The project had thus really assumed the shape of a race for discovery.

The Team of Rocket Women

It is now time for me to introduce the eight women who played a prominent role in this race and often served as its visible face after the successful launch of the satellite, giving TED talks and other interviews. Employed at ISRO as scientists and engineers, in many cases since the 1980s and 1990s, their collective task was to contribute in the work of designing and executing the MOM project.

Here I should be careful in pointing out that my focusing on the work of the women in no way implies any undervaluing of the work of the men who were also parts of the same mission, and who held many of its key positions for design and control. For example, as reported by the Indian newspaper *Hindustan Times*, there were at least eight men who played very indispensable roles in the Mars mission (Srivastava 2013). Among them, I mention here only three – M. Annadurai, who was the Project Director for MOM, in charge of both its budget management and its scheduling. Similarly, S.K. Shivakumar, Director of the ISRO Satellite Center, was the project director for the Deep Space Network Antenna, all indigenously developed in India. Finally, we have A.S. Kiran Kumar, who was responsible for building three of the exploratory payload instruments – the Mars Color Camera, Methane Sensor, and the Thermal Infrared Imaging Spectrometer. Thus, this was a case where both the men and women were jointly involved in a nationalist effort but since my topic in this essay is to explore the epistemic implications of the women's participation in the project, I shall henceforth concentrate only on the women, trying to show how they were beginning to emerge in various leadership roles during the mission.

But, quite obviously, I do not even have space here to cover the work of *all* the women who were associated with this mission, for such a list would surely run into double, even triple, digits. I shall therefore simply focus only on those women scientists who were consistently featured in almost all the reports about the Mars mission and who were perhaps its most visible faces, especially shortly after the launch. Given the fact that almost all published reports celebrated the work of these women, but did not deeply inquire into its epistemic implications, focusing mostly on the entry of a new nation into the elite space club—my chapter can be taken as a philosophical first attempt to construct an analysis of large scale scientific missions like ISRO, and how these missions (slowly) change the gender structure of India's scientific community.

This shorter list now includes eight women, and I list their names below.

1 Nandini Harinath – Project Manager, Mission Design, and Deputy Operations Director, MOM.
2 Ritu Karidhal – served as the Deputy Operations Director of the MOM.
3 Anuradha T.K. – Geosat Program Director, Senior Officer, ISRO.
4 Moumita Dutta – Project Manager for the two payload instruments of the mission.
5 Minal Sampath – Systems Engineer, led a team of 500 engineers.
6 Kriti Faujdar – Computer Scientist, whose job is to keep satellites in their proper orbits.
7 N. Valarmathi – led the launch of the first indigenously developed radar imaging satellite, RISAT-1.
8 Seetha Somasundaram – Program Director, ISRO Space Science Program Office.

Now, in so far as their profiles are publicly available, one can easily see how their career trajectories brought them into their profession at ISRO. It also becomes quickly clear that even though these women came from various parts of the Indian society – and naturally from its various states – a consistent track record of very high educational accomplishment is clearly visible in their family backgrounds. For example, Nandini Harinath, who assisted as the Deputy Operations Director of the project, came from a family where her mother was a teacher and her father an engineer. Having served for nearly 20 years already at ISRO, she had assisted in numerous missions before being inducted into the MOM as its deputy director. Not surprisingly perhaps, it turns out that as a child, she was a fan of the Star Trek.

Similarly, Ritu Karidhal, an aerospace engineer, who also served as a Deputy Operations Director for the project, became something like its public face after the launch, for she gave a TED talk on this mission in 2016. In the course of this talk, Karidhal spoke at length about her own childhood absorption in the stars, and her wonder about what lay beyond the dark night sky. She emphasized her family's great support during her studies as well as during her subsequent employment for which she had to move a great distance away from home. Her entry into the profession took a very time-honored form. After finishing an undergraduate degree in physics from Lucknow University, she joined the Indian Institute of Science, one of the premier science institutions in India, which was founded by Jamsetji Tata in 1909. Having finished her Master's degree in Aerospace Engineering, she then joined ISRO in 1997, and in 2007 she received the Young Scientist Award from the hands of A.P.J. Abdul Kalam, himself a prominent ISRO man as well as a key figure in developing the missile technology for India.

Anuradha T.K., the Geosat Program Director, hails from Bangalore, Karnataka. She completed her Bachelor's degree in Electronics from the Visvesvaraya College of Engineering, Bangalore, and her specialty was communication satellites. She had a long track record in developing different ISRO Communication satellites, e.g. both GSAT-12 and GSAT-10, launched, respectively, in 2011and 2012. Having also served as the Project Manager for the Indian Regional Navigation Satellite System programs, she had held several awards and had been inspired by the image of Neil Armstrong walking on the moon.

Minal Sampath, the next figure in my list, worked as a system engineer for the MOM and played a role in building its infrared camera and the methane detector – two of the five payloads[3] on board.

Kriti Faujdar, whose work consists in keeping and monitoring the satellites in their proper orbits so that they do not veer off from their course by the pull of the sun, has a most interesting story to record. Born in the ancient Buddhist city of Vaishali, Bihar, she completed her bachelor's degree in Computer engineering from Madhya Pradesh, and she is now employed at the Master Control Facility of ISRO at Hassan, Karnataka. During an interview (*The Life of Science*, April 2016), she told the story of how she was initially selected for a job at an IT company, but while she was waiting to join that job, she heard about a recruitment opportunity at ISRO and was successful in getting herself selected. She then moved to Hassan, Karnataka, the location of the Master Control Facility of ISRO. This move was facilitated by the fact that the town is close to a famous Jain pilgrimage site, and her family supported her move. Once again, she is lucky in having supportive colleagues at work as well as a supportive family at home who enthusiastically watches the ISRO launches on television. Her work at ISRO consists of long shifts where the work focuses on monitoring the geostationary satellites that rotate in sync with the earth's orbit, thus allowing weather tracking or communication at particular spots on the earth's surface.

Moumita Dutta, who hails from Calcutta, completed her Masters in Technology in Applied Physics from the University of Calcutta, and joined ISRO in 2004. In 2006, she joined the Space Applications Centre (SAC) of ISRO in Ahmedabad. The development and the testing of the optical and IR sensors being one of her special areas of expertise, she was one of the key figures behind in developing the methane sensor as well the imaging spectrometers. She heads a team which focuses on the theme of "make in India", i.e. designing optical instruments locally, one of the reasons why the project could be made on such a shoestring budget.

N. Valarmathi, hailing from Tamil Nadu, one of the southern states of India, is the first recipient of the Dr. A.P.J. Abdul Kalam Award. She has been working at ISRO since 1984 and has thus been involved in numerous missions, e.g. Insat 2A, a geostationary and communication satellite, IRS-IC, IRS-ID, which are remote sensing satellites, and TES, a technology experiment satellite. After Anuradha T.K., she was the second women to head a satellite project, and India's indigenously developed Radar Imaging Satellite (RISAT-1) was successfully launched under her leadership in 2012.

Finally, we have Seetha Somasundaram, affectionately called Dr. Seetha by her colleagues. She is currently the Director of ISRO's Space Science Program Office, and one of the key figures behind the whole mission, both in charge of the payload characterization and the calibration of the orbit (Venkateswaran and Singh, March 28, 2017). As for her background, she did her MSc from the Indian Institute of Technology, Chennai, and thereafter received her PhD in Physics from the Indian Institute of Science, Bangalore, on the topic of the white dwarf pulsars. She joined ISRO in the early 1980s.

In sum then, one sees a particular kind of profile in these women, which matches quite nicely with that of the ambitious character of the mission itself. Geographically diverse and from all over the country, it is yet quite obvious that they came from families which prized higher education, were mostly middle-class, and who were committed in supporting their family member's continued participation in a high-profile workforce, even when that participation took the form of a race for discovery and many frantic deadlines over a period of 18 months.

Building the Mars Orbiter Vehicle

Let me now briefly consider some of the details of the Mars Orbiter Vehicle itself, pointing out its general design and the engineering challenges. While I shall not be able to go in any depth into the engineering problems or their solutions, a brief survey of some of the main challenges handled by the MOM team will give us an idea of the ambitious, risky, and the innovative character of the mission. After all, until this ISRO venture, no country had ever succeeded in going to Mars at its first attempt.

As per the plan of the mission, a study team was put together in August 2010 to examine all the previous Mars missions undertaken by various international communities, e.g. the United States, the former USSR, Japan, China, and the European Space Agency, and consolidate those insights into a single document. This master document was then circulated to all the different centers of ISRO for feedback, and it covered all possible aspects of the project – the design of the spacecraft, the launch window, and finally the scientific instruments on board for the purpose of which the whole project was being designed. The success rate of the planned mission was not especially encouraging – out of total 55 attempts made, only 26 had been successful. The failures had occurred at multiple points – during the launch, en route, and most important of all, during the insertion of the satellite into the orbit.

Two problems emerged immediately as a result of this overall survey – the first the capability of the required launch vehicle, and the second, the permitted window of time during which a launch was possible. One could indeed plan various kinds of missions to Mars – for example, a fly-by mission vs. a mission that will orbit the planet or will even land on it. While the first type of the mission will produce only a very short window for scientific exploration, and is thus best avoided, it initially seemed that an orbital mission will require a larger transportation and launching capability than what was available at ISRO with its PSLV-C25 launch system – its trusted and regular work horse. However, later calculations showed that by designing a very elliptical path around Mars one could still use the same PSLV launch vehicle and yet do a reasonably good launch. The launch limitation thus turned out to be a window of opportunity for designing a highly elliptical orbit that would place a satellite around Mars on a minimum energy orbit.

Second, the mechanics of the synodic cycles in the sky dictate that a Mars mission from earth could be planned every 2.1 years. The two nearest possible launch windows thus fall either on November 2013 or on January 2016. In spite of its official announcement that India is not in direct competition with any other country, the failure of the Chinese mission in 2011, and the possibility of success in the face of this failure, must have motivated the project to a great extent. The possibility of being the first Asian country to succeed in this venture must have been another equally great motivator. Indeed, such a plan might already have been on the books since 2008, once the lunar satellite had been successfully placed on the orbit.

Next comes the practical business of planning a trajectory for reaching Mars. The journey from earth to Mars takes about 300 days, and it requires the designing of three separate phase trajectories – (i) the geocentric phase when the satellite orbits the earth and gradually gathers enough momentum to escape the gravitational pull of the earth (think of a plane taxiing on the runway before taking flight). (ii) Once it exits the orbit of the earth, it begins orbiting the sun on its way to Mars, and approaches the sphere of the Martian influence following a hyperbolic trajectory. (iii) Finally comes the most challenging phase when the satellite must insert itself into the Mars orbit. This last maneuver consists of quite a few sub-stages which must occur as follows. To enter the Mars orbit it must first fire its 440 N liquid engine to slow down and break the vehicle, rotate the spacecraft to align its thrust vector toward the earth and only then can it be ready to enter the Mars orbit.[4] As this maneuver takes place 10 months (after the launch), when the craft is millions of miles away from the earth, everything must go as per schedule, or else the mission fails.

All this naturally demands an incredible level of precise calculation, for as the satellite leaves the earth, its accuracy must be of the order of .01 arc seconds. A propulsion error of even 1 m/s could result into a massive 200,000 km error at the final trajectory. Any miscalculation thus means that the satellite would either fail to enter the orbit of Mars, and would remain forever trapped in its heliocentric phase. Or worse, it could crash into Mars itself. In either of the two last scenarios, the mission of course fails completely.

In addition to all these trajectory worries, the satellite has to successfully monitor its own inner environment and must be capable of fault management, for on the way to Mars the orbiter has to encounter multiple thermal and radiational environments, any one of which might cause damage to its structure. Regarding the all-important issue of the power on board, all power on MOM

Figure 4.1 The Mars Orbiter Mission spacecraft. Image Credit: www.isro.gov.in

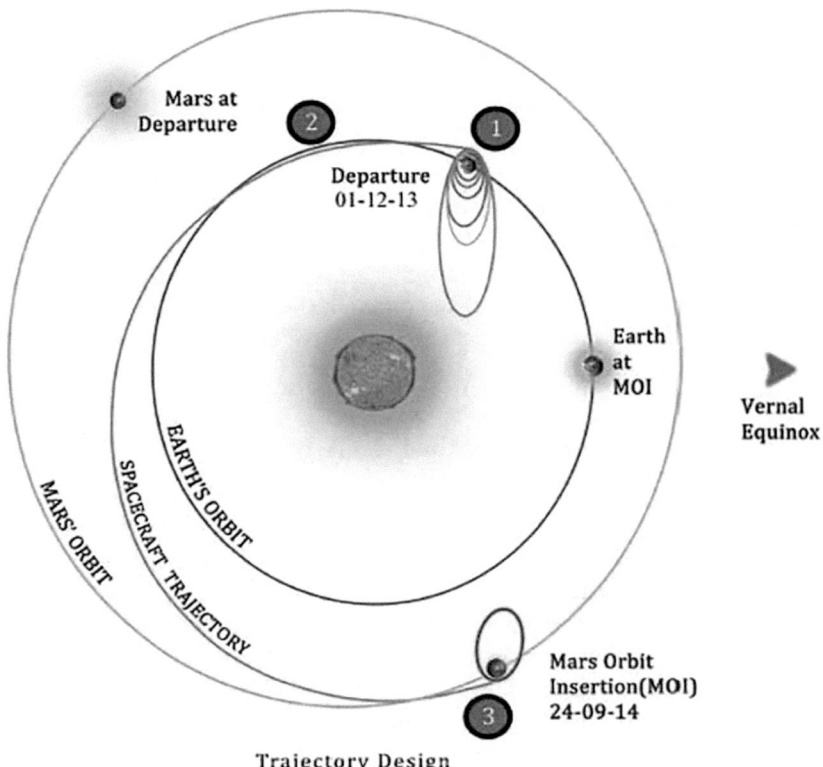

Trajectory Design

Figure 4.2 The elliptical insertion into the orbit of Mars. Image Credit: www.isro.gov.in

came from a single solar array fitted with three panels, as can be easily seen from Figure 4.2. And while the design of this solar array was like that of its previous lunar mission, the major design challenge of the new panel lay the fact that it had to be maximized for generating enough power during its Mars-centric phase where the temperature would be falling much faster compared to the geocentric part of the orbit, for Mars is further away from the sun.

It is now clear that all this calls for an onboard autonomy system. Given the 40-minute time lag of signals between the earth and the Mars, such an autonomy system is absolutely crucial for its success. Not only must the satellite protect itself against all potential thermal and radiation damages, during the last phase it must also wake up, reorient itself, point its high gain antennae toward the earth and only then complete the last orbit-entering phase of the mission. The orbiter must also know how to handle situations when the sun or the Mars comes in between the earth and the satellite and thus disrupts the communication from earth.

The onboard autonomy consisted of two parts – the ability of the system to take decisions about itself and to keep continuous watch on its own various subsystems – the sensors, the transmitter, the solar panel. This was done by giving the system a set of pre-defined stimuli and an ability to re-configure things once it encounters some kind of fault in its operation. The automation part focused on those actions which are known ahead of time, and which can thus be performed chronologically without any intervention from earth.

Finally, a brief description is provided about the instruments on board which were essential for carrying out the mission. Made of composite fiber reinforced plastic and aluminum, the orbiter weighed a total of 1,337.2 kg, and it carried five instruments on board for its scientific explorations, all designed and developed indigenously by ISRO. These were: (i) the Mars Color Camera, (ii) the Methane Sensor (so as to detect any possible signs of life), (iii) Thermal Infra-red Imaging Spectrometer, designed to detect any emitted thermal radiation from the Martian surface, (iv) the Mars Exospheric Neutral Composition Analyzer (MENCA), which, as its name suggests, explores the exosphere, and finally (v) the Lyman Alpha Photometer, a far-ultraviolet sensor that will study the hydrogen isotope ratio in Mars' atmosphere. The instruments thus had three interlinked objectives – to study the surface of the red planet, study its upper atmosphere, and to look for any possible signs of life.

This completes a very quick description of the structure and function of the orbiter vehicle. To sum up quickly, the mission had several challenging aspects that called for innovative thinking – the design of the trajectory, the handling of the different thermal and radiation environments en route to Mars, and finally the insertion of the satellite into the Mars orbit as well as to retain its communication link intact with the earth thereafter.

Success at the First Attempt

The team worked on this ambitious project for 18 months on a breakneck schedule. Finally, on November 5, 2013, the launch took place at 9:08 GMT from the Satish Dhawan Space Centre at Sriharikota, on the east coast of India. On September 24, 2014, ten months afterward, the orbiter was placed flawlessly into Mars orbit. This last event led to a set of iconic pictures in the shape of a group of smiling women congratulating one another in front a row of monitors at ISRO, which were then immediately splashed all over internet and naturally gained considerable international attention (Figure 4.3). It was of course immediately clarified that the smiling women were not the scientists themselves, but rather different ISRO staff but the picture retained its iconic touch. The image of a group of smiling women, dressed in bright silk sarees and flowers in their hair, marked – perhaps for the first time – not a domestic scene, such as a wedding or any other family celebration, but the successful completion of a highly skilled scientific mission.

Since that point of entry, the Mars mission has succeeded in every possible respect. Not only has it entered the orbit as planned, its onboard instruments still continue to be in perfectly good order, returning data as planned. It has also successfully negotiated two cases of blackout (June 6–22, 2015) when the sun came between the earth and the Mars and thus blocked all signals. A whiteout case also took place when there was a sun-earth-mars combination during which the orbiter could not discern any signals emitted from the earth (May 16–29, 2016). In January 2017, it was maneuvered into a new orbit so as to escape the eclipses and thus enjoy a longer life until 2020. In every way then, the orbiter was a great success beyond its originally planned life of six months.

No Longer Just a Modest Practice

What has been the impact of this mission on the Indian space science community? Following the information set forth on ISRO's webpage, I classify this impact under three main headings.

First, with this launch India has become the member of an elite space club, which previously consisted only of the United States, Russia, and the European Space Agency. Also, India became the first Asian nation to join this club. Second, to complete this project, it used a unique launch vehicle trajectory – one of the main innovative features of this mission. Third, it was the most economical interplanetary mission ever, making excellent use of some of the frugal and indigenous engineering concepts as well as the shortest possible project planning. While it is too early to say whether those indigenous strategies amounted to a radical new approach in solving such

Figure 4.3   The iconic picture of the smiling ISRO women. Image Credit: thelifeofscience.com

engineering problems, it surely increased the possibility that such novel contributions might be offered more easily in the future.

Overall then, MOM exhibited undoubted skill in planning and developing the spacecraft, launching it, and then finally maneuvering it successfully on the lowest energy orbit. The project has so far produced 23 peer-reviewed publications, and thousands of color images of the red planet, and in 2015 it received the National Space Society's prestigious Space Pioneer Award.

No longer just confined to Sarabhai's modest philosophy, the future of ISRO now involves a whole string of new projects – *Chandrayaan* 2[5] as well as perhaps a mission in the near future involving human astronauts, which would also include women astronauts. ISRO's general goal seems to be bypassing the idea of land-based networks and leapfrogging straight into a world of space-based technologies. One of their very interesting recent proposals – extremely useful for the common person on the ground – is a solar potential app, which will inform about the monthly or the yearly solar potential of any particular spot or location. The Space Application Center facility (SAC) of ISRO in Ahmedabad has created this app under the directive of the Ministry of New and Renewable Energy.

And while the entire event has been a celebration of India's scientific workforce, it surely included in it a parallel celebration for the work of ISRO's women scientists as well.

Can We Then Draw a Feminist Conclusion?

I started the chapter by raising a question about whether MOM could be taken as showcasing of a stock of feminist expertise. Given the details that I presented earlier, I must now draw a conclusion, and yet I think that the conclusion I can draw will only be a complex one. As I see it, the case can be argued equally well on both sides. For example, it can plausibly be said that the MOM was simply a case of nationalist science, executed along the lines of contemporary Big Science, albeit on a modest scale with some innovative twists. MOM also exhibited nearly all the features of what Kuhn calls normal science, e.g. it was planned with the state-of-the-art knowledge of its day, and the outcome was well-expected, precisely calculated, and so on. And as for the substantial presence of a number of women in the project, who were leaders of several aspects of this mission, it can be interpreted as more of a class phenomenon rather than a case of revolutionary change taking place in society. After all, most of the women scientists came from various middle-class homes where the education of their daughters was already highly prized. Furthermore, they also all seem to be exceptionally fortunate in having supportive families who stood by them during the 18-month breakneck period, when the team was part of a race to the red planet. It then appears from their public profiles that their story had been that of a particularly happy one, where their own upwardly mobile trajectories fitted very nicely within the upward curve of a nation, which had jumped into a space race and thus simply needed as many skilled hands as possible. Indeed, history of science often shows us that when a problem situation arises in science, new members become especially welcome to join the community. Such problem situations then become the entry points into the profession for these new members as well as places for their subsequent foothold. But as a stand-alone event, this might produce very little overall effect in the status of the women in India, especially in those corners where the education of the girl child is not very highly prized.

Against this very plausible, but I think – ultimately erroneous – critique, I present two counter-arguments. First, change, once introduced, tends to spread like ripples on a pond. Second, changes like this often bring in their wake a very visible set of role models – something which this project surely produced in good numbers. Thus, there are likely to be people in the future who will take these women scientists as their role models and will build their dreams around them. Now, in science, as well as in everything else, often imagination opens the doors to a new world, and a dream, once conceived, often has a high probability of being realized in practice. Nandini Harinath's comment that after the success of the Mars mission, both the ISRO and the planners of the mission were put on a different kind of pedestal, leading to a sudden spurt of international and national attention around them captures something of the phenomenon that I am talking about. In her interview with the *Nature India* Blog, Moumita Dutta too records a similar sentiment, saying that the success of the women scientists at ISRO has become something of an attractor, stimulating other women, especially young girls, to think of joining them in the same area of space science. Currently, ISRO has about 20% of women in its workforce, and this percentage is likely to go up in the future. What the project then shows is that a group of highly skilled women were able to offer significant leadership during a crucial nationalist project that involved a big race of skills, time, and prestige, and success in this race had made their own careers look more shining and truly worthy of imagination. The innovative, ambitious, and the frugal character of the mission reinforced their status as reliable practitioners of science, and ability to serve as mentors to a new generation. And since in their own eyes, they were engaged in doing challenging and innovative science, when others follow them in the same path, they would be more than likely to pass this trend on. Thus, using William Blanpied's language, who long ago applied this phrase to describe the first generation of the physicists in India like S.N. Bose, these women too have come to occupy the position of tribal leaders (Blanpied 1972).

I conclude therefore that this project will have significant ripple effects on Indian science, especially on the women scientists who are now in-training, or about to enter their professions, stimulating both their ambitions and their dreams, and through them on the rest of the society. But this significance might not reveal itself immediately in an obvious form. Not all feminist epistemic projects begin with an explicit declaration of feminist methodology or obvious feminist goals. Instead, such projects might begin almost imperceptibly – as forms of routine work – without any explicit announcements of such goals or strategies. Changes like this do not then show up in the form of a violent break, but rather seep into structures and practices in the form of osmosis. There surely exist cases of such subtle, but very pervasive, change. In the story of the proverbial ship of Theseus, for example, the continuous addition of a few new planks in the ship eventually produced an altogether new ship, although no one was sure at what point the old ship had become a new one. The modest addition of a few new planks at a point of time may not even register as an important event, but later on by hindsight this can be seen as the starting point of substantial change by creating a different kind of community. And such a different new community will eventually produce new forms of practice. The doors that had opened once may never again close fully. *And the resulting new forms of practice that usher themselves through those open doors may then – quite plausibly – embody explicit and substantial feminist values.* As evidence of such change consider the increase of interest among the young girls following the newspaper reporting of the successful Mars mission. We can also see the influence in the way the Mars mission entered the pop culture and thus popular imagination by the production of a Hindi movie in 2019, named *Mission Mangal*.

Indeed, India's own entry into science as a nation took precisely this form – a small group of quite peripheral scientists began a modest practice by making a few contributions in the various high-profile problems of European science. Most of those contributions took place within the contexts of creating teaching and research universities in various port cities, e.g. in Kolkata or in Chennai. This, in the space of a couple of generations, led to a modest practice, eventually causing the birth of a new scientific community in a new location. And as I tried to show in the beginning of this chapter, such was the case of ISRO itself – which began its journey as a very modest practice, hoping to gain some small share of the newly-born space science.

To conclude then, I argue that one must always make room for the newcomers in a scientific practice for it gives rise to new paths and new trajectories, even though there may be no quick way of predicting how (or exactly when) those things will lead to new ideas or practices in the future. Such unanticipated turns in the practice then simply re-affirms the open-ended character of scientific inquiry, which Karl Popper long ago dubbed in his autobiography as an unended quest. The Mars mission opened doors to a number of new ambitions and quests for the Indian scientific community, and surely a vision for the future, but one of those open doors clearly revealed the possibility of greater participation of the women scientists in different projects of Indian science.

How long will it be before the change becomes visible all around? We do not know. But a change has been announced no doubt – both in the shape of the entry of a new member into an elite space club and the presence of a new group who emerged as crucial contributors into the Indian space project. And the epistemic significance of the Mars mission seems to be precisely this – that both of these changes took place at the same time, one nicely nested inside the other.

Acknowledgment

I would like to express my thanks to Seetha Somasundaram at ISRO for providing me with valuable details of the Mars Mission in the shape of a special volume released for the occasion. I thank Arnab Rai Chaudhuri for introducing me to Dr. Somasundaram and for his conversations about Indian science. My thanks also go to Paola Castaño, for helping me with a number of necessary documents about ISRO and for her general encouragement.

Related chapters: 5, 31.

Notes

1 "In India, the immediate goal of our space research is modest. We do not expect to send a man to the moon, or put white, pink or black elephants on orbit around the earth" (Singh, 2017, Chap 15).
2 The first Indian satellite named *Aryabhata* was launched by the Soviet Union.
3 A payload is a component of an aircraft or a launch vehicle, measured from the point of view of its carrying capacity and its weight in the overall structure.
4 The satellite's propulsion system consisted of a 440 N liquid engine and eight 22 N thrusters.
5 Contrary to high expectations however, *Chandrayaan 2*, which was launched on July 22, 2019 and reached the lunar orbit on August 20, was only partially successful. While the satellite entered the lunar orbit successfully, the ambitious plan of sending out a lander to explore the surface of the moon ended in total failure. The lander, named *Vikram*, lost contact with the earth, and crashed on the surface of the moon.

References

Blanpied, W. (1972) "Satyendranath Bose: Co-founder of Quantum Statistics," *American Journal of Physics* 40 (4), 1212–1220.

https://www.bbc.com/news/world-asia-india-38253471, accessed on 1/5/2019.

Current Science (2015) Special Issue. : Mars Orbiter Mission. 109 (6).

https://www.diarystore.com/isro-women-scientist-who-took-india-into-space-rocket-women-scientists, accessed on 12/28/2018

Singh, G. (2017) *The Indian Space Programme: India's Incredible Journey from Third World towards the First.* Astrotalkuk Publications.

https://www.amazon.com/Indian-Space-Programme-incredible-journey/dp/0956933734

Mars Orbiter Mission Profile. https://www.isro.gov.in/pslv-c25-mars-orbiter-mission/mars-orbiter-mission-profile, accessed on 1/3/2019.

https://www.isro.gov.in/sites/default/files/AnnualReports/2014/SSPR.html, accessed on 1/4/2019.

https://www.isro.gov.in/second-anniversary-of-mom-launch-celebrated, accessed on 1/4/2019.

https://thelifeofscience.com/2016/04/25/kriti-isro/, accessed on 1/3/2019.

http://blogs.nature.com/indigenus/2018/07/being-rocket-woman.html

Srivastava, B. (2013) "Team-of-11-who-Made-Mangalyaan-Launch-Possible," *Hindustan Times*, November 6.

https://www.thehindu.com/sci-tech/technology/tamil-nadu-woman-stars-in-satellite-launch/article3357654.ece, accessed on 1/4/2019.

http://vigyanprasar.gov.in/isw/drseetha_story.html, accessed on 1/4/2019.

5

CONTRIBUTIONS OF WOMEN TO PHILOSOPHY OF SCIENCE

A Bibliometric Analysis

Evelyn Brister and Daniel J. Hicks

Currently, in 2020, about 20% of philosophers of science are women. This appears to be a lower level of participation by women than in the humanities overall, where over half of faculty members are women (American Academy of Arts and Sciences 2020), and also lower than in the sciences, where women account for 37% of STEM faculty (National Science Foundation 2018). There are also gender differences in subfields of philosophy, with a lower proportion of women participating in philosophy of science than in the discipline at large, since women accounted for 27% of philosophy faculty in 2017 (American Academy of Arts and Sciences 2020). The data presented in this chapter show that since 1980 there has been a relative increase of between 3% and 5% per decade in women's participation in philosophy of science. This represents a steady increase, but the rate of change toward gender inclusion has been slower in philosophy than in many other disciplines, including psychology, history, biology, and chemistry. Although women's participation in philosophy of science is still lower than in most other areas of the humanities and sciences, recent attention to providing institutional support for women and minorities promises to raise the participation of women and other underrepresented groups in the coming decade.

Women have influenced the field of philosophy of science both through their research and via roles with journals and professional societies. When a full range of professional roles is considered, it is clear that women have had a strong influence on the conceptual development of the field in spite of having produced less total scholarship. In this chapter, we: (1) identify some important ways that women have contributed to philosophy of science, (2) describe the contributions of specific women at key moments, (3) examine the social influences that have affected women's participation, and (4) suggest connections between women's participation and the substance of research in philosophy of science.

Many people have long known or suspected that women's participation in philosophy of science has lagged but is slowly improving. Our analysis uses bibliometric data, drawn from publication records, to discover how badly it lagged and how much it is improving. Bibliometric methods allow us to provide an accurate picture of where, when, and how women have contributed to research in philosophy of science. We also draw on traditional historical sources to identify and complete gaps in the data.

Feminist historians and philosophers of science have theorized how women's experiences and social roles have influenced the history of scientific and medical research. Their research shows how gendered metaphors shaped scientific understanding of human gametes (Martin 1991), how class and gender politics influenced debates in physical chemistry (Potter 2001), how assumptions about women's sexuality influenced evolutionary biology (Lloyd 2006), and how the politics of gender continue to influence cognitive science (Fine 2010), along with many other examples of the sway gendered beliefs and norms have had on the development of scientific knowledge and practice. Besides recounting the history of women in philosophy of science, this chapter aims to extend the reflexivity that is common in feminist science studies to urge philosophers to consider how gender has affected the development of our field—that is, how the presence or absence of women in philosophy of science has influenced its content. We hope that this reflexive turn will provide insights relevant to supporting further increases in the social diversity of our profession.

Our Methods

In order to evaluate the participation of women in philosophy of science from 1930 to the present, we used bibliometric analysis of author gender for journal articles, book chapters, and book reviews (see Figure 5.1). Our data show that over this time, women gradually increased from being fewer than 3% of authors in philosophy of science before 1970 to over 18% in 2017.

The bibliometric data for our study were collected from the Crossref database of scholarly publication metadata. Our aim was to collect data on publications and their authors, including author names, article title, journal name, and publication information for the majority of academic articles and reviews written in philosophy of science. In order to compile a core collection of metadata, we targeted 11 journals specializing in the philosophy of science, as well as three book series (*Minnesota Studies in the Philosophy of Science*, *Boston Studies in the Philosophy of Science*, and the *Western Ontario Series in the Philosophy of Science*), focusing on the period from 1930 to 2017. Many philosophers of science publish their research in generalist and interdisciplinary journals, so in order to collect data on these articles while avoiding the inclusion of research that is best identified with other philosophical specializations or as another area of science studies, we designated an author as a "philosopher of science" if two or more articles or reviews by that author had been published in philosophy of science journals. We then added to our publication data set articles and book reviews written by a designated philosopher of science in 16 general or analytic philosophy journals, including *The Journal of Philosophy, Synthese*, and the contemporary *Erkenntnis* (after its refounding in 1975), and in 11 interdisciplinary

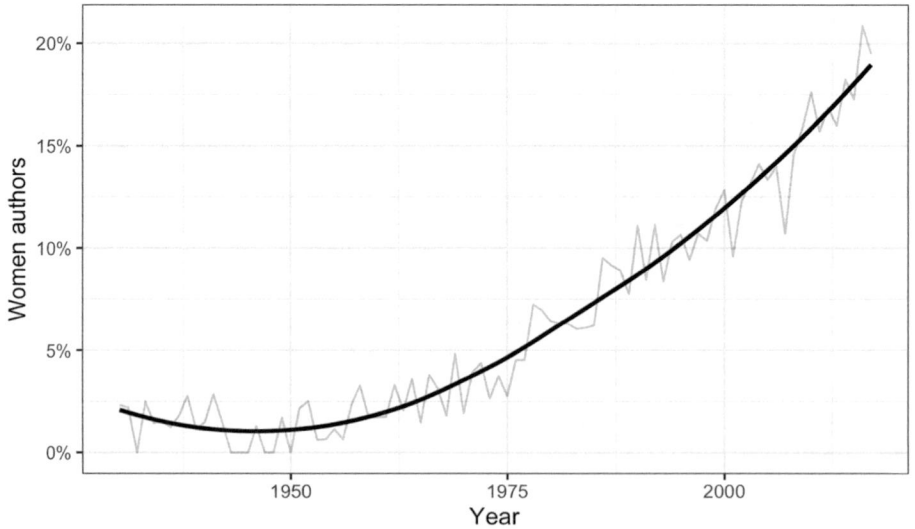

Figure 5.1   Proportion of authors of philosophy of science papers who are women, by year, 1930–2017. Heavy black line is a local regression indicating the trend

science studies journals, including *Foundations of Physics* and the *Studies in History and Philosophy of Science* journals. Because we wanted to be sure to identify contributions to philosophy of science by women—and because feminist work has not always had a positive reception in mainstream philosophy journals—we also included articles and reviews that appeared in the feminist scholarly journals *Hypatia* and *Signs* in a separate feminist category. In total, the data set includes 37,946 journal articles and book chapters by 3,451 authors. The core of metadata from specialized philosophy of science journals and book series accounted for 44% of our final data set; 46% were published in analytic or general philosophy journals; and 10% were published in interdisciplinary science studies. The two feminist journals contributed 0.3% (or 112) of the papers in the data set.[1]

The data set casts a wide net in order to encompass data on areas of scholarship that are relevant to philosophy of science and on authors who contributed to scholarship in philosophy of science, even when those authors primarily wrote in other areas. One drawback with forming the data set from publisher metadata for journal articles and chapters in book series is that it does not include monographs. Thus, philosophers of science who only wrote books or who published articles in generalist journals but not in specialized philosophy of science journals are not included in our analysis. Mary Midgley is an example of someone who, in addition to her work in moral philosophy, published books against scientism as well as books and articles criticizing certain formulations of evolutionary theory. Although these are topics taken up in articles written by others in the data set, Midgley is not included. Another drawback is that the data are more complete for recent years; however, we have addressed this uneven coverage by making liberal use of other historical records and biographies.

The data are available in two formats—one organized by author and the other by individual publication (journal article or book chapter). Since about 8% of publications in the data set have two or more authors, this facilitates analyzing data concerning authors separately from analyzing data about individual publications. We disambiguated names so that variants of a name (such as Carl Hempel and C.G. Hempel) would be classified as the same author, and we used a multi-step algorithmic and manual process to identify and verify author gender. Gender identification was accomplished by combining date-specific information from the US Social Security Administration with data from two online gender attribution services, followed by manual processing that involved looking up biographical information and contemporary personal webpages. We were able to identify author gender of all authors for 99.2% of the 37,946 publications in the data set.[2]

This data set was the basis for further analysis using topic modeling. Topic models are statistical models that analyze the frequency with which words or strings of words appear together (Roberts et al. 2014). Topics represent areas of discourse or subfields that wax and wane over time. The topic model classifies topics and identifies the publications that fit in a topic, which allows dating the initiation of new research areas and identifying the authors or papers that contribute to them. We retrieved text data from JSTOR, Elsevier, and Springer. For each year in the period 1930–2010, approximately half the articles in our data set have text data available in the form of abstracts or full text. We performed topic modeling on text data for 20,827 documents.[3] The bibliometric analysis combined with topic modeling allows us to evaluate the participation of women in scholarly research in philosophy of science, how that participation has changed over time, and specific subfields where women have had a distinctive influence.

Early Decades: 1930–1980

In the twentieth century most scientists and most philosophers were men. Programs to increase the participation of girls and women in science and engineering began to be developed in the 1980s, and by the turn of the century there were national mobilization efforts by the National

Science Foundation, the National Academy of Engineering, and other agencies and professional societies to support women's success in science and engineering education and careers (Bix 2013). Between 1970 and 2010, participation by women increased by a large proportion in most biological and social sciences as well as most areas of the humanities, where women now make up over half of students and faculty. But some disciplines did not see notable increases in gender parity during that period, including engineering, physics, economics, and philosophy.

During the period between 1930 and 2017, the size of the profession grew more than tenfold, from about 40 authors publishing in 1930 to over 500 publishing in recent years. Women made up less than 5% of authors in the years before 1975, and 1969 was the first year when more than ten women published papers in the philosophy of science. Most of the women who published in philosophy of science journals in the early years—before 1950—wrote on logic, probability, and the nature of scientific reasoning.

Only a small number of women published articles on philosophy of science in the 1930s. These are Eleanor Bisbee (1893–1956), who began her career at the University of Cincinnati and then lived and taught in Istanbul before returning to the United States in 1942; Janina Lindenbaum Hosiasson (1899–1942), a Polish logician who was executed by the Gestapo; Maria Kokoszyńska (1905–1981), a Polish logician and epistemologist who presented the views of the Vienna Circle to her colleagues at the Lvov-Warsaw School and published the views of the Lvov-Warsaw School in the Vienna Circle's journal *Erkenntnis* (Brozek 2017); Rose Rand (1903–1980), an Austrian logician and member of the Vienna Circle who immigrated to England and then to the United States; Hilda Geiringer (1893–1973), a German mathematician who immigrated to the United States in 1939; Dorothy D. Lee (1905–1975), an anthropologist and philosopher of anthropology at Vassar; and Susan Stebbing (1885–1943), who was one of the founders of the journal *Analysis* in 1933.

World War II and the political events leading up to it disrupted academic careers. The women who immigrated to the United Kingdom and the United States were dramatically affected by these circumstances, and the names of some well-trained and influential women philosophers and logicians appear in our data set only a small number of times—or not at all— because after immigrating they could only find employment at women's colleges if at all. Their status as women and immigrants was further complicated in some cases by being Jewish. Hilda Geiringer published promising work in the 1930s: she published two articles in *Erkenntnis* and was the first woman to receive an academic appointment in mathematics at the University of Berlin. During and after World War II she taught at Bryn Mawr College and later Wheaton College in Massachusetts, both women's colleges, but she was unable to find work at larger universities closer to her husband's position at Harvard (Friedenreich 2009). Rose Rand is another example of a woman who faced employment challenges in the United States. In spite of her PhD studies with Rudolf Carnap and Moritz Schlick, later support from Susan Stebbing and Karl Popper, and rich correspondence with Otto Neurath, Alfred Tarski, Ludwig Wittgenstein, and others, she primarily published translations rather than original research and does not appear in our data set.

The disruptions to academic philosophy caused by the war and by the wave of immigration from Europe to the United Kingdom and United States that followed, combined with the special difficulty that women had finding academic employment in the United States also explain why fewer women published articles in philosophy of science in the 1940s than in the preceding decade. The only women who began publishing in philosophy of science in the 1940s were May Brodbeck (1917–1983), Marjorie Grene (1910–2009), and Mary Henle (1913–2007). All three were American, and their entry into the field coincided with a shift in the location of cutting-edge work in philosophy of science from Austria, Germany, and Poland to the United States. Mary Henle was a psychologist at the New School for Social Research who found an interested audience among philosophers of psychology. May Brodbeck's first degree was in chemistry before earning a PhD in philosophy, and she spent her career primarily at the University of Minnesota and then the

University of Iowa. As an administrator at the University of Iowa, she oversaw the creation of one of the first women's studies programs (Fine 1997). Marjorie Grene studied zoology before earning a PhD in philosophy from Radcliffe College. As a professor at the University of California at Davis and later at Virginia Tech, she wrote on both existentialism and the philosophy of biology.

In our data set, only five more women began to publish articles in philosophy of science in the 1950s; but 15 were added in the 1960s—most of those joining right at the end of the decade. Social factors limited the research roles that women could assume within academic institutions of philosophy in the 1930s and following decades. Most notably, before 1980 only a few women had professorships in philosophy where they could train doctoral students. Women's modes of influence, therefore, were mostly limited to writing and teaching undergraduate students, usually at women's colleges and often in temporary lectureships and insecure positions. Thus, although there were women writing and teaching philosophy of science, there was not an institutionalized mechanism whereby established women philosophers of science could mentor and support newly minted academics.

Although few women taught graduate students in philosophy during the mid-twentieth century, some individuals did play a role in constructing the institutional framework for philosophy of science—its journals and professional societies. In 1933, Susan Stebbing was on the first editorial board of *Philosophy of Science*; May Brodbeck joined the editorial board in 1959. Mary Hesse presented at the first meeting of the Philosophy of Science Association (PSA) in 1968 (the only woman to do so); and in 1979 she became the PSA's first woman President. In 1971 Marjorie Grene served as President of the Pacific Division of the American Philosophical Association, and in the late 1980s she was involved in the formation of the International Society for the History, Philosophy, and Social Studies of Biology.

Recent Decades: 1980–2010

The participation of women in philosophy of science began to increase slowly in the 1970s and more dramatically through the 1990s. Around 1950, our data identify less than 2% of authors as women. By 1980, that figure had only increased to 7%, but by the turn of the century women made up over 12% of researchers publishing articles on the philosophy of science. By the late twentieth century, there was a growing expectation among women that academic professions be open to them and more awareness of the obstacles that made it difficult to recruit and retain women. Even while there was little organized institutional effort to raise participation by women in philosophy, it became commonly recognized that individuals could benefit from targeted support, especially as undergraduate and graduate students. This was a period when efforts to support women were primarily at the level of individuals, departments, and universities.

The creation of women's studies programs in the 1970s and 1980s fostered an awareness of women's presence—and vulnerability—in the academy, and feminist scholarship began to be published in philosophy of science, drawing on feminist scholarship in related areas of epistemology and history of science. The interdisciplinary feminist journal *Signs* was founded in 1975 to publish scholarship on women, culture, and society, and by the 1980s *Signs* was a vibrant space where feminist scholars addressed how conceptions of gender shape science and technology. *Signs* published formative articles in feminist philosophy of science by Sandra Harding, Evelyn Fox Keller, Helen Longino, Lorraine Code, and others. The first issue of the feminist philosophy journal *Hypatia*, published in 1986, featured an exchange between bioethicist Laura Purdy and philosopher of science Nancy Tuana on biological determinism and the nature-nurture debate, and *Hypatia* ran special issues on feminist philosophy and science studies in 1987, 1988, and 2004.

The opening of a space for feminist scholarship in philosophy of science presented women with the chance to investigate research topics that made sense of their own social and academic experiences, but other scholars often treated feminist scholarship as politicized and dismissed it for that reason. Indeed, feminist scholarship did often challenge existing academic norms. The political nature of these debates forced women to make tactical decisions about how to formulate a research agenda. On the one hand, feminist conferences and "invisible colleges" provided needed social support for women philosophers; on the other hand, feminist scholarship in philosophy was not widely cited outside the intellectual community of feminist scholars. Thus, women who focused primarily on feminist scholarship risked being marginalized by peers in philosophy, including within their own departments, the locus of tenure and promotion decisions.

Feminist scholars in the 1980s and 1990s struggled to establish academic legitimacy, and scholars in philosophy of science were no exception (Gumport 1990).[4] In 1995, Elisabeth Lloyd argued that philosophers of science and analytic philosophers had been violating their own standards of objectivity and legitimate argumentation by ignoring feminist research in epistemology and philosophy of science. She accused the profession of deploying a double standard:

> The vast majority of philosophers still believe that 'feminists' are playing 'out-of-bounds,' in terms of mainstream understandings of the problems of epistemology and philosophy of science; feminist work can, therefore, be safely ignored, set aside, or characterized as of interest only in marginal cases.

> *(Lloyd 1995: 352)*

Twenty years later, many of the research questions that were subjects of feminist investigation in the 1980s and 1990s are central topics in mainstream philosophy of science, metaphysics, and epistemology, including social epistemology, the operation of social power within scientific institutions, the role of social values in scientific research, and the nature of epistemic injustice. A shift to treat these as topics of real philosophical interest rather than beyond the scope of philosophical analysis came about for many reasons, but its initiation by feminist scholars depended on the opportunity to develop such ideas in detail within supportive intellectual circles.

Although the opportunity to pursue feminist philosophy of science was intellectually and professionally valuable to women, the majority of women philosophers of science did not specialize in feminist scholarship, and most researchers who contributed to feminist philosophy of science made additional contributions to mainstream philosophy of science. Of the articles and book chapters in our data set published by women, less than 3% were published in the two feminist journals (*Signs* and *Hypatia*), and two-thirds of the women authors who published some of their work in these journals have published more articles in mainstream philosophy of science than feminist philosophy of science.

These data can be given more context by considering our analysis of topic models constructed from abstracts and full publication texts. Topics represent areas of discourse and are defined according to calculating a probability distribution for words that frequently appear together in articles within the topic while infrequently appearing together in articles outside the topic. Using a fine-grained model, we manually assigned labels to 181 machine-generated topics. Most topics show a pattern of growth over time, such that there is an early period with a few scattered works, followed by a period of rapid increase, or development, as the topic develops into a recognizable subfield. We analyzed the topic models to find contributions by women prior to the completion of the development period. This method permits the discovery of areas of discourse, or subfields, in philosophy of science where women's contributions have had a distinctive and early impact in shaping how an area of inquiry is framed and pursued. The identification of germinal, or pre-development, articles for topics shines light on those areas of philosophy of science where women's contributions have had clear epistemic uptake and formative impact. More than simply showing that women contributed to ongoing conversations, the identification of germinal papers also picks out women who were leaders in shaping new areas of research.

Not surprisingly, the topic with the largest share of women contributing to its development has to do with gender and feminism. At the time it reached development in 1990, 39% of publications in the topic had been authored by women. Although finding that only two-fifths of early authors in the "gender and feminism" topic are women may seem low, this result can be explained by considering the topic modeling method. A topic does not simply contain papers published in feminist journals. Rather, it calculates the frequency of words and combinations of words, and in the period preceding 1990, this topic includes a wide assortment of papers concerning gender in science education, investigating sex and gender in human biological function, and even attacking feminist scholarship, in addition to the papers that represent the beginnings of feminist approaches in philosophy of science. Besides the topic on gender and feminism, a number of other topics also had a notable share of germinal papers authored by women. These topics include (in temporal order of their development dates): continental drift (1977), cognitive science (1991), intentionality (2003), ecology (2005), and model-based science (2008). The women authors with the most germinal papers (in descending order) are Nancy Cartwright, Marjorie Grene, Mary Hesse, Lindley Darden, and Kristin Shrader-Frechette, all of whom contributed germinal papers to dozens of topics.

The scope of women's contributions to philosophy of science is wide and not concentrated in only one or just a few areas. Women made germinal contributions to 71% of the topics we identified and, counting post-development articles, they made contributions to 179 of the 181 topics. (The two exceptions are two topics in logic that peaked in the first few decades of the data set.) When the primary journals in our data set are divided according to whether they cover general philosophy of science or specialize in philosophy of biology, physics, or the social sciences, we find that women have written 15% of articles and reviews in both general philosophy of science and in philosophy of biology, and they have written 9% of articles and reviews in both philosophy of social science and philosophy of physics.[5]

Philosophy of Science Today: 2010–2020

Between 2000 and 2017, the share of women who published articles and reviews in philosophy of science increased at a steady rate from 12% to 20% (again, refer to Figure 5.1). This is comparable to the share of women members of the PSA. About 15% of the members of PSA in 2006 were women, and about 17% in 2014.[6] This figure may be compared to the overall percentage of women who are philosophers, about 27% (American Academy of Arts and Sciences 2020). We are unable to provide a single explanation for the discrepancy between philosophy in general and philosophy of science, but it is likely that there are multiple causes.

The very small number of women who did research and trained graduate students in philosophy of science before 1980 is surely a factor, as this has likely contributed to a slower trend toward inclusion than in other areas of philosophy where women occupied graduate research positions following World War II. It is also likely that the apparent double requirement to train in both philosophy and a science has discouraged the entry of people with less community support and with less opportunity to take longer to train, and this discouragement may combine with the legacy of the sciences being gendered masculine (Nosek et al. 2009). In a bibliographic study using a vast amount of data about the sciences and humanities from JSTOR, West et al. (2013) find differences in publication rates among subfields of philosophy.[7] They find that in subfields related to philosophy of science (such as philosophy of time, confirmation, and scientific methodology), women have published between 3% and 11% of articles, while women have published nearly 20% of articles in moral philosophy (West et al. 2013). A 2017 study of the "Top 25" journals in philosophy found that women publish less in philosophy of science than in history of philosophy, but more than in a category combining philosophy of language, epistemology, mind, and metaphysics (Wilhelm et al. 2018). Studies like these two and ours analyze publication data, and they may find lower participation of women in research outputs compared to employment statistics because not all women who are employed and teach in the field publish journal articles (Wilhelm et al. 2018).

A related question is whether women experience bias when submitting articles for publication. In some male-dominated fields in the sciences and humanities, women publish proportionately less than their male counterparts, but in others they publish on a par. There are a number of explanations for differences in women's publication rates across academic fields, and there is no simple, clear relationship between the proportion of women in a field and their publication rate (Larivière et al. 2013). In some fields where women publish fewer papers proportional to the number of women in the field, this appears to be a direct result of publication bias, but in others it appears to be the result of women's investment strategies and lower submission rates. In our data set, the median woman author published 0.67 papers per year during the span of her career (between her first and last paper in the data set), while the median man author published 0.58 papers per year; however, when we control for the length of career and the year a career began, this difference vanishes. Thus, we did not find evidence of difference in publication rates in our data set, even if they might exist in other disciplines.

We interpret the facts that women and men in philosophy of science have similar publication rates, and that publication rates for women and men are similar to their level of participation in PSA, to be signs that publication bias is not a systematic problem in our subfield. We see this as a positive result indicating that, overall, women researchers in philosophy of science, once established, are not excluded from publication opportunities. However, this finding appears to be in tension with the relatively slow rate of increase for women's participation in the field. One possibility for slow growth during a period of gender equity in the per person publication rate is that some women do not feel sufficiently supported at early educational and career stages to train in philosophy of science, to join the PSA, or to become established researchers. This suggests that early support for undergraduate and graduate students as well as for early career women would be a good investment.

Professional societies have begun to address these concerns and to seek out and advocate for institutional solutions. The PSA Women's Caucus was organized in 2006 to promote networking among women and raise the profile of women's scholarship. The majority of PSA Women's Caucus initiatives provide support by making participation in conferences and the profession easier for all participants, not just for women. These efforts include mentoring scholars from underrepresented groups, providing training on bystander intervention, developing a sexual harassment policy, and distributing dependent-care grants and arranging on-site childcare for conference participants (Bursten 2019). The European Philosophy of Science Association (EPSA) founded a Women's Caucus with a similar purpose in 2011. In 2014, the British Philosophical Association (BPA) and the UK chapter of the Society for Women in Philosophy (SWIP) announced a Good Practice Scheme to assist university departments, professional societies, and journals in adopting policies and practices that encourage the participation of women and underrepresented minorities in philosophy, and in 2015 both EPSA and the British Society for the Philosophy of Science committed to follow the scheme.[8]

Future Paths for Women in Philosophy of Science

Though relatively few in number, women have had a distinctive impact on research in philosophy of science and on the development of its institutions, including journals and professional societies. A goal of this chapter has been to trace the history of women who had a role in developing philosophy of science, to identify the impact of scholarship by women, and to better understand gender dynamics in the discipline. There are now initiatives underway to foster more equitable participation by women and underrepresented groups—initiatives that are motivated by ethical commitments to fairness and inclusion. We would like to raise questions about how the social location and experience of researchers, in the past and future, affect the subject matter and methods of philosophy of science.

An important theme in feminist philosophy of science is the impact that women's experiences and feminist political commitments have had on the development of feminist critiques of science and philosophy. There are at least three ways that this impact occurs. First, feminist scholarship has revived an appreciation of women's contributions to ancient science and to early modern natural philosophy. Women such as Hypatia, Theano, Princess Elisabeth of Bohemia, Margaret Cavendish, and Émilie du Châtelet were influential in their time but have not been appreciated as part of the canon. Feminist historians and philosophers have examined how cultural beliefs and practices made it possible for these women to think of themselves as philosophers and scientists, as well as how social assumptions about women's abilities and their proper social roles caused their contributions to be forgotten (Atherton 1994). Even for a time as recent as the twentieth century, it is important to understand the social forces that affected women's participation and to trace women's impact.

The second way that feminist scholarship has affected science studies and the history and philosophy of science is by showing how women's experiences generate questions about how assumptions about sex, gender, and society figure in scientific theories. Feminist philosophers have used this approach to critique the systems that build scientific knowledge. Consideration of how power operates in academic and research settings is now a mainstream topic in philosophy of science and epistemology. By starting with their own experiences, including with the experience of feminist scholarship being ignored or dismissed by mainstream philosophers, feminists have taken part in analyses of the role of social values in scientific research; how social location and power relations influence the creation of scientific knowledge at the scales of knowers, research institutions, theories, and disciplines; and how scientific communities invite and process criticism. Although this research is often focused on scientific research, questions about values, priorities, and the operation of social power may also be asked of our own discipline (Jenkins 2014).

Feminist philosophers—including philosophers of science—are now developing reflective examinations of the criteria used to establish the credibility of researchers and the priority of research topics in our own discipline, and how these criteria reinforce social power and discourage the inclusion of women and minorities. As Leslie et al. (2015) have shown, in fields such as philosophy where field-specific ability is difficult to identify, cultural associations link men, but not women or underrepresented minorities, with talent. This association may both reflect and trigger bias against women and minorities, including stereotype threat. Identifying and avoiding unsupported mental models and biased linguistic practices (such as the trait of "brilliance") that are common in our field may assist the struggle for greater inclusion in philosophy of science. Leuschner (2015) demonstrates how theories that philosophers of science have developed to explain failures of objectivity in the sciences—bias in peer review, preference for established research topics and methods, resistance to reasoned critique—affect philosophy as well as science. We hope that this analysis of bibliometric data in philosophy of science provides a point of departure for further theory and practical support concerning inclusion and diversity.

Acknowledgments

We are grateful to Rick Morris for providing significant assistance with the collection and processing of data for this project, specifically with disambiguating names. Aaron Crespo assisted with data collection for *Minnesota Studies* book chapters. Many thanks to the dozens of historians of philosophy and philosophers of science who have shared their memories and knowledge about women's activities in developing the field of philosophy of science and its institutions, and especially to Miriam Solomon, Andrea Woody, Jessica Pfeiffer, Julia Bursten, and Inmaculada de Melo Martín.

Related chapters: 30, 31.

Notes

1 The data set is publicly available at https://dx.doi.org/10.5281/zenodo.1400633. Version 2.0 of the data set was used for this chapter.
2 We are not aware of any philosophers of science who were openly nonbinary or trans during the time period of the data set (1930–2017). Two authors who appear in the data set—Rebecca Kukla and Daniel Hicks—currently publicly identify as nonbinary. The Social Security Administration data, as well as the two gender attribution services used for automatic gender attribution, use binary gender.
3 Version 1.1 of the data set was used for topic modeling. Due to logistical issues, we were unable to update the topic models using version 2.0 of the data set in a timely manner.
4 For instance, see the essays collected in Alcoff (2003), which includes many personal narratives about being a woman in philosophy at the end of the twentieth century and demonstrates the tactical and strategic decisions that women made (and still make) to build a research agenda and professional identity for navigating the political pressures of being a woman in a male-dominated field.
5 This is calculated for the years 1986–2017 because *Biology and Philosophy*, the first specialist journal in philosophy of biology, did not begin publication until 1986. Women's participation in research has grown steadily over time; and so, a comparison among specialist areas must cover the same set of years.
6 Demographic reports on PSA membership were compiled in 2006, 2010, and 2014 by Miriam Solomon.
7 An interactive data interface for the results published in West et al. (2013) can be found at http://www.eigenfactor.org/gender/. Because of differences in source data and methods between different studies, one should be wary of comparing women's participation rates between studies.
8 The BPA/SWIP Good Practice Scheme's recommendations can be found at https://web.archive.org/web/20190722053433/https://bpa.ac.uk/resources/women-in-philosophy/good-practice.

References

Alcoff, L. M. (2003) *Singing in the Fire: Stories of Women in Philosophy*, Lanham, MD: Rowman & Littlefield.
American Academy of Arts and Sciences. (2020) *The State of the Humanities In Four-Year Colleges and Universities (2017)*, https://www.amacad.org/sites/default/files/media/document/2020-05/hds3_the_state_of_the_humanities_in_colleges_and_universities.pdf
Atherton, M. (1994) *Women Philosophers of the Early Modern Period*, Indianapolis, IN: Hackett Publishing.
Bix, A. S. (2013) *Girls Coming to Tech! A History of American Engineering Education for Women*, Cambridge, MA: MIT Press.
Brozek, A. (2017) "Maria Kokoszyńska: Between the Lvov–Warsaw School and the Vienna Circle," *Journal for the History of Analytical Philosophy* 5(2), pp. 19–36.
Bursten, J. R. S. (2019) "Field Notes on Conference Climate: A Decade with the Philosophy of Science Association Women's Caucus," *American Philosophical Association Newsletter: Feminism and Philosophy* 19(1), pp. 36–38.
Fine, A. (1997) "Brodbeck, May (1917–1983)," *Jewish Women in America: An Historical Encyclopedia*, New York: Routledge, pp. 187–188.
Fine, C. (2010) *Delusions of Gender*, New York: W. W. Norton.
Friedenreich, H. (2009) "Hilda Geiringer," *Jewish Women: A Comprehensive Historical Encyclopedia*. Jewish Women's Archive. (Viewed on October 20, 2019) https://jwa.org/encyclopedia/article/geiringer-hilda.
Gumport, P. J. (1990) "Feminist Scholarship as a Vocation," *Higher Education* 20(3), pp. 231–243.
Jenkins, F. (2014) "Epistemic Credibility and Women in Philosophy," *Australian Feminist Studies* 29(80), pp. 161–170.
Larivière, V., Ni, C., Gingras, Y., Cronin, B., and Sugimoto, C. (2013) "Global Gender Disparities in Science," *Nature* 504(7479), pp. 211–213.
Leslie, S.-J., Cimpian, A., Meyer, M., and Freeland, E. (2015) "Expectations of Brilliance Underlie Gender Distributions across Academic Disciplines," *Science* 347, pp. 262–265.

Leuschner, A. (2015) "Social Exclusion in Academia through Biases in Methodological Quality Evaluation: On the Situation of Women in Science and Philosophy," *Studies in History and Philosophy of Science* 54, pp. 56–63.

Lloyd, E. (1995) "Objectivity and the Double Standard for Feminist Epistemologies," *Synthese* 104(3), pp. 351–381.

Lloyd, E. (2006) *The Case of the Female Orgasm: Bias in the Science of Evolution*, Cambridge, MA: Harvard University Press.

Martin, E. (1991) "The Egg and the Sperm: How Science Has Constructed a Romance Based on Stereotypical Male-Female Roles," *Signs* 16(3), pp. 485–501.

National Science Foundation. (2018) *Science and Engineering Indicators*, https://nsf.gov/statistics/2018/nsb20181/report/sections/academic-research-and-development/doctoral-scientists-and-engineers-in-academia

Nosek, B., Smyth, F., Sriram, N. et al. (2009) "National Differences in Gender–Science Stereotypes Predict National Sex Differences in Science and Math Achievement," *PNAS* 106(26), pp. 10593–10597.

Potter, E. (2001) *Gender and Boyle's Law of Gases*, Bloomington: Indiana University Press.

Roberts, M. E., Stewart, B. M., Tingley, D. et al. (2014) "Structural Topic Models for Open-Ended Survey Responses," *American Journal of Political Science* 58(4), pp. 1064–1082.

West, J. D., Jacquet, J., King, M. M. et al. (2013) "The Role of Gender in Scholarly Authorship," *PloS One* 8 (7), p. e66212. doi:10.1371/ journal.pone.0066212.

Wilhelm, I., Conklin, S. L., and Hassoun, N. (2018) "New Data on the Representation of Women in Philosophy Journals: 2004–2015," *Philosophical Studies* 175(6), pp. 1441–1464.

PART II

Frameworks

6

FEMINIST EMPIRICISM

Kirstin Borgerson

Feminist Empiricism

Science, it seems, is perpetually under attack. Today the most prominent attackers are probably those who deny the threats posed by climate change. In the past, attacks would have come from other sources, and on other topics: those who denied evolution or thought the sun revolved around the earth (some of these beliefs persist today, of course). In the face of such attacks, it can be tempting to rally uncritically behind scientists, endorsing scientific methods and results and turning one's attention to educating those who disagree. Those who choose, instead (or in addition), to turn a critical eye to science in these moments can be mistaken for sharing views with the attackers. Feminist philosophers of science have had to negotiate this tricky territory – and the misunderstandings it seems to provoke – for many years.

My students are frequently tempted by the view that science must be defended at all costs, perhaps because the media around climate skepticism is so intense, so it may be useful to briefly remind ourselves why feminist epistemologists are not inclined toward this position. Principally, it is because such a position assumes that anything scientific will be good: well-intentioned, well-designed, and aimed purely at knowledge creation for the benefit of all. But this ignores a long history of scientific efforts to further entrench and justify oppression against the members of various minority groups. Janet Kourany surveys some of these efforts, which include attempts by psychologists, biologists, archaeologists, economists, and medical researchers to demonstrate women's inferiority to men "intellectually, socially, sexually, and even morally" (2010: 4–12). Similar track records exist for scientists' efforts to justify a range of racist, ableist, classist, ageist, homophobic, and other oppressive beliefs. Kourany sums up her historical survey with this statement, "Science, in short, has done much to perpetuate and add to the problems women [and oppressed people in general] confront rather than solve them" (2010: 11). (Other chapters in this collection develop these historical examples further.)

At this point some readers will be tempted to argue that "that's all in the past" and such tendencies have since been excised from scientific practice. Even if this were true (contemporary feminist work would suggest otherwise – see for instance ongoing discussions about sexism in neuroscience (Bluhm 2013) and in the continued dominance of male animal models in scientific research (Beery and Zucker 2010)), we should probably still ask how it was that these pernicious beliefs managed not only to persist in science but also to shape research programs and their results over an extended period. If the scientific method is our (human) paradigm for reliable, objective, knowledge production, how is it that it has been (and arguably still is) so easily used for bad ends? Enter feminist empiricism.

What Is Feminist Empiricism?

A neat answer to the question, "What is feminist empiricism?" would call upon the well-accepted definitions of each component concept (feminism and empiricism) and say a bit about how they have been brought together. Because neither term in this pairing is without controversy, let's select the broadest and most inclusive definitions of each term accepted today. The core feminist project involves the identification and elimination of all forms of oppression, and the core empiricist project involves creating knowledge through methods that draw on experience. The feminist project, then, includes both a theoretical orientation (since certain tools and frameworks are useful in identifying subtle, pervasive, and systemic forms of oppression) and an activist orientation (it aims to eliminate oppression once it is identified; this involves making changes to the real world). It's not hard to see how an empiricist methodology might assist in both of these projects – methods from the social and natural sciences may be helpful in identifying oppression, and in proposing and evaluating the effectiveness of responses to oppression. So, for instance, data on domestic violence might be obtained and analyzed by social scientists, and this may inform the development of appropriate tools of intervention, which are then evaluated for efficacy. And yet, while feminist empiricism does include or at least endorse such projects, this analysis falls short of capturing the scope of what is addressed by feminist empiricists in their work.

The deeper project arises because the methods of the sciences themselves require feminist scrutiny, as do the traditional epistemic concepts of objectivity, evidence, rationality, justification, and knowledge. Feminist empiricist epistemologies aren't just a match of subject matter (oppression) and method (scientific); they are full-fledged theories of knowledge in their own right. They offer not only critiques of what's gone wrong in the sciences but also constructive accounts of how scientific communities might do better in their pursuit of knowledge.

The traditional division of feminist epistemology into three camps – empiricism, standpoint theory, and postmodernism – was originally proposed by Sandra Harding (1986). On this early formulation, feminist empiricism "argues that sexism and androcentrism are social biases correctable by stricter adherence to the existing methodological norms of scientific inquiry" (1986: 24). This definition no longer fits feminist empiricism, and a similar evolution is clear with standpoint theory as well. Both standpoint theory and feminist empiricism have responded to early criticisms and become significantly more sophisticated (and some would argue to a large extent converged; see Intemann 2010). But we do see the roots of feminist empiricism here: it originated with concerns about the biases present in scientific inquiry and ultimately seeks to propose a route to preventing idiosyncratic (and oppressive) biases from shaping knowledge. Most feminist epistemologists today – but particularly feminist empiricists – have something to say about the role of values in science. They also likely agree on core commitments to pluralism and the rejection of the possibility and/or desirability of a "view from nowhere." They agree on the importance of diversity and different perspectives in the production of knowledge. And they agree on the contextual and located nature of knowers. Differences remain, some of which are likely the products of the different tools used in different areas. For instance, while empiricists tend to make use of the tools of analytic philosophy of science, standpoint theorists are more likely to draw upon cultural theory and the tools of sociology, and postmodern theorists may draw upon on literary theory, among other sources. In addition, recent work by feminist pragmatists has added the tools of pragmatism to this landscape (Clough 2003). The exact nature of the remaining differences between the theories is a matter of ongoing discussion.

The relationship between feminist and social epistemology has been a productive and interactive one, with many feminist epistemologists identifying as social epistemologists and vice versa. While epistemologists have traditionally focused on the role of individual beliefs and desires in knowledge production, social epistemologists examine the social dimensions of knowledge. The *feminist character of much social epistemology* is evident in the critique of oppressive epistemological practices (through case studies and sociological investigations) – often related to gender, race, and other social categories. Feminists have traditionally valued interconnectedness over isolation, both politically and epistemologically, and this general tendency has also been well integrated with the project of social epistemology, which as noted earlier redirects some attention from the psychological states of individual knowers toward the relationships between and among individuals-in-communities (Grasswick 2004). The *social character of much feminist epistemology* is evident in constructive accounts of knowledge production that require communities (not just individuals) to meet certain epistemic criteria. More on this below.

As the example below will hopefully make clear, feminist epistemologies do not resemble the brutal caricatures sometimes drawn of them by critics (cf. essays by Haack and Almeder in the collection by Pinnick et al. 2003). Contemporary feminist epistemologies, including feminist empiricism, *do not*: valorize or render immune to criticism of sexist ideas about women's intuitive, nurturing, or emotional ways of knowing, hold that a particular privileged viewpoint is held only by people assigned female at birth, retreat uncritically into relativism, or support the wholesale rejection of rationality (see Anderson, 1995, for an overview and rebuttal of these ill-informed critiques). They do not, for the most part, suggest that all of the work done in traditional epistemology is worthless, or believe that entirely new epistemologies need to be developed from scratch without reference to traditional epistemology. Feminist empiricists, in particular, are poorly captured by such descriptions.

Contemporary Feminist Empiricism

As noted earlier, evidence suggests that scientists in a wide range of disciplines and over a long period of history imported sexist and other discriminatory beliefs into their work and even used their research to seek and publicize "explanations" for the inferiority of women and other minority groups. The true political project (justifying inequality) was just beneath the surface of such research, which often purported (as similar projects do today, with similar implausibility) to be simply "interested in differences" between men and women. One of the early tasks of feminist critics of the sciences was to identify these biased areas of research and subject them to scrutiny and critique. Scientists with a feminist orientation were seen as ideally situated to do this work, but those doing so were met with a challenge in return – how is the value-laden position of the feminist critic of science not itself a bias to be eliminated in much the same way as the sexist beliefs of the scientist? In other words, is it the case that the feminist critique of science simply substitutes one bias (sexist bias) for another (feminist bias) in its quest for bias-free science? This is referred to as the *Paradox of Bias* (Antony 1993; Campbell 1998; Rolin 2006).

In keeping with developments in contemporary philosophy of science, feminist empiricists responded to this paradox by arguing that the goal of value-free science is both unachievable and undesirable (for an overview of debates over the role of values in science, see Kincaid et al. 2007). It is unachievable for several reasons: first, epistemic values/virtues (sometimes called constitutive values) are necessarily a part of scientific investigations at all stages of inquiry. Scientists may prize simplicity in explanations, for instance, or value fruitfulness of theories (Kuhn 1977). These values are integral to science at all stages (from choice of research topic to formulation of a hypothesis to dissemination of results). Second, epistemic values will sometimes trade off against each other during the design and conduct of research, and researchers will need to decide which to prioritize – this will itself be a value-laden choice. Third, scientists make risk calculations to determine how much error to allow in their methods and results, and these are value-laden decisions which reflect assessments of potential harm resulting from error (Douglas 2007). (For further arguments, see Chapter 13 of this Handbook.) Once we recognize that values do play a role in the production of knowledge, a natural next question to ask is why it is that certain values are preferred over others (simplicity rather than complexity for instance) (Longino 1993). Ignoring values is clearly a poor choice for scientists since it means they will be adopting fallible value assumptions uncritically, and these might well corrupt their research.

Early attempts to distinguish between these "epistemic" values and other more "contextual" political values (like equality) have given way over time to a recognition that the sciences are infused with values of all types at all stages. This is easily demonstrated in the practically oriented domain of medical research, where we find not only trade-offs between generalizability and precision/accuracy (epistemic values) but also trade-offs made at all stages of research between things like social value (a paradigmatic contextual value) and empirical adequacy (a paradigmatic epistemic value) (Borgerson 2013, 2014). And we can easily recognize the subtle value-driven decisions made by medical researchers in their choice of research topic, comparison treatments, length of study, stopping rules, method of data analysis, and publication plan (including whether to publish or not), even though many of these decisions have implications for the nature and quality of the knowledge produced. Values, then, are pervasive in science. If we are to avoid "anything goes" relativism, some method of managing such values will need to be proposed.

There are many competing feminist empiricist epistemologies in the philosophical literature, several of which have been developed fully in monographs. Each of them attempts, in their own way, to grapple with the value-laden nature of scientific inquiry. Most prominent are Lynn Hankinson Nelson's *feminist empiricism* (1990), Richmond Campbell's *naturalized feminist epistemology* (1998), Miriam Solomon's *social empiricism* (2001), and Helen Longino's *critical contextual empiricism* (1990, 2002). Given limited space in this chapter, I will explain and discuss only this final theory.

Helen Longino's *critical contextual empiricism* (hereafter CCE) is probably the most well-known version of feminist empiricism defended today, achieving what Lorraine Code describes as "landmark status" in the field (2014: 152). While it was first developed almost 30 years ago in her book *Science as Social Knowledge* in the early 1990s, and extended in her 2002 book *The Fate of Knowledge*, it has held up remarkably well under critical scrutiny over time, with only minor modifications proposed over the years (cf. Borgerson 2011; Pinto 2014). I will provide an overview of Longino's theory, highlighting a few points where there have been proposed amendments. I then turn to the key objections to the account, with possible replies.

Longino's key insight, which is preserved at the core of her proposed epistemology, is that open critical discussion is the only way to ensure that the values necessarily pervasive in knowledge-productive endeavors, including paradigmatic scientific investigations, don't operate in secrecy. When values are unacknowledged and unexamined in scientific investigations, and the results of those investigations are then thought to be objective, we are in a dangerous position: biased (and oppressive) results will be touted as fact, and the authority of science will lend these facts more credibility than they are due. This was what happened in the case of sexist, racist, and other biased scientific research in this past, as noted earlier. In order to avoid mistakenly attributing objectivity to the products of closed, and potentially biased, processes, Longino relocates the source of objectivity to the process of community criticism. According to CCE, scientific communities are objective to the extent that they meet four criteria:

1   *Venues*: there must be recognized avenues for criticism of evidence, of methods, and of assumptions and reasoning.
2   *Uptake*: the community as a whole must be responsive to such criticism.

3   *Public Standards*: there must be shared standards that critics can invoke.
4   *Tempered Equality*: intellectual authority must be shared equally among qualified practitioners (Longino 1990: 76, 2002: 128–135).

The four criteria work together to ensure that particular communities have places, times, and locations where critical engagement can take place (venues), that members of communities are responsive to the criticism that arises in those venues (uptake), that members of communities share standards sufficient to provide the common ground for this critical exchange, and that these standards are publicly available (public standards), and that community members are not unjustly denied access to these discussions (tempered equality). Collectively, then, these criteria aim to protect the open critical evaluation of ideas within scientific communities: the theory is in this sense *critical*.

We also see in these criteria Longino's commitment to *social* epistemology. Rather than focusing only on the individual psychological dispositions or commitments required for knowledge production she considers what social conditions must be in place for knowledge to be produced. Knowledge, then, is not something an individual can arrive at independently but is rather something achieved in collaboration with others. It relies on shared commitments and shared standards, and is best thought of as produced by particular communities operating by these shared rules: the theory is in this sense *contextual*. All scientific communities will share a commitment to something like empirical adequacy, though they might differ in some of the other values they hold (e.g. simplicity in theoretical physics vs. complexity in ecology). Different communities will have different background assumptions and criteria for theory choice.

In a paper published in 2011, I offered a few clarifications and minor amendments of Longino's theory, including: (1) an explicit requirement, tied to "venues," that scientist publish or otherwise make the results of their research available, since this facilitates the sort of community-level criticism required by the account and (2) a specification of "shared public standards" that stresses the importance of keeping these expectations to a minimum in order to foster meaningful criticism from relative outsiders to a community, given that outsiders are best positioned to identify hidden background assumptions not shared by those inside the community (Borgerson 2011: 439–443).

The most substantial proposed change arose in my discussion of the requirement of tempered equality of authority, which I argued would be better reconceived as a commitment to "cultivate diverse perspectives" (Borgerson 2011: 443). The key insight in Longino's epistemology is that when values permeate knowledge production it is people with different starting assumptions and values of their own who will be best positioned to expose these otherwise hidden values and assumptions, through critical interrogation. Thus, key to knowledge production is the cultivation and preservation of diversity, where this includes the perspectives of those who actively pursue alternative hypotheses and theories (I refer to this as "strong dissent"). The requirement of tempered equality of authority wasn't adequately protecting strong dissent of the sort most likely to be productive for knowledge, so I revised the account to reflect this core commitment. This shift in focus also helped to resolve any confusion about whether it was the presence of shared standards or the "tempered equality of authority" that earned an individual the right to expect uptake of their criticism of some scientific theory or fact. The explicit commitment to strong dissent makes it clear that the burden of justification is on anyone who wants to exclude a perspective from critical discussion. Regardless of whether this amendment is adopted or not, CCE is, as its name reflects, ultimately a commitment to the view that knowledge is best produced through transparent, critical interrogation of ideas from a wide variety of perspectives.

Four criticisms of CCE stand out in the literature, each with dimensions relevant to other feminist empiricist theories. First, there is a criticism from more-committed naturalistic philosophers, that Longino's theory seems to be too far removed from "real science" (Pinto 2014). According to this critique, the four proposed community norms seem to have been pulled from thin air and granted normative weight without much connection back to the sciences to which they are then meant to be applied. If scientific communities are paradigmatic knowledge-productive communities, wouldn't it make sense to look closely at what they are doing right and incorporate that into any theoretical account?

Two things might be said in response to this sort of concern. First, Longino's account is deliberately context-specific: she aims to leave much of the details of what will count as knowledge for a given community in the hands of that community. (It is, again, critical *contextual* empiricism.) So the fact that she doesn't spend a lot of time pooling the evidence from a range of sciences in her development of the four criteria of objectivity isn't a mistake; it's a feature of the account. Perhaps it is a feature some critics don't like, but it isn't an error: it is a deliberate methodological choice. The standards arise as a result of a shared general concern with the role of values in knowledge production. Thus, the second response to this concern is to say that when particular communities do engage in discussion about, for instance, the content of "shared public standards," they do so facing a common challenge related to the role and place of values in their deliberations. The guidance provided by the four criteria helps to ensure that even these discussions are as open to criticism as possible. A contextual account has to start somewhere – it doesn't mean that the starting point is dogmatically held. In Longino's case, the starting point is the four criteria because her starting concern is the role of values in knowledge claims but the details of how they will be specified in any particular community are left open deliberately, and even these criteria are open to revision. This is a strength of the account.

A second, related worry with CCE is that much of the history of science will appear unscientific, or insufficiently objective in terms of knowledge production, given that the social conditions of knowledge production weren't typically in place before the mid-twentieth century (and not always well-observed even since then). I suspect that historians of science would suggest that more can be said in defense of particular historical scientific endeavors – for instance, about the ways in which communities did in fact subject results to critical scrutiny to the extent possible at the time. But surely not all historical cases will pass this test. Ultimately, I think this is an occasion for biting the bullet – much of what passed as scientific in the past would simply not meet the more rigorous standard defended by Longino today. It would not earn the honorific title "knowledge." And we don't have to look only to the past here: there are domains of scientific research today – medical research for example – which struggle to meet the criteria because of persistent practices such as publication bias (withholding the results of studies from community scrutiny) and other obstacles. While this might seem to lead to skepticism, I think it leads more naturally to criticism of those obstacles and unscientific community standards. This is a productive and useful function of the standards. If all the standards did was rationalize and justify all current science they wouldn't be functioning very well as normative standards. In the historical cases, it can help to contextualize the results of scientific investigations – perhaps they were rigorous in some senses but not in others – and this is useful for appreciating their limits.

The third concern raised about CCE in the literature is that the theory is open to abuse or co-opting by those who "manufacture uncertainty." When the empirical evidence seems to clearly suggest that a product (in the classic example, tobacco) is harmful, one response by manufacturers of the product might be to pursue further research as a stalling and obfuscating technique. Their aim is to cast doubt on the evidence of harm by suggesting that the research on this topic is still being conducted and so results are not yet settled. They aim, in other words, to "manufacture uncertainty" (Michaels 2008). In their book *The Fight against Doubt* (2018), De Melo-Martin and Intemann provide contemporary examples of profit-driven research that seeks to obfuscate rather than generate knowledge, so we know this problem persists and is perhaps even increasing in science (Pinto 2014). Because CCE operates by welcoming new evidence and diverse perspectives, these tactics seem like they would be effective in preventing a community from making a claim to know something (e.g. that smoking causes lung cancer).

I offered a response to this concern in my 2011 paper, which can be summed as follows: there is a difference between assigning the honorific "knowledge" to some claim and providing grounds for policy or action. We might indeed have to await the results of "yet another study" in order to make the strongest possible claim to knowledge in some domain, even when that reflects the use of a stalling technique, but this doesn't stop us from acting. Policy-makers can and should act well before the point where knowledge is decisively labeled as such in domains where risks or harms to humans are high. Once the epistemic and political/action-oriented domains are separated, and a certain comfort with pervasive uncertainty acknowledged, this problem largely disappears. Any remainder can be addressed by using CCE as grounds for a criticism of the trend toward privately funded research (and away from publicly funded research), which seems to be driving this problem.

Finally, there is a longstanding concern that feminist empiricism is somehow less firmly rooted in its moral-political commitment to feminism, because it must be committed to hearing from any and all diverse perspectives, even those of the bigot (Hicks 2011). If feminist values (for example, moral equality) are better than non-feminist values (for example, moral inequality) – and surely feminists of all types believe this if they are feminists – then why think it beneficial to pay attention to the "diverse perspectives" of those with unjustified beliefs about, for instance, the inherent inferiority of women, racial minorities, or members of LGBTQ+ communities? (Intemann 2010: 793). This criticism is an important one – it gets to the heart of what CCE proposes and exposes a vulnerability inherent in any such approach. This discussion mirrors similar debates occurring on university campuses these days, and doesn't give rise to easy or simple answers, but I think two things can be said. First, we can remind ourselves of the nuance in the feminist position: we turn again to Longino, who describes what it means to be a feminist epistemologist. I will quote her passage in full,

> To do epistemology as a feminist is to engage the questions of epistemology with an awareness of the ways in which participation in socially-sanctioned knowledge production has been circumscribed, of the ways in which epistemological concepts like rationality and objectivity have been defined using notions of masculinity (and vice-versa), of the ways women have been derided as knowers, and of the need for alternative theoretical approaches to satisfy feminist cognitive goals. It is to ask how epistemology has participated in or sanctioned these disbursements of privilege and opprobrium and to ask whether the efforts to exclude women from knowledge generating activity has not also resulted in the exclusion from the analysis of knowledge of traits and capacities assigned to women (a shrinking of the conception of knowledge.) What is important for the feminine or the female here is the perspective it affords on the construction of the concept of knowledge and the window it opens on alternatives. But it functions as an object of reflection, not as a subject position.
>
> *(Longino 1994: 475)*

To paraphrase, Longino suggests that the project is not to retreat to claims about "women's ways of knowing" but rather to do epistemology as a feminist – to think about the production of knowledge with an eye to the ways in which epistemology may be related to practices of oppression. Though she doesn't draw this point out here, this approach ought to recognize the ways in which all forms of oppression (racism, classism, ableism, homophobia, transphobia, and others) also influence (and interact to influence) knowledge production. A key feature of the account is awareness – of any and all potentially harmful distortions of our understanding of knowers, knowledge production, and knowledge. Other feminist epistemological theories may explore the subject position that Longino eschews here, but this doesn't make them "more feminist"; it simply makes them differently feminist. (The range of approaches within feminism is, by CCE lights, diversity worth celebrating!)

A second reply is more direct: yes, feminist empiricist theories will allow the expression and consideration of all perspectives including non-feminist perspectives. This will sometimes mean that members of scientific communities will be expected to take the time to respond to yet another claim about the "inherent inferiority" of the members of some group. But this doesn't mean anything goes, since responses can be as thorough and devastating as merited by the original false claim. And it doesn't mean endless time is wasted on such replies, since the repetition of the same argument over and over is not protected by Longino's four norms since it violates the expectation that those *proposing* a new critique also respond to criticisms (they are also expected to engage in "uptake" of ideas). CCE has extensive resources to deal with bigots though this reply will not be entirely satisfying to those who wish to silence them.

Each of the criticisms discussed earlier might arise for other feminist empiricist theories in the literature, with different possible responses. Hopefully this close look at CCE has been helpful at drawing out the resources feminist empiricists have for replies to a few common and persistent concerns.

What's Next for Feminist Empiricism

In reply to the concern about engaging with bigots, feminist empiricism draws on traditional liberal ideas of free speech (feminist empiricists are often explicit in their recognition of ties to liberal theory). I expect there are interesting ways in which discussions about the future of feminist empiricism may echo discussions about the future of liberal feminism. In both cases, it is my hope that critical discussion engages the complex ways in which feminist empiricism connects to contemporary debates rather than relying on simplistic reductions of the view. In her discussion of the relationship between feminist and social epistemology, Lorraine Code identifies some of the ways in which recognizing that values inherent to knowledge-productive processes opens up new avenues for reflection and analysis:

> Now, for many social epistemologists, feminists prominently among them, ethical–political questions about trust, power, epistemic negotiation, advocacy, credibility, communities of inquiry and the ethics of belief enter and shape the discourse yet not, as had been feared, to the detriment of responsible deliberation.
>
> *(Code 2014: 153)*

There has been an explosion of interest in epistemology over the past decade or so on a range of topics that deepen our understanding of the social and community-based practices that underlie knowledge practices of all sorts: epistemic injustice, epistemic oppression, epistemic violence, testimonial quieting and smothering, epistemologies of ignorance, epistemic resilience, and so on (cf. Dotson 2002, 2011, 2014; Fricker 2007; Sullivan and Tuana 2007). Reflecting on recent work on epistemic injustice, in particular, Code points out: "This work would not so readily have claimed a hearing had the way not been paved by developments in social philosophy and feminist philosophy of science, such as Longino's (2002), that contest the tacit effects of the rational/social dichotomy" (Code 2014: 155). I don't want to overstate the point here – much of this literature has independent origins – but the legacy of early work in feminist epistemology and critical science studies is noteworthy as part of the history of where feminist philosophy of science finds itself today. This is in part because what we think about empiricism today has been shaped by feminist critiques over many years. I suspect that the resources of feminist empiricism have not yet been fully engaged and that closer attention to the range of feminist empiricist theories would be fruitful for scholars interested in many of the topics listed earlier.

Conclusion

To review, feminist empiricist epistemologies have a number of strengths: (1) they offer constructive theories of knowledge (not just critiques of particular knowledge practices). These constructive accounts can then be used to identify ways in which scientific communities have been compromised by, for instance, those with sexist, racist, and other oppressive agendas. Whenever communities are rendered secretive, homogenous, and exclusive, there will be an opportunity for feminist empiricists to offer remedies that more carefully preserve open critical discussion and debate; (2) the commitment to diversity reminds us that ensuring scientific communities include a range of perspectives is not only a political goal but also an epistemic goal – it is good for knowledge; (3) they recognize that values are present in all human epistemic endeavors and offer a way to deal with them rather than pretending those values don't exist. This helps to keep us all honest. That's a pretty solid track record, with plenty of promise for the future.

Related chapters: 13, 15, 16, 18.

References

Anderson, E. (1995) "Feminist Epistemology: An Interpretation and Defense," *Hypatia* 10: 50–84.

Antony, L. (1993) "Quine as Feminist: The Radical Import of Naturalized Epistemology," in Antony, L.M. and C. Witt (eds.) *A Mind of One's Own: Feminist Essays on Reason and Objectivity*, Boulder: Westview Press, 185–225.

Beery, A. K. and Zucker, I. (2010) "Sex Bias in Neuroscience and Biomedical Research," *Neuroscience and Biobehavioral Reviews* 35(3): 565–572. doi:10.1016/j.neubiorev.2010.07.002

Bluhm, R. (2013) "New Research, Old Problems: Methodological and Ethical Issues in fMRI Research Examining Sex/Gender Differences in Emotion Processing," *Neuroethics* 6(2): 319–330.

Borgerson, K. (2011) "Amending and Defending Critical Contextual Empiricism," *European Journal for Philosophy of Science* 1(3): 435–449.

Borgerson, K. (2013) "Are Explanatory Trials Ethical? Shifting the Burden of Justification in Clinical Trial Design," *Theoretical Medicine and Bioethics* 34(4): 293–308.

Borgerson, K. (2014) "Redundant, Secretive, and Isolated: When Are Clinical Trials Scientifically Valid?" *Kennedy Institute of Ethics Journal* 24(4): 385–411.Campbell, R. (1998) *Illusions of Paradox: A Feminist Epistemology Naturalized*, Lanham: Rowman and Littlefield Publishers, Inc.

Clough, S. (2003) *Beyond Epistemology: A Pragmatist Approach to Feminist Science Studies*, Lanham: Rowman and Littlefield Publishers, Inc.

Code, L. (2014) "Ignorance, Injustice and the Politics of Knowledge," *Australian Feminist Studies* 29(80): 148–160. doi: 10.1080/08164649.2014.928186

De Melo-Martin, I. and Intemann, K. (2018) *The Fight Against Doubt: How to Bridge the Gap Between Scientists and the Public*, New York: Oxford University Press.

Dotson, K. (2002) "A Cautionary Tale: On Limiting Epistemic Oppression," *Frontiers: A Journal of Women Studies* 33(1): 24–47.

Dotson, K. (2011) "Tracking Epistemic Violence, Tracking Practices of Silencing," *Hypatia* 26(2): 236–257.

Dotson, K. (2014) "Conceptualizing Epistemic Oppression," *Social Epistemology* 28(2): 115–138. doi: 10.1080/02691728.2013.782585

Douglas, H. (2007) "Rejecting the Ideal of Value-Free Science," in Kincaid, H., Dupré, J. and Wylie, A. (eds.) *Value-Free Science? Ideals and Illusions*, Oxford: Oxford University Press, 120–139.

Fricker, M. (2007) *Epistemic Injustice: Power and the Ethics of Knowing*, Oxford: Oxford University Press.

Grasswick, H. (2004) "Individuals-in-Communities: The Search for a Feminist Model of Epistemic Subjects," *Hypatia* 19(3): 85–120.

Harding, S. (1986) *The Science Question in Feminism*, Ithaca: Cornell University Press.

Hicks, D. (2011) "Is Longino's Conception of Objectivity Feminist?" *Hypatia* 26(2): 333–335.

Intemann, K. (2010) "25 Years of Feminist Empiricism and Standpoint Theory: Where Are We Now?" *Hypatia* 25(4): 778–796.

Intemann, K. and de Melo-Martin, I. (2014) "Addressing Problems in Profit-driven Research: How Can Feminist Conceptions of Objectivity Help?" *European Journal of Philosophy of Science* 4: 135–151.

Kincaid, H., Dupré, J. and Wylie, A. (eds.) (2007) *Value-Free Science? Ideals and Illusions*, Oxford: Oxford University Press.

Kourany, J. (2010) *Philosophy of Science after Feminism*, Oxford: Oxford University Press.

Kuhn, T.S. (1977) "Objectivity, Value Judgment, and Theory Choice," in *The Essential Tension*, Chicago: University of Chicago Press, 320–339.

Longino, H. (1990) *Science as Social Knowledge*, Princeton: Princeton University Press.

Longino, H. (1993) "Essential Tensions – Phase Two: Feminist, Philosophical, and Social Studies of Science," in Antony, L.M. and Witt, C. (eds.) *A Mind of One's Own: Feminist Essays on Reason and Objectivity*, Boulder: Westview Press, 257–272.

Longino, H. (1994) "In Search of Feminist Epistemology," *The Monist* 77(4): 472–485.

Longino, H. (2002) *The Fate of Knowledge*, Princeton: Princeton University Press.

Michaels, D. (2008) *Doubt is Their Product: How the Industry's Assault on Science Threatens Your Health*, Oxford: Oxford University Press.

Nelson, L.H. (1990) *Who Knows? From Quine to a Feminist Empiricism*, Philadelphia: Temple University Press.

Pinnick, C., Koertge, N. and Almeder, R.F. (eds.) (2003) *Scrutinizing Feminist Epistemology*, Madison: Rutgers University Press.

Pinto, M. (2014) "Philosophy of Science for Globalized Privatization: Uncovering Some Limitations of Critical Contextual Empiricism," *Studies in History and Philosophy of Science* 47: 10–17.

Rolin, K. (2006) "The Bias Paradox in Feminist Standpoint Epistemology," *Episteme: A Journal of Social Epistemology* 3(1–2): 125–136. doi:10.1353/epi.0.0006.

Solomon, M. (2001) *Social Empiricism*, Cambridge: MIT Press.

Sullivan, S. and Tuana, N. (eds.) (2007) *Race and Epistemologies of Ignorance*, Albany: State University of New York Press.

7

THINKING OUTSIDE-IN

Feminist Standpoint Theory as Epistemology, Methodology, and Philosophy of Science

Catherine Hundleby

Sandra Harding's (1986, 1991) taxonomy of early developments in feminist philosophy of science groups together certain theories in the social sciences and political theory as "feminist standpoint theory." Known sometimes as "feminist standpoint epistemology" or simply "standpoint theory" (herein "FST"), it emerged while Western feminism moved out of its second wave and has since evolved in various directions. As an epistemology of science, FST explains how biased scientific claims, such as the view that human tool use was developed by early men (not women), depend on sexism and patriarchal social structures. It does so by scrutinizing the scientific norms and practices that determine what questions people recognize as scientific and what resources scientists consider.

A feminist standpoint addresses the ideals or norms and attendant practices involved in science and knowledge with a mind to lived experiences of oppression, accounting for sexism but also racism, heterosexism, ableism, and classism. That such matters of social context and awareness of that context influence the ability of individual people to know their worlds constitutes the Situated Knowledge Thesis (Intemann 2016; Wylie 2003), which the first section explains.

Situated knowledges provide the evidence and inspiration for the central *epistemological* tenet of feminist standpoint theory, laid out in the second section: the Epistemic Advantage Thesis. This thesis claims that understanding the world, especially in its social dimensions, benefits from critical reflection on the experiences and interests of marginalized people such as women. Cognitive advances from the history of feminism include the development of new concepts, such as the "glass ceiling" in business and "the male gaze" in art, as well as the concepts of "sexual harassment" and "marital rape." These conceptual tools led to the further development of new types of evidence and knowledge, for example the recognition that rape tends to be committed by acquaintances rather than strangers, and most recently taking the rape of men seriously as a social problem. In response to derogatory stereotypes aiming to control assertive Black women, a Black feminist perspective reveals the power in that assertiveness (Collins 1986, 2000). Currently, popular Western feminism has articulated experiences through concepts such as "mansplaining" (Solnit 2014), "toxic masculinity," and #metoo. LGBTQ2+ people express "pride." Anti-racist movements demand recognition that "Black lives matter" and indigenous peoples fighting oil and gas pipelines on their land proclaim that "water is life."

Individuals and liberatory communities obtain the epistemic advantage of a standpoint through an Outside-In Process of thinking, drawing into science and knowledge the "outside" values and experiences of marginalized people. Versions of this process also go by such names as "theorizing from margin to center" (hooks 1984), "talking back" (hooks 1989), "sciences from below" (Harding 1991, 2015), "oppositional consciousness" (Sandoval 2000), "the new mestiza consciousness" (Lugones 2003), and "oppositional knowledge" (Collins 2013).

The Outside-In Process allows FST to function as a methodology and distinguishes its operation as a philosophy of science. The process has its own logic of tensions between the different roles people occupy and communities that individuals belong to, addressed in section "Thinking Outside-In." Women and other marginalized people may never be wholly outsiders to the dominant culture and often in various ways operate as "outsiders within" (Collins 1986), but individual "epistemic heroes" do not suffice to create the critical perspective of a standpoint (Medina 2013: 225). Developing situated knowledge depends on communities and coalition, as the section "A Standpoint Requires Community, Coalition, and Criticism" explains, and a standpoint becomes most explicit and deliberately involved in the generation of knowledge when it provides a research *methodology*, described in the section "Standpoint Methodology in the Sciences." Finally, the section "Standpoint Philosophy of Science" explains how outside-in thinking manifests both as a theory and a methodology in *feminist philosophy of science*.

Standpoint Epistemology Begins with Situated Knowledge

A feminist standpoint is a social orientation politicized through feminism in which any person – not only women – may participate (Collins 2000; Harding 1991; hooks 1984; Hundleby 1998), but for which women's perspectives provide a nominal starting point. FST's appeal to women's perspectives can essentialize or naturalize the category "woman" and so many philosophers have found it problematic. However, FST need not make any particular assumptions about who counts as a woman and so the category need not be understood as naturally defined or a natural kind. Moreover, as gender intersects with other forms of social identity and feminists engage other liberatory movements in fighting racism, classism, and homophobia, alleviating women's oppression can take many forms, and regularly the lives of people who have been marginalized in those other ways provide the starting place for thinking from a standpoint (Sandoval 2000). Complex social dynamics affect people's knowledge as does their understanding of their situation within those dynamics, according to the Situated Knowledge Thesis. The treatment of one's own perspective as situated within a "matrix of domination" (Collins 2000) turns a perspective into a standpoint.

FST observes how social situations impact the cognitive and material resources of women and others with marginalized social identities. All reasoners live and know through bodies that may be labeled and restricted as part of belonging to a social category, but the systematic restraint of marginalized peoples' social roles benefits people with privilege (Frye 1983). Influences on how people experience the world and the values they develop include social and linguistic practices, but also material economic and political situations (Hennessy 1993). Such ideological conditions affect direct experience when a person experiences sexual harassment, for instance, and that can reveal the need to interrogate social categories such as gender, race, and class. Individual experience from the margins aids the recognition of systematic social problems.

Differently situated experiences of oppression give rise to different movements and differently situated knowledges. What constitutes a standpoint varies according to relevant axes of social privilege and marginalization, such as race, class, and sexuality, plus the sociopolitical basis for a standpoint may change over time and place. A standpoint may also be relevant in different ways to different fields of knowledge. Should there be no gender oppression, then there would be no need for a specifically feminist standpoint theory (Hartsock 1997; Hundleby 1997; Wylie 2003), though other forms of oppression provide a basis for analogous epistemologies.

A Standpoint Provides Epistemic Advantage[1]

The Epistemic Advantage identified by FST emerges from attention to perspectives and experiences that tend to be ignored and even denied, as in "testimonial quieting" that neglects to recognize people as knowers (Dotson 2011) or as in "testimonial injustice" that denies speakers credibility because of social prejudice (Fricker 1998, 2007). Accounting for a range of people's understandings, experiences, and values aids democratic goals of equal representation and it helps to raise fundamental questions about current social norms, practices, and institutions in ways that suggest creative solutions, both of which can be epistemically advantageous. The way cultures and individuals think about such things as sex, assertiveness, employment opportunity, aesthetics, and even death can be transformed by feminism that uncovers gross insufficiencies and deep false-hoods, identifying these ways of thinking, in Nancy Hartsock's words, as "partial and perverse" (1983).

To show epistemic advantage, early theorists drew on Marxist accounts of the working-class standpoint (Hartsock 1983; Rose 1983 Smith 1974), and contemporary standpoint theorists recognize socioeconomic class as one axis of oppression intersecting with others, but they demand no loyalty to that theoretical heritage. In a capitalist economic system, unpaid work receives little credit, but a Marxist feminist analysis demonstrates the value of women's traditional unpaid labor in the home. In many cultures, women's labor caring for children, the elderly, and also able-bodied adults includes meal planning and preparation, gathering resources (including "shopping"), accounting, house cleaning, lay counseling, education, and training. This work facilitates the paid labor performed by members of the household, and so it reproduces the conditions of employment. Also, because this caring labor tends to be performed by the same people, women, who do the work of bearing and nursing children, it can be grouped together as "reproductive labor." A Marxist feminist standpoint reveals further that many women work a "double shift" or "double day" involving both paid labor of various sorts and unpaid reproductive labor. Like the early Marxist feminist views, feminists of color (Collins 2000; hooks 1984, 1989; Lugones 2003) argue for the epistemological significance of women's work in the home, but alongside other standpoint theorists they abandon the problematic universal psychological claims of some early formulations of FST (e.g. Harding 1986; Hartsock 1983), and they also point to a history of women of color caring for other people's children.

The epistemic resources of FST invigorate democratic ideals and validate democratic practices (Harding 2015). In the context of patriarchy, feminist attention to marginalized perspectives compensates for various forms of exclusion. From the dominant perspective, a call for input from marginalized groups may appear to be an inappropriately political "special interest" but FST sees this as a counterbalance to an unrepresentative homogeneity (Harding 1992a; Wieseler 2016). Facilitating marginalized people's contributions to policies and practices that affect their own lives (Mahowald 2005; TallBear 2014) encourages mutual accountability by providing avenues for feedback and makes consensus meaningful by distributing authority (Medina 2013).

Accounting for the value and significance of women's labor and of the labor performed by other marginalized groups also draws attention to the social processes that hide that value. When the neglect of women's knowledge is systematic, addressing it can reveal inconsistencies and even contradictions in social organizations and structures, which provides a distinctive epistemic advantage (Rolin 2009). For instance, thinking from women's lives reveals that schooling does not stop at the boundaries of the institutional property, and that student success depends a great deal on adults in the home providing prior and supplementary education. Adults at home build children's literacy and critical thinking skills, provide nutrition, enforce sleep schedules, and help with homework, all contributions that vary with class and regularly depend on women's work as mothers (Griffith and Smith 1986).

Standpoint criticism of society draws in various ways on personal experiences of subordination and people's efforts to resist it. Self-conscious feminist identity also can spur the self-reflection that provides epistemic advantage, but it is not necessary. Women operating in patriarchal environments may not conceive of themselves as feminist and yet their perspectives foster social criticism from a standpoint insofar as feminist social transformation made their situations possible, for instance the opening up avenues into the sciences for women (Wylie 2003). Such beneficiaries of feminism enjoy a form of epistemic luck (Fricker 2007; Medina 2013). Further, some sorts of community organization, such as block parties, may not even seem political to the organizers and yet may have liberatory significance and affect people's understanding of their social environments. Also, sometimes oppression can be so severe that merely surviving constitutes a form of resistance (Collins 2000).

Knowledge benefits from a reconfigured identity and new ways of reading experience (Haraway 1985; Hennessy 1993). Standpoints initially offer negative insights – they reveal error, personal limitations, and structural social barriers, and it can be easier to offer criticism than to imagine possible improvement (hooks 1984). However, being treated poorly by the current system motivates people to consider alternatives because change suggests improvement to their lives. Thinking creatively about power relationships is vital for keeping feminist consciousness from being coopted (Ludwig 2016).

Some alternatives emerge from the ways people currently live on the margins, from reclaiming or revaluing existing practices. The lived experiences of women provide an understanding of human struggle and flourishing often at odds with popular understanding – this is part of life under patriarchy, and they also exhibit a different set of values – the importance of care, trust, and community. Some standpoint theorists argue that women have a special investment in the politics of peace grounded in their work of bearing and sustaining others' lives, and in the training of girls to complete this work (Hartsock 1983; Rose 1983; Ruddick 1989). Others find sexuality a potent site for resistance and learning (Collins 2000). Another marginalized perspective, that of disability, reveals a higher quality of life experience than people with typical abilities tend to recognize (Scully 2018; Wieseler 2016).

Experiencing one type of oppression provides no guarantee of sensitivity to other forms, but it provides groundwork that creative thinking builds on (Medina 2013). A person's experience of frustration in the development, articulation, and expression of knowledge, which José Medina (2013) calls "epistemic friction," when recognized as part of systematic social oppression, can help familiar social roles come to "seem strange" (Harding 1991; Pohlhaus 2002). Knowers who learn to recognize their own situatedness and their insensitive tendencies, Medina argues, develop "meta-sensitivity" or "meta-lucidity" (2013). This skill allows individuals to recognize how forms of oppression beyond their direct experience may operate.

The epistemic advantage of a standpoint runs counter to the larger systems of sociopolitical privilege, which Alison Wylie (2012) calls FST's "inversion thesis." Social privilege obscures the limits of one's own perspective, making it seem larger and less bounded, a difficulty in understanding one's own situation as a knower that Medina calls "meta-insensitivity" or "meta-numbness," using the example of white people who are "colour-blind" or claim to be. Not seeing or not wanting to see race involves neglecting the realities of racism and the way it harms people, and the myopia comes from the social accommodations that constitute general social privilege: "not ever having the experience of running into cultural limitations that render their experience, problems, and even their lives unintelligible, as a result of not ever feeling severely constrained as speakers and subjects of knowledge" (2013: 75).

Thinking Outside-In

Obtaining epistemic advantage depends logically on a process of thinking from the outside-in, viewing one's own situation in ways that challenge the dominant ideology (Hennessy 1993; hooks 1984; Hundleby 1997; Sandoval 2000), whether coming out as lesbian, proclaiming Black girl magic, or doing feminist research in science or philosophy of science. As Donna Haraway observes, "to see from below is neither easily learned nor unproblematic, even if 'we' 'naturally' inhabit the great underground terrain of subjugated knowledges" (1988: 584), a point so often repeated that Sharon Crasnow (2013) dubs it "the achievement thesis." The complex logic of a standpoint includes operating as an insider in some occasions and in some ways but as an outsider in others, and the strain of juggling multiple perspectives may impede the ability to situate knowledge.

Subordination has greater complexity than simple exclusion because subordinated people must also operate within privileged domains and must understand the basic interests and needs of privileged people. Prototypical "outsiders within" are Black feminist sociologists, being both rooted in the experience of Black women and holding the professional authority of sociologists (Collins 1986). Also, women as caregivers often tend to people of higher social status based on gender, race, or class, whether at home or in paid employment. Analogous insider–outsider orientations have been described and theorized by African American men as "double conscious" (DuBois 1897), by Latinas as "world travelling" (Lugones 2003), by white women academics as "bifurcated consciousness" (Smith 1974), and by others (hooks 1984, 1989; Medina 2013; Narayan 1988).

As feminism and other liberatory movements, such as anti-racist and queer rights movements, advanced over the decades, women and other marginalized people made significant inroads in many fields, becoming increasingly "insiders" with good positions to "work the cracks" within the systems of authority and expertise to transform dominant knowledge structures (Collins 2013: xiii). Yet, these gains remain partial and inconsistent, accompanied by new forms of marginalization that rearticulate oppression, adapting it to the changed environment (Hennessy 1993). For women and visible minorities, the fields in which they find the most success tend to be both associated with stereotypes attached to their social identities and lower in prestige and pay than other fields (England 2010; Ridgeway 2011).[2] For instance, both femininity and low status regularly attach to nursing, food service, education, and secretarial work, fields where women commonly find employment; women also may work more in areas of law and medicine related to the family and community and that have lower status than, say, corporate law; and in science they are more likely to be lab assistants than principal investigators. People in the West often associate ideals of masculinity, such as aggression, leadership, and even certain forms of rationality, with fields of prestigious employment including science and philosophy (Keller 1996; Le Doueff 1990; Lloyd 1984; Moulton 1983). So, participation in science and academic philosophy may conflict with the social norms that partly constitute the identities of women and other similarly marginalized people, making those fields unattractive and difficult to navigate.

The barriers separating people and roles inside from those outside are permeable and shifting (Naples 1996, 1999), and multiple roles yield information that is contradictory or "kaleidoscopic" in Medina's words (2013). So, what lies inside and what outside can be hard to track. Is a white lesbian an outsider to white culture? Often the answer is "yes" because white culture privileges masculinity and heterosexuality. However, she may have cisgender as well as white privilege and that will affect her feminine and sexual identities. Systems of privilege interlock, Collins argues (1986), and women's gendered oppression takes many different shapes. Conscious and unconscious biases, which include sexism against other women, racism, heterosexism, classism, ableism, and so on, also divide women from each other (hooks 1984). In Audre Lorde's words, there is a "piece of the oppressor . . . planted deep within each of us" (1984: 123; Collins 2013; Collins and Bilge 2016).

Thinking from the margins tends to be exhausting. The tension of being both inside and outside can mean feeling unrooted, having nowhere to relax, and may give way to greater uncertainty and despair rather than to critical consciousness. Uma Narayan (1988) explains that "the individual subject is seldom in a position to carry out a perfect 'dialectical synthesis' that preserves all the advantages of both contexts and transcends all their limitations." Similar problems arise for people who speak multiple languages who may as a result lack fluency in any and have significant gaps in vocabulary. Negotiating both environments does not unite them, even under ideal conditions, and they tend to separate and resist full blending, such that María Lugones argues (2003) the various identities a person has are partly mixed together in a way she calls "curdled." Narayan suggests that the difficulties may lead a person either to keep identities separate or to assimilate as much as possible to the dominant norms in the hopes of achieving better access to resources.

A Standpoint Requires Community, Coalition, and Criticism

Developing individual outsider experience and information into a standpoint depends crucially on social connections, including two relationships between individuals and communities: (1) communities foster individuals' understanding of how knowledge is situated; and (2) individual activity provides coordination within particular communities and among communities in flexible coalitions as they pursue social justice (Collins 2000). Coalitions "ebb and flow based on the perceived saliency of issues to group members" (Collins 2000: 248), and the inadequacies of the language produced by cultures of domination, which Fricker describes as "hermeneutical injustice" (2007), make liberatory coalitions inherently unstable (Ruíz and Dotson 2017). Also, people will disagree, challenge, and criticize each other, whether they share fleeting purposes or stable identities, but that friction fosters knowledge from a standpoint (Medina 2013), heightening a standpoint's critical resources.

Communities develop language and practices to show their members' social marginalization and foster situated knowledge. Interacting with other people who have similar identities allows individuals to observe common experiences which helps them to empathize with each other and "become sensitive to those aspects of their experience that have been marginalized, suppressed, and rendered unintelligible" (Medina 2013: 204). Mutual support helps individuals to develop critical politics (Narayan 1988). Even Rosa Parks, heralded civil rights hero, did not act on her own but as part of an orchestrated plan drawing on a history of Black activism (Medina 2013).

A standpoint can also benefit from a degree of separatism: safe and nurturing spaces for marginalized people allow them to reimagine their identities (Medina 2013) and to develop practical skills for resisting oppressive forces. For instance, although recent mainstreaming helps disabled people receive education and employment and participate otherwise in society alongside typically abled people, which provides them with greater opportunities and autonomy, it also can eliminate sources for liberatory knowledge. Jackie Leach Scully observes, "it also means it is harder for disabled people to find the kind of network that holds minority knowledge about living with a particular impairment" (2018: 114).

Although most standpoint theorists retain a traditional Western view that knowledge belongs to individuals, a communal view of epistemic agency has a history in FST as part of Dorothy Smith's (1987) sociology. Communities are so vital to the development of a standpoint that Kristen Intemann (2010) argues they act as the real knowers, as many feminist empiricists hold. FST may be unique in providing a rich empirically based philosophical account of how individuals and communities work together to produce knowledge.

While communities and coalitions can develop in response to a common enemy (Collins 1993; Crasnow 2013; Pohlhaus 2002), Gaile Pohlhaus argues that the active forging of standpoint knowledge requires relationships with other people characterized by "trust, credibility and responsiveness" (2002: 290). Building trust can be difficult for members of subordinate groups whose trust has been frequently violated (Collins 1993), and it requires granting marginalized people credibility and listening to experiences of oppression. Such trust does not mean accepting all reports at face value, and listening can involve questions and challenges. An openness to differences and contestation aids racial solidarity, according to Medina (2013), and the practice has a history in second-wave feminist consciousness-raising. Discussions of experience in consciousness-raising, Alison Wylie (1992) explains, involved critical analysis and so, for instance, internal coherence or consistency could be used to challenge the testimony of someone in a marginalized position. For bell hooks (1984: 62), solidarity involves more than the "shallow sisterhood" of unqualified approval and especially requires criticizing the violence that women perpetrate.

Coalition across different social positions and identities provides extra epistemic friction for developing situated knowledge. For instance, others learn from lesbian lives the complexity of who and what count as lesbian and the importance of identifying oneself. Affectionate behaviors and both emotional and physical intimacies are quite commonplace between heterosexual people of the same sex in many cultures and times while the most frequent representation of lesbians in popular culture may be by actors in pornography playing at sexuality. Such paradoxes encourage Harding (1991) among others to adopt self-identification as the measure of social identity. Admittedly, no method of defining a social category may suffice for all purposes and, notably, religious, ethnic, and racial identities may require shared histories or more formal mechanisms for community membership. However, lesbian experience teaches that self-identification often serves an important liberatory function, providing "an autonomy. . . to name themselves and their world as they wish, an autonomy that women – and especially marginalized women – are all too often denied" (1991: 252).

Learning to understand and trust people despite one's own prejudices and in the absence of shared circumstance requires empathy that may not come easily (Collins 2013; Medina 2013). A lack of emotional experience with particular problems and an inability to transfer what one learns from previous experience, even an experience of oppression, to another context derails many attempts to think from other marginalized people's perspectives (Narayan 1989).

> For instance, men who share household and childrearing responsibilities with women are mistaken if they think that this act of choice, often buttressed by the gratitude and admiration of others, is anything like woman's experience of being forcibly socialized into these tasks and having others perceive them as her natural function in the scheme of things.
>
> *(1989: 265)*

To build empathy, Narayan counsels humility regarding one's understanding of others' experiences.

Standpoint Methodology in the Sciences

As a research methodology, FST looks to perspectives and practices outside the bounds of a discipline, making outside-in thinking especially explicit and deliberate. A methodology is an orientation for selecting research methods and yields different results in different research fields, but FST prioritizes direct community service and leans otherwise toward intersectional analysis. Social justice does not require intersectionality, which can be part of oppressive thinking and action, for instance in targeting Black men (Collins and Bilge 2016), and indeed any method of research may serve oppressive purposes (Harding 1986; Wylie 1992). However, intersectionality tests ideas within practical political contexts in a way that regularly addresses social justice (Collins and Bilge 2016). Intersectional criticism challenges general claims about women that have been basic to certain formulations of FST and counteracts the problematic naturalizing and essentializing tendencies. Also, it multiplies the avenues for outside-in thinking on any topic, increasing by orders of magnitude the potential for valuable epistemic friction and situated knowledge, and strengthening epistemic advantage.

Standpoint methodology gives priority to the personal experiences and concerns of people who tend to be excluded from a discipline and from access to material and cognitive resources of education, training, and technology. That helps to situate the inquirer within social power dynamics, in methodological terms providing "strong reflexivity" (Harding 1992b; Naples 1996). Those who participate in an "outsider" community may be especially able to observe the social dynamics surrounding research institutions and disciplines (Collins 1986).

Researchers must look beyond their institutional walls fortified by private, corporate, and government funding to develop more democratic and oppositional representations of problems to explore and sources for data (Collins 1986; Collins and Bilge 2016). Earnest concern does not suffice, and research will not benefit from outsider experiences without a specific effort to choose methods and modes of inquiry tailored to those experiences and interests. Often open-ended goals work best, Kim TallBear (2014) advises, even if greater certainty must be expressed to funding bodies, and Collins observes too that researchers may have to sacrifice personal income or status (2013, 2016).

Reimagining research problems can start with studying up on the relevant social justice literature or working out ideas and speculating about lines of inquiry in blogs or community-based participatory research (TallBear 2014). Narayan (1988) recommends that when understandings conflict insiders should practice a "methodological humility" by assuming that the insider perspective lacks context, and that in providing criticism a "methodological caution" be employed to avoid denigrating or dismissing the outsider perspective – even seeming to do so. Academic ideas need translation into accessible language (hooks 1984, 1989) and guesting on Black radio call-in shows allowed Collins (2013) to develop a grassroots vocabulary to describe her research.

Because fully engaged community research often proves unworkable (Wylie 1992), intersectionality provides a valuable heuristic for outside-in thinking, a guideline for addressing who informs and benefits from different research topics and questions. Thinking intersectionally about women's lives hedges against the historical feminist tendency to think one set of problems characterizes the lives of all women or other group of people, and to naturalize or essentialize that perspective. Intersectionality can risk fragmenting how feminists understand women, making the concept of "woman" lose meaning, and leaving only very small-scale alliances possible. However, intersectional thinking provides a wealth of details about people's lives that allows feminists to recognize otherwise obscure commonalities, as the commonality in experiences of Mexican and Puerto Rican women created "Latina" identity (Collins and Bilge 2016).

Collins and Bilge (2016) argue that intersectionality functions as a critical praxis when it provides a heuristic for *both* inquiry and social justice. Consider how studies of STEM (science, technology, engineering, and mathematics) education and employment use the logic of "pipelines" that attends separately to girls and women or to people of color and so obscures the specific problems for girls and women of color and allows them to "leak out." Better accounting for girls and women of color in STEM comes from the intersectional logic of structural barriers regarding access to education.

Intersectionality need not even begin with or focus on gender, although "women's studies faculty have been standard bearers for advancing intersectionality as a form of critical inquiry and praxis in the academy" (Collins and Bilge 2016: 103). It may begin thought from a disabled perspective, an indigenous perspective, or a sex workers' perspective. In reasoning about some cultures, where gender is not a central social identity, such as the Yorùbá, imposing gender categories creates deep misunderstanding (Oyěwùmí 1997).

Recent success in mainstreaming outside-in thinking, for instance in Women's and Gender Studies, Indigenous Studies, and Disability Studies, indicates new resources for oppositional knowledge and fresh opportunities for coalition (Collins 2016) and epistemic friction. However, it can also lead to a "testimonial smothering" described by Kristie Dotson (2011), as when Black feminists truncate their discussions to avoid risky exchanges with academics who lack meta-sensitivity about the subject matter's complexity. Also, research on outsider problems can overlook or deny outsider perspectives on those problems:

> For those Black women who confront racial and sexual discrimination and know that their mothers and grandmothers certainly did, explanations of Black women's poverty that stress low achievement motivation and the lack of Black female "human capital" are less likely to ring true.
>
> *(Collins 1986: 528)*

One strategy lies in recruiting actual outsiders into the research community except that tends to involve the expectation that the insider-outsiders perform what Carla Fehr (2011) describes as "epistemic diversity work," which adds to their job burden. Fehr suggests that researchers can avoid being "epistemic free-riders" who simply accept the benefits of a diverse research community by supporting and mentoring socially marginalized colleagues.

Responsibility to a community of people entails that the intellectual products of a standpoint must serve the purposes of the marginalized community from which it draws. As Nancy Naples and Carolyn Sachs put it, "we must recognize our power over those who share their lives, struggles, and visions with us" (2000: 210). For Collins, every project of Black feminist thought, "whether a special issue, a scholarly essay, song, film, or blog," must consider how it benefits Black women and girls (Collins 2016: 141). The outside-in vector must turn back outside to the community in order to evaluate the knowledge generated, especially when standpoint theory operates as a methodology inside powerful academic institutions.

Simply "giving back" to marginalized communities may address symptoms but not the underlying problems (TallBear 2014). FST challenges how some social science methods treat people merely as objects and not as agents who can direct their own futures and decide what they need (Griffith and Smith 1986; hooks 1989). Feminists must fight their own tendencies to treat women and other marginalized people as mere objects (hooks 1984) or even as half-subjects who need to be given a voice (Ortega 2006). For example, to move beyond a form of exchange defined by the settler society in its own terms, researchers must learn to question the cultural context in which they work and their status as researchers in that culture. Kyle Whyte (2017) identifies that in addition to the "supplemental value" scientists gain from adding the content of indigenous knowledge to Western science,[3] they can learn from the knowledge's "governance value." Indigenous knowledge systems account for their own cultivation, transmission, remembrance, and exercise. Understanding and respecting that knowledge sovereignty help scientists as privileged members of the settler society to situate themselves in regard to indigenous cultures.

Harding (2006) recommends the diversifying scientific standards and practice by drawing on science studies and political studies of science. Harding's guidelines for doing so can be vague and uncompelling (Wylie 2008) and her recommendation depends on the diversity of interests and values addressed by different disciplines rather than a diversity of sociopolitical locations, arguments associated with feminist empiricism rather than FST (Intemann 2010). Nevertheless, using the social sciences for self-scrutiny can be justified using FST if one's own discipline has made relatively weak progress, providing little opportunity for liberatory coalition, such that one needs to find like-minded liberatory colleagues and materials wherever one can. Resources for transforming disciplinary standards can come from feminist scholarship in other disciplines and from feminist empiricist reasoning, and that helps FST provide a methodology for feminist philosophy, including in philosophy of science.

Standpoint Philosophy of Science

Philosophers tend to eschew methodology and, as Dotson (2012) argues, Western philosophy tends like Western culture generally to be exceptionalist, only counting its own activities as valuable. Generating questions and theories with input from feminist and other liberatory sources outside of philosophy provides a distinctive epistemology of science and challenges traditional practices in philosophy of science.

Harding (1986, 1991) contrasts FST with other trends in feminist theorizing about scientific knowledge from the late 1970s and early 1980s: "feminist empiricism" and "feminist postmodernism." Ultimately (1991), she makes standpoint theory her own with a socially situated feminist postmodernist view of embodied rationality that rejects the abstract rationality associated with modernism, which helps her to accommodate the ways that gender intersects with sexuality and race. Harding's intersectional FST recognizes multiple feminist standpoints that attend to specific women's experiences of oppression.

Harding (1991, 2015) develops the epistemological norm of "strong objectivity" based on the methodological value of strong reflexivity. Objectivity in its twentieth-century form (Daston 1992; Daston and Galison 2007) requires a situated accounting for the "social values, interests, and assumptions that researchers bring to the research process" (Harding 2015: 33), which strong objectivity detects by starting thought outside the given science. The lives of people in marginalized groups, such as women, give thought starting points that lie outside the dominant frameworks of the sciences, and they involve values and interests that challenge accepted priorities and the values of people with privileged social identities, who are heavily represented among scientists.

Harding argues that the norm of impartiality provides only weak objectivity because it only excludes values of people already on the social margins, failing to exclude the values of people with privileged status. The attractiveness of weak objectivity, she (1991) argues, lies in the promise to provide research that will apply universally regardless of situation. The generality that such science can offer depends on a highly contingent context. It depends on the work of many others, such as subordinates in labs, carers in the home, and service workers in institutions, roles disproportionately filled by women and minorities, and yet those people's interests tend not to drive the theorizing. General applicability can even come at a cost of lives and ecosystems, as Vandana Shiva (2016) exposes in India, and the world must change in painful and sometimes irreversibly destructive ways to conform to serving the priorities of people with social privilege.

The epistemic advantage of thinking from the outside-in applies directly to understanding "the nature of society" (Fricker 1999) and FST makes a clear case that the independence from observers for which traditional objectivity strives provides an inadequate analysis, for instance, of women's underemployment (Rolin 2006) or the absence of Latinas among managers (Pompper 2017). However, critics claim that FST has little relevance to the physical sciences. The worlds of particles or magnetic fields cannot be shared by researchers in the same way as the worlds of women, cultures, or even plants (Keller 1983), and so Kristina Rolin (2006) argues that standpoint theorists must be able to show that the context of patriarchy and forms of oppression that intersect with it are relevant to the subject matter of a particular type of inquiry, that situated knowledge provides epistemic advantage in a specific domain.

Nevertheless, Harding (1991) argues that a feminist social account provides valuable ways for questioning how consensus develops in the physical sciences. Views in physics might possibly, for instance, be affected by patterns in women's unemployment or the causes of that unemployment, regarding which Rolin argues feminists offer significant insights. Different questions than usual might come out of a lab that accommodates women workers or those who come from poor backgrounds. Physicist Barbara Whitten observes how feminism raises questions for her discipline:

We can ask about the relationship between physics and the rest of society. The strong connection between physics and the military, for example, and the relegation of "applied" physics to second-class status, are certainly questions that are amenable to feminist analysis. We can ask about the internal structure of the physics community, how we organize our research groups, educate our students, and allocate our resources.

(1996: 13)

Perhaps in a feminist physics less research would be done that feeds the arms industry. More might address the needs of women, children, and poor people around the world for clean water. Questions about scientific processes and purposes can be scientific questions, and concerns with who science serves were once alien too in the life sciences (Okruhlik 1994) and the social sciences (Collins 2013).

While the Western academy now has models for feminist biology and sociology, Harding (1991) acknowledges the difficulty imagining a feminist physics and it certainly remains distant. Currently, only the most socially privileged people receive the luxury of being able to follow their own curiosity in "pure" or "basic" research.[4] Yet, Whitten argues that "scientists cannot foresee all the applications of their work, and it is not uncommon for apparently esoteric work to have important social consequences" (1996: 9).

Standpoint theorists maintain that ways to devise problems and dream up theories can be better and worse epistemologically, which Miriam Solomon (2009) argues provides an account of creativity that philosophy of science needs. Philosophers tend to invoke a romantic notion of imagination in which "anything goes," treating creative questions of problem choice and theory development as tangential to their work in defining norms with prescriptive force for the testing of theories. Even empirical studies tend to see creativity only as a quality of the isolated individual person.

The standard view of philosophy of science's normative force focuses on testing rather than problem choice and theory development, which goes back to a distinction in twentieth-century philosophy of science between the context of justification where scientists test ideas and the context of discovery that provides inspiration and research questions (Schickore 2018). While many feminists challenge that distinction of justification from discovery (Longino 1990), standpoint theorists especially insist on seeking out descriptions of the norms governing how scientists develop research topics, questions, and theories. Prescriptions for the context of discovery challenge how philosophy of science operates and suggest a new methodology.

The outside-in thinking of FST thus provides theoretical tools for philosophy of science, including situated knowledge, epistemic advantage, strong objectivity, and a methodology for the sub-discipline that reimagines its scope and practices. Starting thought in philosophy of science from the margins involves taking up the results and best practices of feminist and other liberatory scholars, which currently include intersectionality as a research method. Outside-in philosophizing can also involve a commitment to particularism, prioritizing attention to the details of a specific situation over the development of abstract ideals. Setting aside exclusively theoretical debates provides resources for addressing concrete problems, injustices, and systematic disadvantages (Medina 2013) and it creates fresh theoretical challenges for philosophy of science such as developing adequate theories of trust (Whyte and Crease 2010).

Janet Kourany (2009) argues FST provides a methodology needed to complement other feminist philosophies of science. It can motivate the enforcement and reworking of traditional scientific methodologies, as done by "spontaneous" feminist empiricists (Harding 1991). FST also accounts for how social diversity provides the transformative criticism prioritized in Helen Longino's (1990) social empiricism, and it explains how feminist values improve science as feminist naturalists claim.

Treating knowledge as active rather than representational marks FST as a form of pragmatism (Wylie 2008), albeit in a "cynical form" (Dayton 1997). Kristina Rolin (2009) argues that FST reveals how social conditions suppress evidence, which provides an epistemic advantage in understanding how relations of power work. Asking philosophy to, in the same way, equip people in resisting oppression (Collins 2013) demands it provide a "non-ideal theory," perhaps even a naturalist epistemology (Hundleby 2001; Kukla 2006; Mills 2005; Rouse 2009).

FST justifies certain obligations for the practical future in epistemological terms that otherwise might appear to be exclusively ethical or political. Epistemological values also meld with ethical and political values in the way that both pragmatism and FST (Collins 2013; Hundleby 2005) stress the importance of education to democracy. Further, Shannon Sullivan argues that pragmatism provides an appropriate view of knowledge and truth for FST: "putting facts and events to use in the transformation of the world" to provide for human flourishing (2001: 141).

Conclusion: Philosophy of Science from the Outside-in

From a feminist standpoint the questions and answers familiar to philosophers of science – about demarcation, testing, creativity, and so on – seem strange because they fail to inform about the power, promise, and threat of science. Attending to these concerns about human flourishing creates new ways to understand the potential value of scientific knowledge and academic research in general. Science, philosophy, and philosophy of science all can be situated as forms of knowledge, employ the methodology of strong reflexivity, and address the epistemic norm of strong objectivity, each a more specific articulation of the previous.

Reaching outside philosophy to feminist methods in other disciplines and the broad input provided by feminist and other liberatory critiques of science and, in turn, to non-academic liberatory communities distinguishes FST as a philosophy of science. The outward touchstones affect what questions and methods of inquiry count as scientific and philosophical.

Related chapters: 8, 9, 15, 16, 18, 28.

Notes

1 From the standard language, I prefer "advantage" because it seems to admit of degrees, as a feminist standpoint does, whereas "privilege" seems to suggest modularity and uniformity.
2 In some of the discussions, especially by Narayan, the use of the terms "insider" and "outsider" is reversed so that members of a marginalized social group are insiders to that social identity who may also play roles outside their marginalized group.
3 Because the West-East distinction refers to a somewhat dated division of the world between capitalist or democratic and communist, sometimes it is better to distinguish sciences of the industrialized and colonial global North from that of global South. Both distinctions obscure indigenous cultures and sciences, but the alignment of the "Western" characterization with democracy often proves illuminating and it is Whyte's language choice.
4 Harding responded this way to a question from Carla Fehr at the meeting of the International Society for the History, Philosophy and Social Studies of Biology at the Université du Québec à Montréal in 2015.

References

Collins, P.H. (1986) "Learning from the Outsider Within: The Sociological Significance of Black Feminist Thought," *Social Problems* 33(6), Special Theory Issue, pp. S14–S32.
Collins, P.H. (1993) "Toward a New Vision: Race, Class, and Gender as Categories of Analysis and Connection," *Race, Sex & Class* 1(1), pp. 25–45.
Collins, P.H. (2000) *Black Feminist Thought: Knowledge, Consciousness, and the Politics of Empowerment*, 2nd edition, New York: Routledge.
Collins, P.H. (2013) *On Intellectual Activism*, Philadelphia: Temple University Press.

Collins, P.H. (2016) "Black Feminist Thought as Oppositional Knowledge," *Departures in Critical Qualitative Research* 5(3), pp. 133–144.

Collins, P.H. and Bilge, S. (2016) *Intersectionality*, Boston: Polity Press.

Crasnow, S. (2013) "Feminist Philosophy of Science: Values and Objectivity," *Philosophy Compass* 8(4), pp. 413–423.

Daston, L. (1992) "Objectivity and the Escape from Perspective," *Social Studies of Science* 22(4): 597–618.

Daston, L. and Galison, P. (2007) *Objectivity*, New York: Zone Books.

Dayton, E. (1997) "Commentary on Catherine Hundleby's "Where Stands Standpoint Now?' " Canadian Philosophical Society, unpublished commentary.

Dotson, K. (2011) "Tracking Epistemic Violence, Tracking Practices of Silencing," *Hypatia* 26(2), pp. 236–257.

Dotson, K. (2012) "How Is This Paper Philosophy?" *Comparative Philosophy* 3(1), pp. 3–29.

DuBois, W.E.B. (1897) "Strivings of the Negro People." *Atlantic Monthly* 80, pp. 194–198.

England, P. (2010) "The Gender Revolution: Uneven and Stalled," *Gender & Society* 24(2), pp. 149–166.

Fehr, C. (2011) "What Is in It for Me? The Benefits of Diversity in Scientific Communities," in Grasswick, H.E. (ed.) *Feminist Epistemology and Philosophy of Science: Power in Knowledge*, Dordrecht: Springer, pp. 133–155.

Fricker, M. (1998) "Rational Authority and Social Power: Towards a Truly Social Epistemology," *Proceedings of the Aristotelian Society*, New Series 98, pp. 159–177.

Fricker, M. (1999) "Epistemic Oppression and Epistemic Privilege," *Canadian Journal of Philosophy*, 29(suppl. 1), pp. 191–210.

Fricker, M. (2007) *Epistemic Injustice: Power and the Ethics of Knowing*, New York: Oxford University Press.

Frye, M. (1983) *The Politics of Reality*, Trumansberg: The Crossing Press.

Griffith, A.I. and Smith, D.E. (1986) "Constructing Cultural Knowledge: Mothering as Discourse," in Gaskell, J. and McLaren, A. (eds.), *Women and Education: A Canadian Perspective*, 1st edition. Calgary: Detselig Enterprises, Ltd., pp. 87–103.

Haraway, D. (1985) "A Manifesto for Cyborgs: Science, Technology, and Socialist Feminism in the 1980s," *Socialist Review* 15(80), pp. 65–105.

Haraway, D. (1988) "Situated Knowledges: *The Science Question in Feminism* and the Privilege of Partial Perspective," *Feminist Studies* 14(3), pp. 575–599.

Harding, S. (1986) *The Science Question in Feminism*, Ithaca: Cornell University Press.

Harding, S. (1991) *Whose Science? Whose Knowledge? Thinking from Women's Lives*, Ithaca: Cornell University Press.

Harding, S. (1992a) "After the Neutrality Ideal: Science, Politics, and 'Strong Objectivity,' " *Social Research* 59(3): 567–87.

Harding, S. (1992b) "Rethinking Standpoint Epistemology: What is 'Strong Objectivity'?" *The Centennial Review* 36(3), pp. 437–470.

Harding, S. (2006) *Science and Social Inequality: Feminist and Postcolonial Issues*, Chicago: University of Illinois Press.

Harding, S. (2009) "Standpoint Theories: Productively Controversial," *Hypatia* 24(4), pp. 192–200.

Harding, S. (2015) *Objectivity and Diversity: Another Logic of Scientific Research*, Chicago: University of Illinois Press.

Hartsock, N.C.M. (1983) "The Feminist Standpoint: Toward a Specifically Feminist Historical Materialism," in Harding, S. and Hintikka, M.B. (eds.), *Discovering Reality: Feminist Perspectives on Epistemology, Metaphysics, Methodology, and Philosophy of Science*, Boston: Kluwer, pp. 283–310.

Hartsock, N.C.M. (1997) "Standpoint Theories for the Next Century," *Women & Politics* 18(3), pp. 93–101.

Hennessy, R. (1993) "Women's Lives/Feminist Knowledge: Feminist Standpoint as Ideology Critique," *Hypatia* 8(1), pp. 14–34.

hooks, b. (1984) *Feminist Theory: From Margin to Center*, Boston: South End Press.

hooks, b. (1989) *Talking Back: Thinking Feminist, Thinking Black*, Boston: South End Press.

Hundleby, C. (1998) "Where Standpoint Stands Now," *Women & Politics* 18(3): 25–43.

Hundleby, C. (2001) *Feminist Standpoint Theory as a Form of Naturalist Epistemology*. PhD thesis, University of Western Ontario. Available at https://www.researchgate.net/publication/320563907_Feminist_Standpoint_Theory_as_a_Form_of_Naturalist_Epistemology.

Hundleby, C. (2005) "The Epistemological Evaluation of Oppositional Secrets," *Hypatia* 20(4), pp. 44–58.

Intemann, K. (2010) "25 Years of Feminist Empiricism and Standpoint Theory: Where are We Now?" *Hypatia* 25(4), pp. 778–796.

Intemann, K. (2016) "Feminist Standpoint," in Disch, L. and Hawkesworth, M. (eds.), *The Oxford Handbook of Feminist Theory*, New York: Oxford, pp. 261–282.

Keller, E.F. (1983) *A Feeling for the Organism: The Life and Work of Barbara McClintock*, New York: Henry Holt and Co.

Keller, E.F. (1996) *Reflections on Gender and Science*, New Haven: Yale University Press.

Kourany, J. (2009) "The Place of Standpoint Theory in Feminist Science Studies," *Hypatia* 24(4) pp. 209–218.

Kukla, R. (2006) "Objectivity and Perspective in Empirical Knowledge," *Episteme: A Journal of Social Epistemology*, 3(1–2), pp. 80–95.

Le Doueff, M. (1990) *The Philosophical Imaginary*, C. Gordon (trans.), Stanford: Stanford University Press.

Lloyd, G. (1984) *The Man of Reason: "Male" and "Female" in Western Philosophy*, New York: Methuen and Co. Ltd.

Longino, H.E. (1990) *Science as Social Knowledge: Values and Objectivity in Scientific Inquiry*, Princeton: Princeton University Press.

Lorde, A. (1984) *Sister, Outsider*, Trumansberg: Crossing Press.

Ludwig, D. (2016) "Overlapping Ontologies and Indigenous Knowledge. From Integration to Ontological Self-determination," *Studies in History and Philosophy of Science* 59, pp. 36–45.

Lugones, M. (2003) *Pilgrimages/Perigrinajes: Theorizing Coalition against Multiple Oppressions*, New York: Rowman and Littlefield.

Mahowald, M.B. (2005) "Our Bodies Ourselves: Disability and Standpoint Theory," *Social Philosophy Today* 21, pp. 237–246.

Medina, J. (2013) *The Epistemology of Resistance: Gender and Racial Oppression, Epistemic Injustice, and Resistant Imaginations*, New York: Oxford University Press.

Mills, C.W. (2005) "'Ideal Theory' as Ideology," *Hypatia* 20(3), pp. 165–184.

Moulton, J. (1983) "The Adversary Method: A Paradigm of Philosophy," in Harding, S. and Hintikka, M.B. (eds.), *Discovering Reality: Feminist Perspectives on Epistemology, Metaphysics, Methodology, and Philosophy of Science*, Boston: Kluwer, pp. 149–164.

Naples, N.A. (1996) "A Feminist Revisiting of the Insider/Outsider Debate: The 'Outsider Phenomenon' in Rural Iowa," *Qualitative Sociology* 19(1), pp. 103–124.

Naples, N.A. (1999) "Towards Comparative Analyses of Women's Political Praxis: Explicating Multiple Dimensions of Standpoint Epistemology for Feminist Ethnography," *Women and Politics* 20(1), pp. 29–57.

Naples, N.A. and Sachs, C. (2000) "Standpoint Epistemology and the Uses of Self-Reflection in Feminist Ethnography: Lessons for Rural Sociology," *Rural Sociology* 65(2), pp. 194–214.

Narayan, U. (1988) "Working Together across Difference: Some Considerations on Emotions and Political Practice," *Hypatia* 3(2), pp. 31–47.

Narayan, U. (1989) "The Project of Feminist Epistemology: Perspectives from a Non-western Feminist," in Jaggar, A. (ed.), *Gender/Body/Knowledge: Feminist Reconstructions of Being and Knowing*, Rutgers: Rutgers University Press, pp. 256–272.

Okruhlik, K. (1994) "Gender and the Biological Sciences," *Canadian Journal of Philosophy*, supplementary volume 20, pp. 21–42.

Ortega, M. (2006) "Being Lovingly, Knowingly Ignorant: White Feminism and Women of Color," *Hypatia* 21(3), pp. 56–74.

Oyěwùmí, O. (1997) *The Invention of Women: Making an African Sense of Western Gender Discourses,* Minneapolis: University of Minnesota Press.

Pohlhaus G. (2002) "Knowing Communities: An Investigation of Harding's Standpoint Epistemology," *Social Epistemology* 16(3), pp. 283–293.

Pompper, D. (2017) "The Gender-ethnicity Construct in Public Relations Organizations: Using Feminist Standpoint Theory to Discover Latinas' Realities," *Howard Journal of Communications* 8(4), pp. 291–311.

Ridgeway, C. (2011) *Framed by Bender: How Gender Inequality Persists in the Modern World*, New York: Oxford University Press.

Rolin, K. (2006) "The Bias Paradox in Feminist Standpoint Epistemology," *Episteme: A Journal of Social Epistemology* 3(1–2), pp. 125–136.

Rolin, K. (2009) "Standpoint Theory as a Methodology for the Study of Power Relations," *Hypatia* 24(4), pp. 21–42.

Rose, H. (1983) "Hand, Brain and Heart: A Feminist Epistemology for the Natural Sciences," *Signs* 9(1), pp. 73–90.

Rouse, J. (2009) "Standpoint Theories Reconsidered," *Hypatia* 24(4), pp. 200–209.

Ruddick, S. (1989) *Maternal Thinking: Towards a Politics of Peace*, Boston: Beacon Press.

Ruíz, E. and Dotson, K. (2017) "On the Politics of Coalition," *Feminist Philosophical Quarterly* 3(2), pp. 1–15.

Sandoval, C. (2000) *Methodology of the Oppressed*, Minneapolis: Minnesota University Press.

Schickore, J. (2018) "Scientific Discovery," in E.N. Zalta (ed.), *The Stanford Encyclopedia of Philosophy* (Summer 2018 Edition). Available at https://plato.stanford.edu/archives/sum2018/entries/scientific-discovery/

Scully, J.L. (2018) "From 'She Would Say That, Wouldn't She?' to 'Does She Take Sugar?' Epistemic Injustice and Disability," *International Journal of Feminist Approaches to Bioethics* 11(1), pp. 106–124.

Shiva, V. (2016) *Who Really Feeds the World? The Failures of Agribusiness and the Promise of Agroecology*, Berkeley: New Harvest.

Smith, D. (1974) "Women's Perspective as a Radical Critique of Sociology," *Sociological Inquiry* 44(1), pp. 7–13.

Smith, D. E. (1987) *The Everyday World as Problematic*, Toronto: University of Toronto Press.

Solnit, R. (2014) *Men Explain Things to Me*, Chicago: Haymarket Books.

Solomon, M. (2009) "Standpoint and Creativity." *Hypatia* 24(4), pp. 226–237.

Sullivan, S. (2001) *Living across and through Skins: Transactional Bodies, Pragmatism, Feminism*, Bloomington: Indiana University Press.

TallBear, K. (2014) "Standing with and Speaking as Faith: A Feminist-indigenous Approach to Inquiry," *Journal of Research Practice* 10(2), pp. 1–7.

Whitten, B. (1996) "What Physics is Fundamental Physics? Feminist Implications of Physicists' Debate over the Superconducting Supercollider," *NWSA Journal* 8(2), pp. 1–16.

Whyte, K.P. (2017) "What Do Indigenous Knowledges do for Indigenous Peoples?" in Nelson, M.K. and Shilling, D. (eds.), *Traditional Ecological Knowledge: Learning from Indigenous Methods for Environmental Sustainability*, Boston: Cambridge University Press, pp. 57–82.

Whyte, K.P. and Crease, R.P. (2010) "Trust, Expertise, and the Philosophy of Science," *Synthese* 177, pp. 411–425.

Wieseler, C. (2016) "Objectivity as Neutrality, Nondisabled Ignorance, and Strong Objectivity in Biomedical Ethics," *Social Philosophy Today* 32, pp. 85–106.

Wylie, A. (1992) "Reasoning about Ourselves: Feminist Methodology in the Social Sciences," in Harvey, E. and Okruhlik, K. (eds.), *Women and Reason*, Ann Arbor: University of Michigan Press, pp. 225–244.

Wylie, A. (2003) "Why Standpoint Matters," in Figueroa, R. and Harding, S. (eds.), *Science and Other Cultures: Issues in Philosophy of Science and Technology*, New York: Routledge, pp. 26–48.

Wylie, A. (2008) "Social Constructionist Arguments in Harding's *Science and Social Inequality*," *Hypatia* 23(4), pp. 201–211.

Wylie, A. (2012) "Feminist Philosophy of Science: Standpoint Matters," *Proceedings and Addresses of the American Philosophical Association* 86(2), pp. 47–76.

Further Reading

Harding, S. (ed.; 2004) *The Feminist Standpoint Theory Reader: Intellectual and Political Controversies*, 1st edition, New York: Routledge. (A collection of classic works from the first two decades of feminist standpoint theory.)

Linker, M. (2014) *Intellectual Empathy: Critical Thinking for Social Justice*, Ann Arbor: University of Michigan Press. (This philosophy undergraduate textbook employs an intersectional method for critical thinking education from a feminist standpoint.)

Neely, B. (1993) *Blanche on the Lam*, New York: Penguin (A mystery illustrating many aspects of feminist standpoint epistemology. Wylie (2003) outlines how it represents important aspects of the theory.)

Wylie, A. and Sismondo, S (2015) "Standpoint Theory, in Science," in Wright, J.D. (ed.), *International Encyclopedia of the Social and Behavioral Sciences*, 2nd edition. Amsterdam: Elsevier, pp. 324–330.

8

LATIN AMERICAN DECOLONIAL FEMINIST PHILOSOPHY OF KNOWLEDGE PRODUCTION

Sandra Harding and Breny Mendoza

Introduction: Institutional Space for Latin American Feminist Decolonial Philosophy?[1]

Latin American feminist philosophy of science is currently producing innovative contributions to international philosophies of science and their feminisms. Yet both its history and its present projects remain mostly invisible in international contexts (Rivera Berruz 2018). This first section describes this invisibility and its sources, and very briefly points to a few key moments in the more than five centuries of Latin American thought about women and the production of knowledge. The invisibility of present and past Latin American concerns with women and the production of knowledge has several origins. An important one is the fact that much of this work cannot be read by English-only readers because it is published in Spanish or Portuguese. Another reason for its invisibility is that some of its most innovative contributions have their origins in the oral and spiritual traditions of indigenous women, as current feminist debates about gender are revealing. Moreover, a lot of scholarship is produced outside academic contexts, involving political activism and therefore eschewing traditional academic conventions. Furthermore, while women's or gender studies programs and centers became institutionalized in Latin American universities in the mid to late 1980s and 1990s, they were for a long time focused mostly on gender and development scholarship (Mendoza 2012). Thus, contexts for the articulation of feminist issues in philosophy or other disciplines in Latin America were scarce or occurred in non-academic and non-conventional formats (Femenias and Oliver 2007; Mendoza 2012; Schutte and Femenias 2013).

However, equally important here is the long history of colonial domination of Latin America which has involved also epistemic violence and the westernization of universities. Thus Latin American philosophy students are still taught that philosophy is a unique European invention, and that their own philosophical thoughts are folklore. Latin American decolonial scholars call this the coloniality of knowledge. Of course, formal Spanish and Portuguese rule of Latin American had almost completely ended by 1830. Yet its effects and reinventions linger into the present, including Roman Catholic and, especially Jesuit assumptions. The coloniality of knowledge is powerfully enacted still today in the North through its insistence on defining the terms philosophy and science in ways that exclude Latin American thoughts and practices (Stehn 2013: 16). Restricting "real philosophy" to analytic traditions, and "science" to Northern" high sciences," such as physics and chemistry, has had the result of disvaluing and obscuring rich traditions of Latin American thought. That is, these have been the institutional consequences of such restrictive definitions.[2] Interestingly, modern Western science itself was co-produced and co-constituted with Iberian colonialism in the Americas (Dussel 1995; Giraldo 2016; Mignolo 2011). Its definition was part of the initial institutionalization of the colonial project, which included very early on the foundation of the universities.[3] The twentieth-century restriction of philosophy to analytic assumptions and practices has been among the more recent reinventions of such coloniality.

Finally, the term "Latin America" has also been contentious. It was not chosen by Latin Americans to refer to themselves and it excludes indigenous and Afro-descendant peoples (ibid.). Rather, it emerged in nineteenth-century French intellectual circles as a way to refer to that geographic region, and subsequently became a site of debate about the identity of those cultures and thinking (Stehn 2013: 16). Today many prefer calling the region, *Abya Yala*, which means land in full maturity in the language of the Kuna of Panama and Colombia.

That all said, in fact Latin American feminist philosophies and their thinking about knowledge production do have rich histories. Many observers insist that for centuries before 1492, Amerindian women participated in the production of useful agricultural, environmental, and technical knowledge, as well as in the speculative ontology and cosmovision known to have been created by the Aztecs, Incas, and Mayans. After 1492, the first Latin American feminist thought visible to Europeans would have been the writings of the Catholic nun, Sor Juana Inés de la Cruz 1651–1695, who succeeded in reaching a significant readership for her arguments on behalf of women's abilities and rights to participate in international intellectual life, to the dismay of her Church's critics. Interestingly, Sor Juana was writing one hundred years before Mary Wollstonecraft (de la Cruz 2009; Gargallo 2004; Goodale 2008; Mendoza 2007; Oliver 1998; Rivera Berruz 2018; Sousa 2017).

Yet it would not be until the late nineteenth century that a significant first wave of feminist activist projects appeared. This lasted into the 1940s. Here women (mostly mestiza/criolla, upper middle-class women) were mobilized for liberal and socialist political movements, and they demanded the rights to higher education, to vote, to administer their own property, for labor rights, and for family rights — the latter including official registration of children born out of wedlock. And they criticized the sexual double standard. The first International Feminist Congress was held in Buenos Aires in 1910. Within the various movements of this first wave, disputes broke out about whether women should be struggling for recognition of the superiority of feminine practices, or whether the goals should be specifically feminist ones. As is the case today, advocates of feminist agendas argued that restricting calls for change only to valuing women's practices within male-dominated societies does not deeply enough challenge such male-domination. Women are capable of far more ambitious social participation than only what male-dominated societies have permitted (Miller 1991; Schutte and Femenías 2013: 397–398).

The 1950s–1960s were decades low in feminist activism in the United States and Europe, as well as in Latin America. The institutionalization of women's/gender studies programs in Latin America in the 1980s and 1990s ran parallel to a series of biennial (and, later, triennial) meetings of Latin American women: *Encuentros Feministas Latinoamericanos y del Caribe*. The first was held in 1981 in Bogota, Colombia. These have provided ongoing connections for the debates and struggles over valuably conflicting positions of women activists, scholars, and the feminist policy makers of the international so-called development agencies (Alvarez et al. 2002; Stehn 2013: 17).

While Latin American Studies programs and courses in North American universities began in 1960s after the Cuban revolution with US foreign political interests in mind, they have increasingly proliferated and become more populated with Latin American scholars in recent years. This has occurred in part due to globalization's tendency to open more space to formerly excluded groups of people in North American universities, such as women, peoples of color, and international students, on both sides of the podium, who would not have been present two decades ago (Hess 2011). Consequently, a number of helpful monographs and collections of papers on feminist philosophy, and in particular on philosophy of knowledge production, have appeared in English.

Significant in organizing and contributing to these are scholars today located in US universities who were born in Latin America or are of Latin American origins. These include, for example, Linda Alcoff (2006, 2015), Gloria Anzaldua (1987), Maria Lugones (2008), Susana Nuccetelli and Seay (2003), Nuccetelli et al. (2010), and Ofelia Schutte (1994, 1998b). At the same time, in Latin America indigenous and Afro-descendent peoples are also entering Latin American universities in significant numbers or are creating their own centers of knowledge, which are changing profoundly the terms of the debates around philosophy and gender. They are creating what they call the "anthropology of us" (Pérez Moreno 2019). Indigenous feminist scholars transcribe the debates taking place in their communities into writing and into Spanish.

The effervescence of knowledge production in and about Latin America occurs in the context of a return to anti-colonial thought. The Anglo academic is more familiar with those postcolonial criticisms that stem from British colonial experiences. Yet, Latin Americans have a long tradition of anti-colonial scholarship based on their experiences of Spanish and Portuguese colonialism, which predates British colonialism by several hundred years. One of its latest iterations is the Decolonial theory founded by the modernity/coloniality/decoloniality (MCD) group. Decolonial/anti-colonial thinking in general today is very influenced by this approach, although it is by no means the only strand of such thinking in the region. However, it is important to have clarity about the distinctiveness of Latin American anti-colonial thought. Section "The Decolonial beyond the Postcolonial" will briefly identify the differences between these Latin American approaches to anti-colonial thinking and those characteristic of the more familiar postcolonial projects. Section "Decolonizing Gender: Contexts for Debates" focuses on examples of the distinctively new questions that Latin American feminists are raising about knowledge-production issues through their debates about gender.

The Decolonial beyond the Postcolonial[4]

Decolonial analyses occur in significantly different historical contexts than those in which the more familiar postcolonial accounts were generated. Here the focus is on four such different contexts: different chronologies or moments in history, encounters with different physical geographies, different cultures and their particular projects from which the colonizers set out, and different cultures in which lived the peoples whom they colonized. First, there are important chronological differences marked especially by the MCD scholars. Colonial relations in the Americas began in 1492 – more than two and a half centuries before the British began to establish their colonies in India and the Middle East. For the Decolonial scholars, it is no accident that the so-called discovery of the Americas coincides with the emergence of modernity in Europe, though standard Northern histories tend not to link these two phenomena. "Modernity appears when Europe organizes the initial world-system and places itself at the center of world history over against a periphery equally constitutive of modernity" (Dussel 1995: 9–10). So, for Latin American theorists, modernity and Iberian colonialism co-produce and co-constitute each other. This not only shifts the beginning of modernity to a much earlier date but also inserts Iberian colonialism centrally into the history of modernity, which is something that has been largely denied by North Atlantic scholars.

Another chronological difference is that formal independence from European rule began much earlier in the Spanish, Portuguese, and French colonies in the Americas than in the British colonies (with the exception of the United States). Most of the other colonies in the Americas achieved formal independence from Spain, Portugal, and France by 1830, except Cuba, which gained independence in 1898.[5]

Moreover, for the anti-colonial scholars, 1492 is the starting date of anti-colonial thinking. The Amerindians whom Cortes encountered, as well as Nahua and Quechua intellectuals in the early sixteenth century, clearly resisted both the idea and the reality of Iberian colonization (Brotherston 2008; Todorov 1984). Anti-colonial arguments by Nahua intellectuals had already appeared *in Spanish* in the Americas by 1538. Quechua author Felipe Guamán Poma de Ayala (1615) wrote between 1606 and 1615 what is considered the longest sustained critique of Spanish colonial rule produced by an indigenous subject in the entire colonial period. (And it was written in Spanish and Quechua.) So, while the postcolonial theory that began to appear in the British Empire after World War II preceded the development of the Latin American Decolonial theory of the MCD group by several decades, anti-colonial thought has a longer and different history in Latin America than the familiar British postcolonial accounts.

Second, the origins of the Scientific Revolution are broader than assumed in conventional philosophies and histories of science, and they have roots in colonialism. Colonization of the Americas required that the conquerors interact effectively with physical worlds different from those familiar to them. Yet they lacked an astronomy of the Southern Hemisphere with which to navigate back to Europe across the South Atlantic. The cartography of the South Atlantic and of their environments in the Americas had to be created. They also needed climatology, oceanography, and better engineering to secure the safe travels of their crews and their "precious cargoes." In the Americas they needed knowledge of the unfamiliar flora and fauna that they encountered. They needed better geology, mining, and engineering, even though they soon appropriated from the Amerindians sophisticated forms of these technologies which they improved in order to extract the gold and silver that they found in Mexico and Peru. In 1492, the Europeans were behind the Amerindians in these kinds of scientific and technical knowledge: they were the backward ones. Moreover, several science and technology research and teaching academies were established in both Iberia and the Americas in the sixteenth century. These were focused on training future colonizers into the new challenges of piloting, navigation, cartography, oceanography, climatology, mining, and engineering. Europe's colonial projects in the Americas turned a huge part of the globe into a laboratory for European sciences (Cañizares-Esguerra 2006; Saldana 2006).

Third, in addition to the scientific and technical needs created by the different chronologies and geographies, the Iberian colonizers lived in social worlds different from those that shaped the coloniality of the British Empire. For the Europeans, the "discovery" of new lands across the Atlantic appeared as a solution to some of their most vexing social problems. Europeans welcomed the thought of being able to leave behind the economic and political challenges of the continual religious and political wars, as well as of overpopulation and famines. The Europeans imagined that they could start over in the "Garden of Eden" that had been "discovered" across the Atlantic.

Fourth, yet those peoples that the Spanish and Portuguese colonized were culturally different from those the British colonized centuries later. For the Amerindians the arrival of the Europeans was a cataclysmic event. It meant the destruction of their cultural and physical worlds, the loss of sovereignty over their lands, the loss of their freedom, and the destruction and devaluation of their forms of knowledge and spirituality.

It is only relatively recently that demographic, historical, and environmental research undermined long-held assumptions that the Americas were only sparsely inhabited in 1491, and that those inhabitants were at a much more primitive stage of social and scientific development than were Europeans. In 1491 there were probably more people living in the Americas than in Europe (e.g. Denevan 1992; Mann 2005). Estimations of the actual numbers in the Americas vary hugely, from 10 million to over 100 million. Some of the world's largest cities at the time were in the Americas (Mann 2005). Inca, Aztec, and Mayan architecture, engineering, and road systems were among the most advanced of ancient civilizations, and in some respects superior to those of the Europeans. Amerindians had extensive agricultural techniques, such as controlled fires to clear the land and increase the nutrients in the soil, and were able to preserve food that could last for years through processes of freezing, dehydration, and rehydration. Aztecs used the technique of the *chinampa* which allowed them to grow crops on shallow lake beds. The Incas developed resilient breeds of crops such as potatoes, quinoa, and corn. They built cisterns and irrigation canals that snaked and angled down and around the mountains, some of them still in use today. In ecologists' terms, Amerindians were also the keystone species of the Amazonian ecosystem. Some observers have referred to Amazonia as a cultural artifact since its trees, plants, and soils were themselves cultivated by the Amerindians, as recent studies have revealed (Hecht 2004).

By 1620, the Americas did appear almost empty. Over 90% of the Amerindian population had been eliminated. This was arguably thanks to pandemics, but perhaps more significantly due to the persistent efforts of the colonialists to kill Amerindians and take possession of their lands, and to the Europeans' superior armaments. The pandemics would not have created such high mortality figures if the Iberians had not systematically captured the Amerindians for forced labor, and then overworked, underfed, and refused to provide medical care for them, decimated their provisioning environments, and otherwise treated them as disposable slaves (Pratt 2008).

What did the Amerindians know in 1491 in addition to their agricultural, environmental, and spiritual-philosophical knowledge? The Nahua, Incas, and Maya had produced highly sophisticated knowledge systems that in 1491 were superior in many respects to those of the Europeans. The Nahua effectively mined silver and gold, as indicated, and drained the swamps and then engineered the hanging gardens of the town that became Mexico City. Moreover, the Europeans had no way to project dates into BC eras, and no precise way to measure a solar year. The Nahua, Mayans, and Incas could do both. Amerindians also learned that they could locate their own calendars on the European Christian calendars; Aztec and Inca events could be celebrated to coincide with Christian events, unbeknownst to the Europeans. And there was more knowledge production in the realm of medicine, pharmacopeia, and botany.

Yet the "invading Europeans heaped up and burned books in New Spain and *quipus* in Peru, . . .whole 'libraries' of both, to use the Spaniards' own term" (Brotherston 2008: 25). The Europeans could not understand what was in them and saw them as products of dangerous infidel practices. The destruction was massive, and it caused great disorientation and chaos among the Amerindians, but European colonization was not able to eradicate completely indigenous knowledge forms. Today indigenous knowledges are being reconstructed and are experiencing a boom perhaps never seen since the conquest.

To conclude this section, philosophy of science and technology was produced in the Americas, along with the knowledge itself, for centuries before the Spanish and Portuguese arrived in 1492. Above are brief accounts of four distinctive characteristics of the encounters in the Americas that generated kinds of knowledge different from that produced by the British in India and Africa that has been designated "postcolonial" knowledge. These characteristics are first, the Americas were colonized more than two centuries before the British got to India, and this colonization co-constituted Europe's modernity. Second, the colonizers encountered needs for different kinds of physical, geographical, and astronomical knowledge in order to successfully colonize. Third, the colonizers themselves were different European peoples with different projects driving their colonization efforts. And finally, the indigenous peoples encountered in the Americas and the Far East were also different peoples, with different kinds of existing knowledge and different subsequent knowledge projects.

Decolonizing Gender: Contexts for Debates

Indigenous philosophies appeared dormant or invisible to the non-indigenous outsider until very recently and are still largely unknown to Anglo academics. Yet they have existed underground and persisted throughout the centuries inside indigenous communities. Today, indigenous peoples see as their task not only to reconstruct ancestral knowledges for their own survival but also for the survival of the rest of humanity that seems unable to halt the most pernicious aspects of modernity such as infinite economic growth, the destruction of the planet or "*Pachamama*" – the earth mother, and a modern science and technology at the service of profit and constant wars (Mendoza 2019). What is especially remarkable about this process is that for the first time in colonial history, indigenous women's voices can now be heard. Indigenous women had in the pre-intrusion era occupied important social and political positions that were undermined with colonization. Equally important was the place women occupied in indigenous cosmogonies and ontologies. These positioned women in a parallel, but not always equal position with men. It is this last point that not only defines the particularity of indigenous epistemologies, cosmogonies, and ontologies but also gives rise to one of the most contentious points in today's feminist debates around gender.

The following section is an account of these debates around gender that are led by indigenous women and mestiza/criolla feminists in the region. These debates have the potential of changing the terms of the Northern debates around gender. Yet they also provide evidence for the need to assess the degree to which contemporary indigenous knowledge production, led mainly by indigenous men, can accurately incorporate or give access to indigenous women's knowledge forms in the latter's philosophical as well as political projects.

These debates are not monolithic throughout the subcontinent. There are important differences between debates taking place in Mexico and Guatemala and those taking place in the Andean region. Mestiza/criolla feminists in Latin America are also paying much more attention to indigenous knowledge forms and are engaging in an unprecedented dialogue with indigenous women. Many mestiza/criolla feminists, such as Sylvia Marcos (2019) in Mexico and Silvia Rivera Cusicanqui (2004) in Bolivia, serve as mediators between the indigenous and mestiza/criolla cultures of knowledge. And yet, it is not only indigenous and mestiza/criolla feminists who are debating gender, Afro-descendant feminist voices from Brazil, Colombia, and other places are also debating gender and participating in the dialogues. What brings them together are a fatigue with ethnocentric gender ideologies and Western feminist theories, a project of reconstruction of ancestral knowledges, and a necessity to defend their territories and their bodies from the assaults of a new phase of conquest and colonialism by capitalist forces, both local and foreign.

It seems important therefore that we understand the context in which these debates occur before we delve into them. For the Mayan and Xinca feminists of Guatemala it is the genocide that took place between 1960 and 1996 and the subsequent occupation of their lands for projects of extractivism, the reorganization of the military into organized crime, and the upsurge of femicides that accompanies it that brought forward an epistemic uprising. For the indigenous peoples of Mexico, it was the NAFTA Treaty of 1994, the onslaught of neoliberalism, organized crime, and the extremely high rates of femicide that reawakened Zapatismo, creating one of the most important epistemic revolutions of our times. In the Andean region in places like Bolivia, the epistemic revolution has its roots in the 1952 revolution, which still showed strong influence from Marxism. It also has roots in the 1970s boom of "theological studies of the Andeans" that claimed a radical departure from the Western world (Alvizuri 2017), but gained precedence with the election of Evo Morales, the first indigenous president of Bolivia, and a so-called "Pink Tide" government. Afro-descendant feminists gained force in Colombia in particular when the decades long war against the FARC[6] started unwinding and a massive displacement of rural and Afro-descendant populations from their territories began to make room for national and transnational extractivist projects. In Brazil, Afro-Latin feminism took center stage under PT, Brazil's Labor Party governments of Lula da Silva and Dilma Rousseff which opened the doors of the universities to Afro-descendants, a population long excluded from mainstream institutions. Mestiza/criolla debates occur not only in response to indigenous or Afro-descendant debates. The phenomenon of femicide that has spread throughout the region that affects all women independently of race and ethnicity and the economy of extractivism are equally of great concern for mestiza/criolla feminists. They also want to make sense of the pandemonium of violence that has been unleashed against women in the region. They are forced to dig deeper into history for answers.

The search for the keys of history is what leads all these distinct groups to conclude that what women need is not equality or recognition, but decolonization and depatriarchalization. So instead of the slogan of Western feminists, "the personal is political," which exudes its connection to individualism and liberalism, their slogan is "there is no decolonization without depatriarchalization" (Galindo 2013). This reveals not only their awareness of the role that patriarchalization played in the process of colonization but also the treacherous effects of liberalism and individualism in contexts where communitarian ideals have been at the forefront. Indeed, many of these debates are conceptualized in terms of decolonization and the reconstitution of communality. One must note, however, that decolonization here has different sources that are not limited to the debates of the Decolonial feminism linked to the MCD group that is more known to Anglo academics, and must therefore not be conflated with them. Many of the debates around decolonization in Latin America have little or no connection to the MCD group. In fact, many of these debates often feel at odds with the MCD group which can be perceived as of Northern provenance and/or often not recognizing feminists' decade long research on coloniality (Cusicanqui 2012).

Since innovative gender debates take place in parts of the region other than the ones highlighted here, this section must be seen as a partial account of the gender debates in *Abya Yala*/Latin America. What follows is a sketch of some of these indigenous and mestiza/criolla debates, and not others such as Afro-descendant feminist debates. Also included is the work of Maria Lugones (2008). Lugones' work on the coloniality of gender, which was first published in the United States, has its locus of enunciation in the region and is in conversation with the gender debates referred to here. In particular, Lugones' differences with the work of Rita Segato (2003, 2016, 2018), who is based in Latin America, are highlighted at the end to mark the very complex distinct interpretations of gender and colonization that exist within the Latin American gender debates.

Gender and the Resurrection of Ancestral Knowledges

The political reorganization and advancement of indigenous peoples in *Abya Yala*/Latin America since the nineties began with a project of recovery and reinvention of ancestral knowledges. This has meant the resurrection of ancestral concepts, philosophies, and ways of life that had been forgotten, transformed, or destroyed both in the colonial process, and also recently with the genocide in Guatemala and the recolonization of indigenous territories. Indigenous people understand this process as a return to their cosmovisions and "*cosmovivencias*," or the ways the cosmos is lived by experience (Marcos 2019). In many ways, we can call this the return of the colonized.

As expected, the recuperation of ancestral knowledges is necessarily a contested terrain, not only between indigenous and non-indigenous peoples but also within the indigenous communities themselves. Unsurprisingly, the internal conflicts inside indigenous communities are often between indigenous men and women, and as such are a gendered process. Many of these conflicts get transferred into the feminist arena and provoke a feminist response from both indigenous and mestiza/criolla feminists. In this context, claims of essentialism and conservatism abound, especially in places where decolonization has been elevated to state policy, as in Bolivia.

The difficulty of recovering ancestral knowledges lies not only in their fragmented and dispersed state after centuries of colonization; they also have fused with Western, Christian elements which have altered not only the collective memory but also their existence in the present. It is not always clear what remains of the past and what is a recent invention. To complicate matters, the process of recuperation is often manipulated by present-day political interests of both indigenous and mestiza men, but also of women. Yet no matter how important it is to keep in mind these contradictions in the process of recuperation of ancestral knowledges, ancestral knowledges do pose serious challenges to Western totalitarian knowledge that sees itself as the only valid knowledge. It is perhaps in the discussions about gender where the disparities seem to be the greatest. Gender permeates the entire recomposition of indigenous cosmovisions.

The reconstitution of indigenous conceptions of gender has not been a task solely of indigenous women or indigenous feminists. Mestiza/criolla feminists have also been involved in the process, as well as indigenous men. Still there are others, such as Maria Lugones, who would dispute the existence of gender altogether in indigenous societies. She interprets this project of resurrection of an indigenous concept of gender as futile and even colonialist, since she considers gender a Western, colonial construct that has nothing to do with ancestral social arrangements. This is a point of contention that has not been resolved. Yet, it reveals how tempestuous the debates around gender can be.

Indigenous conceptions of gender in both the Mesoamerican and Andean regions are based on a cosmic vision of life that is entirely different from the West. Cartesian dichotomies that separate mind and body, humans and nature, nature and society are foreign in these cultures. In their cosmic vision, all of these elements are interdependent; they must maintain an equilibrium for a harmonious existence. There is a fluidity that runs through the earth, heavens, water, wind, and the humans and non-humans (Marcos 2019) that fuses them together. The cosmos is itself constituted by dualistic forces that are fluid, but not hierarchical as in Cartesian precepts, nor gendered. Thus, the feminine and masculine forces are complementary, of equal importance to the cosmos, and must maintain an equilibrium to guarantee the perpetuity of life.

Sociologically, this gendered division of the cosmos translates into gender complementarity, gender parallelism, or what the Aymara call *chachawarmi*. Man-woman constituted a unit of pairs. The married couple of man and woman were the basic unit of the community. Their work in tandem, although differentiated, was of equal worth. Women were not economically dependent on men. In gender parallel structures women constituted a lineage where inheritance was passed down to their daughters.

And yet, historically, we can see that elements of gender hierarchies were present. Gender differentiation increased as empire and state building advanced both among the Mexicas and the Incas (Kellog 1995; Silverblatt 1987). Men as soldiers and warriors had a public face that women lacked. Men were the representatives of the community before the ruler. While noble women had class privilege and could occasionally occupy positions of power, the highest positions of power were still reserved for men. War, although understood to be as important as women's power of child birthing, constituted the center of power of indigenous *real politik*.

But it is the elements of complementarity, parallelism, and reciprocity between the genders that many indigenous men and women and their mestiza/criolla allies want to claim as either still existent or in need of resurrection. This position encounters many criticisms.

First is the fact that this gender regime did not survive colonialism intact. Colonialism itself involved a social pact between colonized and colonizer men based on the acceptance of the subordination of women to men in exchange for a limited access for colonized men to power inside the community. Indigenous men while emasculated in the public sphere were granted the control of women, children, and the elderly in the household and the community. These gender colonial norms have in time installed gender violence, something that was unknown to them in the era of pre-intrusion. As Argentinean anthropologist Rita Segato has maintained, the separation of the public and private spheres not only privatized and minoritized indigenous women; it had lethal consequences for them (Segato 2016). More recent experiences of genocide, such as the one in Guatemala where the state forced indigenous men to rape, kill, and mutilate indigenous women, have increased violence against indigenous women dramatically, and thereby led to some of the highest femicide rates in the world.

The presence of hierarchical elements in the gender imaginaries of indigenous societies and present-day gender violence has led some Mayan and Aymara women to resist the essentialization and ethnic fundamentalism that they perceive motivates the resurrection of gender complementarity or *chachawarmi*, in particular among indigenous men. Lorena Cabnal, a Xinca communitarian feminist scholar from Guatemala, compels us to question indigenous heteronormative philosophies that in her view build and sacralize a cosmogenic heteroreality that is oppressive to women and gender non-conforming persons. From her perspective, there is such a thing as ancestral patriarchies, based as are most patriarchies on the principles of power and war. Under colonization, ancestral patriarchies were defeated, but they were re-functionalized and became more oppressive for indigenous women. Indigenous women must, therefore, not only strive for cosmogenic equity and the end of ancestral patriarchies; they must also fight against Western patriarchies (Cabnal 2010). Communitarian feminists in Bolivia say something similar and speak of the juncture of patriarchies. Julieta Paredes, an Aymara communitarian feminist who coined the term, appeals to a longer memory and to the acceptance of the fact that indigenous histories and patriarchies do not begin with European colonization (Paredes 2010). However, Aura Cumes, a Maya-*kaqchikel* anthropologist, does not agree with the notion of the juncture of patriarchies of communitarian feminists. According to her, patriarchy as we know it is based on the exclusive authority of the white man from which indigenous men are not only excluded but, like indigenous women, are also oppressed. Based on an analysis of the *Popol Vuh*, which is a book of origins transcribed into Spanish by three *Kiche'* men between 1554 and 1558, 25 years after the conquest, Cumes reconstructs Mayan notions of gender complementarity. She recognizes tenuous elements of hierarchy between men and women in them and admits that women may have had parity with the men at the symbolic level. However, at the level of human society, women had a secondary position, and this had the potential of constituting a patriarchy (Cumes 2019). Yet this was never developed fully because it was interrupted by the European intrusion. So, it is not a full-blown ancestral patriarchy that the Spaniards encounter, but one that is in the process of development, and that becomes hardened with the introduction of European conceptions of gender binarism.

Mestiza/criolla interventions into these debates have been varied. Mexican anthropologist of religions and sociologist Sylvia Marcos (2019), for instance, has a much more benevolent view of indigenous notions of gender. She is not so much preoccupied with the existence of ancestral patriarchies as she is with the potentiality for indigenous cosmovision to accomplish an epistemic rupture with Western philosophy. As a companion of Zapatista indigenous women's struggles, Marcos prefers to "walk with them" in their fight for gender justice within their communities in the present, based on ancestral notions of gender complementarity. Marcos, as well as Pérez Moreno, a *Tseltal* indigenous scholar, is focused on the recovery of those parts of the culture that, as Pérez Moreno says, refuse to die (Pérez Moreno 2019). In so doing they rescue women's ways of connecting spirituality and knowledge. Marcos speaks of a form of incarnated spirituality that should not be confused with religion. Incarnated spirituality derives from indigenous cosmic visions in which everything – nature, land, bodies – are so fused that spirituality is experienced in the body. One constructs knowledge from the body as fused with the cosmos. Spirituality as a bodily experience becomes thus the basis of knowledge and worship. In fact, "prayer" or knowledge is expressed by the bodily movements of dance. "We pray dancing" they say (Marcos 2019: 121). As spirituality is experienced in the body, it does not need the priest as mediator nor a masculine god that is omnipotent. In fact, the masculine God should be conceived as a colonial concept (ibid.: 120).

Pérez Moreno alludes to other practices of indigenous women that rupture Western binary thinking of the genders such as the conception of the unity of *abuelas/abuelos*. Here the sexes are a unity and can take either shape. She focuses on an indigenous philosophy which is based on the heart, or the *o'tanil,* as the center of thought (Pérez Moreno 2019). So, it is not the head from which thoughts emanate, but the heart. The heart is not just an organ; it is the site of thought, feeling, emotions, dreams, and knowledge. All living beings have a heart and are interdependent. It is from the heart that one is one with the world and the cosmos; one thinks, feels, and acts with the heart. Thinking and feeling are not separated. We do not just think; we "*senti-pensamos*," feel-thinking. In this respect, Western knowledge "breaks hearts"; it is divorced from emotions and the sacred and is based on the coldness and separation of the mind, of reason, from the world and the cosmos.

Segato vs. Lugones

It is worth considering the dispute between feminist scholars Rita Segato and the late Maria Lugones, both of whom are mestiza/criolla and Argentinean. Segato, like Marcos, has a long history of working with indigenous peoples, and she is central to the theorizing of femicide. In the case of Segato, her work has been centered on the Yoruba of Brazil.

Segato is, however, not a supporter of the idea of a pre-colonial past that is free of patriarchies. According to her, all human societies have functioned with the mandate of masculinity, which requires the subordination of women. Resonating with ideas of Western radical feminism, for Segato patriarchy is universal; it is "the cornerstone and center of gravity of all forms of power" (Segato 2016, 2018). Indigenous societies were communitarian and collectivist, and each gender had its own ontology. Men in indigenous cultures were not a universal reference; and in fact, men and women could transit from one gender to another with impunity. But this did not preclude hierarchy. Prestige was not equally distributed; it was distributed according to patriarchal principles. Yet the ancestral patriarchies were of low intensity in comparison to the high intensity patriarchies of Europeans.

It was colonialism that introduced gender binarism and destroyed what she calls the village world. Colonialism separated the public and private spheres, which was unknown by indigenous societies, and it domesticated, privatized, and expelled women from the political world. It also introduced sexual violence. This created an explosive combination of patriarchies that has been lethal for indigenous women. Sexual violence became not only a *modus operandi* of the state but also masked as private the violence that it had introduced in state affairs. As we can gather, Segato's position is congruent with indigenous communitarian feminists who also envision a juncture of patriarchies.

Maria Lugones disagrees. The non-existence of gender binarism in indigenous societies leads her to think that whatever form of power differentials existed between men and women, it cannot be described with the imaginary of gender. Not even after the colonial imposition of gender binarism does she see the presence of gender in colonial societies. This is because, in her view, the gender binarism only applied to those deemed human and civilized – that is, European men and women. Coloniality did not engender indigenous societies; instead it dehumanized and bestialized both men and women (Lugones 2008). This is why it is inappropriate to look for gender where there is none. So here we have a conception of gender that is encapsulated in the European cultural and historical relations between men and women. This is a conception that is simply further expanded in the development of colonialism and capitalism, but that was totally unknown to indigenous cultures. To understand, then, power differentials between men and women in indigenous societies, we would have to come up with another concept. Yet, many indigenous women today respond "We do have gender!"

Conclusion

Debates about gender might have seemed unlikely sites of innovative philosophies of science. Yet in the Latin American/*Abya ayalense* feminist challenges to the standard modern Western ontological and epistemological tendencies, one can see the parochiality of the latter, still rooted in their imperial histories. Long after formal colonial rule in the Americas has ended, its residues and reinventions persist. The new debates around gender led by indigenous women and mestiza/criolla feminists in Latin America give us the opportunity to engage in a dialogue about how reliable knowledge should be produced that perhaps should have occurred 500 years ago!

Related chapters: 7, 9.

Dedication

We want to dedicate this article to our dear friend, the late Margie Waller. Margie read the first versions of this article and gave us invaluable feedback. She was unable to see the finished version. Margie this is for you!

Notes

1 For helpful comments on this manuscript we thank Jane Bayes, Jane Jaquette, Shu-mei Shih and the late Margie Waller, and the editors of this volume. The errors that remain are our own.
2 By "institutions," we refer to libraries and their book-purchasing and journal subscription policies; universities and their standards for hiring and promotion, PhD admissions, and doctoral topic standards; disciplines and their selections for prizes and conference plenaries; journal and book publishers' and their criteria for publishable topics.
3 The first colonial universities were founded in Santo Domingo, Lima, and Mexico City as early as 1538 and 1551.
4 An expanded version of section "The Decolonial beyond the Postcolonial" may be found in Harding (2018). See also Harding (2017), Morana, Dussel, and Jauregui (2008), and Rivera Cusicanqui and Bragan (1997). We capitalize "Decolonial" here to distinguish this group of MCD Latin American theorists from the more general anti-colonial projects described in section "Decolonizing Gender: Contexts for Debates."
5 However, some would argue that Puerto Rico in fact remains a colony of the United States: residents of Puerto Rico are residents of the United States, but they are not citizens of the United States. They cannot vote in national elections (Noriega, 2020).
6 FARC is Spanish for Revolutionary Armed Forces of Colombia – People's Army, a guerrilla movement involved in Colombia's internal conflict between 1964 and 2017.

References

Alcoff, L. (2006) *Visible Identities: Race, Gender and the Self*. Oxford: Oxford University Press.
———. (2015) *The Future of Whiteness*. Cambridge: Polity Press.
Alvarez, S. et al. (2002) "Encountering Latin American and Caribbean Feminisms," *Signs* 28(2): 537–579.
Alvizuri, V. (2017) "Indianismo, Política, Religión en Bolivia (2006–2016)," *Caravelle* 108: 83–98.
Anzaldua, G. (1987) *Borderlands/La Frontera*. San Francisco: Spinsters/Aunt Lute.
Brotherston, G. (2008) "America and the Colonizer Question: Two Formative Statements from Early Mexico," in Moraña, M., Dussel, E. and Jáuregui, C. A. (eds.) *Coloniality at Large: Latin America and the Postcolonial Debate*. Durham: Duke University Press: 23–42.
Cabnal, L. (2010) "Acercamiento a la Construcción de la Propuesta de Pensamiento Epistémico de las Mujeres Indígenas Feministas de Abya Yala," retrieved from https://porunavidavivible.files.wordpress.com/2012/09/feminismos-comunitario-lorena-cabnal.pdf
Cañizares-Esguerra, J. (2006) *Nature, Empire and Nation: Explorations of the History of Science in the Iberian World*. Stanford: Stanford University Press.
Cumes, A. (2019) "Cosmovisión Maya y Patriarcado: Una Aproximación en Clave Crítica," in Muñoz, K. O. (ed.) *Miradas en Torno al Problema Colonial*. Mexico: Akal: 73–89.
Cusicanqui, S. R. (2012) "Ch'ixinakax utxiwa: A Reflection on the Practices and Discourses of Decolonization," *South Atlantic Quarterly* 111(1): 95–109.
De la Cruz, S. J. I. (2009) *The Answer/La Respuesta* (ed. and trans. E. Arenal and A. Powell), New York: Feminist Press at the City University of New York.
Denevan, W. M. (1992) "The Pristine Myth: The Landscape of the Americas in 1492," *Annals of the Association of American Geographers* 82(3): 369–385.

Dussel, E. (1995) *The Invention of the Americas: Eclipse of 'the Other' and the Myth of Modernity*. (trans. Michael D. Barber). New York: Continuum.

Femenias, M. L. and Oliver, A. (eds.) (2007) *Feminist Philosophy in Latin America and Spain*. New York: Rodopi: 197–207.

Galindo, M. (2013) *No se Puede Descolonizar, sin Despatriarcalizar*. La Paz, Bolivia: Mujeres Creando.

Gargallo, F. (2004) *Las Ideas Feministas Latinoamericanas*. Mexico City: Universidad de la Ciudad de Mexico.

Giraldo, I. (2016) "Coloniality at Work: Decolonial Critique and the Postfeminist Regime," *Feminist Theory* 17(2): 157–173.

Goodale, M. (2008) *Dilemmas of Modernity: Bolivian Encounters with Law and Liberalism*. Stanford: Stanford University Press.

Guamán Poma de Ayala, F. (1615) *El Primer Nueva Coronica y Buen Gobierno (The First New Chronicle and Good Government)*. John Murra and Rolena Adorno (eds.) Mexico City: Siglo XXI.

Harding, S. (2017) "Latin American Decolonial Studies: Feminist Issues," *Feminist Studies* 43(3): 624–636.

———. (2018) "State of the Field: Latin American Decolonial Philosophies of Science," *Studies in History and Philosophy of Science* 78: 48–63 https://doi.org/10.1016/j.shpsa.2018.10.001.

Hecht, S. (2004) "Indigenous Soil Management and the Creation of Amazonian Dark Earths: Implications of Kayapo Practices," *Amazonian Dark Earths: Origins, Properties and Management of Fertile Soils in the Humid Tropics*. Dordrecht: Kluwer: 355–373.

Hess, D. J. (2011) "Science in an Era of Globalization: Alternative Pathways," in Harding, S. (ed.) *The Postcolonial Science and Technology Studies Reader*. Durham: Duke University Press.

Kellog, S. (1995) "The Woman's Room: Some Aspects of Gender Relations in Tenochitlan in the Late Pre-Hispanic Period," *Ethnohistory* 42(4): 563–576 (Women, Power and Resistance in Colonial Mesoamerica).

Lugones, M. (2008) "The Coloniality of Gender," *Worlds and Knowledges Otherwise* Spring 2(2): 1–17.

Mann, C. C. (2005). *1491: New Revelations of the Americas before Colombus*. New York: Knopf.

Marcos, S. (2019) "Espiritualidad Indígena y Feminismos Descoloniales," in Muñoz, K. O. (ed.) *Miradas en torno al Problema Colonial*. Mexico: Akal: 119–134.

Mendoza, B. (2007, Fall) "Juxtaposing Lives: Mary Wollstonecraft and Sor Juana Inés de la Cruz," *Women's Studies Quarterly* 35: 287–291.

———. (2012) "The Geopolitics of Political Science and Gender Studies in Latin America," in Bayes, J. (ed.) *Gender and Politics: The State of the Discipline*. Berlin, Germany: Barbara Burdich Publishers.

———. (2019) "Can the Subaltern Save Us?" *Tapuya: Latin American Science, Technology and Society* 1(1): 109–122.

Mignolo, W. D. (2011) *The Darker Side of Western Modernity: Global Futures, Decolonial Options*. Durham: Duke University Press.

Miller, F. (1991) *Latin American Women and the Search for Social Justice*. Chicago: University Press of New England.

Morana, M., Dussel, E. and Jauregui, C. A. (eds.) (2008) *Coloniality at Large: Latin America and the Postcolonial Debate*. Durham: Duke University Press.

Noriega, C. (2020) "Not Inconceivable: Knowledge-Production, the Arts, and the Pre-History of a Puerto Rican Artist, 1934–1882" (sic), *Tapuya: Latin American Science, Technology and Society*. https://doi.org/10.1080/25729861.2020.1763687.

Nuccetelli, S. and Seay, G. (2003) *Latin American Philosophy: An Introduction with Readings*. Upper Saddle River: Pearson.

Nuccetelli, S., Schutte, O. and Bueno, O. (eds.) (2010) *A Companion to Latin American Philosophy*. Malden: Blackwell Publishing.

Oliver, A. A (1998) "Feminist Thought in Latin America," in *Routledge Encyclopedia of Philosophy*. Oxford: Taylor and Francis. doi:10.4324/9780415249126-ZA007-1

Paredes, J. (2010) *Hilando Fino Desde el Feminismo Comunitario*. Querétaro: Mexico: Grietas.

Pérez Moreno, M. P. (2019) "*O'tanil*: Corazón. Una Sabiduría y Práctica de Sentir, Pensar, Entender, Explicat y Vivir el Mundos Dedse los Mayeas Tseltales de Bachajón, Chiapas, Mexico," in Muñoz, K. O. (eds.) *Miradas en Torno al Problema Colonial*. Mexico: Akal: 157–173.

Pratt, M. L. (2008) *Imperial Eyes: Travel Writing and Transculturation*. New York: Routledge.

Rivera Berruz, S. (2018, Winter) "Latin American Feminism," in Zalta, E. N. (ed.) *The Stanford Encyclopedia of Philosophy*, https://plato.stanford.edu/archives/win2018/entries/feminism-latin-america/

Saldana, J. J. (ed.) (2006) *Science in Latin America: A History* (trans. Bernabe Madrigal). Austin: University of Texas Press.

Schutte, O. (1994) "Special Cluster on Spanish and Latin American Feminist Philosophy," *Hypatia* 9(1): 147–194.

———. (1998a) "Cultural Alterity: Cross-Cultural Communication and Feminist Thought in North-South Dialogue," *Hypatia* 13(2): 53–72.

———. (1998b) "Latin America," in Jaggar, A. M. and Young, I. M. (eds.) *A Companion to Feminist Philosophy*. Malden: Blackwell: 87–95.

Schutte, O. and Femenias, M. L. (2013) "Feminist Philosophy," in Nuccetelli, S., Schutte, O. and Bueno, O. (eds.) *A Companion to Latin American Philosophy*. New York: Wiley-Blackwell: 397–411.

Segato R. L. (2003) *Las Estructuras Elementales de la Violencia*. Buenos Aires: Prometeo and Universidad Nacional de Quilmes.

———. (2016) "Patriarchy form Margin to Center: Discipline, Territoriality, and Cruelty in the Apocalyptical Phase of Capital," *South Atlantic Quarterly* 115(3): 615–624.

———. (2018) "A Manifesto in Four Themes," *Critical Times* 1: 198–211.

Silverblatt, I. (1987) *Moon, Sun and Witches*. Princeton: Princeton University Press

Sousa, L. (2017) *The Woman Who Turned Into a Jaguar, and Other Narratives of Native Women from Archives of Colonial Mexico*. Stanford: Stanford University Press.

Stehn, A. (2013) "From Positivism to Anti-Positivism: Some Notable Continuities," in Gilson, G. D. and Levinson, I. W. (eds.) *Latin American Positivism: New Historical and Philosophical Essays*. New York: Lexington Books.

Todorov, T. (1984) *The Conquest of America* (trans. R. Howard). New York: Harper and Row.

9

SCIENCES OF CONSENT

Indigenous Knowledge, Governance Value, and Responsibility

Kyle Whyte

Introduction

Peoples' right to consent/dissent to actions of others that affect them is a major norm in the ethics of science. The right to free, prior, and informed consent (FPIC) is one example. Specifically, science refers to some practice authorized as research and sponsored by anglophone universities or non-profit organizations, agencies of industrial nations like Canada, or multinational corporations. Consent—i.e. the right to consent/dissent—is used as a norm that restricts research to reduce risks and prevent harm. The consequences of negating consent are appalling. Research and science-policy practices have allowed Indigenous persons, including children, to be malnourished (Mosby 2013), tortured (George 2019), and poisoned (Brugge et al. 2006; Hoover 2017). Far too often the outcomes were lethal.

As a restriction, consent may be perceived as a norm that is not directly related to the procedures of empirical inquiry. Procedures understood by some as *research itself*, including data gathering and analysis, can be conceived separately from consent. Then consent must be applied later to mitigate risk and harm, including disallowing certain research projects. Empirical inquiry is cordoned off from being considered a consent-based practice. Instead, science is mainly to be regulated by consent when its procedures are risky or harmful. It is sometimes the case that scientists' identities as researchers are not tied to a sense of responsibility to honor consent beyond their the norm's restrictive or regulatory functions.

Some scientific practices and philosophies of science do *not* treat consent *only* as a restriction and regulation. Feminist philosophers have articulated philosophies of science that center diversity in design and procedures of empirical research, including feminist standpoint and feminist empiricist theories (Harding 1995; Intemann 2010; Longino 1990). These theories do not always invoke the specific words *consent* or *dissent*. Yet the theories do typically emphasize that the quality of empirical inquiry improves when its procedures involve wide ranging human perspectives and knowledge. The perspectives and knowledge bring to bear different life experiences, generations, genders, cultures, abilities, and epistemological systems.

Such diversity of perspectives and knowledge increases the degree to which procedures of empirical inquiry are endorsable and revisable by multiple constituencies of humans. To me, this last point suggests a profound commitment to consent in terms of consensus in data gathering and analysis. On the flipside, such diversity increases the opportunities for and the acceptability of constructive criticism (i.e. dissent) from multiple voices. In other areas of feminist philosophy of science and science studies, scholars have theorized the significance of consent by showing how humans are not the only self-determining agents in procedures of empirical inquiry (Barad 2007; Haraway 1997). Plants, insects, animals, ecosystems, among others, have their own expressiveness, are unpredictable, and can act independently in ways that substantially impact procedures like data gathering and analysis.

The subject I will focus on here is that some Indigenous scientific traditions understand consent as involved in every part of empirical inquiry. This essay attempts to dialogue with feminist philosophy science by addressing a shared philosophical issue—science and consent. But I will do so by writing from a standpoint grounded in Indigenous philosophy of science. That is, I will not write about *Indigenous content* by interpreting it through concepts developed by scholars in different philosophical fields. In Indigenous philosophy of science, science means some practice authorized as research and sponsored by institutions emanating from Indigenous peoples, where these institutions range from families, to ceremonial lodges and societies, to First Nations government agencies, to Tribal colleges and enterprises. In future writing by myself and others, I hope there is more circling back to compare side by side some of the aforementioned feminist scholarship with what I'm about to share in the essay.

In my experience, practitioners of Indigenous philosophy of science are not terribly interested in understanding how to improve the objectivity of research claims or how to explain normatively how science might work best. In my own reflection, I see Indigenous philosophy of science as first focused on understanding the *governance value of science*. That is, science or knowledge refers to a set of actions taken by diverse agents, not exclusively humans. The actions aim to exercise responsibilities that improve the relationships of interdependence within collectives of diverse beings and entities who must be responsive together to the constantly changing world around them.

Consent is not mainly articulated as a right. Rather, it is an integral quality of the responsibilities that diverse agents exercise in scientific practices. Consent means that responsibilities are acted out in ways that are accountable to the animacy of diverse beings and entities that makes the world a place that presents itself through unending motion. Here, consent is not a restriction or regulation on science, nor is consent a characteristic that increases the objectivity or rigor of empirical inquiry. Quite differently, it would not be possible for anyone to authorize a practice *as research* unless it is consensual. For non-consensual practices are irresponsible, and some Indigenous scientific traditions do not separate—even for analytic purposes—responsibility from empirical inquiry. The identities of Indigenous scientists, knowledge-keepers, learners, and culture-bearers honor consent as an integral quality of how they exercise their responsibilities across every scientific activity.

In this essay, I will first cover a little bit more information about what I am choosing to call Indigenous philosophy of science and the governance value of science. I will then move to discuss consent and animacy through three entry points into Indigenous philosophy of science: a philosophical tradition, an individual philosopher, and a case. In this analysis, I will both draw on Indigenous traditions I am more personally associated with (Anishinaabe ones) and contribute to inter-Indigenous dialogue across traditions that I engage with through reading or coalitional advocacy. While there isn't really such an accepted English language phrase as *Indigenous philosophy of science*, I am using it to denote the numerous efforts out there in the world to creatively and critically reflect on Indigenous scientific traditions—most often undertaken by Indigenous persons. Usually the word *science* isn't really used, and specific Indigenous language words are used instead, or English language words are used like *traditional ecological knowledge* (TEK) or *Indigenous knowledge*.

Indigenous Philosophy of Science

I am routinely asked by philosophers of science for cases of Indigenous peoples engaging with various sciences. Sometimes my colleagues' interest stems from their hopes of finding examples that support their particular theories of how values and science are related in research, funding, or policy. Perhaps a particular Tribe's struggle with a university doing genetic sampling reveals racism in that very research paradigm; perhaps an Indigenous community organization's grass-roots testing of pollution exposure represents an instance of democratic science; perhaps the TEK of a group of Native elders offers more refined environmental knowledge than the research of climate scientists.

I have no doubt that cases of Indigenous peoples' engagement with science are relevant for the lines of inquiry of philosophers of science working in anglophone institutions of research and higher education. At the same time, there are entire worlds of Indigenous scientific traditions that are practiced in their own right throughout the planet. Indigenous sciences are grounded in particular philosophical assumptions. Indigenous persons also philosophize about these sciences.

Indigenous philosophies of science have not had to struggle with certain scientific issues in quite the same ways as philosophies of science have had to in anglophone institutions of research and higher education. Few Indigenous scientific traditions, for example, have ever been founded philosophically on cultural aspirations to objectivity or value-free inquiry privileged in some European and European diasporic societies. In some places, Indigenous sciences do not have histories in which research practices are rooted in patriarchy, anthropocentrism, ableism, homophobia, and racism.

There is a difference between (A) Indigenous case studies used within non-Indigenous philosophies of science and (B) Indigenous philosophies of science in their own right. This essay is about (B). Regarding (B), *non*-Indigenous cases certainly figure as part of the content of Indigenous philosophies of science. For example, Indigenous scholarship has analyzed certain sciences in the United States or Canada as cases of non-kinship-based knowledge systems that have little respect for consent, trust, or reciprocity (Battiste 2000; Pihama et al. 2002; Smith 1999; TallBear 2013). Indigenous scholarship has also shown how certain environmental sciences, such as ecology, reveal alternative accounts of systems of interdependence that can be put in dialogue with Indigenous science under conditions determined by Indigenous peoples to be safe (Ford and Martinez 2000; Kimmerer 2013; Shilling and Nelson 2018). In this spirit, this essay is about philosophies of science with starting points that do not begin with the analysis of scientific traditions emanating from European and European diasporic societies.

Governance Value of Science

The governance value of science is one of the concepts I have used to make sense of what appears to me as a different starting point for some of the Indigenous philosophy of science with which I'm most familiar. I use *governance* for thinking about the *coordination* of the interdependent mesh of relationships that connect different beings and entities together. The relationships make up a collective. The collective is some group of related members, whether described as a community, network, society, or nation. The members are beings or entities that have diverse intentions, motivations, goals, types of awareness, agency, and learning styles. In some Indigenous philosophies, the members include flows (e.g. water, wind), fishes, insects, mountains, landscapes, animals, forests, and plants, among others.

The members are interdependent because their behavior impacts one another systematically. For example, water temperature affects fish health; fish health affects animal, bird, and human nutrition; human economies affect water temperature through land use; and this example can keep expanding. Depending on the tradition, they are also interdependent because their relationships are determinative of who they are to one another. Many Indigenous persons I know readily acknowledge that depending on the perspective, a plant, flow, animal, landscape, or human is a very different being or entity depending on who or what is perceiving it.

When collectives are *coordinated*, it means that the members' behavior tends to be feasibly conducive to the well-being of all members. Here, *well-being* is not to be taken statically. For in any collective, well-being can be multifaceted, as in the case of humans where it involves psychological, economic, cultural, and moral aspects. Or the meaning of well-being is unknown and contested, as in the steep learning curve for empathizing with the experiences and suffering of animals. The concept of governance refers to a normative field where humans try to figure out what responsibilities they have for performing their part to maintain or promote coordination in a collective.

Responsibilities are actions that are taken with the explicit awareness that the actor intends to contribute to coordination. Responsibilities are better able to serve their intended contributions if they have certain qualities attached to them. Qualities are features of responsibilities that—when present— facilitate their contributory capacity. They include consent, trust, reciprocity, and transparency, among others. In the traditions I am going to discuss here, coordination is always needed because the world itself—from the environment, to culture, to animals—is animate. That is, the beings and entities in the world are all constantly in motion, which means there is complexity, unpredictability, and perpetual change.

Governance, on this understanding, is also a field seeking to understand how qualities of responsibilities bear on maintaining or promoting coordination. In many Indigenous traditions, science is inseparable from its value for governance—that is, the governance value of science. Consider Robin Kimmerer's Indigenous description of empirical inquiry:

> Because we can't speak the same language, our work as scientists is to piece the story together as best we can. We can't ask the salmon directly what they need, so we ask them with experiments and listen carefully to the answers. We stay up half the night at the microscope looking at the annual rings in fish ear bones in order to know how the fish react to water temperatures. So we can fix it. . . We measure and record and analyze in ways that might seem lifeless but to us are the conduits to understanding the inscrutable lives of species not our own. Doing science with awe and humility is a powerful act of reciprocity with the more-than-human world.
>
> *(Kimmerer 2013: 252)*

Using examples of fish and water temperature, Kimmerer understands the procedures of empirical inquiry as exercises of responsibilities. The responsibilities facilitate the coordination of interdependent members of a collective. Beings like salmon are animate (not "lifeless"), as are all the other beings and entities they relate to. The responsibilities are carried out with qualities of consent and reciprocity. Consent involves careful listening, and respect for humans' inability to communicate clearly with fish. Reciprocity involves doing research in good faith since fish support human life and the many other lives that humans depend on (and that, in turn, depend on humans).

I will focus on the quality of consent. Consent has diverse formal and informal meanings that express an overall honoring of the self-determination of any being or entity that is a member of a collective. Honoring self-determination means respecting the animacy or unique agencies of all beings and entities. It means ensuring consensus across beings and entities affected by similar issues—especially where cooperative coordination can improve everyone's situation. Consensus requires that communication across diverse beings and entities is attentive to their languages, behaviors, and needs as best they can be understood or empathized with in good faith.

Within some Indigenous scientific traditions, there are not really tradeoffs that are associated with acquiring knowledge while neglecting responsibility. Deborah McGregor writes that

> Native understandings of [traditional ecological knowledge] tend to focus on relationships between knowledge, people, and all of Creation. TEK is viewed as the process of participating

(a verb) fully and responsibly in such relationships, rather than specifically as the knowledge gained from such experiences. For Aboriginal people, TEK is not just about understanding relationships, it is the relationship with Creation. TEK is something one does.

(McGregor 2008: 145)

For McGregor, there is not really such a thing as responsible or irresponsible science. Without responsibility there can be no science. Science can be philosophized about in relation to its governance value.

The governance *value* of knowledge systems, when taken as a starting point for philosophizing about science, has implications for our historical and thematic orientations. Indeed, Indigenous science already starts with assumptions that topics like consent or animacy are key beginnings for investigating the philosophical challenges of science. In what follows, I share some examples of how consent and animacy may be interpreted within diverse areas of Indigenous scientific traditions and philosophies of science, where governance is a significant value of science.

Anishinaabe Science

Late grandmother and knowledge keeper Josephine Mandamin led the Mother Earth Water Walk as a movement seeking to protect water quality in the Great Lakes and North America by honoring Indigenous knowledge of water. Mandamin spoke to the responsibility that "people of all walks of life" have to water at the same time she emphasized traditions of Anishinaabe women's responsibilities to water. Mandamin spoke about our interdependence and responsibility relating to water. "We have to think of our relatives. . . At a time when we needed them, they were there for us. Take care of them, like they took care of us" (Mandamin 2015: xix). As a Potawatomi person, the philosophy of Anishinaabe persons, with special respect paid to Mandamin, has influenced how I think about concepts of interdependence, coordination, responsibility, consent, and animacy as parts of governance value.

McGregor, in their work with Mandamin, discusses the nature of knowledge in the Mother Earth Water Walk.

> We must look at the life that water supports (plants / medicines, animals, people, birds, etc.) and the life that supports water (e.g., the earth, the rain, the fish). Water has a role and a responsibility to fulfill, just as people do. We do not have the right to interfere with water's duties to the rest of Creation. Indigenous knowledge tells us. . . that water itself is considered a living entity. . .
>
> *(McGregor 2009: 37–38)*

For McGregor, in light of Mandamin's work, knowledge or science is described as disclosing the animacy ("living entity") of water. It is through animacy that the critical coordination of relationships is protected, that is, governance.

There is an important principle of consent at work in this account of knowledge. It is about non-interference of water—but so that water can exercise responsibilities it has for others and in coordination with others in a collective of interdependent members. For animacy is also about a being or entities' being able to self-determine their own actions for the sake of contributing to others' well-being. It is not about brute non-interference. Beings and entities, whether considered as human or non-human, are supposed to protect the conditions of their being able to express consent. Knowledge, or science, relating to water, is inseparable from responsibilities laden with this quality of consent.

Concepts in Anishinaabe science that are tied to science often speak to animacy and consensuality. Margaret Noodin discusses this in relation to *ganawendamaw*, which is a verb that can be translated into something like "sustainability" and the knowledge and practices associated with sustainability. She defines the verb as referring to the "spectrum of animacy for all life, allowing rocks, water, and humans to be described as coequal partners in the creation, maintenance, and evolution of a place. This basic concept conveys the idea of observation, protection, and preservation" (Noodin 2017: 247). For Noodin, scientific conceptions like observation are tied to animacy and "coequal" partnerships that promote coordination ("maintenance, and evolution of a place") through the self-determination/consent of all members of the collective.

Christine Sy's work on Indigenous knowledge and land shares Noodin's focus on animacy and its relevance to consent. Based on her interviews, she writes that "Anishinaabe language theorist Helen Roy Fuhst, translating the word *aki*, contends that land is everything physical—the earth and universe that we live in—and that this physicality is animated with constant motion and movement." Sy goes on to state, paraphrasing Roy Fuhst, that "Seeing relationships as animated through persons, inclusivity, reciprocity, and mutual reciprocity must shape then how humans carry out their relationship with land" (Sy 2018: 227–228).

Citing James Dumont, Sy describes four principles of relationships:

> First, all beings are persons (e.g., human, tree), and therefore all relationships are personal. Second, all relationships are inclusive. Third, all relationships are familial (i.e. kinship, relatives). And fourth, relationship is reciprocal and mutually reciprocal, meaning that relationships are maneuvered through a back and forth between two or more persons, and each person in a relationship has the ability to enact their volition.
>
> *(Sy 2018: 227–228)*

For Sy, in their work with elders, this conveys a theory of knowledge relating to land that is at heart about interdependence, animacy, and consent ("enact their volition"). Similar to Kimmerer, Sy references other qualities beyond consent, like reciprocity, which I am not focusing on here in this essay.

Basil Johnston describes an Anishinaabe story that discusses one of the origins of humans. In the story, the animals invite humans to the earth, and it is clear at that time that there are relationships of interdependence: "Without the animals the world would not have been; without the animals the world would not be intelligible" (Johnston 1990: 49). Animals provided nourishment, "shelter," "joy," and voluntary "labor" on behalf of humans. Humans and animals could understand each other communicatively. Yet humans abused the consent of the animals by harshly treating them as if humans were owed these services, which animals had previously performed "without complaint." Humans were irresponsible, failing to perform actions that are feasibly conducive to the coordination of the collective.

Johnston writes that

> At last, weary of service, the animals convened a great meeting to gain their freedom. All came at the invitation of the courier. The bear was chosen to be the first speaker and to act as chairman of the session. He explained the purpose of the meeting. 'We are met to decide our destiny. We have been oppressed far too long by [humans]. He has taken our generosity and repaid us with ingratitude; he has taken our labors and repaid us with servitude; he has taken our friendship and fostered enmity among us'.
>
> *(1990: 50)*

One of the lessons of the story then is to not abuse consent, as such abuse can ruin the coordination of interdependent relationships. In fact it is really animals who have the knowledge. So science is more like humans acting responsibly to animals, including honoring animals' animacy through respecting consent, in order to benefit from their knowledge.

I see these examples of Anishinaabe philosophy as critical reflections on science. They are part of what I have no problem calling a philosophy of science. In them, governance value is a major starting point. For the examples are about how knowledge relates to the coordination of interdependent members of a collective, and in ways where responsibility is tied to qualities—like consent—in an animate, constantly changing world.

Jeanette Armstrong: Salish Science

Individual Indigenous persons globally have offered detailed philosophies of science across their publicly available bodies of work. Jeanette Armstrong offers significant articulations of such traditions in Salish philosophy. This section considers their contributions as a philosopher, culture-bearer, institution-builder, knowledge-keeper, facilitator, creative writer, and leader from the Okanagan Valley in what is currently called British Columbia, Canada. While I have benefited greatly from learning from Salish persons over the years, I approach the interpretation of Armstrong in the spirit of dialogue, not the spirit of being knowledgeable according to the criteria of that tradition.

For philosophers like Armstrong, the intellectual traditions about knowledge and science have multiple origins, and here I will focus on just a few. Knowledge arises from systems of governance, and their political philosophies, that were designed to be coordinated to seasonal and interannual environmental changes. Knowledge was always in the context of a mobile society, one in which people moved during the year or changed their behaviors annually to adjust to the dynamics of constant ecological change. Enduring questions in such philosophies ask about the challenges of how to organize a society to be as coordinated as possible to constant change—changes could be environmental in nature, or social (e.g. pertaining to trade), or biological, among others. In the work of Armstrong, consent and animacy, as I am understanding the concepts, are key vectors through which to understand science's governance value.

As a prefacing remark to a discussion of knowledge, Armstrong writes about mobility, change, responsibility (caretaking), and coordination:

> We simply move around on the territory at different seasons, and different times of the year, but we always return to our villages in the winter months after all the harvesting is done. So it's like harvesting a huge garden and it's like taking care of a huge garden. Think of the garden as being vertical, rather than flat, then you have some idea of the different seasons and the different levels of growth patterns. Migration patterns of the deer and the moose and the elk and other sources of food that live off the other relatives, and occupy our land and take care of us.
>
> *(Armstrong 1998: 175)*

Knowledge is directly tied to movement and responsibility. She writes that

> I have heard elders explain that the language changed as we moved and spread over the land through time. My own father told me that it was the land that changed the language because there is special knowledge in each different place.
>
> *(Armstrong 1998: 175)*

Armstrong's account of science is tied to concepts of land that emphasize animacy and consent. She writes that

> . . .language was given to us by the land we live within. . . All my elders say that it is land that holds all knowledge of life and death and is a constant teacher. It is said in Okanagan

that the land constantly speaks. It is constantly communicating. Not to learn its language is to die. We survived and thrived by listening intently to its teachings—to its language—and then inventing human words to retell its stories to our succeeding generations. It is the land that speaks *N'silxchn* through the generations of our ancestors to us.

(Armstrong 1998: 176)

Here, in dialogue with Anishinaabe traditions, Armstrong emphasizes that it is the land which is animate, and that humans must listen to it, which is a way of articulating consent relationships with animate beings and entities.

Consider Armstrong's description of the word *naw'qinwixw* as an account of scientific knowledge. Here I will primarily focus on relationships connecting humans in Armstrong's account, for the sake of brevity. Armstrong describes the word in relationship to governance coordination. For the word has to do with being "able to sustain community, and to be able to transfer that knowledge, and that ethic to each succeeding generation, and to be able to bring the community continuously in balance with all of the other living life forms" (Armstrong 2007). More specifically within this context of governance, *naw'qinwixw* is about coordinating responsibilities. It refers to a particular

a tool or methodology that can be used for finding out what the best solution to any question might be. . . it's thought about as a dialogue tool. Again, that word has a series of images that are attached to it.

(Armstrong 2007)

For Armstrong, something like knowledge is referenced when they write that *naw'qinwixw* is about a "solution to any question." Knowledge is about the bonds conducive to dialogue. But the meaning of knowledge, and *naw'qinwixw* more broadly, is best explainable through discussing images. Here I will quote at length some of the images that Armstrong provides.

The first part of the word, "*naw'qin,' -'aw*" has to do with water dripping in a really slow, one drop at a time, that kind of action. So that would be an image. And "*naw'qin*" the meaning of "*qin*" always has to do with the top of the head, or the top of a mountain. So there is water dripping one drop at a time in the top of the head. The last part of the word, "*wixw*," means we do that for each other. I do it for you, you do it for me. That's what the "*wixw*" at the end means. If we "*tkwinsenwixw*" and "*wixw*"—you hear the "*wixw*" on the end?—that means we are shaking each other's hands.

(Armstrong 2007: 8)

Their interpretation of the imagery concerns knowledge.

So what the symbol is speaking about is being able to put into the mind the knowledge, to be able to let it drop in. You know how if you were to take a drop of water and put it on say cotton, and you'd see that the drop slowly permeate the cotton. . . If you were to give knowledge in that way, then knowledge becomes integrated into the whole person: into their mind, and their spirit, and their emotions, every part of them. It becomes integrated into their family, and into the work that they do, they way that they live and think. It becomes a part of them. So when we are making decisions, unless a person can receive information and knowledge in that way, it doesn't become a part of them, it becomes something that's external to them and remains external to them. So knowledge must be brought in in a way that takes into consideration the feelings of a person, the level of knowledge or information or facts a person might have, the background that the person might have been exposed to, the understanding and the status that the person's role might have in the community itself.

(Armstrong 2007: 8)

Knowledge is inextricable from governance ("the person's role might have in the community itself"). Consent is a key part of this, for unless people are involved in a consensus process that engages their communicative capacities and needs appropriately, then knowledge that supports governance cannot be achieved. Honoring people's animacy, including their differences and uniqueness, motivates the importance of consent.

So, you wouldn't speak to say a teenager and tell them in the same way you would tell my grandmother about something. So that difference in terms of diversity in the community is the primary request in our dialogue, that says, "you have to respect diversity, you have to respect that the other person never is going to think like you, be like you, know what you know, because they're not you." They have all of these different experiences. That is, in dialogue, really necessary for you to know, so you can't assume anything. So, you have to try and clarify for the person in the way that they can take it, in a way that they can understand. And that's your role whenever there is something in-between us that we don't understand.

(Armstrong 2007: 8)

Moral qualities of consent, trust, transparency, accountability, among others, are present here given that knowledge exchange must recognize and respect difference.

It's not your role to come and convince me, "this is what it looks like; this is how it is because that's the way I see it." Which is very disrespectful and destructive because you're not seeking clarity, you're seeking to be aggressive, you're seeking to dominant, and that's not acceptable. What you should be doing, very clearly saying, 'we have this problem, clearly, one of us doesn't understand it, and so I'll try and tell you how I see it, what I know about it, how I think about it, how I feel about it, how I feel it might affect me, or effect things that I know about, and that will help inform you. But I'm requesting the same things from you. I want you to tell me how you feel about it, how it affects you, the things you know about how it affects you. Then we'll have a better understanding; we'll have a chance at a better understanding of what it is we need to do. We can only do that by giving as much clarity from our diverse points of view.' So, to seek the most diverse view is what *naw'qinwixw* asks for.

(Armstrong 2007: 8–9)

For Armstrong, it is impossible to speak about knowledge or science without—at the same time—speaking about difference, dialogue, and community in ways that prioritize consent. Since knowledge is for the community, it has to be relevant to the members of the collective so that their animacy can be honored and they can consent to and invest in the knowledge. Taking this approach, for Armstrong, is tied to the conditions that need to be established for coordinated action, or governance, that is aimed at being responsive to a constantly changing world composed of many animate beings and entities.

Haudenosaunee Science

Haudenosaunee peoples living in what are currently called the United States and Canada have published widely on their scientific work to address environmental concerns. Their communities have faced industrial pollution, land dispossession, and climate change. By science, of course,

I mean both Haudenosaunee knowledge systems and Haudenosaunee adaptations of scientific methods stemming from Canadian and US institutions. Haudenosaunee people keep some of the oldest traditions of Indigenous governance that are tied to consent, self-determination, animacy, and science—among many other dimensions. I want to discuss some of these governance relationships as told through Haudenosaunee scholarship and writing. In particular, Haudenosaunee have endeavored to pursue science as always involving cooperative partnerships and coordination with other groups in shared regions.

Mary Arquette writes about the governance value of the knowledge and science conveyed in "Our original instructions." She discusses how the instructions

> tell us that we must acknowledge, respect, and give thanks to the Animal Nations. . . animals are our ancestors. . . In the Skyworld, they were acknowledged as people and for that reason, we continue to see them as such today. . . When we think about our elders, the Animals, we realize that from the start, we owe our very survival to their existence.
>
> *(Arquette 2000)*

Arquette talks about animacy through the unique personalities of different animals and their mutual responsibilities in relation to human and non-human beings and entities like water and air. She discusses that animals that threaten to harm humans should be treated with respect, and that human societies must respect the exercise of animals' own self-determination.

Arquette emphasizes how humans depend on animals, both for nutrition and ecosystem services, but also for spiritual and emotional health. But animals and plants, as animate beings, can better exercise these responsibilities if humans treat animals in ways that respect their self-determination (consent).

> Unlike plants, we cannot directly use the energy of the sun to make food. . . we are the most dependent of all species. . . There have been times when humans have thought they knew more. . . we have become manipulative, trying to make the Creation suit our needs, wants, and desires. . .
>
> *(Arquette 2000)*

For Arquette, the world is an interdependent collective of interdependent beings and entities that have responsibilities to be coordinated to respond to change. When humans violate animals' consent, then interdependence is less supportive of well-being.

Another Haudenosaunee governance tradition is the *Kaswentha* treaty relationship, which Susan Hill describes as follows:

> Within the oral record of the Haudenosaunee, it is noted that the relationship was to be as two vessels travelling down a river—the river of life—side by side, never crossing paths, never interfering in the other's internal matters. However, the path between them, symbolized by three rows of white wampum beads in the treaty belt, was to be a constant of respect, trust, and friendship. Some might say that this is what kept the two vessels apart, but in fact, it is what kept them connected to each other. Without those three principles, the two vessels could drift apart and potentially be washed onto the bank (or crash into the rocks). This agreement was meant to provide security for both sides. In essence, they agreed to live as peaceful neighbors in a relationship of friendship, predicated on an agreement to not interfere in each other's internal business. The contemporary oral record of the treaty also notes that individuals could choose which boat to travel in with the understanding that one must be clear in one's choice and avoid "having a foot in both." The premise of non-interference,

within the concept of brotherhood, demonstrates the desire to be allies rather than to have one side be subjects of the other.

(Hill 2008: 31)

Hill's description of the *Kaswentha* philosophy highlights "respect, trust, and friendship" as qualities of relationships that create mutually accountable connections. Consent is woven throughout the proposed relationship, including the "[provision of] security," an "agreement to not interfere in each other's internal business," the freedom of individuals to "choose which boat to travel in," and the "desire to be allies rather than to have one side be subjects of the other." These are fine-grained fabrics of consensual relationality that demonstrate respect to partners in diverse ways.

The *Kaswentha*, among many other traditions, grounds how some Haudenosaunee initiatives approach their own sciences and knowledge systems and their interactions with settler and other non-Indigenous peoples. James Ransom and Kreg Ettenger write that

According to Haudenosaunee tradition, indigenous knowledge and western science are analogous tools developed and used by their respective societies. The former is used by Native people to help them fulfill their responsibilities given to them by the Creator. It is their science, and part of how the canoe navigates the river. Western science is the tool used by the environmental agencies responsible for achieving the goals of the ship.

(Ransom and Ettenger 2001: 225–226)

Ransom and Ettenger's article discusses numerous Haudenosaunee environmental projection initiatives that seek to put the different knowledge systems in dialogue and coordination. In the different cases, the Haudenosaunee proposed the *Kaswentha* as a basis of knowledge exchange but also of consensual science that brings different groups together by respecting their self-determination. The very idea of a knowledge system is already grounded in the *Kaswentha* philosophy, which has the basics of consensual relationality that Hill describes.

Haudenosaunee people have created scientific research aimed at governance across regions of many interdependent members. Ransom and Ettenger describe some cases in their article, including "collaborative partnerships which respect and support Haudenosaunee sovereignty and cultural identity while addressing critical environmental problems." They include "(1) a regional organization representing traditional Haudenosaunee Nations on environmental issues; (2) an action plan for environmental restoration; (3) a proposal to create culturally-based environmental protection processes; and (4) guidelines for environmental research in Haudenosaunee territories" (Ransom and Ettenger 2001: 222). They write that the work of these projects

reflects the fundamental concept of the *Kaswentha*—the ability to preserve one's identity and autonomy while working with allies in response to common interests. . .. The development of community environmental processes to protect traditional ways of life, the natural world, and future generations while being consistent with the sovereignty of the Haudenosaunee.

(Ransom and Ettenger 200: 223)

In this way, consent and governance are completely woven throughout the basic activities of empirical inquiry.

Haudenosaunee peoples and their allies have developed a portfolio of scientific work studying the relationship among pollution, health, self-determination, and cultural vitality in the Saint Lawrence River watershed. They designed this research to respond to widespread industrial pollution burdening Mohawk communities on both the US and Canadian sides, including toxicants like polychlorinated biphenyls, which has been widely documented by Mohawk scholars, scientists, and leaders (Tarbell and Arquette 2000). The work had to be based on coordination of the interdependent members of the collective of humans and non-humans in the region. The work has been about governance value. Henry Lickers, speaking of planning and environmental science in relation to consensus, writes that

> Planning is then seen as a long-term activity with a time period of many generations. Characteristics which help to maintain place are beneficial. Time is sacrificed for understanding, sensitivity, and consensus. A place is better protected, preserved and enhanced when everyone knows why.
>
> *(Lickers 1993)*

Understanding interdependence and coordination was truly crucial to the Mohawk's efforts. Elizabeth Hoover has written that, in Akwesasne, "the relationship between fish–whose duty it is to cleanse the water and offer themselves as food–and humans–whose role it is to respectfully harvest these fish–has been interrupted by environmental contamination" (Hoover 2017). Those most at risk from pollutants include women of childbearing age, pregnant and nursing women, and children under 15, especially given the bioaccumulation of some toxicants in breast milk. Indigenous environmental scientists Alice Tarbell and Mary Arquette estimate that 50% of the economy used to be based on fishing before the pollution started. Beyond fish, they tell how the contamination of medicinal plants leaves traditional health care providers unable to recommend natural remedies that some elders in Mohawk communities rely on (Tarbell and Arquette 2000).

Arquette and their collaborators at Akwesasne have famously discussed how the consent of Mohawk people is essential to acceptable science. When non-Indigenous scientists did not involve Mohawk persons in the determination of risks of contamination, they came up with some problematic and irresponsible conclusions. They argued irresponsibly that when people avoid eating contaminated foods, there is no risk. Yet Mohawk people later pointed out that "alternative diets are consumed that are often high in fat and calories and low in vitamins and nutrients," which produces additional negative health outcomes that affect Mohawks acutely, including diabetes (Arquette et al. 2002). Mohawk advocate Katsi Cook, through the Mother's Milk Project, has worked to make environmental health science accessible to affected communities. Importantly, Cook sees the scientific work as motivated by how the project participants "bless the seeds, pray to corn, and continue a one-on-one relationship with the earth" (LaDuke 1999: 20).

Haudenosaunee actions have made powerful contributions to what I am calling the governance value of science. Knowledge cannot be detached from consent as a way of honoring animacy, whether that is the self-determination of non-human beings and entities, or the consensus of humans collaborating to protect the environment. Consent is integral to the governance value of science and is generative of coordination of members of collectives.

Conclusion

Some Indigenous scientific traditions and philosophies of science, I hope to have shown, understand science as a responsibility. Empirical inquiry, no different from any other dimension of science, represents a domain of actions that must be understood as responsibilities to uphold and promote the coordination of interdependent members of a collective. Consent, as an honoring of animacy, which includes honoring self-determination, is a quality that makes it possible for responsibilities to be exercised for the sake of governance. Consent is not only a right, and not merely a restriction or regulation. Indigenous peoples value consent and animacy *not* because these concepts merely represent assumptions of Indigenous worldviews. Rather, consent and animacy are critical concepts and qualities of responsible action that contribute to the governance of interdependent members of collectives.

I would argue that some Indigenous scientific traditions and philosophies of science would not deem a practice to be counted as research if consent is not pervasive in the procedures of empirical inquiry. For any irresponsible actions—e.g. science without consent—cannot have governance value. At the beginning of the essay, I covered some well-known violations of consent within research ethics. Insofar as these cases have generated deep-seeded distrust among Indigenous peoples that their consent will be honored by scientists, is it really acceptable to say that the scientific studies should still be articulated as research? The unethical studies affected relationships negatively among interdependent members of shared and overlapping collectives (e.g. settler Canadian and Indigenous communities). The same could be said for scientific cases involving studies that fail to honor the challenges of having consensual relationships with animals, plants, insects, flows, and landscapes (recall Kimmerer's description of listening to salmon in science).

Indigenous philosophy of science should be a partner with the feminist philosophy of science that occurs in European and European diasporic societies. Partnership does not refer to philosophical practices that delimit *Indigenous content* to case studies or examples within books and papers written according to the standards of, for example, settler Canadians or settler Americans. As Indigenous philosophies of science reflect on scientific traditions that have their own histories, the philosophical treatment of issues like diversity in empirical inquiry or patriarchy in science arise from different sources of experience, culture, politics, and knowledge. As these sources of experience, culture, politics, and knowledge are rarely (if at all) countenanced in anglophone institutions of higher education, the well of possibilities for partnership and dialogue is poisoned.

Related chapters: 7, 8, 12, 15, 16.

References

Armstrong J. (1998) "Land Speaking," in Ortiz, S. (ed.), *Speaking for the Generations: Native Writers on Writing. Sun Tracks: An American Indian Literary Series.* Tucson: University of Arizona Press, pp. 174–194.

Armstrong J. (2007) "Native perspectives on sustainability: Jeannette Armstrong," (Syilx)[Interview transcript]. *Native Perspectives on Sustainability* (www. nativeperspectives. net).

Arquette M. (2000) "The Animals," in Haudenosaunee Environmental Task Force (ed.), *Words That Come Before All Else.* Akwesasne: Haudenosaunee Environmental Task Force, pp. 82–101.

Arquette M, Cole M, Cook K, et al. (2002) "Holistic Risk-Based Environmental Decision-making: A Native Perspective," *Environmental Health Perspectives* 110, pp. 259–264.

Barad K. (2007) *Meeting the Universe Halfway: Quantum Physics and the Entanglement of Matter and Meaning.* Durham: Duke University Press.

Battiste M. (2000) *Reclaiming Indigenous Voice and Vision.* Vancouver: University of British Columbia Press.

Brugge D, Benally T and Yazzie-Lewis E. (2006) *The Navajo People and Uranium Mining.* Albuquerque: University of New Mexico Press.

Ford J and Martinez D. (2000) "Traditional Ecological Knowledge, Ecosystem Science, and Environmental Management," *Ecological Applications* 10, pp. 1249–1250.

George J. (2019) "Igloolik Inuit Sue Feds Over Medical Experiments that They Saw Them Treated Like 'Monkeys'," *Nunatsiaq News.* Nunavut, Quebec, Canada: Nortext Publishing Corp., June 21.

Haraway DJ. (1997) *Modest_Witness@Second_Millennium. FemaleMan__Meets OncoMouse.* New York: Routledge.

Harding S. (1995) "Strong Objectivity": A Response to the New Objectivity Question," *Synthese* 104, pp. 331–349.

Hill S. (2008) "Travelling Down the River of Life Together in Peace and Friendship, Forever: Haudenosaunee Land Ethics and Treaty Agreements as the Basis for Restructuring the Relationship with the British Crown," in Simpson, L. (ed.), *Lighting the Eighth Fire: The Liberation, Resurgence, and Protection of Indigenous Nations.* Winnipeg: Arbeiter Ring, pp. 23–45.

Hoover E. (2017) *The River Is in Us.* Minneapolis: University of Minnesota Press.

Intemann K. (2010) "25 Years of Feminist Empiricism and Standpoint Theory: Where Are We Now?" *Hypatia* 25, pp. 778–796.

Johnston B. (1990) *Ojibway Heritage*. Lincoln: University of Nebraska Press.

Kimmerer R. (2013) *Braiding Sweetgrass: Indigenous Wisdom, Scientific Knowledge and the Teachings of Plants*. Minneapolis: Milkweed Editions.

LaDuke W. (1999) *All Our Relations: Native Struggles for Land and Life*. Cambridge: South End Press.

Lickers H. (1993) "Talking to My Granddaugther," in Ahenakew, F., Gardipy, B., and Lafond, B. (eds.), *Native Voices*. Toronto: McGraw-Hill Ryerson

Longino H. (1990) *Science as Social Knowledge*. Princeton: Princeton University Press.

Mandamin J. (2015) "Water Is Life," in Scott, D.N. (ed.), *Our Chemical Selves: Gender, Toxics, and Environmental Health*. Vancouver: UBC Press, pp. xi–xix.

McGregor D. (2008) "Linking Traditional Ecological Knowledge and Western Science: Aboriginal Perspectives from the 2000 State of the Lakes Ecosystem Conference," *The Canadian Journal of Native Studies* XXVIII, pp. 139–158.

McGregor D. (2009) "Honouring Our Relations: An Anishnaabe Perspective on Environmental Justice," in Agyeman, J., Cole, P., Haluza-DeLay, R., and O'Riley, P. (eds.), *Speaking for Ourselves: Environmental Justice in Canada*. Vancouver: UBC Press, pp. 27–41.

Mosby I. (2013) "Administering Colonial science: Nutrition Research and Human Biomedical Experimentation in Aboriginal Communities and Residential Schools, 1942–1952," *Histoire Sociale/Social History* 461, pp. 45–172.

Noodin M. (2017) "Ganawendamaw: Anishinaabe Concepts of Sustainability," in Haladay, J. and Hicks, S. (eds.), *Narratives of Educating for Sustainability in Unsustainable Environments*. East Lansing: Michigan State University Press, pp. 245–260.

Pihama L, Cram F and Walker S. (2002) "Creating Methodological Space: A Literature Review of Kaupapa Maori Research," *Canadian Journal of Native Education* 26, pp. 30–43.

Ransom JW and Ettenger KT. (2001) "'Polishing the Kaswentha': A Haudenosaunee View of Environmental Cooperation," *Environmental Science & Policy* 4, pp. 219–228.

Shilling D and Nelson M. (2018) *Traditional Ecological Knowledge: Learning from Indigenous Methods for Environmental Sustainability*. Cambridge: Cambridge University Press.

Smith LT. (1999) *Decolonizing Methodologies: Research and Indigenous Peoples*. London: Zed Books.

Sy WC. (2018) "Relationship with Land in Anishinaabeg Womxn's Historical Research," in: Gallagher, J. and Winslow, B. (eds.), *Reshaping Women's History*. Urbana-Champagne: University of Illinois Press, pp. 227–228.

TallBear K. (2013) *Native American DNA: Tribal Belonging and the False Promise of Genetic Science*. Minneapolis: University of Minnesota Press.

Tarbell A and Arquette M. (2000) "Akwesasne: A Native American Community's Resistance to Cultural and Environmental Damage," in Hofrichter, R. (ed.), *Reclaiming the Environmental Debate: The Politics of Health in a Toxic Culture*. Cambridge: MIT Press, pp. 93–111.

10

QUEER SCIENCE STUDIES/QUEER SCIENCE

Kristina Gupta and David A. Rubin

This chapter makes two simple claims: first, feminist science studies has always already been queer science studies. Second, critiquing science and medicine has always already been central to the projects of queer, trans, and intersex scholarship and activism. In the West, scientific and medical practices and discourses have played a key role in constructing ideas about "natural" gender and sexual expression in human and non-human animals. The ideas about sex, gender, and sexuality arising from science and medicine have most often reflected and served to reinforce regressive norms and institutions that marginalize and oppress women, sexual minorities, and gender non-conforming and trans people, particularly poor, queer, and trans folks of color, among others. Throughout their respective histories, feminist science studies and queer, trans, and intersex scholarship and activism have critiqued and challenged these regressive scientific understandings of binary sex/gender, female sexuality, gay and lesbian sexuality, transsexual and transgender identity and embodiment, and intersex identity and embodiment, along with intersecting regressive understandings of race, disability, and class, among other categories of difference.

In *Queer Feminist Science Studies: A Reader*, we argued that queer critique, at its best, ". . .contests the naturalization of the categories of normal and deviant sexuality and binarized notions of sexed anatomy, gender identity, sexual desire and sexual identity" (Cipolla et al. 2017: 7). To repeat, it has been scientific and medical practices and discourses that have played a key role in naturalizing social categories of normal and deviant sexuality, sex, and gender, and in reinforcing the idea that sexed embodiment, gender identity and expression, and sexual identity and expression each exist in only two forms and must align in certain ways. As we will show below, often these uneven processes occurred through the elevation of whiteness and Western scientific and cultural values into the unmarked regulatory ideals through which sexual dimorphism, binary gender, and heteronormativity were naturalized. It is our first argument that, in critiquing and challenging the understandings of sex, gender, sexuality, and other social categories produced in scientific and medical practice and discourse, feminist science studies has always already been doing the work of queer science studies. At the same time, because science and medicine have played such a key role in regulating the lives of sexual and gender minorities, by necessity, queer, trans, and intersex scholarship and activism has long been engaged in the practice of analyzing and critiquing science. Finally, we contend that recent queer science studies scholarship that draws on queer and trans of color critique, among other traditions, has critically interrogated the racial, geopolitical, multispecies, and ecological implications of histories and practices of Western science in nuanced and capacious ways.

In what follows, we briefly describe some of the foundational feminist science studies work and queer, trans, and intersex scholarship and activism that we argue should also be considered queer science studies. We then suggest that some of the work in feminist science studies on methodologies and epistemologies also performs the work of queer science studies. Finally, we review some recent work in the field of queer/feminist science studies that further queers—arrests, transforms, and taxes—assumptions about what counts as difference in both the sciences and in feminist and queer scholarship. This work reaffirms the centrality of race to the modern production of ideas about sex, gender, and sexuality—and thereby challenges scholars to not only persistently critique science, technology, and biomedicine from diverse perspectives but also develop queer, feminist, and antiracist sciences ourselves.

Foundational Work

In one of her foundational contributions to queer theory, *Tendencies* (1993), Eve Sedgwick points out that, in dominant understandings of "sexual identity," a variety of elements are expected to "align" in certain ways: your biological sex (male or female), your gender assignment (masculine or feminine), your gender identity and expression (masculine or feminine), the biological sex of your preferred partner, the gender assignment of your preferred partner, the gender identity and expression of your preferred partner, your sexual identity (gay or straight), your preferred partner's sexual identity, your procreative choices (yes or no), your preferred sexual acts (insertive or receptive), your most eroticized sexual organs, your sexual fantasies, your main locus of emotional bonds, your enjoyment of power in sexual relations, and so on. According to Sedgwick, one of the things that *queer* can refer to is "the open mesh of possibilities, gaps, overlaps, dissonances and resonances, lapses and excesses of meaning when the constituent elements of anyone's gender, or anyone's sexuality aren't made (or *can't be* made) to signify monolithically" (Sedgwick 1993: 8). In other words, for Sedgwick, one meaning of the verb *to queer* is to expose the ways in which these elements of sexual identity don't always (or can't be made to) coherently line up according to social norms. By exposing and critiquing scientific belief in the naturalness of "monolithic" sex/gender/sexuality, feminist critiques of the science of sex, gender, and sexuality have long been engaged in the work of *queering*. At the same time, because science and medicine have played such a key role in naturalizing and normalizing the alignment of these elements, by necessity queer, trans, and intersex scholarship and activism has always been centrally concerned with challenging and reconfiguring the assumptions of scientific research and medical practice.

Here we don't provide an exhaustive overview of this large body of scholarship, but instead highlight a few key works from the history of (queer) feminist science studies and queer, trans, and intersex analyses of science and medicine. Some of the earliest work in (queer) feminist science studies critiqued scientific work on sex/gender for relying on and thus reproducing a binary understanding of sex. For example, in *Myths of Gender: Biological Theories about Men and Women* (1985), Anne Fausto-Sterling challenges, among other things, the binary model of genetic sex (Fausto-Sterling 1985). Fausto-Sterling and others have built on this early work. Nelly Oudshoorn, for example, argues that the field of endocrinology naturalized certain understandings of testosterone and estrogen as male and female "sex" hormones, respectively (Oudshoorn 1994). In *Brainstorm: The Flaws in the Science of Sex Differences* (2010), Rebecca Jordan-Young critiques scientific claims that pre-natal exposure to "sex" hormones "hardwires" the brain as male or female (also called "human brain organization theory") (Jordan-Young 2010). In *Sex Itself: The Search for Male and Female in the Human Genome* (2013), Sarah Richardson further deconstructs the influence of binary conceptions of sex/gender on "sex" chromosome research (Richardson 2013). This impressive body of feminist work has queered scientific understandings of sex and gender as binary and of gender as directly produced by and unalterably tied to "biological" sex.

Relatedly, feminist science studies scholars have long critiqued scientific and medical efforts to disappear and/or manage those who do not fit into the scientifically naturalized sex/gender binary. In regards to intersex people, scholars including Fausto-Sterling, Suzanne Kessler, and Alice Dreger offered foundational critiques of scientific and medical efforts to theoretically (through definition and diagnosis) and practically (through medical intervention) reduce various forms of sexed embodiment into a binary sex model (Dreger 1998; Fausto-Sterling 1993, 2000a, 2000b; Kessler 1998). At the same time, because medicine has played such a key role in the oppression of people with intersex forms of embodiment, critiques of medicine have played a central role in the history of intersex scholarship and activism. For example, in the foundational piece "Hermaphrodites with Attitude: Mapping the Emergence of Intersex Activism" (1998), Cheryl Chase exposes the assumptions about sex/gender informing medical responses to the birth of intersex infants and advises intersex people and their allies to reject the medical paternalism of non-consensual surgical and hormonal normalization treatments (Chase 1998).

In regards to transgender experience, the record of feminist scholarship is less laudable; early feminist critiques of "transsexual medicine" were either explicitly anti-trans (Raymond 1994) or positioned transsexual individuals largely as dupes of medicine's sexist and heterosexist norms (Hausman 1995). Importantly, these versions of anti-trans and trans exclusionary essentialism and technodeterminist feminism were powerfully critiqued in early 1990s trans scholarship (Stone 1992; Stryker 1994), which drew explicitly on both feminist science studies and queer theory (Haraway 1991). Moreover, as in the case of intersex scholarship and activism, because medical practice has played such an important role in the lives of many trans individuals, by necessity, "queer science studies" has always been a central part of trans scholarship and activism. For example, in "Resisting Medicine, Re/modeling Gender" (2003), Dean Spade exposes the ways in which transgender medicine demands the performance of traditional masculinity and femininity from transgender individuals, while also acknowledging the agency of transgender individuals in sometimes collaborating with and sometimes resisting the heteronormative demands of medicine (Spade 2003).

In addition to queering scientific and medical research and practice on sex/gender (conformity and nonconformity), feminist science studies has also long performed the work of queering scientific work on male and female sexuality. Early feminist science studies work challenged the naturalization of male sexuality as "aggressive" and "promiscuous" and female sexuality as "passive" and "choosy" in scientific research. For example, in some of her earliest work, Donna Haraway critiqued the work of primatologists, revealing the influence of sexism on their interpretation of male and female primate sexuality. According to Haraway, female primatologists eventually challenged conventional theories about female primate sexuality, which revolutionized the world of behavioral biology (Haraway 1989).

Relatedly, there is a significant body of scholarship in feminist science studies critiquing over a century of scientific and medical research on homosexuality. Jennifer Terry, for example, analyzes heterosexist understandings of homosexuality in scientific research since the mid-nineteenth century (Terry 1999). According to Terry, it is clear that much early scientific research on homosexuality viewed homosexuality as abnormal and deviant, and justified its study of homosexuality by arguing that the research could be used to shore up "normal" heterosexuality. In addition, according to Terry, early scientific research on homosexuality equated gender and sexuality, and thus scientists studying homosexuality found not only sexual deviance but gender deviance as well (Terry 1999). Terry and others have queered more contemporary scientific research on homosexuality as well, including research on the prenatal hormone/brain theory of homosexuality (the hypothesis that prenatal hormone exposure leads to the feminization of the brains of gay men and the masculinization of the brains of gay women) and the genetic theory of homosexuality. For example, both Terry and Jordan-Young point out that in studies on homosexuality in non-human animals (which form part of the prenatal hormone/brain theory of homosexuality), non-human male animals are categorized as engaging in homosexual behavior if they engage in "receptive" sex with either male or female conspecifics (e.g. if they "present") and non-human female animals are categorized as engaging in homosexual behavior if they engage in "insertive" sex with either male or female conspecifics (e.g. if they "mount"). Scientists then equate a particular kind of sexual activity engaged in by non-human animals (presenting in males and mounting in females) with a particular kind of sexual orientation or identity in humans (preferring a sexual partner of the same gender), despite the clear dissonance between the two phenomena (Jordan-Young 2010; Terry 2000). In general, both (queer) feminist science studies scholars and queer theorists and activists have been suspicious of efforts by scientists and gay rights activists to discover "the" biological basis of homosexuality and to base gay rights claims on the argument that homosexuality is "natural" and "innate" (see Halley 1994; Hegarty 1997; Stein 1994; Terry 1999). In these ways, critiques of scientific research on homosexuality by both feminist science studies scholars and queer activists and scholars have performed the work of queer science studies by challenging the naturalization of heterosexuality, the conflation of sexual acts with sexual identities, and the conflation of sexual nonconformity with gender nonconformity.

Methodologies, Epistemologies, and Human/Technological Relationships

We also argue that early work in feminist science studies on objectivity and epistemology along with work that articulates a complicated relationship between humans and technology performs the work of queer(ing) science studies. According to many queer theorists, queer critique extends beyond the domains of gender and sexuality per se. Harper et al. explain that they conceive of queer critique as "a means of traversing and creatively transforming conceptual boundaries, thereby harnessing the critical potential of queer theory while deploying it beyond the realms of sexuality and sexual identity" (Harper et al. 1997: 1). It is in this broader sense—of *queering* as traversing and transforming conceptual boundaries—that feminist science studies engagements with the epistemology and philosophy of science perform the work of queer(ing) science studies. Haraway's research, for example, undermines the distinction between nature and culture through concepts such as "material-semiotic" and "naturecultures." These concepts highlight the productive interactivity of human and non-human phenomenon, thereby not only denaturalizing foundational binaries in Western thought but also bringing out the queerness—the infinite variations, strange patterns, and often unnoticed nuances—of naturecultural worlds, cyborg politics, companion species relations, the chthulucene, and their deeply interconnected and messy histories and biopolitics (Haraway 2008, 2016). As an example of the use of these concepts in practice, anthropologist Agustín Fuentes uses the term "natureculture" to reflect upon the multispecies interface between humans and monkeys in Bali, Indonesia, demonstrating the ways in which the two species are "simultaneously actors and participants in sharing and shaping mutual ecologies" (Fuentes 2010: 600). Similarly to "naturecultures," Haraway's concept of "situated knowledges" also undermines the binary opposition between objectivity and relativism, by framing knowledge and vision as embodied, socially situated practices, while also maintaining a "no-nonsense commitment to faithful accounts of the real world" (Haraway 1988).

Haraway also offers a queer vision of the relationship between humans and technology. Carrie Sandahl describes "queering" as follows: "Queering describes the practices of putting a spin on mainstream representations to reveal latent queer subtexts; of appropriating a representation for one's own purposes, forcing it to signify differently; or of deconstructing a representation's heterosexism. . ." (Sandahl 2003: 37). In her image of the cyborg, Haraway offers a vision of human-technological intervention that includes the potential for both deconstructive critique and creative appropriation. According to Haraway, cyborgs (both an enabling fiction and an accurate description of human and many, though not all, non-human ontologies under late capitalism) are creatures of hybridity, blurring the boundaries between humans and non-human animals, between humans and machines, and between the material and the immaterial (or digital). Crucially, this blending of the organic and the technical is a place of both critique *and* opportunity for Haraway. She writes,

> It is not just that science and technology are possible means of great human satisfaction, as well as a matrix of complex dominations. Cyborg imagery can suggest a way out of the maze of dualisms in which we have explained our bodies and our tools to ourselves. This is a dream not of a common language, but of a powerful infidel heteroglossia.
>
> *(Haraway 1991: 181)*

For Haraway, then, biomedical technology actually provides chances to reconfigure and challenge current structures of domination, and to transform and perhaps overcome dualisms between man/woman, nature/culture, and mind/matter. Haraway's work has inspired a number of feminists to adopt a "queer" attitude of creative appropriation toward science, medicine, and technology. Deboleena Roy's work, discussed later in this chapter, provides an example of this creative appropriation of science and technology.

Queer Science Studies: Contemporary Contributions

Drawing and expanding on the earlier feminist traditions of queer(ing) science studies, recent work in a number of interdisciplinary fields has opened up a wide range of critical vistas for queer inquiries and analyses. While many scholars continue to focus on critiquing the production of racial and sexual biases in scientific research, some femi-queer scholars have moved toward developing *queer sciences* themselves. In this section, we review and reflect on some contemporary contributions to both trajectories of research. Rather than attempting to be comprehensive, we focus instead on what we view as some especially generative and transformative areas of queer research and queer science.

Critical Intersex Studies and Trans of Color Critique

Recent scholarship in critical intersex studies by Lisa Downing, Iain Morland, and Nikki Sullivan (2014), Zine Magubane (2014), Georgiann Davis (2015), David A. Rubin (2017), Hilary Malatino (2019), and others has more rigorously historicized the modern medical paradigm of intersex treatment. Pushing beyond the field's earlier critical focus on the sexological legacies of the gender identity clinic of the 1950s, these scholars have attended to the roles of settler colonialism, Western imperialism, chattel slavery, eugenics, racial capitalism, and neoliberalism in the medical normalization of atypically sexed bodies and the scientific reification of the twinned ideologies of sexual dimorphism and binary gender. Mobilizing a wide range of theoretical frameworks, critical intersex studies continues to critique the stigma, shame, and trauma that intersex medicalization produces, but also attends to the centrality of intersex lives and bodies to the genealogy of gender as a universalizing category and regulatory technology of subjectivity and sociality in late modernity. In these ways, critical intersex studies creates important opportunities for rethinking body politics, health, bioethics, reproductive justice, and human rights from the perspectives of a variety of subjugated knowledges.

Alongside this work, recent scholarship in trans studies and particularly trans of color critique has acknowledged and critically grappled with the politics of transnormativity, or the ideological accountability structure which the medical model subjects trans, gender non-conforming, and some intersex individuals to (Johnson 2016), the unmarked whiteness and ableism of the dominant medical model of gender transition (Puar 2017), the centrality of histories of antiblack racism to genealogies of trans and intersex medicalization (Gill-Peterson 2018; Snorton 2017), and the implications of the cultural and regional contingency of Western conceptions of trans, gender nonconforming, and intersex life in a transnational world (Aizura 2018; Rubin 2017; Swarr 2012). Taken together, critical intersex studies and trans of color critique powerfully extend queer, feminist, and antiracist commitments to critiquing histories of medical violence and help us to imagine more holistic, decolonial approaches to intersex, trans, and gender nonconforming healthcare and well-being.

Queer and Trans Zoologies and Ecologies

Another growing body of work that contributes to queer(ing) science studies might be called queer and trans zoologies and ecologies. According to Myra J. Hird (2006), the so-called natural world is not only more diverse and queerer but also more transitive and transversal than previous scientific research has assumed. Expanding on works by evolutionary biologists Joan Roughgarden (2013) and Bruce Bagemihl (2000), Hird argues that "trans exists in non-human species" and councils that we should "extend feminist interest in trans as a specifically sexed enterprise (as in transitioning from one sex to another), but also in a broader sense of movement across, through, and perhaps beyond traditional classifications" (Hird 2006: 37). To support these claims, Hird details the trans-materialities of intersex barnacles, gay penguins, cross-species queer interactions between butterflies and rove beetles, intersex plants, multisex fungi, amphibians that lack sex chromosomes and determine sex environmentally (via temperature), transsex fish such as the coral goby, and bacteria that completely defy the logics of sexual dimorphism altogether. "In some families of fish," Hird notes, "transsex is so much the norm that biologists have created a term for those 'unusual' fish that do not change sex: *gonochoristic*" (2006: 41). The point here is not to exoticize the somatic diversity and variation of non-human animals and organisms; indeed, Hird explicitly cautions against "the temptation to name certain species as queer" (2006: 45). Rather, Hird explains,

> given the diversity of sex amongst living matter, and the prevalence of transsex more specifically, it does not make sense to continue to debate the authenticity of trans when this debate necessarily relies on a notion of nature that implicitly excludes trans as a non-human phenomenon.
>
> *(2006: 45)*

While Hird's usage of the terms "transsex" versus "trans(gender)" might seem to preserve the sex/gender distinction and the attendant presumption that sex is biological while gender is cultural and therefore uniquely human, one could also argue that her account of sex diversity among living matter troubles and destabilizes the assumed coherence of that distinction by decentering the human as the measure of sexological categorization altogether. Recognition of the heterogeneity of transsex in the non-human world, then, provincializes the politics of human debates about transgender authenticity by disarming and displacing dominant anthropomorphic assumptions about the relative stability and/or plasticity of sex/gender, the nature of embodiment, and the organization of organic life more broadly.

A related body of work on queer ecologies (Mortimer Sandilands and Erickson 2010) considers the inter- and intra-actions and relations among sexualities, species, waterways, landscapes, atmospheres, and environments as dynamically interdependent developmental systems. Situated at the intersections of sexuality and environmental studies, research in queer ecologies investigates diverse topics, including "queer animals" as subjects of environmental and other popular fascination (Alaimo 2010), political interrogations of colonial discourses organizing sex (especially sex between men) as ecological threat (Gosine 2010), histories of lesbian and gay creations of natural space (Unger 2010), and hybrid nature writing as a source of queer futurity (Chrisholm 2010). By focalizing the queer agencies and relations of multispecies interdependence and vulnerability that enfold naturecultural phenomena such as hurricanes, earthquakes, blight, watershed management, environmental racism, and environmental preservation, queer ecological research helps us to rethink the layered materialities and semiotics of environmental conditions that frequently blur the lines between nature and culture.

Queer(ing) Biology

A more specific trajectory of research within feminist science studies centers on rethinking and queering biology. For instance, Banu Subramaniam's (2014) research on flower color variation in morning glories led her to rethink the genealogies of (cultural) diversity and (natural) variation in evolutionary biology. While it may seem relatively benign to most, the study of color variation in a common North American flower helped Subramaniam to connect debates over genetic variation with larger histories of power.

> Questions of genetic variation. . .are deeply linked to questions of diversity and difference in human populations steeped in tortured histories of slavery, colonialism, and genocide. *A naturecultural analysis reveals that the question of variation is fundamentally about power—the politics of life and death.*
>
> *(2014: 7, emphasis in original)*

Subramaniam argues for "a feminist reconstructive project for experimental biology" (2014: 4), one that brings feminist situated knowledges and postcolonial ethical sensibilities together with the curiosity and spirit of scientific experimentation to discover new modes of interdisciplinary and perhaps even anti-disciplinary knowing—including not only traditional lab work in the academy but also indigenous knowledge systems, kitchen science, and DIY science—queer projects indeed!

Angela Willey also analyzes the politics of science and the possibilities of biology in her book *Undoing Monogamy* (2016). At the outset, Willey emphasizes the import of disarming and displacing inherited sexological and eugenic paradigms in order to call into question the assumed naturalness, whiteness, and status of monogamy as a moral good in both heteronormative and homonormative Western contexts. According to Willey, "feminist engagements with embodiment as a naturecultural or material-discursive phenomenon have radical potential for transforming science and the worlds it helps materialize" (2016: 4). In a powerful rereading of the seminal writing of Audre Lorde, Willey casts Lorde as a *queer feminist critical materialist* whose theory of the erotic reshapes critical understandings of "biopossibility," or the complexly mediated capacity to embody certain socially salient traits and differences (2016: 138). Analyzing the work of scientists who have named a "monogamy gene" in the prairie voles they study, as well as popular and scholarly polyamory writers who similarly look for *non*monogamy in our biology, Willey declines to position monogamy or nonmonogamy on either side of the nature/culture division. Instead, Willey takes a patently naturecultural attitude toward understanding how both humans and their laboratory proxies, prairie voles, "pair bond," have sex, and/or form attachments.

Willey further argues that conflating sex with romantic love and then prioritizing it over nonsexual love enable monogamy to exist. After explaining the cultural entrenchment of the ideal of romantic love, Willey analyzes the scientific study of monogamy, which views it in evolutionary terms as a "mating system or strategy" (2016: 11). The similarities between both approaches couldn't be clearer. Scientific narratives, like social narratives, view males as more promiscuous in their mating strategies (that is, as "sperm spreaders") whereas they cast females as more "passive" and "domestic" (as "egg protectors"). In that regard, scientific assumptions regarding monogamy's evolutionary advantage (e.g. two parents nurturing offspring = an increase in chances of survival) figure monogamy not as "an object of inquiry itself [but] an a priori assumption informing scientific research" (2016: 16). In these ways and others, Willey demonstrates how assumptions about human monogamy are naturalized and shape notions of the normal (sexual/social/pair-bonded) and abnormal (asexual/asocial/promiscuous)—both of which are thoroughly racialized, as Willey demonstrates through archival research—that are then waylaid into the modeling of gene-brain-behavior connections in contemporary neuroscience.

Capping off her argument, Willey's epilogue "Dreams of a Dyke Science" to *Undoing Monogamy* (2016) uses an encounter with the science file at the Lesbian Herstory Archives in Brooklyn, New York to reimagine what it might mean to claim and to craft feminist and queer sciences:

> Queer feminist activist, "ecosexual," and porn star Annie Sprinkle famously called feminists to make our own pornography. We can make our own sciences in the same spirit. . .When we claim sciences, instead of "engaging" them, the terrain shifts from one of how un/friendly feminists are to Science to one of what a world of sciences has to offer, where so much is at stake.
>
> *(Willey 2016: 146)*

Where so much is at stake, Willey provides a timely call to shifting the frame from "engagement" to "practice" in order to create feminist and queer sciences. To be sure, such a shift may require different institutional, economic, and disciplinary arrangements than those offered by the contemporary neoliberal university. Nonetheless, Willey's work demonstrates that science and biology in particular hold many possibilities for current and future queer, feminist, and antiracist research.

Queering the Gut, Molecular Biology, and Beyond

Willey is one of a small but growing number of scholars who question the epistemic authority typically granted to neuroscience as the patronymic or most superior form of science in the early twenty-first century. Elizabeth Wilson takes a similar tack in her book *Gut Feminism* (2014), which decenters neuroscience and the brain in favor of a more gut-centered approach. Pursuing a model of feminist theory that thinks innovatively and organically at the same time, Wilson puts critical pressure on longstanding feminist assumptions that biology is flat, sovereign, static, and deterministic. What if, Wilson queries, biology is queerer than both conventional biologists and feminists have typically assumed? What if there is no intrinsic orthodoxy to biological matter, what if it can be "as perverse and wayward as any social, textual, cultural, affective, economic, historical, or philosophical arrangement" (2014: 27)?

Building on her previous endeavors to bring feminist theory and the neurosciences into uneasy critical intimacy and nonconsilient alliance, in *Gut Feminism* Wilson shifts her focus from the cerebral cortex to the neurological periphery: the enteric nervous system that encases the gut. Analyzing research on antidepressants, placebos, transference, phantasy, eating disorders, and suicidality, Wilson explores what cultivating curiosity about the biochemistry and psychopharmacology of the gut can do to and for feminist theory. Principally an effort to dislodge the commonplace (Wilson calls it "instinctual" [2014: 1]) antibiologism of feminist theory, *Gut Feminism* also offers a much-needed critique of the unbridled enthusiasm for flat and totalizing biologizing explanations that has been naturalized within the recent neurological turn that spans the hard sciences, social sciences, and even the humanities.

Wilson's project is successful partly owing to her keen attention to issues of data and method. She carefully reads biological data—its myriad complexities, contradictions, and errancies—using a unique methodology that combines biochemical, physiological, and pharmacological analysis, psychoanalysis, and deconstructive close reading. This capacious interdisciplinary methodological routine is well suited to Wilson's task, which is not to settle matters, but precisely to unsettle them. In the spirit of "a different mode of interdisciplinarity" (2014: 171)—a mode that is nonlinear, non-totalizing, disjunctive, and wondrous, one that exemplifies rather than prescribes (2014: 170)—Wilson argues that a more robust engagement with biological and pharmaceutical data can be "helpful" for feminist theory (2014: 2). Perversely—indeed queerly—Wilson uses the adjective *helpful* not to mean productive, ameliorative, or reparative, but "arresting, transforming, taxing" (2014: 2). In these ways, Wilson disarms, displaces, and reconfigures the epistemological and ontological assumptions about depression, biology, and aggression that feminist theory, queer studies and related disciplines have come to take for granted.

Wilson importantly acknowledges that some feminist theorists, particularly feminist science studies scholars—including Ruth Bleir, Donna Haraway, Sandra Harding, Evelyn Fox Keller, Ann Fausto-Sterling, Susan Squire, Vicky Kirby, Karen Barad, Catherine Waldby and Robert Mitchell, and a handful of others—have been working for decades to rethink the life sciences through feminist and queer frameworks. These scholars' influential interventions notwithstanding, "one thing feminist theory still needs," Wilson contends, "is a conceptual toolkit for reading biology" (2014: 3). The crucial word in this quotation is not biology per se but *reading*. Wilson does not argue for the supremacy of "biological data" (2014: 13) over and above interpretation. On the contrary, she shows that biology is not a static thing or all-determining substance opposed to hermeneutics but a dynamic, ruminative substrata in constant, variable intra-action with other co-entangled interpretive systems, including environment, atmosphere, psyche, culture, intersubjectivity, and companion species, both friendly and hostile. Wilson thus provides an indispensible account of how queer feminist thinkers can *read* the hermeneutics of biological matter—not merely scientific *discourses* about biology, but "entanglements of affects, ideations, nerves, agitation, sociality, pills, and synaptic biochemistry" (2014: 1). Taking biology seriously, Wilson shows, enervates and transforms the scope and critical capacities of feminist and queer inquiry in significant, unexpected ways.

Key to Wilson's argument is her insight that the neurological periphery—the viscera—constitutes an interpretive system that has been obscured and underanalyzed in feminist theory and the neuroscientific turn alike. She argues "not that the gut *contributes* to minded states, but that the gut *is* an organ of mind: it ruminates, deliberates, comprehends" (2014: 5). By analyzing the mindedness of the gut, Wilson exposes the limitations of isolating "brain from body, psyche from chemical, neuron from world" (2014: 13). In the process, she provides a radically decentered and decentering account of subjectivity, sociality, and biology alike. At the same time, *Gut Feminism* also makes "a strong case for the necessary place of aggression (bile) in feminist theory" (2014: 5). This secondary argument is actually the underpinning—the underbelly, if you will—of Wilson's primary project of tracing how biology challenges and reconfigures feminist (and queer) theory. Drawing on "the so-called antisocial thesis in queer theory" (2014: 5), especially the now canonical work of Leo Bersani and Lee Edelman, Wilson contends that "negativity is intrinsic (rather than antagonistic) to sociality and subjectivity" (2014: 6). Wilson does not mean this claim figuratively. Rather, she uses biological and pharmaceutical data on the treatment of depressive states to reveal that there are literally aggressive forces within and between each of us that bring the ideas of coherent selves, communities, *and* biologies capable of seamlessly orienting themselves exclusively toward the good into crisis. "Politics always involves hostility against the objects that we love" (2014: 86). This argument effectively counters frequent misreadings of the antisocial thesis. Such misreadings endeavor to redeem negativity, to turn it toward the good, to transform the destructive into the productive, and thereby conceal their own aggression toward negativity's irrevocably and irreducibly *negative* character in subjective, social, and biological life.

The implications of Wilson's entangled arguments about the gut, depression, biology, and aggression are multifarious. Wilson suggests that there is no ingestion, digestion, expulsion, or regurgitation without bile (hostility), no depression without inward *and* outward aggression, no pill without placebo, no cure without harm, no reparation without paranoid hostility. As Wilson puts it in one of her particularly gorgeous Sedgwickian reformulations of a familiar Derridian axiom, the center is always "vitally dependent" on the periphery (2014: 14), and "this makes a world of difference politically" (2014: 6). Our ecological, zoological, and bacterial inter- and intra-dependence makes a world of difference politically, in other words, because it refutes the orthodoxy of "bifurcated terms," revealing yet again that "biology and culture are not separate, agonistic forces" (2014: 8). For Wilson, there is no originary division between these domains. Rather, it is the very act of division that constitutes an underinterrogated ground of potential affinity *and* hostility. In this way, Wilson theorizes biology and culture and gut and mind as dynamic systems that are always co-entangled, co-constitutive, and co-troubling.

We recognize that some might be reluctant to label Wilson's work *science*. After all, Wilson's arguments do not rest on her own original experimental designs and data, but rather emerge from her close readings of other researcher's experiments, case studies, data, findings, and texts. Deboleena Roy's (2018) recent work, however and on the other hand, follows Willey's charge directly and fully lays claim to the radical interdisciplinary possibilities of claiming feminist, queer, postcolonial, and decolonial sciences. Roy investigates science as feminism at the lab bench, engaging an interdisciplinary conversation between molecular biology, Deleuzian philosophies, posthumanism, and postcolonial studies. In chapters on bacteria, subcloning, and synthetic biology, Roy shows how using feminist, queer, and postcolonial insights to conduct scientific research can produce new forms of scientific knowledge that challenge taken for granted assumptions in both the sciences and the humanities and social sciences. Roy writes,

> Practice-oriented feminist STS approaches can help. . .by providing us with the everyday knowledge and tools to conduct our experiments. Doing science in our backyards could include setting up local science shops where experts from a diverse range of knowledge bases come together, through a shared common interest, to solve local problems. It could involve creating feminist, postcolonial, and decolonial technoscience salons where academics learn to bring their research into interdisciplinary conversations. It could involve creating shared community maker spaces that prioritize the involvement of typically marginalized groups. It could even involve setting up interdisciplinary mentoring structures that support the radical act of having feminist sciences practice both their science and their feminism at the lab bench.
>
> *(2018: 206)*

By thinking molecularly, using grasses and stolons as her prime examples, Roy reveals that "being and becoming coexist and even work to coproduce each other" (2018: 21). Rather than providing blueprints or roadmaps for how to do queer science, Roy does offer, to borrow from a central theme of her book, "grass roots" (2018: 206) for growing queerer—as well as more critically feminist, postcolonial, and decolonial—laboratories of our own.

Conclusion

In this chapter, we have argued that feminist science studies scholars have developed a significant tradition of queering science studies and even sciences themselves. We have also suggested that critiquing science, medicine, and technology has long been central to queer, intersex, and trans activism and scholarship. Both of these claims are supported not only by the genealogies we have traced of these fields but also by new interdisciplinary research (some of which we reviewed earlier) that is pushing us to think in new, dynamic, and transformative ways about queerness across a wide range of scientific domains.

To be sure, we recognize that there are significant institutional mechanisms that constrain the potential of queer(ing) science and science studies in the contemporary moment. Prominent among them is the reluctance of institutions to properly fund genuinely interdisciplinary collaborations across disciplines such as those Roy imagines and invents. While we do not believe that disciplinarity must always be inherently anti-queer, the hard realities of current funding structures and administrative priorities in the academy in the Global North and Global South mean that resources are rarely allocated to scholars working at the intersections of queer research and STS. More importantly, these disciplinary problems are compounded by the ongoing structural dispossession of multiply marginalized queer, intersex, and trans communities, especially people of color, indigenous peoples, disabled folks, undocumented folks, and poor folks. Under such complexly stratified circumstances, access to the fruits of scientific knowledge production continues to be highly policed and usually limited to relatively privileged populations. For these reasons, as we move further into the twenty-first century, we hope future scholarship will stay attuned to the power dynamics of efforts to queer the sciences. Remembering how central and generative feminist science studies and queer, intersex, and trans studies and activism have been to the work of queering the sciences, we also want to recall how many scientific fields and areas of knowledge we have yet to investigate, reanimate, and transform with our queerest of inquiries, analytics, and energies.

Related chapters: 9, 14, 20.

References

Aizura, A. (2018) *Mobile Subjects: Transnational Imaginaries of Gender Reassignment*, Durham: Duke University Press.

Alaimo, S. (2010) "Eluding Capture: The Science, Pleasure, and Culture of 'Queer' Animals," in C. Mortimer Sandilands and B. Erickson (eds.) *Queer Ecologies: Sex, Nature, Politics, Desire*, Bloomington: Indiana University Press, 51–72.

Bagemihl, B. (2000) *Biological Exuberance: Animal Homosexuality and Natural Diversity*, New York: St. Martin's Press.

Chase, C. (1998) "Hermaphrodites with Attitude: Mapping the Emergence of Intersex Political Activism," *GLQ: A Journal of Lesbian and Gay Studies* 4 (2): 189–211.

Chrisholm, D. (2010) "Biophilia, Creative Involution, and the Ecological Future of Queer Desire," in C. Mortimer Sandilands and B. Erickson (eds.) *Queer Ecologies: Sex, Nature, Politics, Desire*, Bloomington: Indiana University Press, 359–382.

Cipolla, C., K. Gupta, D. Rubin, and A. Willey (eds.) (2017) *Queer Feminist Science Studies: A Reader*, Seattle: University of Washington Press.

Davis, G. (2015) *Contesting Intersex: The Dubious Diagnosis*, New York: NYU Press.

Downing, L., I. Morland, and N. Sullivan (2014) *Fuckology: Critical Essays on John Money's Diagnostic Concepts*, Chicago: University of Chicago Press.

Dreger, A. (1998) *Hermaphrodites and the Medical Invention of Sex*, Boston: Harvard University Press.

Fausto-Sterling, A. (1985) *Myths of Gender: Biological Theories about Women and Men*, 2nd ed. New York: Basic Books.

———. (1993) "The Five Sexes," *The Sciences* 33 (2): 20–24.

———. (2000a) *Sexing the Body : Gender Politics and the Construction of Sexuality*, 1st ed. New York: Basic Books.

———. (2000b) "The Five Sexes, Revisited," *The Sciences* 40 (4): 18–23.

Fuentes, A. (2010) "Naturalcultural Encounters in Bali: Monkeys, Temples, Tourists, and Ethnoprimatology," *Cultural Anthropology* 25 (4): 600–624.

Gill-Peterson, J. (2018) *Histories of the Transgender Child*, Minneapolis: University of Minnesota Press.

Gosine, A. (2010) "Non-White Reproduction and Same-Sex Eroticism: Queer Acts Against Nature," in C. Mortimer Sandilands and B. Erickson (eds.) *Queer Ecologies: Sex, Nature, Politics, Desire*, Bloomington: Indiana University Press, 149–172.

Halley, J. (1994) "Sexual Orientation and the Politics of Biology: A Critique of the Argument from Immutability," *Stanford Law Review* 46 (3): 503–568.

Haraway, D. (1988) "Situated Knowledges: The Science Question in Feminism and the Privilege of Partial Perspective," *Feminist Studies* 14 (3): 575–99.

———. (1989) *Primate Visions: Gender, Race, and Nature in the World of Modern Science*, New York: Routledge.

———. (1991) "A Cyborg Manifesto: Science, Technology, and Socialist-Feminism in the Late Twentieth Century," in *Simians, Cyborgs and Women: The Reinvention of Nature*. New York: Routledge, 149–181.

———. (2008) *When Species Meet*, Minneapolis: University of Minnesota Press.

———. (2016) *Staying with the Trouble: Making Kin in the Chthulucene*, Durham: Duke University Press.

Harper, P.B., A. McClintock, J.E. Muñoz, and Tr. Rosen (1997) "Queer Transexions of Race, Nation, and Gender: An Introduction," *Social Text* 52/53: 1–4.

Hausman, B. (1995) *Changing Sex: Transsexualism, Technology, and the Idea of Gender*, Durham: Duke University Press Books.

Hegarty, P. (1997) "Materializing the Hypothalamus: A Performative Account of the 'Gay Brain'," *Feminism & Psychology* 7 (3): 355–372.

Hird, M. (2006) "Animal Transex," *Australian Feminist Studies* 21 (49): 35–50.

Johnson, A. (2016) "Transnormativity: A New Concept and Its Validation through Documentary Film about Transgender Men," *Sociological Inquiry* 86 (4): 465–491.

Jordan-Young, R. (2010) *Brain Storm: The Flaws in the Science of Sex Differences*, Boston: Harvard University Press.

Kessler, S. (1998) *Lessons from the Intersexed*, New Brunswick: Rutgers University Press.

Magubane, Z. (2014) "Spectacles and Scholarship: Caster Semenya, Intersex Studies, and the Problem of Race in Feminist Theory," *Signs: Journal of Women in Culture and Society* 39 (3): 761–785.

Malatino, H. (2019) *Queering Embodiment: Monstrosity, Medical Violence, and Intersex Experience*, Lincoln: University of Nebraska Press.

Mortimer Sandilands, C. and B. Erickson (eds.) (2010) *Queer Ecologies: Sex, Nature, Politics, Desire*, Bloomington: Indiana University Press.

Oudshoorn, N. (1994) *Beyond the Natural Body: An Archaeology of Sex Hormones*, New York: Routledge.

Puar, J. (2017) *The Right to Maim: Debility, Capacity, Disability*, Durham: Duke University Press.

Raymond, J. (1994) *The Transsexual Empire: The Making of the She-Male*, New York: Teachers College Press.

Richardson, S. (2013) *Sex Itself: The Search for Male and Female in the Human Genome*, Chicago: University of Chicago Press.

Roughgarden, J. (2013) *Evolution's Rainbow: Diversity, Gender, and Sexuality in Nature and People*, Berkeley: University of California Press.

Roy, D. (2018) *Molecular Feminisms: Biology, Becomings, and Life in the Lab*, Seattle: University of Washington Press.

Rubin, D. (2017) *Intersex Matters: Biomedical Embodiment, Gender Regulation, and Transnational Activism*, Albany: SUNY Press.

Sandahl, C. (2003) "Queering the Crip or Cripping the Queer? Intersections of Queer and Crip Identities in Solo Autobiographical Performance," *GLQ: A Journal of Lesbian and Gay Studies* 9 (1): 25–56.

Sedgwick, E. (1993) *Tendencies*, Durham: Duke University Press.

Snorton. C. R. (2017) *Black on Both Sides: A Racial History of Trans Identity*, Minneapolis: University of Minnesota Press.

Spade, D. (2003) "Resisting Medicine, Re/Modeling Gender," *Berkeley Women's Law Journal* 18: 15–39.

Stein, E. (1994) "The Relevance of Scientific Research about Sexual Orientation to Lesbian and Gay Rights," *Journal of Homosexuality* 27 (3–4): 269–308.

Stone, S. (1992) "The Empire Strikes Back: A Posttranssexual Manifesto," *Camera Obscura*. 10 (2 (29)): 150–176.

Stryker, S. (1994) "My Words to Victor Frankenstein above the Village of Chamonix: Performing Transgender Rage," *GLQ: A Journal of Lesbian and Gay Studies* 1 (3): 237–54.

Subramaniam, B. (2014) *Ghost Stories for Darwin: The Science of Variation and the Politics of Diversity*, Chicago: University of Illinois Press.

Swarr, A. (2012) *Sex in Transition: Remaking Gender & Race in South Africa*, Albany: SUNY Press.

Terry, J. (1999) *An American Obsession: Science, Medicine, and Homosexuality in Modern Society*, 1st ed., Chicago: University of Chicago Press.

———. (2000) "'Unnatural Acts' in Nature: The Scientific Fascination with Queer Animals," *GLQ: A Journal of Lesbian and Gay Studies* 6 (2): 151–193.

Unger, N. (2010) "From Juke Joints to Sisterspace: The Role of Nature in Lesbian Alternative Environments in the United States," in C. Mortimer Sandilands and B. Erickson (eds.) *Queer Ecologies: Sex, Nature, Politics, Desire*, Bloomington: Indiana University Press, 173–198.

Willey, A. (2016) *Undoing Monogamy: The Politics of Science and the Possibilities of Biology*, Durham: Duke University Press.

Wilson, E. (2014) *Gut Feminism*, Durham: Duke University Press.

11

NATURALIZING AND DENATURALIZING IMPAIRMENT AND DISABILITY IN PHILOSOPHY AND FEMINIST PHILOSOPHY OF SCIENCE

Shelley Lynn Tremain

Two Spheres of Influence

Over the past several years, my research on the social constitution of disability has expanded to encompass two distinct, but interrelated, spheres: (what I call) a reconstructive–conceptual sphere and a metaphilosophical sphere. The work in the reconstructive–conceptual sphere has been two-fold: it (1) has endeavored to show how philosophers in various subfields of philosophy naturalize disability as a non-accidental and disadvantageous biological human characteristic (attribute, property, or difference) that ought be corrected or eliminated, that is, how disability is individualized and medicalized in (for instance) mainstream bioethics, cognitive science, feminist philosophy, and political philosophy and (2) has advanced an alternative conception of disability that construes it as an apparatus, in Foucault's sense, that is, as a heterogeneous ensemble of discourses, institutions, scientific statements, laws, architectural forms, administrative measures, and philosophical and moral propositions that responds to an urgent need in a given historical moment (see Foucault 1980a: 194). Normalization, I have argued, is the urgent requirement to which the apparatus of disability responds. In the metaphilosophical sphere, I have used empirical data and theoretical argument to identify and interrogate mechanisms, strategies, and policies within the profession of philosophy—including discriminatory hiring practices, exclusionary conference organizing, and biases in publication—that produce the underrepresentation of disabled philosophers and the marginalization of critical philosophical work on disability. Throughout this research on the social constitution of disability and its naturalized foundation, impairment, I have steadily come to recognize that the phenomena of the two spheres are mutually constitutive, entangled and entwined.

Given the predominance in philosophy of an individualized and medicalized understanding of disability, philosophers generally do not regard disability as pertinent to a form of social and political philosophy—namely, feminist philosophy—which is fundamentally about asymmetrical relations of social power and domination. Most philosophers assume, rather, that disability is appropriately and adequately addressed in the domains of medicine, science, and bioethics. Hence, the assumptions about disability that condition conceptual and analytical work in the discipline also condition the judgments that philosophers make about (for instance) faculty searches, peer review, graduate school applications, fellowship applications, and tenure and promotion. In other words, the prevalence in philosophy of the seemingly self-evident assumption that disability is a natural disadvantage—that is, the idea that people "with disabilities" are "naturally" disadvantaged—has feedback effects for the careers of disabled philosophers (and disabled philosophers of disability in particular), for the composition of the profession, and for the very content of the discipline. Indeed, the prevalence in philosophy of an individualized and medicalized understanding of disability is co-constitutive with the exclusion of disabled philosophers (especially disabled philosophers of disability) from adequate employment in the profession and contributes to the marginalization of philosophy of disability in the discipline. My work in the two spheres has thus been designed to show that the conception of disability that predominates in philosophy, according to which disabled people are naturally flawed or defective, is *causally related* to the underrepresentation of disabled philosophers in the profession.

One of the central aims of these inquiries has been to compellingly demonstrate that disability is not a natural, transhistorical, and transcultural phenomenon; that is, I have endeavored to denaturalize and de-biologize disability by combining feminist philosophy and critical philosophical work on disability to produce a "feminist philosophy of disability" that draws upon Foucault's insights about (for example) apparatuses, biopower, the constitution of subjects, governmentality, and the problematization of phenomena. Foucault's work, I maintain, offers feminist philosophy of disability the most philosophically sophisticated and politically astute tools with which to denaturalize and de-biologize disability, doing so without appeal to spurious claims about ideology and an allegedly pristine realm of truth that such claims necessarily presuppose.

In the metaphilosophical sphere of my research, I have pointed out, for instance, that PhilPapers (N.d.), the increasingly influential database for philosophical research, is one of the institutionalized mechanisms of the discipline that both reproduces the individualized and medicalized conception of disability that prevails in philosophy and contributes to the marginalization of disabled philosophers and disabled philosophers of disability especially. Indeed, the architectural framework of PhilPapers (which its spin-off, PhilJobs, replicates) subordinates an array of areas of inquiry that underrepresented philosophers produce insofar as the content of the database is organized into predetermined and hierarchically arranged areas of specialization, subfields, and topics in philosophy according to dominant ideas in the tradition of Euro-American, Western philosophy about which areas, subfields, and topics: (1) have the most/less philosophical import, (2) have the most/less explanatory power, and (3) should be endowed with the most/less authoritative status. Thus the so-called core or fundamental areas of this tradition—"Metaphysics and Epistemology," "Value Theory," "Science, Logic, and Mathematics," "History of Western Philosophy," and "Philosophical Traditions"—are designated as the supreme categories on the database and other areas of inquiry are designated as subcategories of these supreme categories, or subcategories of the subcategories of the supreme categories, or "leaf" subcategories of the subcategories of the subcategories of the supreme categories, where a category's distance from the supreme categories implicitly announces its diminished philosophical import, explanatory power, and authoritative status.

Feminist philosophical work on disability is situated under a "leaf" subcategory of PhilPapers—namely, "Feminism: Disability"—that is subordinate to the subcategory of "Topics in Feminist Philosophy," a subcategory of "Philosophy of Gender, Race, and Sexuality," which is itself a subcategory of the supreme category of "Value Theory." In the schema of the PhilPapers database, that is, feminist philosophy of disability is represented as on par with "topics" in feminist philosophy such as "Autonomy," "Love," "Identity Politics," and "Reproduction" rather than represented as on par with and in relationship with other apparatuses of identity and subjection in a more comprehensive superior category of "Philosophy of Gender, Race, Sexuality, and Disability," to which the subcategory of "Topics in Feminist Philosophy" would, in turn, be subordinate. Although the superior category of "Philosophy of Gender, Race, and Sexuality" comprises subcategories of "Philosophy of Gender," "Philosophy of Race," and "Philosophy of Sexuality," it does not encompass an analogous subcategory of "Philosophy of Disability." There is of course no objective and value-neutral explanation that can be offered for why feminist philosophy of disability has been so categorized in the PhilPapers database. The relegated status that feminist philosophy of disability occupies in the database is rather a value-laden and interested political decision that, among other things, precludes and even prevents the incorporation of disability into intersectional or other integrated analyses, thereby reinforcing depoliticized conceptions of disability in philosophy and contributing to the marginalization and diminution of critical work on disability within the subfield of feminist philosophy and the discipline more generally (see Tremain 2013, 2017: ix–xi).

The rationale for the subordinated status that feminist philosophical work on disability occupies in PhilPapers becomes even more evident when one considers how disability is classified elsewhere in the database. For example, "leaf" sub-sub-sub-categories with respect to disability can be found under the broader sub-sub-category of "Biomedical Ethics," alongside and on par with items such as "Drugs," "Death and Dying," and "Neuroethics," as well as under the sub-sub-category of "Social Ethics," alongside and on par with items such as "Deception" and "Friendship." These "leaf" sub-sub-sub-categories of "Disability" are derivatives of the superior (sub-)sub-category of "Applied Ethics," itself a subcategory of the supreme category of "Value Theory." In short, the inclusion of "Disability" under the rubric of "Biomedical Ethics" both reinforces individualized and medicalized conceptions of disability that prevail in philosophy and minimizes the social, political, and discursive significance of disability, especially in light of how subordinate the positioning of "Disability" is within the database as a whole (see, e.g., Tremain 2013, 2017, 2019; also see Tremain 2015).

Let me underscore that the classification of subfields in philosophy and the questions and concerns that these subfields comprise are no mere value-neutral representation of objective differences, relations, and similarities that await discovery and recognition. Rather, classification and classification systems in philosophy (and everywhere else) are performative insofar as they contribute to the constitution of the value-laden resemblances, distinctions, and relationships between phenomena and states of affairs that they put into place. Although the formula of PhilPapers represents philosophy as a value-neutral, detached, disinterested, and impartial enterprise, as my critiques of the database have emphasized, political, social, economic, cultural, and institutional force relations influence every aspect of the discipline (and profession) of philosophy. My critiques of the PhilPapers database have emphasized, furthermore, that every philosophical question and concern, as well as every subfield that these questions and concerns comprise, is a politically potent artifact of historically contingent and culturally specific discourse. Insofar as every philosophical question, every philosophical subfield, and every specialization in philosophy is a contingent artifact of discourse, these questions, subfields, and specializations in philosophy have histories that can be traced genealogically (see Tremain 2013, 2017: ix–xi).

Genealogy as a Practice of Denaturalization

Genealogy is, I maintain, the most effective philosophical practice with which to denaturalize and de-biologize disability (and its naturalized foundation, impairment). Genealogy can, for instance, reveal how certain domains and subfields of philosophy such as bioethics have emerged and been configured in ways that reproduce the naturalization and biologization of disability. The pathologization of Phineas Gage, which enabled the emergence and consolidation of the fields of cognitive science and neuroscience, is a case in point. In 1848, Gage, a railroad supervisor, was impaled by a tamping iron that entered his left cheek and exited the back of his skull. Over the course of more than a century and a half, an almost mythical narrative has been elaborated within psychology and medical textbooks about the aftermath of Gage's injury, a mythology to which philosophers and cognitive scientists have subscribed and promoted. As Allan Ropper (a neurologist at Harvard

Medical School) puts it, Gage's "famous case" helped to "establish brain science as a field." Says Ropper, "If you talk about hard core neurology and the relationship between structural damage to the brain and particular changes in behavior, this is ground zero." Ropper explains that Gage's brain injury offered scientists and medical practitioners "an ideal case" because it involved one region of the brain, was very evident, and the resulting changes in personality were "stunning" (Ropper quoted in Hamilton 2017).

The contested status of assumptions about the modularity of the brain aside (see, for instance, O'Donovan 2013), these sorts of claims about the impact of Gage's injury have led philosophers of mind and cognitive scientists (among others) to use the story of Gage as a springboard to advance arguments, develop experiments, and formulate positions on, among other things, personal identity, the true self, and the moral self (Strohminger 2014; Knobe 2016; Tobia 2016). My argument is, however, that the uses to which philosophers of mind and cognitive scientists routinely put the story of Gage naturalize the phenomenon of brain injury and are highly contestable insofar as they simultaneously contribute to the social significance attributed to brain injury and tendentiously isolate brain injury from the social, cultural, and political contexts in which it is embedded and in which its social significance is materialized (see my discussion of Gage in Tremain 2017).

Foucault, in his work on the history of the modern prison and the history of sexuality, adapted the genealogical method that Nietzsche famously introduced in studies of the descent of Western morals, variously referring to his own incarnations of genealogy as "histories of the present" and "historical ontologies of ourselves." For Foucault, the genealogist asks: of what is given to us as universal, necessary, and obligatory, how much is occupied by the singular, the contingent, the product of arbitrary constraints? In other words, Foucault used genealogy to critically inquire into the history of necessity on a given topic and the historical emergence of the necessary conditions of a certain state of affairs, pointing out that a critical ontology of ourselves must not be considered as a theory, doctrine, or permanent body of knowledge, but rather as a "limit-attitude," that is, an ethos, a philosophical life in which the critique of what we are is at the same time the historical analysis of the limits imposed on us. Genealogy is designed to identify how historically contingent practices, encounters, events, and accidents have enabled the emergence of current modes of thinking and acting and the limits that they impose (see Foucault 1982: 216–219). As Ladelle McWhorter (2010) puts it, genealogies help us to make sense of how we are now, in this historical moment, by looking at how we got here and how this, here, now, is historically possible.

Genealogy is thus especially apt for an examination of the ways that disability is naturalized in philosophy of science in general and feminist philosophy of science in particular. For example, one can use genealogy to trace the emergence of modern notions of normality within philosophy of science and the role that these historical artifacts have played in the constitution of disability within philosophy of science, including feminist philosophy of science. As Foucault remarked, genealogies are not positivistic returns to a form of science that more accurately represents phenomena. Rather, genealogies are "antisciences." Genealogies, Foucault wrote, "are about the insurrection of knowledges. . ... An insurrection against the centralizing power effects that are bound up with the institutionalization and workings of any scientific discourse organized in a society such as ours." Genealogy is an "attempt to desubjugate historical knowledges . . . to enable them to oppose and struggle against the coercion of a unitary, formal, and scientific theoretical discourse" (Foucault 2003: 9). Thus, genealogies require the excavation and articulation of subjugated knowledges, knowledges that "have been disqualified as inadequate to their task or insufficiently elaborated: naïve knowledges, located low down on the hierarchy, beneath the required level of cognition or scientificity," knowledges of the psychiatrized individual, of the delinquent, and of the personal care worker (Foucault 1980b: 82). Historical ontologies—that is, genealogies—exhume these phenomena, these subjugated knowledges, in order that the historically contingent character of the self-understandings and self-perceptions that we hold in the present can be understood.

Few feminist philosophers of science have challenged the alleged self-evidence of accepted philosophical and scientific assumptions about the putatively natural character of disability, nor have more than a few feminist philosophers of science made a noticeable effort to incorporate critical philosophical work on disability and the subjugated knowledges of disabled people into their own analyses of science, their critiques of mainstream philosophy of science, or feminist philosophy of science itself. Hence the discussion of this chapter goes some distance to correct the omission of (feminist) philosophy of disability from feminist philosophy of science by offering a cutting-edge understanding of disability that denaturalizes disability, that is, a historicist understanding of disability that feminist philosophers of science (and feminist philosophers more generally) can fold into their research and teaching, thereby increasing the likelihood that philosophy of disability will play a more formative role in shaping the future of feminist philosophy of science than it does at present.

To motivate the discussion, I draw upon insights that Foucault introduced in his critical ontologies (genealogies) of abnormality, perversion, madness, and other discursive objects that intellectuals and non-intellectuals alike generally associate with disability. My discussion suggests ways in which feminist philosophers of science can extend the inquiries that Foucault made in these critical ontologies and could, therefore, be aptly characterized as a feminist philosophical treatise on what he referred to as the "problematization" of phenomena in the present, that is, characterized as a critical feminist ontology of the problematization of disability. Foucault did not advance normative proposals in his critical ontologies of abnormality, perversion, madness, and other phenomena; rather, Foucault's critical ontologies were designed to uncover how these phenomena became thinkable in the first place, that is, how they emerged as problems to which solutions were produced (see Tremain 2017). Following Foucault, my discussion in this chapter is designed to (1) persuade feminist philosophers of science that a historically contingent and culturally specific regime of power—namely, biopower—has constituted certain acts, practices, subjectivities, bodies, relations, and so on as a problem for the present, (2) convince feminist philosophers of science that philosophy has played and continues to play a considerable role in the elaboration of this problem, and (3) point out to feminist philosophers of science that in all likelihood their own theoretical analyses and discursive practices contribute to the persistence in philosophy of the simultaneous problematization of disability and exclusion of critical work about it, for feminist philosophers themselves do indeed continue to naturalize disability, with deleterious consequences.

One might think that the persisting naturalization and problematization of disability in feminist philosophical contexts is surprising given how much work feminist philosophers have done to denaturalize and de-biologize other social categories such as sex, gender, and sexuality, with some of the most powerful feminist arguments to denaturalize sex and gender articulated by feminist philosophers of science themselves. Indeed, feminist philosophy of science emerged in large part as a critical response to the essentialist assumptions about sex and gender that have conditioned Euro-American thinking in general and Euro-American science and philosophy of science in particular, that is, emerged as a critical response to the essentialist assumptions about sex and gender that continue to both limit the kinds of questions that mainstream scientists and philosophers of science regard as worthy of investigation and circumscribe the kind of responses to the questions that they will seriously consider. Feminist scientists and feminist philosophers of science have, among other things, thrown into relief the biased nature of taken-for-granted assumptions about reproductive processes (Martin 1991), undermined the gendering of the brain (Fine 2011), and subverted the very idea of two natural binary sexes (Fausto-Sterling 2000). Furthermore, some of the earliest work that feminist philosophers of science designed to denaturalize and de-biologize sex and gender made associations between essentialist arguments about sex-gender and, among other things, degradation of the environment, subjugation of nonhuman animals, and colonial projects worldwide (see Harding 1986, 1991). I want to point out, however, that although some feminist philosophers of science have not confined themselves solely to critique of the philosophical assumptions that underpin traditional philosophical and scientific claims about sex and gender, critical questions and concerns about disability, race, indigeneity, and nonhuman animals (among other things) nevertheless remain sidelined in feminist philosophy of science.

Many feminist philosophers, including many feminist philosophers of science, continue to implicitly construe gender (and sex) as prior to, more fundamental than, and indeed separable from disability and other apparatuses of subjecting power, even though they explicitly claim to endorse and uphold the political, theoretical, and discursive value of intersectionality. In other words, many feminist philosophers continue to presume that insofar as "women" share so many experiences in virtue of their gender an analytic focus on gender in isolation from, say, disability, race, ethnicity, class, sexuality, age, and nationality constitutes a legitimate project. I want to point out, however, that the analytical purity of the conception of gender that these feminist philosophers employ is achieved only by obscuring other apparatuses of power with which gender is complicit and mutually constitutive, usually through the implicit institution of a nondisabled white norm. Nevertheless, the omission of critical philosophical work on disability from feminist philosophy of science and feminist philosophical analyses more generally is in largely due to the endurance in philosophy of an individualized and medicalized conception of disability according to which disability is a natural, transhistorical, and transcultural disadvantage, that is, a deleterious and non-accidental human characteristic or property that some individuals possess or embody rather than a contingent apparatus of power in which everyone is implicated and positioned. In the terms of the individualized and medicalized understanding of disability that many feminist philosophers (like almost all mainstream philosophers) assume, disability is indistinguishable from its naturalized foundation, impairment, is a prediscursive and hence philosophically uninteresting human characteristic or attribute, a transhistorical and transcultural defect or difference that medicine and science can accurately represent.

An especially worrisome example of the naturalization and problematization of disability in feminist philosophical contexts can be found in *Perfect Me: Beauty as An Ethical Ideal*, feminist philosopher and bioethicist Heather Widdows's recent book on the ethics of feminine beauty standards. One of the central, if not *the* central, theses of the book is that norms with respect to beauty and appearance more generally are social constructs, never ethically or politically neutral, always value laden (Widdows 2018: 122, 131–137). Yet the brief remarks that Widdows makes about disability *Perfect Me* actually undermine these claims and the overall argument of the book, as well as reproduce an individualized and medicalized conception of disability in feminist philosophy and philosophical bioethics. Although Widdows's explicit remarks about disability span only a page-and-a-half of her 341-page book, uncritical endorsement of notions that rely upon individualized and medicalized conceptions normality and disability—"health," "healthy," "healthy functioning," "disorder," "risk," "abnormal bodies," and "suffering"—peppers the analysis of beauty throughout the book. Now, I surmise that Widdows's training in and implicit allegiance to analytic philosophy and in particular analytic bioethics likely precluded from her discussion opportunities for critical analysis of disability and of the relations between normality, normalization, beauty, and disability. Were, however, Widdows to have engaged with (for instance) my work on disability and normality and the incremental normalization of force relations that neoliberal bioethics enables, she might have recognized how the genealogies of impairment, normality, and normalization that I have provided undercut the outdated claims about disability that she ultimately makes in her book, claims that variously serve to re-naturalize and re-medicalize the apparatus of disability and that, hence, run counter to some of the most philosophically and politically perspicacious work currently done in philosophy and theory of disability. In *Foucault and Feminist*

Philosophy of Disability (2017) and elsewhere, for example, I advance arguments according to which the ideas of normality and impairment are artifactual mechanisms and effects of biopower, which is a complicated matrix of social power relations that facilitates population management and control in the service of neoliberalism.

I have asserted that the most effective and philosophically tenable way to throw into relief the artifactual and contingent character of normality or any other seemingly natural and objective practice, state of affairs, identity, and value is to trace its genealogy, that is, genealogy is the most effective way to trace the historical accidents and contingent circumstances from which normality or any other putatively prediscursive value, practice, state of affairs, or identity emerged. I have also asserted that Foucault's work demonstrates how feminist philosophers (of science) can advance a philosophically up-to-date critique of the naturalization and biologization of disability that avoids spurious appeals to ideas about ideology (see Tremain 2015, 2017). I want to suggest, therefore, that if Widdows were to have consistently drawn upon Foucault's insights throughout in her book, she would likely not have taken argumentative recourse to contestable ideas about ideology and historically specific notions of normality. Insofar as appeals to ideology always presuppose the existence of a realm of truth that ideology obfuscates and to which it prevents or withholds access, Widdows's claims that ideology plays a role in the production and naturalization of ideals of beauty necessarily presuppose that there exists a realm of truth with respect to beauty that ideology about it distorts or obscures (Widdows 2018: 43). Yet an argument, such as Widdows's, that presupposes an unfettered realm of truth about beauty and closely associates the value of beauty with the value afforded to normality cannot, without contradiction, also assume that there is nothing objectively true about normality nor assume that the idea of normality is never value neutral. Indeed, although Widdows repeatedly asserts that normality is never neutral, always value laden, the remarks that she explicitly makes about disability actually belie these claims. Widdows writes:

> [T]hose who fall outside the normal range [of acceptable appearance] fall into two broad groups. The first are those who are disfigured by disability or accident or have physical features that are dramatically outside the normal range. No matter what those in these groups do they will never be able to attain normal. The second group are those who fall only a little outside the normal range, who could bring themselves within the normal range using products or procedures. For the first group there is no possibility of attaining normal, and therefore as appearance matters more, it is likely that discrimination against this group will increase. In addition, those who fail to measure up to appearance standards will become rarer, as appearance issues become regarded as disabilities.
>
> *(Widdows 2018: 150)*

This passage in Widdows's relatively short discussion of disability is both troubling and, I want to argue, virtually self-contradictory. Notice that in order to distinguish between people who can never become normal and people who can become normal with some effort, Widdows installs a stable and unchanging conception of normality, despite her subsequent suggestion that the category of disability itself shifts in dimensions. For Widdows, no measure of social change could bring perceptions of the people in the first group within the normal range of appearance standards. In other words, Widdows seems to suggest that some people are *prediscursively* outside of the normal range of appearance standards. In short, some people are, it seems, objectively, naturally, and ontologically, abnormal. Normal and abnormal aren't historically contingent, culturally specific, and value-laden social categories after all, at least not for everyone, at least not when they are applied to some people. Although Widdows's final remarks in the cited passage seem to suggest that she regards disability as a historically contingent and culturally specific phenomenon, subsequent remarks that she makes on p. 151 indicate that her ideas about the ontology of disability are far more undeveloped than an account that characterizes disability in this way would be. In short, Widdows's remarks on p. 151 rely upon a rather inaccurate representation of the British social model of disability, an early model of disability whose distinction between impairment (construed as a biological characteristic) and disability (construed as a form of social disadvantage imposed upon "people with impairments") naturalizes and medicalizes impairment and, hence, naturalizes and (re)medicalizes the apparatus of disability. As I have argued in various places, the distinction between impairment and disability that the British social model of disability institutes is in fact a chimera that collapses upon scrutiny. Impairment is as socially constructed as disability.

In the reconstructive-conceptual sphere of my work, I have argued that insofar as (feminist) bioethicists, philosophers of medicine, and philosophers of cognitive science (among others) assume individualized and medicalized conceptions of disability, they take for granted that the (social) disadvantages that disabled people confront are the inevitable consequences of natural (biological, physiological, or structural) disadvantages that should be prevented, eliminated from the human gene pool, or cured through scientific and technological discovery and medical intervention. Indeed, most feminist philosophers uncritically and implicitly accept the aforementioned assumptions about what disability is, though they may not derive the same normative conclusions from this ontology as mainstream philosophers do. Even most philosophers of disability do not rigorously question the metaphysical and epistemological status of disability, but rather advance ethical and political positions that largely assume the self-evidence of that status. That this cluster of assumptions is contestable, that it might be the case that disability is a historically and culturally specific and contingent social phenomenon, a complex apparatus of power rather than a natural attribute or property that certain people possess, is not considered, let alone seriously entertained.

Feminist Philosophy of Science and the Apparatus of Disability

An astute feminist philosophy of science would denaturalize and de-biologize disability by arguing that disability is a historical artifact, drawing upon Foucault's insights in order to do so. As I have indicated, I recommend that feminist philosophers of science adopt Foucault's idea of an apparatus to argue that disability is a historically specific and dispersed system of force relations that produces and configures practices toward certain strategic and political ends. As an apparatus, disability is a historically contingent aggregate that comprises, constitutes, and is constituted by and through a complex and complicated set of discourses, technologies, identities, and practices that emerge from medical and scientific research, government policies and administrative decisions, academic initiatives, activism, art and literature, mainstream popular culture, and so on. Although some of the diverse elements of the apparatus of disability seem to have different and even conflicting aims, design strategies, and techniques of application, the elements of the apparatus are nevertheless co-constitutive and mutually reinforcing. The apparatus of disability is expansive and expanding, differentially subjecting people on the basis of constructed perceptions and interpretations of (inter alia) bodily structure, appearance, style and pace of motility, mode of communication, emotional expression, size, mode of food intake, and cognitive character, all of which phenomena are produced and understood within a culturally and historically contingent frame and shaped by place of birth, place of residence, gender, education, religion, years lived, and so on.

Philosophical analysis that understands disability as an apparatus treats these phenomena as the outcomes of contextually specific and performative relations of biopower rather than as transcultural and transhistorical objective and determined facts about humans. Biopower, Foucault wrote, is "what brought life and its mechanisms into the realm of explicit calculations and made knowledge-power an agent of transformation of human life" (Foucault 1978: 143). Foucault used the term *biopower* to refer to the convergence of these immanent strategies and techniques for population management from the eighteenth century onward. Biopower, Foucault remarked, is

> the set of mechanisms through which the basic biological features of the human species became the object of a political strategy, of a general strategy of power, or, in other words, how, starting from the eighteenth century, modern [W]estern societies took on board the fundamental biological fact that human beings are a species.
>
> *(Foucault 2007: 1)*

Sex, Foucault pointed out, has been central to the economic and political problem of "population" that emerged with biopower and to which biopolitical strategies have responded; that is, with the emergence of biopower came the problematization of sex, including surveillance practices and policies designed to manage and improve relations between two supposedly natural sexes, birthrates, fecundity, sexual practices, and so on. Biopower, Foucault stated, is "what brought life and its mechanisms into the realm of explicit calculations and made knowledge-power an agent of transformation of human life" (Foucault 1978: 143). As with the apparatus of sex, so too with disability: a whole array of scientific, administrative, and social techniques and strategies have emerged from within the apparatus of disability to calculate and manage the ostensibly natural embodiment and experience of disability.

The conception of disability as an apparatus that I have developed does not rely upon some variation of the assumption that impairment and disability could be taken up as politically neutral and value-neutral objects of inquiry were it not for disabling practices and policies of exclusion that the ideological requirements of power place upon them. This assumption is, however, fundamental to the British social model of disability, which construes impairment as a politically neutral and value-neutral human characteristic or trait on which disability (construed as social oppression) is imposed. With the conception of disability as an apparatus, no domain of impairment or disability exists apart from relations of power, that is, there is no politically neutral and value-neutral description or definition of impairment and disability. Impairment and disability can never be freed from power, nor, furthermore, can there be a phenomenology that articulates these supposedly prediscursive domains. On the Foucauldian understanding of disability that I have developed, power relations are neither external to impairment and disability nor to their nexus in the apparatus of disability, but rather are integral to this relationship, constituting the knowledge and objects that these historical artifacts affect, as well as the artifacts themselves. Indeed, there is no exteriority between techniques of knowledge and strategies of power. Rather, knowledge-power relations are constitutive of the objects that they affect.

Efforts to denaturalize disability and its naturalized foundation, impairment, variously presuppose certain assumptions about the relation between biology and society—that is, between nature and nurture—while rejecting certain other assumptions about how biology and society are related to each other. Feminist sociologist and legal scholar Dorothy E. Roberts (2016) has schematized these assumptions by vividly describing two approaches to inquiry in the social sciences: the "old biosocial science" and the "new biosocial science." Roberts notes that the old way of doing biosocial science posits that biological differences produce social inequality, whereas the new way of doing biosocial science posits that social inequality produces biological differences. Roberts explains that the biological determinism of the former approach is achieved in several ways: first, nature is separated from nurture and the origins of social inequalities are located in inherent traits rather than in imposed societal structures; second, the old biosocial science postulates that social inequalities are reproduced in the bodies of socially disadvantaged people rather than reinvented through unjust ideologies and institutions; third, problems that stem from social inequality are claimed to derive from the threats that oppressed people's biology itself poses to society rather than from structural barriers and state violence imposed upon oppressed people; and fourth, the old bioscience endeavors to intervene and fix the perceived biological deficits of oppressed people themselves rather than end the structural violence that dehumanizes them and maintains an unjust social order. Roberts, whose own position on human biology is nonessentialist and constructivist, writes that, by contrast, the new way of doing biosocial science posits that every single biological element, every single biological process in the human body, every human cell, and everything that happens to a human cell is affected by society. In short, all of life is both biological and social. There is, Roberts notes, no natural body. Biology is not a separate entity that interacts with the environment; rather, these interactions constitute biology (Roberts 2016; see also Roberts 1998, 2012; Prinz 2012; Gilman and Thomas 2016). With Roberts, I have argued, furthermore, that critical analyses of biosocial science must consider how claims about the social construction of biological phenomena are produced, in what contexts they are mobilized, and for what political purposes (for example, Tremain 2015, 2017, 2019).

Feminist Resistance to Foucault

I have asserted that feminist philosophers of science should incorporate a denaturalized conception of impairment and disability into the field of feminist philosophy of science, drawing upon the work of Foucault in order to do so. Nevertheless, I predict that many, perhaps most, feminist philosophers are likely to resist my conviction that Foucault's insights suggest ways in which feminist philosophers of science might contribute to the philosophical denaturalization of impairment and disability and, by doing so, extend critical inquiry into the problematization of disability. Feminist philosophers have long disparaged and dismissed Foucault's work because of its alleged sexism and masculinism. In particular, feminist philosophers (and theorists) have argued that Foucault's history of sexuality ignored the significance of ("social") gender. Such feminist criticisms are vexing, however, especially given that one of Foucault's central aims in his genealogy of sexuality was to denaturalize "sex." In the first volume of *The History of Sexuality*, for example, Foucault introduced a reversal of the causal relation between sex and sexuality that the feminist sex-gender distinction institutes. Recall that the feminist sex-gender distinction posits that "sex" is a self-evident fact of nature and biology from which gender and sexuality follow. Foucault argued, however, that "sex is the most speculative, most ideal, and most internal element in a deployment of sexuality organized by power in its grip on bodies and their materiality, their forces, energies, sensations, and pleasures." The notion of sex, Foucault maintained, "made it possible to group together, in an artificial unity, anatomical elements, biological functions, conducts, sensations, and pleasures, and it enabled one to make use of this fictitious unity as a causal principle, an omnipresent meaning." With this innovative reversal of the terms of debate about sex, gender, and sexuality, in other words, Foucault posited sexuality as the "real historical formation [that] gave rise to the notion of sex" (Foucault 1978: 157), that is, as the apparatus that produces the ideal of "sex" as the foundational property that sexuality seeks to evoke or express. For Foucault, in other words, "sex" is an *effect* of force relations that has come to pass as the *cause* of a naturalized human desire. Sex, although it initially emerged as a mechanism of the apparatus of sexuality, has itself become an apparatus of force relations with which the apparatus of sexuality (among other apparatuses) is inextricably intertwined, mutually constitutive and reinforcing.

Let me draw attention to remarks that Foucault made in the closing pages of *The History of Sexuality, Volume 1* in order to indicate why the charges of androcentrism and masculinism that have been directed at him should not be accepted. In part 4 of this contribution to Foucault's genealogy of sexuality, he identified "four great strategic unities which, beginning in the eighteenth century, formed specific mechanisms of knowledge and power centering on sex" (Foucault 1978: 103). One of the "four great lines of attack" (p. 146) was "the hysterization of women's bodies," which Foucault described as a process in which the feminine body was analyzed and understood as "thoroughly saturated with sexuality," was integrated into the sphere of medical practices, by reason of a pathology intrinsic to it, and was placed in "organic communication" with: (1) "the social body (whose regulated fecundity it was supposed to ensure)," (2) "the family space (of which it had to be a substantial and functional element)," and (3) "the life of children (which it produced and had to guarantee, by virtue of a biologic-moral responsibility that extended through the entire period of the children's education)" (p. 104). Through the process of the "hysterization" of women (and their bodies), Foucault wrote, "sex" became defined in three ways: first, as an intrinsic property that men and women have in common; second, as an intrinsic property that men possess par excellence and that women lack; and third, as that property which, by itself, constitutes woman's body, ordering it entirely in terms of its reproductive functions and keeping it in a state of constant agitation with respect to the effects of that function (p. 153).

Notice that Foucault's remarks with respect to the three-fold process of women's hysterization precisely describe what feminists who endorse the (dated) sex-gender distinction identify as women's gendering in a specific historical moment and cultural context, that is, late eighteenth-century France. Foucault pointed out that, from the eighteenth century, women became increasingly subjected to a distinct medical-psychiatric discourse and to a discourse on motherhood that rendered them responsible for the lives of their own children and for the life of society at large. Thus, it seems evident that Foucault accounted for the discursive and epistemological space that feminist discourse refers to as "gender," even though he did not actually use the term *gender* to signify this space in his analyses. Furthermore, Foucault's transposition of sex and sexuality laid the groundwork for both the reversal of the causal relation between the categories of sex and gender and the pathbreaking argument about gender performativity that Butler (1999) makes some years later. In short, feminists who argue that Foucault fails to account for gender in his history of sexuality, that is, who argue that Foucault's history of sexuality incorporates masculinist and androcentric biases, seem to disregard the substance of his remarks about women in the first volume of the history of sexuality series. Indeed, I think that there is ample reason to dispose of the claim that Foucault's history of sexuality and his body of work in general are androcentric, masculinist, and rely upon sexist biases.

A Way Forward

The continued reluctance, if not refusal, on the part of many feminist philosophers, to seriously entertain Foucault's body of work and the work of feminist philosophers who draw upon his insights imposes conceptual and discursive limits on feminist philosophy of disability in particular and feminist philosophy more generally, as well as enables the continued marginalization of feminist philosophy of disability within feminist philosophy itself, including within feminist philosophy of science. The lessons that feminist philosophers of disability could derive from sustained engagement with Foucault's claims are undermined when feminist philosophers (and feminist theorists), relying on the legitimacy of unjustified and unsubstantiated objections to Foucault's work, disparage and dismiss his claims, refusing to countenance them. As I have indicated in this chapter and elsewhere, much of Foucault's work represents, among other things, a significant challenge to the self-evidence of assumptions about the ontological status of disability (see Tremain 2001, 2006, 2010, 2015, 2017).

Indeed, Foucault was the first philosopher, and in fact the first disabled philosopher, to persuasively expose the historical specificity and contingent character of the category of the normal and its cognates abnormal and pathological. In his writing on punishment (1977) and the history of sexuality (1978), in particular, Foucault described how knowledges produced about the "normal" case become vehicles for the exercise of disciplinary force relations that target certain people.

As philosopher and historian of science Ian Hacking puts it, "the benign and sterile-sounding word 'normal' [became] one of the most powerful ideological tools of the twentieth century" (see Hacking 1990: 169; see also Tremain 2019). Thus, Foucault's inquiries into the problematization of abnormality, racism against the abnormal, perversion, and madness were importantly groundbreaking, suggesting a host of avenues of investigation along which feminist philosophers of disability can proceed. Feminist philosophy of science that draws upon Foucault would open doors to critical feminist philosophical analyses of disability. Furthermore, the philosophical investigations about the apparatus of disability that could potentially follow from Foucault's critical ontologies and other inquiries would likely expand and enrich discussions in feminist philosophy of science about a range of other philosophical, political, professional, and social concerns and questions, including concerns and questions related to race, gender, settler colonialism, neoliberalism, exclusion, and underrepresentation. Despite calls from within feminist philosophy (and the broader discipline) for greater attention to be paid to the epistemologies, ontologies, concerns, and perspectives of previously marginalized and excluded groups in philosophy, the work of Foucault, a disabled gay male philosopher, remains systematically neglected. Feminist philosophers and other philosophers who are genuinely committed to increasing the heterogeneity of the discipline should want to change this state of affairs.

Foucault's insight according to which knowledge-power relations are constitutive of the very objects that they are claimed to merely represent and affect dissolves the binary distinctions between (for instance) description and prescription, fact and value, and form and content. Among other things, the insight indicates that any given description is indeed a prescription for the formulation of the object (person, practice, or thing) to which it is claimed to refer. Knowledge-power relations have brought impairment into being, brought it into being as a certain kind of thing, that is, as negative, as a natural disadvantage, as a problem to be corrected or rectified. Impairment has emerged as an area of investigation only because productive relations of power established it as a particular kind of object of a particular kind of inquiry, inquiry that has been possible and remains possible only because techniques of knowledge-power have been able to invest it as such.

If feminist philosophers of science were to assume the conception of disability as an apparatus in their analyses, they would help to move philosophical discussion about disability away from restrictive conceptualizations of it as (for instance) a personal characteristic or attribute, a property of given individuals, an identity, or a difference. As I have indicated, on the understanding of disability as an apparatus, disability is not a metaphysical substrate, a natural, biological category, or a characteristic that only certain individuals embody or possess, but rather is a historically contingent network of force relations in which everyone is implicated and entangled and in relation to which everyone occupies a position. Hence, if feminist philosophers of science were to assume that disability is an apparatus, in Foucault's sense, they would help to move philosophical discussion of disability toward a more comprehensive conceptualization of it than other conceptions of disability provide, a conceptualization of disability that is (among other things) historicist and culturally sensitive in ways that other conceptions of disability are not.

Related chapters: 20.

References

Butler, J. (1999) *Gender Trouble*, Tenth-anniversary edition. New York: Routledge.

Fausto-Sterling, A. (2000) *Sexing the Body: Gender Politics and the Construction of Sexuality*, New York: Basic Books.

Fine, C. (2011) *Delusions of Gender: How Our Minds, Society, and Neuroscience Create Difference*, New York: W.W. Norton & Company.

Foucault, M. (1977) *Discipline and Punish: The Birth of the Prison* (trans. Alan Sheridan), New York: Vintage Books.

———. (1978) *The History of Sexuality*, Vol. 1, *An Introduction* (trans. Robert Hurley), New York: Vintage Books.

———. (1980a) "The Confession of the Flesh," in C. Gordon (ed.) *Power/Knowledge: Selected Interviews and Other Writings, 1972–1977*, New York: Pantheon Books, pp. 194–228.

———. (1980b) "Two Lectures," in C. Gordon (ed.) *Power/Knowledge: Selected Interviews and Other Writings, 1972–1977*, New York: Pantheon Books, pp. 78–108.

———. (1982) "The Subject and Power," in H. L. Dreyfus and P. Rabinow (eds.) *Michel Foucault: Beyond Structuralism and Hermeneutics*, 2nd ed., Chicago: University of Chicago Press, pp. 208–226.

———. (2003) *"Society Must Be Defended": Lectures at the Collège de France, 1975–1976*. M. Bertani and A. Fontana (eds.). Translated by David Macey, New York: Picador, pp. 1–21.

———. (2007) *Security, Territory, Population: Lectures at the Collège de France, 1977–1978*. M. Senellart (ed.) Translated by Graham Burchell, New York: Palgrave Macmillan.

Gilman, S. L. and Thomas, J. M. (2016) *Are Racists Crazy? How Prejudice, Racism, and Antisemitism Became Markers of Insanity*, New York: New York University Press.

Hacking, I. (1990) *The Taming of Chance*, Cambridge: Cambridge University Press.

Hamilton, J. (2017) "Why Brain Scientists Are Still Obsessed with the Curious Case of Phineas Gage," *Shots: Health News from the NPR* (blog). NPR. May 21.

Harding, S. (1986) *The Science Question in Feminism*, Ithaca: Cornell University Press.

———. (1991) *Whose Science? Whose Knowledge? Thinking from Women's Lives*, Ithaca: Cornell University Press.

Knobe, J. (2016) "Personal Identity and the True Self," *Flickers of Freedom* (blog).

Martin, E. (1991) "The Egg and the Sperm: How Science Has Constructed a Romance Based on Stereotypical Male-Female Roles," *Signs: Journal of Women in Culture and Society* 16 (3), pp. 485–501.

McWhorter, L. (with S. Tremain). (2010) "Normalization and Its Discontents: An Interview with Ladelle McWhorter," *Upping the Anti: A Journal of Theory and Practice* 11. https://uppingtheanti.org/journal/article/11-normalization-and-its-discontents-an-interview-with-ladelle-mcwhorter

O'Donovan, M. (2013) "Feminism, Disability, and Evolutionary Psychology: What's Missing?" *Disability Studies Quarterly* 33 (4). http://dsq-sds.org/article/view/3872

PhilPapers: Online Research in Philosophy. N.d. https://philpapers.org/

Prinz, J. J. (2012) *Beyond Human Nature: How Culture and Experience Shape the Human Mind*, New York: W. W. Norton.

Roberts, D. (1998) *Killing the Black Body: Race, Reproduction, and the Meaning of Liberty*, New York: Vintage Books.

———. (2012) *Fatal Invention: How Science, Politics, and Big Business Re-create Race in the Twenty-First Century*, New York: The New Press.

———. (2016) "The Ethics of the Biosocial: The Old Biosocial and the Legacy of Unethical Science," Tanner Lectures on Human Values, Mahindra Humanities Center, Harvard University, November 2.

Strohminger, N. (2014) "The Self Is Moral." *Aeon*, November 17.

Tobia, K. (2016) "The Phineas Gage Effect." *Aeon*, December 21.

Tremain, S. (2001) "On the Government of Disability," *Social Theory and Practice* 27 (4), pp. 617–636.

———. (2006) "Reproductive Freedom, Self-Regulation, and the Government of Impairment in Utero," *Hypatia: A Journal of Feminist Philosophy* 21 (1), pp. 35–53.

———. (2010) "Biopower, Styles of Reasoning, and What's Still Missing from the Stem Cell Debates," *Hypatia: A Journal of Feminist Philosophy* 25 (3), pp. 577–609.

———. (2013) "Introducing Feminist Philosophy of Disability," *Disability Studies Quarterly* 33 (4), http://dsq-sds.org/article/view/3877/3402

———. (2015) "This Is What a Historicist and Relativist Feminist Philosophy of Disability Looks Like," *Foucault Studies* 19, pp. 7–42.

———. (2017) *Foucault and Feminist Philosophy of Disability*, Ann Arbor: University of Michigan Press.

———. (2019) "Feminist Philosophy of Disability: A Genealogical Intervention," *The Southern Journal of Philosophy* 57 (1), pp. 132–158.

Widdows, H. (2018) *Perfect Me: Beauty as an Ethical Ideal*, Princeton: Princeton University Press.

12

EPISTEMIC VICES AND FEMINIST PHILOSOPHIES OF SCIENCE

Ian James Kidd

Introduction

This chapter surveys some points of contact – conceptual, topical, and methodological – between contemporary vice epistemology and feminist epistemology and philosophy of science. There are three fundamental shared features: first, the *motivating conviction* that our epistemic practices and systems are typically suboptimal, defined relative to the goals and values of inquiry; second, the *methodological conviction* that investigation of these suboptimalities must attend to epistemic agents, construed as subjects situated in social systems and practices of inquiry; third, the *ameliorative conviction* that rectification of the suboptimalities of our projects and communities of inquiry is a main task of philosophy.

Such motivational, methodological, and ameliorative convictions are not exclusive to vice epistemologies and feminist epistemologies. One also finds them, too, in pragmatist science studies (see Clough 2003). It's also clear that these convictions can be articulated in many ways, depending on the specific agenda, concerns, and objectives of different projects in vice and feminist epistemologies. Once those convictions start to play out within specific disciplinary and intellectual contexts, they variegate and, often, diverge. Moreover, epistemology looks very different depending on the sorts of epistemic practices and institutions being studied: for instance, much feminist epistemological work has been on science, whereas most current vice epistemology focuses on educational and political systems. So, let's think through feminist vice epistemologies of science.

Character epistemology

The systematic study of epistemic vices and failings has a long history (Kidd 2019a). Some 30 years ago, virtue epistemology emerged, initially as a response to frustrations about the limitations of abstract, individualist analyses of epistemic agency – the main bugbears being coherentism, foundationalism, and Gettier problems: some date the decisive break to Ernest Sosa's 1980 paper, 'The Raft and the Pyramid', which suggested a role for epistemic virtues in the resolution of puzzles about justified true belief.

A more decisive thickening of conceptions of epistemology was Lorraine Code's 1987 book, *Epistemic Responsibility*, which argued for a 'responsibilist epistemology' – one that has at its heart the concept of responsibility as the core attribute of an epistemic agent, 'from which [all] other virtues radiate' (1987: 44). For a responsibilist, inquirers are embodied subjects, working with (and sometimes against) other agents, with whom they exist in complex, often conflicting interdependencies as a result of their sharing (or failing to share) in social practices and institutions of inquiry. Although Code's work was not explicitly feminist, nor able to deliver a worked-out account of epistemic responsibility, its ideas came to inform work in feminist epistemologies (Grasswick 2003). But as Code also subsequently noted, another merit was that 'thinking about epistemic responsibility moves close to the realm of virtue epistemology' (2014: 19).

The Emergence of Virtue Epistemology

The discipline of virtue epistemology gradually evolved during the 1990s, finding its own distinctive voice with Linda Zagzebski's *Virtues of the Mind* (1996). Situated within a broadly Aristotelian framework, she argued that virtue theory could help revitalise epistemology, as it had done for ethics. 'Given that almost everyone thinks the virtues are important in our moral life', Zagzebski presciently remarked ode, 'I see no reason why they would deny that they are important in our cognitive life' (1996: 74). Subsequently, a new project of studying the virtues of the mind emerged, even if much of it challenged or rejected several of her theses – her claim, for instance, that an epistemic virtue must have a success component, a strong claim given the phenomenon of epistemic luck (Lackey 2007).

By the late 1990s, one could distinguish two main ways of thinking about virtue epistemology in relation to earlier projects in analytical epistemology (cf. Baehr 2011: 12). A *conservative* conception sees the concept of epistemic virtue as useful for addressing one or more traditional problems in epistemology (such as justified true belief, externalism, and the like). An *autonomous* conception sees epistemic virtue as interesting in their own right, independently of any contributions to those 'classic' problems. Around this time, virtue epistemologists started to focus on analysis of specific epistemic virtues or of clusters of virtues such as the 'maps' of certain virtues offered by Roberts and Wood (2007: Part II; see, further, the essays in Battaly 2019: Part II).

Three claims should be made about the development of virtue epistemology that set up my discussion of vice epistemology in relation to feminist epistemology of science. First, none of these foundational studies were explicitly feminist in their methods, aims, or inspirations: the books by Baehr, Roberts and Wood, and Zagzebski do not refer to work in feminist epistemology, even if they dutifully cite Code's on epistemic responsibility. Roberts and Wood's brief history of twentieth-century epistemology mentions post-Kantianism, postpositivist philosophies of science, and Gettier problems, but not feminist or social epistemology (2007: ch. 1). More generally, those three monographs have little to no engagement with issues of intersectional social identity, systems of social power, patriarchy, and other core concepts of feminist analysis.

Second, virtue epistemologists engage with the history and philosophy of science as sources of case studies of epistemically virtuous practice and to support their calls for attention to the character of inquirers (see, e.g., Roberts and Wood 2007: ch. 7). But those case studies are usually presented and discussed without due sensitivity to historical and social contexts of scientific practice. Moreover, use of the relevant philosophies of science typically was decoupled from critical discussions internal to the philosophy of science, such as well-known work by such influential feminist philosophers of science, such as Evelyn Fox Keller (1985), Helen Longino (1990), Carolyn Merchant (1980), and the essays in Ann Garry and Marilyn Pearsall's influential 1989 collection, *Women, Knowledge, and Reality*. Indeed, when virtue epistemologists do engage with philosophy of science, they tend to focus almost exclusively on mid-twentieth-century Anglo-American male philosophers of science, such as Kuhn, engaged primarily in epistemological debates about rationality, theory change, or the semantics of scientific language.

A third feature of virtue epistemology is that it has tended, until quite recently, to neglect epistemic vices. The focus is almost entirely on virtues, flourishing, and other positive concepts, broadly mirroring the focus of much virtue ethics, which also focuses on the sunnier side of character. Unsurprisingly, some of the most lucid critiques of the frontloading of virtue and flourishing and the occlusion of vice and corruption in ethical and epistemological theorising came from feminist critics like Claudia Card (1996), Robin Dillon (2012), and Lisa Tessman (2005). The emergence of vice epistemology has therefore been a welcome, timely corrective to the one-sidedness of what we should really call *character epistemology*, whose constituent sub-disciplines are virtue epistemology and vice epistemology.

Vice Epistemology: A Primer

The current interest in vice epistemology was really initiated by the work of Heather Battaly (2010a, 2010b, 2014) and, as a discipline, was formally named by Quassim Cassam, who characterises it as 'the philosophical study of the nature, identity, and epistemological significance of intellectual vices' (2016: 159). Most contemporary work in vice epistemology falls into one or more of three kinds:

A   *Foundational*: analyses of the concept(s) of epistemic vice, their relation to the concept(s) of epistemic virtue, their relation to ethical virtues and vices, normative issues about the badness of vices, ontological issues about the nature of epistemic vice(s), the classification of epistemic vices, the significance of epistemic vices to other philosophical topics (e.g. Battaly 2014, 2016; Cassam 2016, 2019; Kidd 2016).

B   *Studies of specific vices*: the analysis and articulation of specific epistemic vices, such as the *familiar epistemic vices* – arrogance (Tanesini 2016), closed-mindedness and dogmatism (Battaly 2018) – and the *esoteric epistemic vices* – epistemic hubris (Kidd 2019a), epistemic insouciance (Cassam 2019: ch. 4), epistemic self-indulgence (Battaly 2010), and epistemic timidity (Tanesini 2018).

C   *Applied vice epistemology*: the application of vice-epistemic resources to the analysis and amelioration of epistemological problems arising in specific social and practical domains, such as education, healthcare, political deliberation, and scientific inquiry, as well as normative and methodological questions about effective and appropriate uses of vice-epistemic resources in applied contexts (e.g. Battaly 2013; Biddle, Kidd, and Leuschner 2017; Cassam 2019: chs. 2–4; Kidd 2019b).

There are much stronger feminist themes flowing through vice epistemology, much of it due to the influence of Miranda Fricker's 2007 book, *Epistemic Injustice*. A striking feature of this book is that the theoretical framework is a feminist-inspired virtue epistemology, despite its primary critical target being a vice – *testimonial injustice*. Commentators tend to neglect the vice-epistemic framing of epistemic injustice, which is fine, if that suits their purposes: nothing *requires* that epistemic injustices be analysed in terms of character epistemology. Still, it's worth our being reminded *testimonial injustice* itself is an epistemic vice and most fully understood in those terms (Battaly 2017).

An overtly feminist exercise in applied vice epistemology, informed by Fricker's work among much else, is José Medina's 2013 book *The Epistemology of Resistance*. It presents a complex analysis of gendered and racial oppressions through the lens of systematic social-structural intersectional epistemic injustices. Starting from the insight of liberatory theorists that 'oppression involves corrupted and distorted forms of social relationality', Medina defines an epistemic vice as 'a set of corrupted attitudes and dispositions that get in the way of knowledge' (2013: 41, 30). This is a feminist analysis in at least two key senses: there is, first, a built-in sensitivity to intersectional social identities and situatedness relative to systems of power, and, second, an explicit liberatory aspiration to find effective strategies to help mitigate the vices that motivate, enact, and sustain patterns of social oppression (2013: ch. 4). Similar feminist themes and sensibilities flow through the work of Alessandra Tanesini on the vices accompanying the virtue of humility: she has analysed such vices of the privileged as arrogance and *superbia* and the oppressive vices of timidity and servility, construed as contributions to a study of the psychology of oppression (e.g. Tanesini 2016, 2018).

Granted, these feminist themes and influences are not universal among all work in contemporary vice epistemology. Not all work on epistemic vices will obviously require any distinctively feminist approach, of course, and feminist theorists could take a lot from vice epistemological work even if it wasn't developed with feminist concerns in mind.

Vice Ontology

Before we can engage with those sorts of issues, though, there is one last abstract issue to address – *vice ontology* (Cassam 2020). I focus on two ontological questions: *what kind of thing is an epistemic vice* (the Kind Question) and *what are the bearers of epistemic vice* (the Bearer Question). Concerning the Kind Question, Cassam distinguishes two positions: *vice-monism* is the view that epistemic vices are one kind of thing (character traits, say, as with Aristotle) while *vice-pluralism* is the view that epistemic vices can be many kinds of things (Cassam is a pluralist, since, for him, epistemic vices can be *character traits*, *attitudes*, and *ways of thinking* such as wishful thinking). I think vice-pluralism is more useful for feminist purposes, since it's natural to speak of *sexist epistemic attitudes* or *sexist ways of thinking* – bell hooks, for one, refers to 'sexist thinking' as 'ways of knowing' that tend systematically to create 'distortions', 'misinformation', and 'false assumptions' (2013: ch. 16).

The Bearer Question asks to what sorts of things can we attribute epistemic vices, in the sense of talk of their being epistemically vicious being descriptive of their properties or qualities, rather than being figurative, as when one talks of 'the cruel sea'. The traditional answer within Western philosophical character theory has been *bearer-monism*: the sole bearers of epistemic vices (and epistemic virtues) are individual epistemic subjects, and everyone who does character epistemology agrees that there are epistemically vicious individuals. More recently, however, some epistemologists have argued for a more pluralist picture. Consider two options, each with interesting feminist precedents:

A   *Epistemically vicious collectives*: a collective can have one or more specific epistemic vices, V, whether or not some or all of its constituent members also possess V. We can therefore talk about arrogant committees, dogmatic research teams, and so on, without this simply meaning that the entirety or majority of their members are arrogant or dogmatic: the position called *non-summativism* (Fricker 2010; Lahroodi 2019).

B   *Epistemically vicious abstracta*: abstract objects – such as doctrines, policies, or conceptions or ideologies – are epistemically vicious if they tend to (a) *promote* epistemically vicious attitudes, actions, motives, or desires in the subjects and environments structured by them or (b) if their enactment tends to require the *exercise* of one or more epistemic vices when enacted or (c) if they could only be regarded as *attractive* to agents with at least a latent disposition to epistemic vice (Battaly 2013; Kidd 2018).

An interesting source for epistemically vicious abstracta is the criticisms of androcentric science as they have contingently developed historically – histories, of course, more deeply influenced by women than was once supposed, thanks to a generation of historical research on occluded women scientists, technicians, craftspersons, and others, which shows that the 'masculine profile of the sciences, as they have developed in Euro-American contexts in the last 300 years, was by no means monolithic or inevitable' (see Crasnow, Wylie, Bauchspies, and Potter 2018: §2).

Sandra Harding offers several candidates of epistemically vicious abstracta in her 1986 book, *The Science Question in Feminism*. She specifically targets positivist conceptions of science, especially those modelled on the physical sciences, arguing they're epistemically dogmatic in the sense that they incorporate conceptual and methodological presuppositions which structurally militate against critical self-appraisal:

> Their non-social subject matter and the paradigmatic status of their methods appear to preclude critical reflection on social influences on their conceptual systems; indeed, prevalent dogma holds that it is the virtue of modern science to make such reflection unnecessary.
>
> *(1986: 34)*

A philosophical conception of science (or some specific science) whose constitutive values and assumptions preclude critical self-evaluation is one at strong risk of promoting such vices as dogmatism and closed-mindedness – two vices described by Battaly (2018), which I show how to attribute to *epistemic stances* in Kidd (2018).

Another vice-based critique of abstracta is María Lugones and Elizabeth Spelman's famous attack on androcentric 'empirical, philosophical, and moral theories', according to which 'theories appear to be the kinds of things that are true or false; but they also are the kinds of things that can be, e.g., useless, arrogant, disrespectful' (1983: 578). Arrogance and disrespectfulness are vices, so there is precedent for attribution of epistemic vices to sexist and androcentric theories. Whether this could be cashed out in ways useful to feminist aims is a task for future work.

A feminist epistemologist can be ontologically pluralistic about epistemic vices: they can be many *kinds of thing* (character traits, attitudes, ways of thinking) and have many *bearers* (individual agents, collective agents, abstracta). Such pluralism can accommodate the range of targets of feminist critiques of sciences – the arrogantly sexist professor of physics, the closed-mindedness of a research team closed to the possibility of gendered politics in their laboratory, or dogmatic conceptions of science as an epistemically impeccable enterprise performed by genderless 'objective' inquirers.

Conceptions of Epistemic Vice

There are two main normative conceptions of what makes vices bad – one appeals either to their *effects* or to their *motives* – and this section sketches them and assesses their pertinence to feminist epistemology and philosophy of science against its main thematic concerns (see Anderson 2019: paragraph 1):

1   The exclusion of women from practices and communities of scientific inquiry.
2   The denial of epistemic authority (and other epistemic goods and attainments) to women.
3   The denigration of putatively 'feminist' cognitive styles and epistemic traits.
4   The generation of theories of women that are derogatory or subordinatory.
5   The occlusion of women's experiences from epistemic projects and practices.
6   The production of knowledge that is irrelevant or injurious to the interests of women.

Clearly enough, these themes need some clarification, at least when stated in this schematic way – for instance, to guard against the tendency to avoid inadvertently essentialising talk of women as a putatively homogeneous category, or, even worse, as a gloss that disguises a background focus on the concerns and experiences of relatively privilege white women in the developed world (the classic statement of this criticism is Lugones and Spelman 1983). It also seems clear that each of these epistemological themes refers to assumptions, attitudes, and actions that are epistemically vicious, at least intuitively.

Consider closed-mindedness, characterised by Battaly (2018) as an inability or unwillingness to engage with relevant epistemic options. We see this vice in the attitudes and actions of those who try repeatedly to exclude women from collaborative inquiry and try to deny them epistemic authority. Such persons are closed to such options as *including women in inquiry, affirming the epistemic authority of women, revising inherited assumptions about gendered-coded cognitive styles*, and so on.

Contemporary vice epistemology offers two main answers to the question of what it is that makes epistemic vices *bad* or *negative* in a way that can justify their classification as vices. *Consequentialism* regards the vices as tending to systematically cause epistemically bad effects such as overlooking evidence, ignoring explanatory possibilities, or blocking the pursuit of truth and other epistemic goods. This is the opposite of *virtue-reliabilism*, the idea that epistemic virtues are traits that tend to make one a more reliable agent, in the sense of one who tends successfully to come up with the epistemic goods.

Motivationalism is the view that epistemic vices are traits that express or manifest epistemically bad motives or desires, which their possessor may endorse and be aware of, or not. Typical examples might include a desire to interfere with the epistemic agency of other subjects, or a lack of concern for the evidential basis of one's beliefs, or a failure to appropriately value or desire epistemic goods. Motivationalism is the position opposed to *virtue-responsibilism*, the position introduced by Code (1987), on which the epistemic virtues are those traits that make on a responsible epistemic agent. An epistemically vicious agent lacks the motivations constitutive of an epistemically responsible subject, including what Zagzebski regards as the fundamental motivation for all the epistemic virtues, a desire for 'cognitive contact with reality' (1996: 167).

Actually, there is a third option, too, which Battaly (2014, 2015) dubs *pluralism*: a vice is a trait that (a) tends to cause a preponderance of epistemically bad effects *or* (b) expresses epistemically bad motives/desires *or* (c) both. Sometimes, the focus on either consequences or motives is enough for an account of why some epistemic trait is vicious, whereas in other cases one might think that a vice is only satisfyingly described attends to both its bad effects and motives. In the latter case, consider *epistemic malevolence*, which Jason Baehr (2010) argues has two forms: *impersonal epistemic malevolence* is (roughly) opposition to knowledge as such, while *personal epistemic malevolence* is (roughly) opposition to the epistemic well-being or development of one or more other persons (2010: 204). This vice surely has special pertinence to feminist epistemologists, since, intuitively, it is epistemically malevolent to have or endorse a desire to exclude women *qua* women from systems of inquiry by denying them epistemic authority – these being forms of *epistemic violence* (see Dotson 2011; Spivak 1988).

We therefore have three options – *consequentialism, motivationalism*, and *pluralis*m – and in the following sections I present fuller accounts of them and ask how they might be profitably appropriated by feminist epistemologists and philosophers of science.

Consequentialist Conceptions of Epistemic Vice

A consequentialist conception of epistemic vice focuses on the *epistemically bad effects* of traits, attitudes, and ways of thinking – what Heather Battaly dubs *effects-vices*. An epistemic vice is an epistemic failing because it tends to have, as Cassam puts it, 'a negative impact on our intellectual conduct' (2019: 2). The fullest statement of a consequentialist conception of this sort is *obstructivism*, which Cassam develops in his book, *Vices of the Mind*. Its guiding claim is that:

> An epistemic vice is a blameworthy or otherwise reprehensible character trait, attitude, or way of thinking that systematically obstructs the gaining, keeping, or sharing of knowledge.
> *(2019: 23)*

Note three features of this definition. First, vices must be 'reprehensible', in the sense that they are objects of *criticism*, of which *blame* is only one type. Cassam emphasises that we're not always justified in *blaming* people for their vices, many of which are acquired as a result of our environments. Our evaluative responses to vices therefore are, and ought to be, broad, and blame is only one sort (see Cassam 2019: 22f).

Second, obstructivism is ontologically pluralistic, since vices can be, at the least, character traits, attitudes, or ways of thinking; later in the book, Cassam introduces the terms *character-vice*, *attitude-vice*, and *thinking-vice* (2019: 79, 38, 56). Third, epistemic vices are bad due to their epistemically obstructive tendencies, which Cassam qualifies in his remark that epistemic vices must '*systematically* obstruct the gaining, keeping, or sharing of knowledge *in the actual world*' (2019: 12, my emphasis). I flag this point because feminist epistemologists, alert to the huge variations in the epistemic circumstances of differently socially situated subjects, will rightly point out that our unjust socio-epistemic environments are *already* typically epistemically obstructive, in all sorts of ways. Moreover, the tendency for epistemic character traits to be epistemically *obstructive* varies depending on the social identities of their possessors – a point to which I'll return.

An obstructivist conception of epistemic vice therefore relies on a form of epistemic consequentialism, specifically their tendencies to systematically obstruct three generic sorts of epistemic activity – the *gaining*, *keeping*, or *sharing* of knowledge, which includes self-knowledge about one's own possession of the self-concealing vices Cassam dubs *stealthy vices* (2019: ch. 7). We should also add other sorts of epistemic good, such as true belief, insight, or understanding, and also epistemic attainments such as epistemic virtues (2019: ch. 8). Within those three generic activities, one can make further distinctions: an *insensible* subject fails to *perceive* certain things as epistemically good, owing to a false or deficient conception of the epistemic good: think of a radical instrumentalist, maybe a Secretary of State for Education, who CAN only perceive things as epistemically valuable if they are also economically valuable (Battaly 2013: §2).

Cassam argues that epistemic vices can obstruct knowledge in at least three ways:

1   By reducing the likelihood that the vicious individual's beliefs will be true.
2   By getting in the way of belief.
3   By undermining one's right to be confident in one's beliefs (2019: 11).

Consider closed-mindedness: a person with this vice may be epistemically closed off to true beliefs, and unlikely to form beliefs about topics or issues to which they are closed, and the confidence they have in their beliefs is likely to be insecure, since they are unlikely to have openly explored certain aspects of those beliefs or salient critical perspectives on them (Cassam 2019: ch. 2).

Obstructivism may appeal to feminist epistemologists and philosophers of science for several reasons. First, the focus on the gaining, keeping, and sharing of knowledge and other epistemic goods is obviously consistent with the broad epistemic aims of scientific inquiry. Cassam (2019: 7, 86) actually frames obstructivism in the context of what Christopher Hookway called *enquiry epistemology*, a pragmatist-inspired conception of epistemic life as consisting of individual and collective efforts 'to find things out, to extend our knowledge by carrying out investigations directed at answering questions, and to refine our knowledge by considering questions about things we currently hold true' (1994: 211).

Second, obstructivism shares with feminist epistemology and philosophy of science an upfront sensitivity to the manifest suboptimalities of our epistemic activities and systems. Although Cassam focuses on the epistemic defects of the media and the political systems in the developed world, one can plausibly find epistemic vices across individuals, institutions, and cultures, including the sciences. Third, obstructivism is *ameliorative* in the sense that its analyses are intended to guide corrective measures, if and where possible. For Cassam, 'epistemic vices of various kinds have been at least partly responsible for a series of political and other disasters', and that our 'only hope of avoiding such disasters in the future is to improve our thinking, our attitudes, and our habits of thought and inquiry', with the alternative 'too ghastly to contemplate' (2019: 187).

A final feature of obstructivism with appeal to feminist theorists is an insistence that 'vice-explanations' of epistemic failures – ones invoking the epistemic vices of agents – are only apposite in certain cases and only then to a certain degree. Depending on the particular case, we may also need *structural explanations, ideological explanations, political-rational explanations, situational explanations,* and *sub-personal explanations,* like appeals to cognitive bias. Sometimes, a fuller story needs to be told about the causes of epistemic failures and vices may play only a marginal, secondary role in some of them. Obstructivism is thus committed only to the conviction that 'there are cases where [other] explanations don't do the job and where it is difficult to understand a person's intellectual conduct other than by reference to their epistemic vices'. Sometimes, says Cassam, 'the explanation *is* personal' (2019: 27).

Still, certain feminist epistemologists and philosophers of science may demur on at least three other features of obstructivism, each of which requires changes to its normative or methodological components. To start with, obstructivism, at least as it is presented in *Vices of the Mind*, has no built-in sensitivity to gender or social identity. There is recognition of the role of degrees of social power, as when Cassam argues that occupying 'positions of power and privilege can easily result in intellectual overconfidence or a cognitive superiority complex' (2019: 97). In such cases, vice-explanations and structural-explanations reveal the phenomenon of *epistemic corruption,* a form of characterological damage that occurs when a subject interacts with individuals, groups, structures, or processes that tend to facilitate the development and exercise of epistemic vices. Such epistemically corrupting conditions also underscore the need for *aetiological sensitivity* to the contingencies and complexities of the development of an agent's vices and character (see Kidd 2016, 2020).

A second concern with obstructivism is the *perception* of epistemic character traits, attitudes, and ways of thinking as epistemic vices can be distorted by sexist and racial biases, spoiling our capability to detect epistemically vicious agents. What may be needed to offset this concern is what Robin Dillon calls a critical sensitivity to 'the politics of character assessment' (2012: 100). Otherwise, charges of epistemic vice risk doing what they have been used for historically – the denigration, marginalisation, and oppression of women and other subordinated groups (Superson 2004). Think of the systematic tendencies within the sciences to exclude women by attributing to them – individually or collectively – such 'feminine' vices as foolishness, stupidity, or distorting sentimentality, obstructive relative to the demands of objective, rational inquiry (some classic statements are Anderson 1995, Lloyd 1984, Longino 1987).

A final feature of obstructivism that requires amendment relative to the aims and sensibilities of feminist analyses of science concerns its *axiological monism*: the reliance on epistemic values in the normative appraisal of vices, which assigns either a secondary role or no role at all to non-epistemic values – ethical, social, political. Cassam typically notes the bad practical, social, and political effects of epistemically vicious conduct, using case studies like the US invasion of Iraq, the Yom Kippur attack on Israel, and Brexit. But epistemic values are the normative core of obstructivism, and a feminist can response in two ways.

First, there is a weaker claim that normative accounts of at least *some* epistemic vices must necessarily appeal to both epistemic and non-epistemic values, insofar as any satisfying analysis of the badness of, say, the vice of testimonial injustice must attend to its ethical and political effects, too. Analyses of gendered epistemic injustices in science, for instance, must of necessity attend to the ethical and political consequences of the unjust silencing of the members of socially marginalised groups – otherwise, the ethically and politically charged aspects of occurrences of epistemically vicious conduct within science come to seem secondary, incidental features of what is primarily an epistemically problematic state of affairs (Grasswick 2018, which also contains a valuable survey of relevant work in feminist histories and philosophies of science and healthcare).

A second, stronger response is to reject axiological monism by denying the putative distinction between epistemic and non-epistemic domains of value. Nancy Daukas, for instance, urges us to reject any separation of epistemic and socioethical domains of value. Given the inseparability of knowledge, power, and agency, it's better to appreciate that the goals of feminist analysis are 'inseparably socio-ethical and epistemic goals' (2019: 379). But as a consequence, talk of *epistemic vices* goes out of the window. An authentically liberatory character epistemology sensitive to the *socially situated* nature of epistemic agency as structured by unjust inequities in power and privilege cannot endorse an axiological monism that construes the epistemic as separable from the socioethical – the epistemic *is* political and this demands an axiology undistorted by the misleading idea that a separate epistemic domain could be neatly cut free of its socioethical and political aspects. Indeed, if Kristie Dotson (2012) is right, that axiological assumption may itself help perpetuate systems of oppression.

Whether or not one can make obstructivism more axiologically pluralistic is unclear. Much depends upon the extent to which it is defined by its epistemic consequentialism. I think obstructivism could be made more alert to social positionality and the politics of character assessment and that doing so would complicate and deepen our understanding of the epistemically obstructive effects of the vices of the mind. If that's possible, then one can also develop obstructivism into a liberatory vice epistemology, which construes the vices of the mind as both socially oppressive and epistemically obstructive.

Motivationalist Conceptions of Epistemic Vice

The second main normative conception of epistemic vices involves their *motivations*. This is the position endorsed by Zagzebski: a virtue or vice has 'a component of motivation, which is a disposition or tendency to have a certain motive', the latter being 'an action-initiating and directing emotion', directed towards some end (1996: 134). There can be a fundamental motivation common to all epistemic virtues and vices and specific motivations that serve to individuate specific virtues and vices from one another.

Within vice epistemology, there are three main sorts of motivational account (see Crerar 2018: §§2 and 3).

A   *Presence accounts*: epistemic vice requires the presence of *bad epistemic motivations*, such as the motivations not to believe certain things, to be and remain ignorant about a certain topic, for beliefs that are comfortable or convenient rather than well-justified (cf. Battaly 2016: 105). Such epistemically bad motivation can be rooted in deeper, non-epistemic motivations such as maintaining one's privilege (cf. Crerar 2018: 757). Vices get 'some (or all) of their dis-value from the dis-valuable motivations they require' (Battaly 2016: 106).

B   *Absence accounts*: epistemic vice involves an absence of *good epistemic motivations*, such as the motivation to form justified true beliefs, to acquire and retain knowledge about salient topics, to establish and maintain cognitive contact with reality (cf. Zagzebski 1996: 167). Consider vices like epistemic laziness, incuriosity, or negligence (cf. Crerar 2018: 757). An epistemically vicious person may manifest either an *absent* or an *insufficient* motivation for epistemic goods relative to other goods (cf. Baehr 2010: 209).

C   *Compatibility accounts*: certain epistemic subjects can exemplify epistemic vices even though they are actually and ultimately motivated by a love of and desire for epistemic goods. A conspiracy theorist may be viciously closed-minded and gullible, while still being motivated by a genuine desire for the truth about certain events (cf. Cassam 2016: 162–163). Epistemic viciousness is thus compatible with the presence of epistemically good motives in at least certain cases (cf. Crerar 2018: 759–761).

The fullest development of a motivational analysis of epistemic vice if the presence type is Alessandra Tanesini's work on the vices opposed to humility. The excess-vices opposed to humility include arrogance and *superbia* and the deficiency-forms include timidity and servility (Tanesini 2016, 2018). She argues that epistemic vices have behavioural, emotional, and motivational components, and the latter two are causally efficacious in bad epistemic conduct. Arrogance includes behaviours like bragging, motivated fundamentally by a desire for ego-inflationary self-enhancement, and characterised by dispositions to frequently, situationally experience anger. Such specific motivations individuate epistemic vices, although all of them also share a more fundamental motivation to 'turn away from epistemic goods', the opposite to the desire for cognitive contact with reality arguably characteristic of epistemic virtues.

Motivational conceptions of epistemic vice resonate with key themes in feminist epistemologies and philosophies of science, reflecting Tanesini's earlier work in those very areas (see Tanesini 1999). Three points stand out. To start with, there's the emphasis on the affective and motivational dimensions of epistemic character and activity and a consequent appreciation of the significance of anger, fear, and other affective states to epistemology. Second, – the sensitivity to the ways that social identities shape the motivational psychology of epistemic subjects – Tanesini focuses on the ways that patterns of oppression and social privilege affect the development of the vices of humility. A final point connecting motivationalist conceptions of epistemic vice and feminist philosophies of science is the emphasis on the *messiness* of the motives informing scientific inquiry. Such laudable epistemic motivations as a 'love of truth' or the desire for knowledge do play a role within scientific inquiry, but these motivations are (a) often only partially expressed and are (b) perpetually accompanied by other bad epistemic motivations. A survey of feminist studies of science reveals an enormous variety of invidious motivations evinced by scientists. Elizabeth Potter, for instance, mentions:

- Being motivated to deny the significance of gendered assumptions to scientific practice.
- Being motivated to ignore evidence about the role of sexist assumptions and values in scientific theory and practice.
- Being motivated to resist claims about the social nature of scientific knowledge.
- Being motivated to exclude women and the members of other marginalised groups from scientific training, communities, and systems of advancement.

(Potter 2006, 19, 111, 119)

These are all epistemically bad motivations, since they involve what Tanesini calls a turning away from epistemic goods – for instance, from the evidence of persistent and systematic gendered inequalities in historical and contemporary scientific communities (a good survey and discussion is Schiebinger 1999: chs. 2–3, which also documents some absurd strategies used by apologists for androcentric science). I think there are good prospects for analyses of these strategies of denial and resistance in terms of motivated epistemic vice (cf. Kidd 2018: §6.3).

There are two issues about motivationalism when it comes to feminist analyses of science. The first is the *normativity question*: must the bad motivations be *epistemic* or can they be *ethical, social, or political* – another form of the issue about axiological pluralism. An array of epistemic vices evident in science are plausibly explained fundamentally in terms of misogynistic motivations, especially in medical and healthcare sciences (some recent studies include Criado Perez 2017: Part IV; Dusenbery 2017; Freeman and Stewart 2019). A vice can still be *epistemic* if it primarily affects epistemic activities, such as inquiring, even if the operative motivations are ethical, social, or political. Second, how do motivational accounts of vice fare relate to the emphasis by many feminist epistemologists and philosophers on the fundamentally collective character of scientific inquiry? We might argue that certain motivations can be shared by a collective (a research team, say), which could mean that we can talk about *collective epistemic vices* grounded in collective motivations. But this needs conversations between vice epistemology, social metaphysics, and feminist philosophies of science.

Related chapters: 6, 7, 8, 15, 16, 18, 17, 19.

References

Anderson, E. (1995) "Knowledge, Human Interests, and Objectivity in Feminist Epistemology," *Philosophical Topics* 23 (2): 27–58.

Anderson, E. (2019) "Feminist Epistemology and Philosophy of Science," in E.N. Zalta (ed.), *The Stanford Encyclopedia of Philosophy* (Summer 2019 Edition), forthcoming URL https://plato.stanford.edu/archives/sum2019/entries/feminism-epistemology/.

Baehr, J. (2010) "Epistemic Malevolence," *Metaphilosophy* 41 (1–2): 189–213.

Baehr, J. (2011) *The Inquiring Mind: On Intellectual Virtues and Virtue Epistemology* (Oxford: Oxford University Press).

Battaly, H. (2010a) "Epistemic Self-Indulgence," *Metaphilosophy* 41 (1–2): 214–234. Reprinted.

Battaly, H. (2010b) "Introduction: Virtue and Vice." *Metaphilosophy* 41 (1–2): 1–21.

Battaly, H. (2013) "Detecting Epistemic Vice in Higher Education Policy: Epistemic Insensibility in the Seven Solutions and the REF," *The Journal of Philosophy of Education* 47 (2): 263–280.

Battaly, H. (2014) "Varieties of Epistemic Vice," in J. Matheson and R. Vitz (eds.), *The Ethics of Belief* (Oxford: Oxford University Press), 51–76.

Battaly, H. (2015) *Virtue* (Cambridge: Polity).

Battaly, H. (2016) "Epistemic Virtue and Vice: Reliabilism, Responsibilism, and Personalism," in C. Mi, M. Slote, and E. Sosa (eds.), *Moral and Intellectual Virtues in Chinese and Western Philosophy: The Turn towards Virtue* (New York: Routledge), 99–120.

Battaly, H. (2017) "Testimonial Injustice, Epistemic Vice, and Vice Epistemology," in I.J. Kidd, G. Pohlhaus, Jr. and J. Medina (eds.), *The Routledge Handbook of Epistemic Injustice* (New York: Routledge), 223–231.

Battaly, H. (2018) "Closed-mindedness and Dogmatism," *Episteme* 15 (3): 261–282.

Battaly, H. (ed.) (2019) *The Routledge Handbook to Virtue Epistemology* (New York: Routledge).

Biddle, J., I.J. Kidd, and A. Leuschner (2017) "Epistemic Corruption and Manufactured Doubt: The Case of Climate Science," *Public Affairs Quarterly* 31 (3): 165–187.

Card, C. (1996) *The Unnatural Lottery: Character and Moral Luck* (Philadelphia: Temple University Press).

Cassam, Q. (2016) "Vice Epistemology," *The Monist* 99 (3): 159–180.

Cassam, Q. (2019) *Vices of the Mind: From the Intellectual to the Political* (Oxford: Oxford University Press.

Cassam, Q. (2020) "The Metaphysics of Epistemic Vice," in I.J. Kidd, H. Battaly, and Q. Cassam (eds.), *Vice Epistemology* (London: Routledge), forthcoming, 37–52.

Clough, S. (2003) *Beyond Epistemology: A Pragmatist Approach to Feminist Science Studies* (Oxford: Rowman & Littlefield Publishers).

Code, L. (1987) *Epistemic Responsibility* (Andover: University Press of New England).

Code, L. (2014) "Feminist Epistemology and the Politics of Knowledge: Questions of Marginality," in M. Evans, C. Hemmings, M. Henry, H. Johnstone, S. Madhok, A. Plomien, and S. Wearing (eds.), *The SAGE Handbook of Feminist Theory* (London: SAGE), 9–25.

Crasnow, S., A. Wylie, W.K. Bauchspies, and E. Potter (2018) "Feminist Perspectives on Science," in E.N. Zalta (ed.), *The Stanford Encyclopedia of Philosophy* (Spring 2018 Edition), https://plato.stanford.edu/archives/spr2018/entries/feminist-science/.

Crerar, C. (2018) "Motivational Approaches to Intellectual Vice," *Australasian Journal of Philosophy* 96 (4): 753–766.

Criado Perez, C. (2017) *Invisible Women: Exposing Data Bias in a World Designed for Men* (London: Chatto & Windus).

Daukas, N. (2019) "Feminist Virtue Epistemology," in H. Battaly (ed.), *The Routledge Handbook of Virtue Epistemology* (New York: Routledge), 379–391.

Dillon, R. (2012) "Critical Character Theory: Toward a Feminist Perspective on 'Vice' (and 'Virtue')," in S.L. Crasnow and A.M. Superson (eds.), *Out from the Shadows: Analytical Feminist Contributions to Traditional Philosophy* (New York: Oxford University Press), 83–114.

Dotson, K. (2011) "Tracking Epistemic Violence, Tracking Practices of Silencing," *Hypatia* 26 (2): 236–257.

Dotson, K. (2012) "A Cautionary Tale: On Limiting Epistemic Oppression," *Frontiers: A Journal of Women Studies* 33 (1): 24–47.

Dusenbery, M. (2017) *Doing Harm: The Truth About How Bad Medicine and Lazy Science Leave Women Dismissed, Misdiagnosed, and Sick* (New York: HarperOne/HarperCollins).

Freeman, L. and H. Stewart (2019) "Microaggressions in Clinical Medicine," *Kennedy Institute of Ethics Journal* 28 (4): 411–449.

Fricker, M. (2007) *Epistemic Injustice: Power and the Ethics of Knowing* (Oxford: Oxford University Press).

Fricker, M. (2010) "Can There Be Institutional Virtues?," in T.S. Gendler and J. Hawthorne (eds.), *Oxford Studies in Epistemology*, vol. 3 (Oxford: Oxford University Press), 235–252.

Garry, A. and M. Pearsall (eds.) (1989) *Women, Knowledge, and Reality* (Boston: Unwin Hyman).

Grasswick, H. (2003) "The Impurities of Epistemic Responsibility: Developing a Practice-oriented Epistemology," in H. Nelson and R. Fiore (eds.), *Recognition, Responsibility and Rights: Feminist Ethics and Social Theory* (Lanham: Rowman and Littlefield), 89–104.

Grasswick, H. (2018) "Understanding Epistemic Trust Injustices and Their Harms," *Royal Institute of Philosophy Supplement* 84: 69–91.

Harding, S. (1986) *The Science Question in Feminism* (Ithaca: Cornell University Press).

hooks, bell (2013) *Teaching Critical Thinking: Practical Wisdom* (London: Routledge).

Hookway, C. (1994) "Cognitive Virtues and Epistemic Evaluations," *International Journal of Philosophical Studies* 2: 211–227.

Keller, E.F. (1985) *Reflections on Gender and Science* (New Haven: Yale University Press).

Kidd, I.J. (2016) "Charging Others with Epistemic Vice," *The Monist* 99 (3): 181–197.

Kidd, I.J. (2018) "Is Scientism Epistemically Vicious?," in J. de Ridder, R. Peels, and R. van Woudenberg (eds.), *Scientism: Prospects and Problems* (Oxford: Oxford University Press), 222–249.

Kidd, I.J. (2019a) "Deep Epistemic Vices," *Journal of Philosophical Research* 43: 43–67.

Kidd, I.J. (2019b) "Epistemic Corruption and Education," *Episteme* 16 (2): 220–235.

Kidd, I.J. (2020) "Epistemic Corruption and Social Oppression," in I.J. Kidd, H. Battaly, and Q. Cassam (eds.) *Vice Epistemology* (London: Routledge), forthcoming, 68–85.

Lackey, J. (2007) "Why We Don't Deserve Credit for Everything We Know," *Synthese* 158: 345–361.

Lahroodi, R. (2019) "Virtue Epistemology and Collective Epistemology," in H. Battaly (ed.), *The Routledge Handbook of Virtue Epistemology* (New York: Routledge), 407–419.

Lloyd, G. (1984) *The Man of Reason: "Male" and "Female" in Western Philosophy* (London: Methuen).

Longino, H. (1987) "Can There Be a Feminist Science?," *Hypatia: A Journal of Feminist Philosophy* 2: 51–64.

Longino, H.E. (1990) *Science as Social Knowledge: Values and Objectivity in Scientific Inquiry* (Princeton: Princeton University Press).

Lugones, M. and E. Spelman (1983) "Have We Got a Theory for You! Feminist Theory, Cultural Imperialism, and the Demand for 'the Woman's Voice'," *Women's Studies International Forum* 6: 573–581.

Medina, J. (2013) *The Epistemology of Resistance: Gender and Racial Oppression, Epistemic Injustice, and Resistant Imaginations* (Oxford: Oxford University Press).

Merchant, C. (1980) *The Death of Nature: Women, Ecology, and the Scientific Revolution* (New York: Harper and Row).

Potter, E. (2006) *Feminism and Philosophy of Science: An Introduction* (London: Routledge).

Roberts, R.C. and W.J. Wood (2007) *Intellectual Virtues: An Essay in Regulative Epistemology* (Oxford: Oxford University Press).

Schiebinger, L. (1999) *Has Feminism Changed Science?* (Cambridge: Harvard University Press).

Sosa, E. (1980) "The Raft and the Pyramid: Coherence Versus Foundations in the Theory of Knowledge," *Midwest Studies in Philosophy* 5 (1): 3–26.

Spivak, G.C. (1988) "Can the Subaltern Speak?," in C. Nelson and L. Grossberg (eds.), *Marxism and the Interpretation of Culture* (London: Macmillan), 271–313.

Superson, A. (2004) "Privilege, Immorality, and Responsibility for Attending to the 'Facts about Humanity'," *Journal of Social Philosophy* 35 (1): 34–55.

Tanesini, A. (1999) *An Introduction to Feminist Epistemologies* (Oxford: Wiley-Blackwell).

Tanesini, A. (2016) "Calm Down, Dear: Intellectual Arrogance, Silencing and Ignorance," *Aristotelian Society Supplementary Volume* 90 (1): 71–92.

Tanesini, A. (2018) "Intellectual Servility and Timidity," *Journal of Philosophical Research* 43: 21–41.

Tessman, L. (2005) *Burdened Virtues: Virtue Ethics for Liberatory Struggles* (New York: Oxford University Press).

Zagzebski, L. (1996) *Virtues of the Mind: An Inquiry into the Nature of Knowledge and the Ethical Foundations of Knowledge* (Cambridge: Cambridge University Press).

13

"WHERE ARE ALL OF THE PRAGMATIST FEMINIST PHILOSOPHERS OF SCIENCE?"

Sharyn Clough and Nancy Arden McHugh

Introduction

As pragmatist feminists we have been committed to the integration of theory with practice, of thinking with doing, since the earliest stages of our careers in philosophy of science. However, the more our pragmatist feminist inclinations influenced the trajectory of our academic careers in philosophy of science, the more we recognized and began to push back against the troublesome separation between the expectations and reward systems of an academic life of the mind, and the practices and habits informing the material worlds with which we were engaged as scholars, as community members, as citizens. Increasingly, we both found ourselves working outside the confines of the academy, actively engaged with the public sphere—territory that would be familiar to the earliest of our pragmatist role models, John Dewey, W. E. B. Du Bois, and Jane Addams. Inspired by their work, we have each over the years moved our feminist philosophical research on science into closer alignment with our identities as public philosophers.

We draw on this journey in our entry for this *Handbook*, examining the communities within which our pragmatist feminist approaches to philosophy of science have developed, the similarities and differences in those approaches, the historical precedents that created space for philosophy of science to be approached in these ways, and the future lines of inquiry that this history allows us to imagine.

We first began engaging with each other's work in 2004 at the inaugural meeting of the *Association for Feminist Epistemologies, Methodologies, Metaphysics, and Science Studies* (FEMMSS) at the University of Washington, where Catherine Hundleby had organized a panel discussion of Clough's book *Beyond Epistemology: A Pragmatist Approach to Feminist Science Studies* (Clough 2003). This discussion was published in 2006 as a set of essays in *Metascience*. We return to that panel discussion, below. For now we highlight another panel discussion, eight years and many conferences later, at the 2012 FEMMSS meeting at Penn State. For this meeting, McHugh had invited Clough to participate in a plenary discussion "Where Theory and Practice Meet: Pragmatist Feminism as a Means of Knowing and Doing Scientific Practice." At this meeting, Clough began by establishing some common ground among the panel participants about what we might mean by "feminism," and "pragmatism." Recovering some of the terms and conditions that provided that common ground seems a good place to start.

Terms and Conditions

Clough circulated some definitions ahead of the 2012 panel discussion and they formed the basis of a fruitful exchange that eventually made their way into Clough's essay "Pragmatism and

Embodiment as Resources for Feminist Interventions in Science" published in the journal of *Contemporary Pragmatism* (Clough 2013). We reproduce and build from those definitions here.

With respect to feminism, Clough first addressed the social conditions to which she took feminism to be responding:

> Power, including the power associated with the practices and knowledge claims of science, is differentially defined, produced, and distributed according to complex social hierarchies. These hierarchies are calibrated in terms of a normative matrix of embodied markers, such as (presumed) primary and secondary sex characteristics, gender roles, ethnic, and national backgrounds, disability, and social/economic status.
>
> *(Clough 2013: 122)*

There are many notions of feminism available to us, but of relevance to this discussion is the kind of feminism that involves "a normative charge of inequity directed at these social conditions" (Clough 2013: 122). The normative feminist charge of inequity is, she argued, well-supported by a variety of broadly empirical claims such as the following:

1   The embodied markers according to which the complex social hierarchies are calibrated (as earlier) are socially and historically contingent features of human lives, always and already in play, even within feminist theorizing.
2   These embodied markers are irrelevant and arbitrary when used as criteria for considering the limits and possibilities of human flourishing.
3   When power is distributed under these social conditions, those who (have less access to power) are actively discouraged, via a variety of social and psychological mechanisms, from investigation of, and commitment to, points 1 and 2 (Clough 2013: 122).

We note in addition that one result of (3) is that the arbitrary embodied markers can become reified and made relevant as criteria for considering human flourishing, and that science practices have often been complicit in this reification.

Feminism, then, as we are using the term, involves

> not only the normative charge of inequity directed at the social conditions [above], but a call to respond to that inequity – a call to the institutions and persons benefiting from these social conditions to work for change: to more equitably define, produce, and distribute power, including the power associated with the practices and knowledge claims of science.
>
> *(Clough 2013: 123)*

Another part of the feminist response to that inequity involves genealogical digging to reveal the historical pathways that helped reify current relations between power and embodied markers, that helped those relations seem obvious and necessary rather than contingent. Because science has not been a neutral force in the reification process, obscuring contingencies, and reifying inequitable social conditions, science becomes an important site of critical feminist response.

Turning now to our definition of "pragmatism," we build on the definitions discussed at our 2012 panel and focus in particular on pragmatism as an epistemic outlook or methodological orientation that is problem-focused and naturalistic, and at the meta-epistemic level, pluralist and committed to the detection of contingency. Pragmatism helps us recognize the social and historical contingencies that underlie distinctions believed to be universal and necessary (e.g. the distinction between facts and values, and even the distinction between what counts as natural and non- or super-natural). Pragmatists are devoted to working with and through communities, and to the amelioration of oppressive social conditions. In short, pragmatism focuses on the relationship between knowers and the world—including relationships between scientists and the world—using the empirical, naturalized practices and knowledge claims of science to investigate the inequitable distributions of power associated with those very same practices and knowledge claims.

The ameliorative, community, and problem-oriented focus of pragmatism is part of what we think makes it such a helpful approach for feminist interventions in science and philosophy of science, but, as we will discuss later, this focus also often puts feminist pragmatists at odds with the reward systems of the academy, and of analytic philosophy of science more particularly.

Keeping in mind these definitions of feminism and pragmatism, we turn to a brief discussion of our American pragmatist role models and their relationship to science, especially social science, beginning around the turn of the twentieth century and moving into a review of some of the more contemporary feminist work in the philosophy of science and science studies that draws from or is informed by the American pragmatist tradition.

Historical Overview

In many ways, contemporary pragmatist feminist philosophy of science had its start in pragmatist feminist philosophy of the social sciences through the work of Jane Addams and W. E. B. Du Bois in the late nineteenth and early twentieth centuries. As scholar-activists, Addams's and Du Bois's work moved between, on the one hand, theorizing and researching about the causes of oppression, and, on the other hand, politically and materially working to ameliorate the social and political conditions that caused gender, class, race, and ethnicity-based oppression. Many of the traits we have characterized as features of contemporary pragmatist feminist philosophy of science—the commitment to epistemic pluralism, naturalism, focused on engagement with and through communities, and the amelioration of oppressive social conditions—were developed through the social science research and activism of Addams and Du Bois, as well as the work of John Dewey, G. H. Mead, and William James. These are among the traits that drew us to the pragmatist tradition.

Understanding how these traits are developed through the work of Du Bois and Addams is key to understanding the development of pragmatist feminist philosophy of science in the late twentieth century. It also helps to explain why early twenty-first-century pragmatist feminists have tended to move away from philosophy of science into other areas of philosophy that are more actively engaged with social justice in communities.

Both Addams and Du Bois took epistemic pluralism as a consistent starting point for their theorizing and their practice. By pluralism in this context, we mean the recognition of the epistemic strength that comes from attending to a plurality of voices, especially those from marginalized social positions. In her essay "Widening the Circle of Enlightenment" (2017 [1930]), and her writing about Hull House, "Twenty Years at Hull House" (1990 [1910]) Addams argues for both the epistemic and social necessity of pluralism. Her view was that immigrants had a significant amount to share that would enhance the social life and social structure of American culture, and that the dominant modes of forced assimilation as well as a lack of integration prevented these contributions from being realized. Epistemically, she viewed the voices of immigrants to be necessary for knowing how to shape policy and practice that would, in turn, shape and improve the experiences of immigrants in the United States. Addams argues that "no one so poignantly realizes the failures in the social structure as the man [sic] at the bottom, who has been most directly in contact with those failures and has suffered the most" (1990: 137). Contemporary pragmatist feminist philosophers, Marilyn Fischer and Charlene Haddock Seigfried, emphasize the importance of pluralism in Addams's work, with Seigfried arguing that Addams was "more explicit than [that of the white] male pragmatists about the value of insights of [white] women and of [men and women from] disadvantaged ethnic groups" (Seigfried, *Pragmatism and Feminism: Reweaving the Social Fabric*, 1996: 76).

Du Bois was also explicit in his commitment to epistemic pluralism. This is evident in one of the grounding arguments in his 1903 essay collection *The Souls of Black Folk* (Du Bois 1968 [1903]). Du Bois's formulation of double consciousness and the value of Black consciousness and knowledge for the advancement of progress in the United States, socially, politically, and intellectually, is an argument for epistemic pluralism. Du Bois argues that the goal for Black consciousness should not be an erasure of African Americans' epistemic and cultural links to African identity or a loss of their American identity and experience; instead the wish is for

> neither of the older selves to be lost. He would not Africanize America, for America has too much to teach the world and Africa. He would not bleach his Negro soul in a flood of white Americanism, for he knows that Negro blood has a message for the world.
>
> *(Du Bois 1968: 4)*

The valuing of these perspectives and the uneasy epistemic tension they create are, for Du Bois, a type of gift, a "second sight," that can allow a more clear view of the state of knowledge, politics, and social systems (1968: 3).

Because of Du Bois's and Addams's deep commitments to epistemic pluralism, they were both guided to a methodological pluralism that lent itself toward a naturalist approach. Both Du Bois and Addams initiated their research from a pluralist and naturalist starting point through empirical methodologies. For example, in his extensive research project, *The Health and the Physique of the American Negro* (1906) Du Bois researched the lives of rural Blacks and rural whites and urban Blacks and urban whites from multiple regions in order to understand the interaction of health and environment. He ultimately found that health status was tied to location and economic class, not biologically fixed categories of poor or robust "racial health." As Boles et al. argue in "The Dance between Addams and Du Bois: Collaboration and Controversy in a Consequential 20th Century Relationship" (2016), in his research to develop evidence regarding the social construction of race, Du Bois's "methodological dimensions focused on his use of multiple methodologies: surveys (interviews), archival work, and field observation" (Boles et al. 2016: 41). This allowed Du Bois to develop an extensive body of research that enabled him to make the argument that race is a social construct 50 years before anthropologists began making this claim, countering claims of the growing eugenics movement. This was the case even though his study received little uptake by white researchers or the white political structure.

Du Bois's commitment to feminist causes is less well known, but explicit in his work. One could think of him as an early intersectional feminist. For example, in the chapter "The Damnation of Women" from his 1920 book *Darkwater*, Du Bois early on notes how all women had historically been put in a position in which "[t]hey existed not for themselves, but for men" (Du Bois 2004 [1920]: 162), and that in order to be mothers, many women have done so at the "sacrifice of intelligence and the chance to be their best work" (2004: 164). He moves on to argue that all women must be enabled to have meaningful work and economic freedom from men. Du Bois then begins to explain the unique experience of what it means to be a Black person and a Black woman in the United States by stating "All of this woman—but what of black women?" (2004: 165). He points to the ways that racism, the legacy of slavery, and sexism uniquely impact Black women such that they are never really seen as women in white society. He states:

> I cannot forget that it is such Southern gentlemen into whose hands smug Northern hypocrites of today are seeking to place our women's eternal destiny—men who insist upon

withholding from my mother and wife and daughter those signs and appellations of courtesy and respect elsewhere he withholds only from [white] bawds and courtesans.

(Du Bois 2004: 172)

Using a similarly pluralistic methodology, Addams worked to develop an extensive body of data from interviews, detailed maps of the Hull House community. In her book *Jane Addams's Evolutionary Theorizing* (2019) Marilyn Fischer's close and nuanced reading of Addams's *Democracy and Social Ethics* (Addams 2002 [1902]) reveals the ways in which Addams utilized the social evolutionary theory of her time to shape knowledge, policy, and practice for the benefit of those most oppressed, while at the same time critiquing those with power. Fischer argues that in Addams's appeals to the science of the time she was able to pull on the moral sentiments of powerful people in order to work to meet the needs of oppressed members of her community. What is evident in Addams's and Du Bois's pluralism and in their naturalism, which started from the material and historical conditions of communities, is their joint commitment to the evidence of everyday experience as a site for knowledge.

Because Du Bois and Addams took seriously everyday experience as a site for knowledge, they both worked extensively in and through communities with the goal of social amelioration. "The Health and the Physique of the American Negro" is just one of many examples of Du Bois's research that started from his interaction with communities, with the goal of improving their material conditions. As Boles et al. argue, Du Bois and Addams were both invested in the settlement model of social development. Addams co-founded Hull House in 1889 and Du Bois visited Addams there numerous times, which led to his work in housing settlements in Philadelphia from 1896 to 1898 and his book-length project *The Philadelphia Negro* (1899), and to portions of the research for the report "The Health and the Physique of the American Negro" (1906). Du Bois's goal with this extensive research was two-fold—illuminating the life and health of African Americans and in doing so to provide the impetus for city, state, and federal government as well as aid groups to change policy and practice to positively shift those conditions and to direct funds toward those communities.

Addams's work is marked by her commitment to engaging in and through immigrant communities in Chicago. This was the foundation of her work at Hull House, but also of her work on other pertinent social issues that she continued to approach using pluralist, naturalistic methods. For example, Addams was instrumental in founding The Juvenile Protective Association and advocated to reform the justice system by arguing for separate juvenile courts and housing for juveniles. The first juvenile court and juvenile detention center was established in Chicago across the street from Hull House in 1899 by the Juvenile Court Act. Children who committed crimes were no longer prosecuted as criminals, but were viewed as children who needed help to learn skills and to live a better and more productive life. The Hull House reformer Julia Lathrop was the first chair of the Juvenile Court Committee, and Alzina Stevens, a Hull House resident, was the court's first probation officer. Addams and the Hull House reformers were invested in ameliorating both adult and juvenile crime and their research helped show that social conditions, primarily poverty, immigrant and minority status, and social marginalization, played a significant role in the causes of crime and, importantly, who was convicted of crime.

Du Bois and Addams also worked together on key national projects to ameliorate frequently deadly social conditions such as the lynching of African American men. To this end, Du Bois and Addams (along with fellow pragmatist John Dewey) helped to found the National Association for the Advancement of Colored People (NAACP).

It is important to note at this point that although her association with Du Bois and the NAACP illustrates the ways Addams was much more progressive on social issues than many of her white colleagues, we don't want to ignore or excuse her racism. There were a number of ways in which

Addams was inattentive to race or outright mistaken on matters related to race, even while she was highly attentive to the social conditions attending white ethnicity. For example, although she vehemently and publicly argued against lynching, Addams continued to hold on to a patently false and racist view that was commonly accepted among whites, viz., that the reason Black men were lynched was because they had raped white women (Boles et al. 2016). Du Bois provides an important contrast here because he seemed to be able to think across difference in a more significant and intentional way than Addams did. This may be because he was starting his work from a significantly more marginalized position, which allowed him to be attentive to gender in way that Addams appeared not to be as thoroughly able to with race. Du Bois's attention to gender was unusual for a man of any race at this time. He actively worked with women and actively worked to promote not just individual women's work, but the cause of women over all. And he did this in ways that showed he understood how women's intersectional identities shaped their experiences in the world, certainly better than Addams seemed to be able to do with respect to her understanding of Black experience.

Because of their own subject positions, Du Bois as an African American man, and Addams as a white woman, and because of their scholar-activist focus—with Du Bois challenging racism and its companion injustices of class, politics, health, and education, and with Addams challenging ethnocentrism and sexism and their companion injustices of class, politics, health, and education—neither Du Bois or Addams was accepted into the academy in any meaningful way. Both were always working on the periphery of academic life. While their research was welcomed in the growing field of sociology, their presence was not. And neither their research nor their presence was welcomed in academic philosophy, which participated more heavily in racialized and gendered boundary patrolling, and was in its disciplinary and professional identity formation becoming increasingly separate from the concerns of daily living—a point we return to below.

Addams and Du Bois were well aware of the boundary patrolling and gatekeeping of the academy in their experiences of sexism and racism from all levels of university life. They were also both well aware of the ideological challenges that were developing in the early part of the twentieth century, even in sociology, that would move the scholar-activist model so integral to pragmatism, out on to the periphery of the academy. For example, Addams who "was committed to reform which she understood required activism, including political action," saw the shift in the academy in the 1920s toward a more protective mode (Boles et al. 2016: 44). Addams ascribed this shift to the "fear of change" which, she argued, led many academics to "play safe" and stay away from the pragmatist tendency to combine the scholar-activist role (Boles et al. 2016 quoting Addams [1930]: 155). This boundary patrolling of the academy, along with the rise of the analytic tradition in academic philosophy in the late 1930s, led to a decrease in the role of academics in the shaping of the social fabric beyond the campus. While the analytic tradition had some of its roots in the socially progressive ambitions associated with the left-leaning Jewish émigrés who brought logical empiricism to the United States, these ambitions were censured in the political climate of the Cold War (Reisch 2005). This meant that the analytic tradition in the United States tended to prioritize questions and arguments that were internal to philosophy, such as "what is knowledge?" over those that raised questions about social ideals or social progress such as "why is affirmative action in college admissions valuable?" By the mid-to-late twentieth century, the move away from the concerns of the material conditions of daily living pushed academic involvement in social change to the very margins of the US intellectual tradition.

Pragmatism as Methodology, or How to be a Feminist Pragmatist Philosopher of Science

Despite the conceptual and practical parallels between the work of early American pragmatists and contemporary feminist philosophers, up until the late 1980s surprisingly few feminist philosophers had made use of pragmatist resources in their work. The counter-intuitive nature of this situation led Charlene Haddock Seigfried to ask "Where Are All the Pragmatist Feminist Philosophers?" (1991). A number of feminist philosophers answered her call, forming a robust conversation among feminist scholars, especially in feminist social and political philosophy.

Siegfried did not ask "Where are all the Pragmatist Feminist Philosophers of Science?" but she might have. The conceptual tools of pragmatism have had less uptake within feminist philosophy of science and science studies than they have within feminist social and political philosophy. While it is somewhat arbitrary to draw boundaries separating social, political, and epistemological topics in feminist philosophy (especially pragmatist feminist approaches to these topics!), if we take science broadly as an object of social, political, and epistemological investigation, the boundaries of our topic become a little clearer.

As we noted, by the mid-twentieth century in the United States, philosophical approaches to science were increasingly informed by the analytic tradition. Despite its roots in the ameliorative projects and naturalistic methods promoted by some of the scholars associated with the Vienna Circle (Cartwright et al. 1996), philosophy of science in the analytic tradition arose in a particular post-war climate that encouraged a more conservative task: to preserve the authority of science by providing rational, and often formal and highly idealized reconstructions of scientific theorizing. The analytic approach to philosophy of science tended to discourage attention to the messier details associated with social science methods, and highlighted physics instead as the model (and materially embedded) organism.

The decline in available pragmatist models and mentors meant that the few feminists in philosophy who had an interest in science and were attracted to naturalistic methods that attended to empirical (and often social) details of how science was actually practiced found themselves isolated and confined to the relatively unhelpful canonical works associated with the analytic philosophy of science. There were some bright spots in the canon however, around which feminists could orient their work—pockets of resistance in the tradition that began to unravel the idealized models of science. It is no surprise that the bright spots that attracted feminist attention often took as their starting point some feature of pragmatism as we have defined it, especially epistemic pluralism, naturalism, and holism with respect to the fact/value distinction. For example, feminists recovering from the inevitable letdown that typically accompanied the discovery that Hilary Putnam was not a woman could find some comfort in his fact-value holism. Thomas Kuhn's naturalism and Nelson Goodman's recognition of epistemic pluralism also provided important tools for feminist thought-in-action.

Some of the earliest links between feminist philosophy of science and pragmatism were routed through the work of W. V. O. Quine. A number of feminist philosophers interested in epistemology and the philosophy of science did their dissertations on Quine, including Sandra Harding, Nancy Tuana, and Miriam Solomon. Quine's links to pragmatism are not straightforward to be sure (Godfrey-Smith 2013); however, in the 1990s, feminists began to publish work highlighting pragmatist features of Quine's ideas, especially his naturalism and his fact/value holism, showing them to be helpful both for dismantling the received view in analytic philosophy of science and for building a positive program that aimed to set science in its social context (see, for example, Nelson 1990, 1993; Antony 1993; Campbell 1994, 1998).

The most comprehensive treatment of Quine in feminist science studies remains Lynn Hankinson Nelson's groundbreaking book *Who Knows? From Quine to a Feminist Empiricism* (1990); however she did not focus on pragmatism, Quinean or otherwise. On those occasions when she addressed pragmatism, she distinguished between the question "does this theory or hypothesis work?" as distinct from the question "is this theory or hypothesis true?" and expressed concern that the former pragmatist question could be construed as inviting skepticism. Insofar as Quine could be associated with pragmatism of this sort, she aimed to defend him from this skeptical charge (Nelson 1990: 92–93).

In Clough's 1998 essay "A Hasty Retreat from Evidence: The Recalcitrance of Relativism in Feminist Epistemology" and later in her book *Beyond Epistemology* (2003) she argued that defending Quine from these skeptical charges was a tricky epistemological task that feminist philosophers of science ought to avoid (see also Clough 2004). She offered a way to avoid the epistemic task by rethinking the definition of pragmatism at issue, and by highlighting the ameliorative target at which she took feminist science studies to be aiming. She argued against epistemic definitions of pragmatism, regardless whether those definitions were set in terms of truth, evidence, or usefulness. She focused instead on a notion of pragmatism that involved local, naturalistic investigations of evidence in particular science communities, aimed at ameliorating particular problems such as sexism and racism in scientific theories of evolution. This focus is closer to the definition of pragmatism we have offered here.

She traced her particular understanding of pragmatism through the work of Richard Rorty, another scholar associated with the analytic canon whose links to pragmatism, as with Quine's, are often questioned. Rorty was himself inspired by Dewey, and by the pragmatist influences he detected in Quine's student Donald Davidson (e.g. Rorty 1999 [1994]). Using a pragmatist reading of Davidson, Clough argued that epistemic debates about skepticism and the nature of truth might not be the best focus for feminists interested in the amelioration of oppressive social conditions and in the role of science in perpetuating those conditions.

We feminist pragmatist philosophers of science encourage a refocusing of the debate in more explicitly pragmatist terms back on to scientific practice itself, which means, among other things, bringing the evaluative political and cultural commitments of scientists within the purview of empirical investigation. We urge scientists (and philosophers of science) who are trained to consider the evidential support of their descriptive judgments about the world to consider also the evidential support for their value judgments about the world.

Elizabeth Anderson makes exactly these points in her essay "Feminist Epistemology: An Interpretation and a Defense" (1995). Anderson describes feminist epistemology as naturalized social epistemology (a thread she followed from Harding 1986; Nelson 1990; Duran 1991; Tuana 1992; Antony 1993; Potter 1993). Picking up on the naturalized empiricist approach of Nelson in particular, Anderson gives a now familiar nod to Quine:

> Naturalized epistemologists consider knowledge production as an activity in which inquirers are subject to the same causal forces that affect their objects of study (Quine 1969). They ask of science that it provide an account of its own activity.
>
> *(Anderson 1995: 54)*

A decade later in 2004, Anderson published "Uses of Value Judgments in Science: A General Argument, with Lessons from a Case Study of Feminist Research on Divorce" where she offered the most explicit articulation of the methodological commitments we have identified with pragmatism, that is, commitments that are naturalized, holistic, and pluralistic with an ameliorative, community, and problem-oriented focus.

As with Nelson, neither Anderson's 1995 nor 2004 essays gave explicit mention to Dewey nor did she focus on pragmatism more generally, though Anderson's intellectual debt to Dewey became more explicit as time went on. In 2004 she was asked by the editors of the *Stanford Encyclopedia of Philosophy* to write the entry on John Dewey's moral philosophy. In 2013 when she was given the title of Distinguished University Professor of Philosophy and Women's Studies at the University of Michigan, she was allowed to choose a philosopher for which she wanted the position named; Anderson chose Dewey. In 2014, feminist pragmatists everywhere cheered when she was invited to give the Dewey Lecture at the Central Division of the *American Philosophical Association*—the title of her talk: "Journeys of a Feminist Pragmatist." In this talk, Anderson described the route by which she had come to see that Dewey provided the critical conceptual tools needed for feminist work in the epistemic, social, and political realms. She wrote:

> Feminist philosophers of science were showing how social and political values could be embedded in the practice of good science—and indeed, how the claims of science to objectivity could be enhanced by such engagement with non-epistemic values. I argued that this could work only if moral and political values were also subject to empirical testing. I discovered a powerful theoretical account of how to do this in John Dewey's philosophy.
>
> *(Anderson 2014: 82)*

Building upon these historical and contemporary insights, we have been self-consciously defining feminist pragmatist approaches in philosophy of science less in terms of any expressed allegiance to particular members of the American pragmatist tradition, and more in terms of the commitment to epistemic pluralism, naturalism, and fact/value holism, focused on ameliorating oppressive social conditions with and across communities. We have by design cast a wide net. While we can't list the important work of all the scholars who meet some or even most of these criteria, we want to at least give a flavor of the wide range of approaches that can be traced to the pragmatist themes we've highlighted. These include Sharon Crasnow's work on objectivity and feminist values (e.g. 2006, 2013) as well as naturalism and pluralism in political science (e.g. 2015, 2019), Naomi Zack's work on pragmatism, science, and racial injustice (e.g. 2016, 2002) and community disaster preparedness (2009), Nancy Tuana's work on climate change, gender, and race (e.g. Tuana 2019; Tuana and Scott 2018), Maya Goldenberg's engaged feminist philosophical work with healthcare practitioners, especially her work on the difficult issue of vaccine hesitancy (Goldenberg 2016, 2019), and Inmaculada de Melo-Martin and Kristen Intemann's engaged research on a range of biomedical health issues (e.g. de Melo-Martín and Intemann 2011, 2007) and especially their new book *The Fight Against Doubt: How to Bridge the Gap Between Scientists and the Public* (2018).

We note also the work of feminist philosopher of science Evelyn Brister who researches the framing of environmental management decisions and how communities can impact environmental decision-making through epistemic feedback (see for example her essay on forest service management, "Distributing Epistemic Authority: Refining Norton's Pragmatist Approach to Environmental Decision-Making" Brister 2012). Alison Wylie's naturalized approach bridges philosophy and archaeology focused on pluralism and feminist values (e.g. 2012, 2017) often in collaboration with indigenous communities (e.g. 2014, 2015). Kyle Powys Whyte brings an intersectional and ameliorative lens to his extensive research program concerning indigenous communities and climate change (e.g. Whyte 2012, 2013, 2018) as does Liam Kofi Bright in his work in the social epistemology of science (e.g., Bright 2017); including a shout out to Du Bois in "Du Bois' Democratic Defence of the Value Free Ideal" (Bright 2018).

Danielle Lake's work provides a helpful example of engagement with pragmatist feminism inside and outside of the academy focused on "wicked problems." In her essay "Jane Addams and Wicked Problems: Putting the Pragmatic Method to Use" (2014) Lake argues that engagement with wicked problems—social problems, such as climate change, food access, and healthcare, that are "characterized by intense disagreement between fragmented stakeholders, multiple and often conflicting objectives, as well as high levels of uncertainty, variability and risk"—is an inherently pragmatist feminist methodology (Lake 2014: 77). Like many pragmatist feminists, Lake's work takes her outside of the academy where she is working with communities on capacity building to address a range of issues such as barriers to K-12 attendance and degree completion, and developing housing kits for Grand Rapids residents (Lake 2019; see also her earlier essay "Restructuring Science, Re-Engaging Society" 2012).

We end our discussion at the beginning, by returning to the inaugural meeting of FEMMSS at the University of Washington in 2004, where McHugh contributed to the panel discussion on Clough's book *Beyond Epistemology*. Here McHugh expressed concern that Clough had overstated the case, insofar as Clough's attack on certain kinds of epistemology still seemed to reject much that feminists could find useful in it. McHugh argued that pragmatist approaches to epistemology, grounded in case studies, were still possible and important. In her remarks later published as an essay in the *Metascience* review, McHugh (2006) cited the work of Tuana (2004), Charles Mills (1999), and Robert Proctor (1985) as exemplars of feminist epistemological work that withstood Clough's criticisms of epistemology.

McHugh's own commitment to feminism, pragmatism, and science; to epistemic pluralism, within and across communities; and to ameliorating oppressive social conditions is represented in much of her academic research, including her essay "More than Skin Deep: Situated Communities and the Case of Agent Orange in Viet Nam" (2011) which she then expanded into her book *The Limits of Knowledge: Generating Pragmatist Feminist Cases for Situated Knowing* (2015). McHugh's essay was prompted by a research trip she took to Viet Nam in 2004 to study how the transition in Viet Nam to a market economy ("Doi Moi") affected the daily lives and health of Vietnamese rural women. On this trip she was asked to visit one of the Peace Villages, part of the Tu Du Hospital in Ho Chi Minh City where child victims of Agent Orange were housed and cared for. After meeting the director of the Tu Du Hospital to discuss the health effects of Agent Orange and US policy related to Agent Orange, and then meeting the children and infants in the Peace Village, McHugh began to ask a different set of research questions than those that had originally motivated her trip. The new questions she asked helped her to recognize a gap between what many scientists and lay people want from science—that is the generation of knowledge that improves human flourishing—and what science actually provides. In *The Limits of Knowledge* McHugh argued that many of our current scientific methods do not and cannot meet the needs of communities that are situated outside of dominant culture and that experience multiple negative impacts, e.g. from poverty, poor access to medical care, environmental contaminants, physical and social isolation, stress, globalization, war, racism, colonialism, and sexism. She offered "transactionally situated" research methods as alternatives that could meet these needs. These research methods build from the insights of pragmatism, feminism, and critical race philosophy, and embed and start research from the oppressed communities whose health needs they seek to address. Her case-based analysis led finally to an argument for rethinking the job of philosophy as one that should exist primarily within the academy. McHugh concluded that philosophy should take up the challenge set out by Du Bois, Addams, and Dewey, viz., we should be out in the world, engaging in social change.

In her review of *The Limits of Knowledge* for *Hypatia Reviews Online* Clough was so inspired by McHugh that she issued a challenge:

> If you think, as McHugh does, that one of the best ways for philosophers to make change in the world is to connect knowing and doing—to rethink knowledge as a practice that is engaged and situated—then you're going to have to do philosophy in a way that is self-consciously engaged and situated. There are lots of ways to do this. McHugh. . . provides numerous examples of philosophically robust engagement. So what are you waiting for?
>
> *(Clough 2016)*

So What Are You Waiting For?

In Anderson's 2014 Dewey Lecture, she noted: "I have long thought, . . . that pragmatists spend far too much time talking about methodology. The key is to put it into practice and display its attractions by its fruits" (2014: 82). This important reminder is of course easier said than done. Analytic philosophy of science constrains the kinds of things that count as legitimate problems for scholarly attention. If you're a feminist *and* a pragmatist interested in science, then it's hard to practice engaged philosophy and make it fit. The forming of the *Consortium for Socially Relevant Philosophy of/in Science and Engineering* (SRPoiSE) was one kind of response to these constraints and its meetings attract feminists, especially those interested in work on marginality, oppression, and needed social change. The *Public Philosophy Network* (PPN) is another resource. We imagine that Rorty would have fit in well at the PPN. For his time he was a very publicly engaged philosopher—an identity, we note, that left him marginalized from the analytic philosophy community.

McHugh takes up her commitment to practice philosophy outside the traditional walls of academia by engaging with oppressed communities through teaching in prisons, detention centers, and reentry programs, as well as running restorative and transformative justice projects with a county juvenile court. She has also been co-writing with people who are incarcerated. The LoCI-Wittenberg University Writing Group, of which she is a member, published an article in the feminist philosophy journal *philoSOPHIA* entitled "An Epistemology of Incarceration: Constructing Knowing on the Inside" (LoCI 2016), which builds upon work in epistemic injustice to argue for a type of resistant epistemology that the group refers to as "carceral consciousness." The group worked intentionally to end the paper with a set of recommendations for change in the carceral system.

Clough's current research interests have shifted away from conceptual arguments about the nature and ineliminability of political values in science, and toward the practice of deliberating about these values, especially in controversial science policy contexts (Clough 2020). She focuses on the importance of skills like empathy and epistemic humility in these deliberations, building on feminist pragmatist themes in virtue epistemology, such as Alessandra Tanesini's work (2016a, 2016b), and explicated in terms of peace literacy—a new kind of phronesis or practical wisdom, borrowed from public philosopher Paul K. Chappell (2012, 2017). Conceiving of peace pragmatically as a set of skills or habits, she also works with Chappell to design curriculum on Peace Literacy for K-12 and higher education, and to train teachers and students in the effective use of the curriculum in the service of social justice for a sustainable future (http://www.peaceliteracy.org).

While we, Clough and McHugh both, engage in these activities because we find them inherently valuable, we also see our work as part of the pragmatist tradition that seeks to challenge the boundaries of the academy. Like the early American pragmatists, we believe the role of philosophy is best and most productive when it expands beyond the walls of academia and into and through communities, to engage in just, cooperative, and responsive social change. Our roots in feminism and pragmatism incline us to use the empirical, naturalized practices and knowledge claims of science to investigate inequitable distributions of power, even or especially as the inequitable distributions of power inflect these very same practices and knowledge claims—in science, and in the academy more generally. Feminist pragmatist philosophy of science challenges all of us to help build communities of knowers that are self-consciously and reflexively struggling with ameliorative projects; with engaged philosophy-in-the-world. There is work to be done.

Related chapters: 1, 9, 22, 32, 33.

References

Addams, J. (2017 [1930]) "Widening the Circle of Enlightenment," In *On Education*. New York: Routledge Press. https://doi.org/10.4324/9781315125541

———. (2002 [1902]) *Democracy and Social Ethics*, Urbana, IL: University of Illinois Press.

———. (1990 [1910]) *Twenty Years at Hull House*, Urbana, IL: University of Illinois Press.

———. (1930) *The Second Twenty Years at Hull House*, New York: McMillan.

Anderson, E. (2014) "Journeys of a Feminist Pragmatist," *Proceedings and Addresses of the American Philosophical Association* 88, pp. 71–87.

————. (2004) "Uses of Value Judgments in Science: A General Argument, with Lessons from a Case Study of Feminist Research on Divorce," *Hypatia: A Journal of Feminist Philosophy* 19(1), pp. 1–24.

————. (1995) "Feminist Epistemology: An Interpretation and a Defense," *Hypatia: A Journal of Feminist Philosophy* 10(3), pp. 50–84.

Antony, L. (1993) "Quine as Feminist: The Radical Import of Naturalized Epistemology," in Antony, L. and Witt, C. (eds.) *A Mind of One's Own: Feminist Essays on Reason and Objectivity*, Oxford, UK: Westview Press.

Boles, D., Hopps, J., Clayton Jr., O. and Brown, S. L. (2016) "The Dance between Addams and Du Bois: Collaboration and Controversy in a Consequential 20th Century Relationship," *Phylon* 53(2), pp. 34–53.

Bright, L. K. (2018) "Du Bois' Democratic Defence of the Value Free Ideal," *Synthese* 195(5), pp. 2227–2245.

————. (2017) "Logical Empiricists on Race," *Studies in the History and Philosophy of Science: Part C* 65(1), pp. 9–18.

Brister, E. (2012) "Distributing Epistemic Authority: Refining Norton's Pragmatist Approach to Environmental Decision-Making," *Contemporary Pragmatisim* 9(10), pp. 185–203.

Campbell, R. (1998) *Illusions of paradox: A Feminist Epistemology Naturalized*, New York: Rowman and Littlefield.

————. (1994) "The Virtues of Feminist Empiricism," *Hypatia: A Journal of Feminist Philosophy* 9(1), pp. 90–115.

Cartwright, N., Cat, J., Fleck, L., and Uebel, T. E. (eds.) (1996) *Otto Neurath: Philosophy Between Science and Politics*, New York: Cambridge University Press.

Chappell, P. K. (2017) *Soldiers of Peace: How to Wield the Weapon of Nonviolence with Maximum Force*, Norwalk, CT: Easton Studio Press.

————. (2012) *Peaceful Revolution*, Easton Studio Press.

Clough, S. (2020) "Charity, Peace, and the Social Epistemology of Science Controversies," in Ashton, N., Kusch, M., McKenna, R., and Sodoma, K. (eds.) *Social Epistemology and Relativism*, New York: Oxford University Press.

————. (2016) "Nancy McHugh's *The Limits of Knowledge: Generating Pragmatist Feminist Cases for Situated Knowing*," *Hypatia Reviews Online*, https://www.hypatiareviews.org/reviews/content/68

————. (2013) "Pragmatism and Embodiment as Resources for Feminist Interventions in Science," *Contemporary Pragmatism* 10(2), pp. 121–134.

————. (2004) "Having It All: Naturalized Normativity in Feminist Science Studies," *Hypatia: A Journal of Feminist Philosophy* 19(1), pp. 102–118.

————. (2003) *Beyond Epistemology*, New York, NY: Rowman & Littlefield.

————. (1998) "A Hasty Retreat from Evidence: The Recalcitrance of Relativism in Feminist Epistemology," *Hypatia: A Journal of Feminist Philosophy* 13(4), pp. 88–111.

Consortium for Socially Relevant Philosophy of/in Science and Engineering, http://srpoise.org

Crasnow, S. (2019) "Political Science Methodology: A Plea for Pluralism," *Studies in History and Philosophy of Science Part A*, 78, pp. 40–47. https://doi.org/10.1016/j.shpsa.2018.11.004

————. (2015) "Natural Experiments and Pluralism in Political Science," *Philosophy of the Social Sciences* 45(4–5), pp. 424–441.

————. (2013) "Feminist Philosophy of Science: Values and Objectivity," *Philosophy Compass* 8(4), pp. 413–423.

————. (2006) "Activist Research and the Objectivity of Science," *APA Newsletter on Feminism and Philosophy* 6(1), pp. 3–5.

de Melo-Martín, I. and Intemann, K. (2018) *The Fight Against Doubt: How to Bridge the Gap Between Scientists and the Public*, Oxford: Oxford University Press.

————. (2011) "Feminist Resources for Biomedical Research: Lessons from the HPV Vaccines," *Hypatia: A Journal of Feminist Philosophy* 26(1), pp. 79–101. https://doi.org/10.1111/j.1527-2001.2010.01144.x

————. (2007) "Can Ethical Reasoning Contribute to Better Epidemiology? A Case Study in Research on Racial Health Disparities," *European Journal of Epidemiology* 22, pp. 215–221. https://doi.org/10.1007/s10654-007-9108-3

Du Bois, W. E. B. (2004 [1920]) *Darkwater*, New York, NY: Washington Square Press.

————. (1968 [1903]) *The Souls of Black Folk: Essays and Sketches*, Chicago, IL, A. G. McClurg, 1903. New York: Johnson Reprint Corp.

————. (1906) *The Health and the Physique of the American Negro*, Atlanta, GA: Atlanta University Press.

————. (1899) *The Philadelphia Negro*, Philadelphia: The University of Pennsylvania Press.

Duran, J. (1991) *Toward a Feminist Epistemology*, Totowa, NJ: Rowman and Littlefield.

Fischer, M. (2019) *Jane Addams's Evolutionary Theorizing*, Chicago, IL: University of Chicago Press.

Godfrey-Smith, P. (2013) "Quine and Pragmatism," in Harman, G. and Lepore, E. (eds.) *A Companion to W.V.O. Quine*, New York, UK: Wiley, pp. 54–68. https://doi.org/10.1002/9781118607992.

Goldenberg, M. (2019) "Vaccines, Values, and Science," *Canadian Medical Association Journal* 191(14), pp. E397–E398.

———. (2016) "Public Misunderstanding of Science? Reframing the Problem of Vaccine Hesitancy," *Perspectives on Science* 24(5), pp. 552–581.

Harding, S. (1986) *The Science Question in Feminism*, Ithaca, NY: Cornell University Press.

Lake, D. (2019) Interview in *Engaged Philosophy*, https://www.engagedphilosophy.com/2019/08/25/danielle-lake/

———. (2014) "Jane Addams and Wicked Problems: Putting the Pragmatic Method to Use," *The Pluralist* 9(3), pp. 77–94.

———. (2012) "Restructuring Science, Re-Engaging Society," *The Pluralist* 7(3), pp. 51–56.

LoCI-Wittenberg University Writing Group. (2016) "The Epistemology of Incarceration: Constructing Knowing on the Inside," *philoSOPHIA Special Issue: Queer, Trans, and Feminist Responses to Mass Incarceration* 6(1), pp. 9–26.

McHugh, N. (2015) *The Limits of Knowledge: Generating Pragmatist Feminist Cases for Situated Knowing*, Albany: State University of New York Press.

———. (2011) "More than Skin Deep: Situated Communities and the Case of Agent Orange in Viet Nam," in Grasswick, H. (ed.) *Feminist Epistemology and Philosophy of Science: Power in Knowledge*, Dordrecht, The Netherlands: Springer.

———. (2006) "On the Very Idea of a Feminist Epistemology for Science," *Metascience* 15(1), pp. 15–21.

Mills, C. (1999) *The Racial Contract*, Ithaca, NY: Cornell University Press.

Nelson, L. H. (1993) "A Question of Evidence," *Hypatia: A Journal of Feminist Philosophy* 8(2), pp. 172–189.

———. (1990) *Who Knows? From Quine to a Feminist Empiricism*, Philadelphia, PA: Temple University Press.

Potter, E. (1993) "Gender and Epistemic Negotiation," in Alcoff, L. and Potter, E. (eds.) *Feminist Epistemologies*, New York: Routledge.

Proctor, R. (1985) *Cancer Wars: How Politics Shapes What We Know and Don't Know about Cancer*, New York: Basic Books.

Public Philosophy Network, https://www.publicphilosophynetwork.net.

Quine, W. V. O. (1969) *Ontological Relativity and Other Essays*, New York, NY: Columbia University Press.

Reisch, G. (2005) *How the Cold War Transformed Philosophy of Science: To the Icy Slopes of Logic*, Cambridge, MA and New York: Cambridge University Press.

Rorty, R. (1999 [1994]) "Truth Without Correspondence to Reality," in *Philosophy and Social Hope*, New York, NY: Penguin Books.

Seigfried Haddock, C. (1996) *Pragmatism and Feminism: Reweaving the Social Fabric*, Chicago, IL: University of Chicago Press.

———. (1991) "Where Are All the Pragmatist Feminists?" *Hypatia: A Journal of Feminist Philosophy* 6(2), pp. 1–20.

Tanesini, A. (2016a) "Intellectual Humility as Attitude," *Philosophy and Phenomenological Research* 96(2), pp. 399–420. doi: 10.1111/phpr.12326.

———. (2016b) "Teaching Virtue," *Logos and Episteme* 7(4), pp. 503–527.

Tuana, N. (2019) "Climate Apartheid: The Forgetting of Race in the Anthropocene," *Critical Philosophy of Race* 7(1), pp. 1–31. https://doi.org/10.5325/critphilrace.7.1.0001

———. (2004) "Coming to Understand: Orgasm and the Epistemology of Ignorance," *Hypatia: A Journal of Feminist Philosophy* 19(1), pp. 191–232.

———. (1992) "The Radical Future of Feminist Empiricism," *Hypatia: A Journal of Feminist Philosophy* 7(1), pp. 100–113.

Tuana, N. and Scott, C. (2018) "Border Arte Philosophy: Altogether Beyond Philosophy," *Journal of Speculative Philosophy* 32(1), pp. 70–91.

Whyte, K. P. (2018) "Critical Investigations of Resilience: A Brief Introduction to Indigenous Environmental Studies and Sciences," *Daedalus: Journal of the American Academy of Arts and Sciences*, 147(2), pp. 136–147.

——— . (2013) "On the Role of Traditional Ecological Knowledge as a Collaborative Concept: A Philosophical Study," *Ecological Processes* 2(7), pp. 1–12.

———. (2012) "Indigenous Peoples, Solar Radiation Management and Consent," in Preston, C. (ed.) *Engineering the Climate: The Ethics of Solar Radiation Management*, Lanham, MD: Rowman and Littlefield, pp. 65–78.

Wylie, A. (2017) "Feminist Philosophy of Social Science," in Garry, A., Khader, S. J., and Stone, A. (eds.) *Routledge Companion to Feminist Philosophy*, New York: Routledge.

———. (2015) "A Plurality of Pluralisms: Collaborative Practice in Archaeology," in Padvani, F., Richardson, A., and Tsou, J. Y. (eds.) *Objectivity in Science: New Perspectives from Science and Technology Studies*, New York: Springer, pp. 189–210.

———. (2014) "Cognitive/Social Norms in Community Based Collaborative Research," in Cartwright, N. and Montuschi, E. (eds.) *Philosophy of Social Science: A New Introduction*, Oxford: Oxford University Press, pp. 68–84.

———. (2012) "Feminist Philosophy of Science: Standpoint Matters," *Proceedings and Addresses of the American Philosophical Association* 86(2), pp. 47–76 [Pacific APA Presidential Address].

Zack, N. (2016) *Applicative Justice: An Empirical Pragmatic Approach to Correcting Racial Injustice*, Lanham, MD: Rowman and Littlefield.

———. (2009) *Ethics for Disaster*, Lanham, MD: Rowman and Littlefield.

———. (2002) *Philosophy of Science and Race*, New York: Routledge.

PART III

Key Concepts and Issues

14

IS SEX SOCIALLY CONSTRUCTED?

Catherine Clune-Taylor

Introduction

The sex/gender distinction was an important theoretical and political tool for second-wave feminists in their fight against sexist oppression and the biologically determinist arguments used to justify it. According to this distinction, sex is a natural, biological fact of the body comprised of multiple variables including external genitalia, internal reproductive structures, gonads, hormones, chromosomes, and (more recently) neurocognitive structures. Gender, on the other hand, refers to one's identification and expression as masculine or feminine, and is socially constructed. Individuals are taught socio-culturally and temporally specific norms regarding masculinity and femininity, and come to internalize, identify with, and express themselves in relation to these norms. This distinction – first employed by Anne Oakley in 1972's *Sex, Gender, and Society* – elegantly, and potently, undermined biologically determinist claims that one's male or female sex naturally gave rise to roles traditionally gendered as masculine or feminine – particularly those which supported the oppression of women. While the sexual difference of the body may be natural (and thus immutable), gender norms are socially constructed, and thus they could be otherwise.

In the late 1980s and throughout the 1990s, feminist scholars across the sciences and the humanities began to argue that, in addition to gender, sex was also socially constructed, challenging both the sex/gender distinction that had been so politically efficacious for their predecessors, and nature/culture binary on to which it maps. Sex, they argued, was socially constructed in at least two senses. First, our concepts regarding sex, itself – that is, our understanding of biological sex and its multiple material components – are shaped by socio-culturally and temporally specific meanings, and material arrangements. Many of these scholars turned to both contemporary and historical management of folks with intersex conditions in order to reveal the discursive and practical "sexing" of the body, as well as the persistent instability of definitions of sex. The second sense in which sex is socially constructed emphasizes what I will refer to as the "sociomaterial constitution of sex," pointing to the ways which the environment socio-culturally and materially influences the development of biological sex. That is, while we may find evidence of sex differences within and across bodies, those differences may be better characterized as material effects of development within a gendered environment than evidence of naturally binary sex. For example, differences in muscularity between men and women that are frequently seen as indicative of natural sex dimorphism depend heavily on exercise opportunities, such that they could be reduced with increased access to activity, and different social norms regarding gendered embodiment (Fausto-Sterling 1985: 218).

Interdisciplinary research revealing the social construction of sex across an array of dimensions has provided stark evidence for many of the central tenants of feminist philosophy of science. Analysis of the social and material (re)construction and (re)naturalization of sex (and gender) via the biosciences (and before) has been, and remains, revelatory with regard to the social nature of knowledge, knowers, and knowledge communities, as well as deficiencies (and outright flaws) in our conceptions of knowledge and knowers, objectivity, and scientific methodology. In the following sections, I shall explore both the social construction, and the sociomaterial constitution of sex. However – as shall become clear in the course of this exploration – the distinction between these two senses of the social construction of sex is quite fuzzy, and becomes fairly difficult to maintain at various points. This is because sex is not merely the additive products of natural and cultural influences. Rather, it is a dynamic, contextually dependent assemblage – an instantiation of "natureculture" in Haraway's sense of the term, insofar as the natural and the environmental cannot be easily disentangled (Haraway 2003). For example, according to the now-dominant biomedical model of sex difference known as brain organization theory, differences between the sexes with regard to gendered cognition, behavior, sexual and non-sexual desires, and identity are the result of the "organization" of the brain into feminine or masculine patterns by prenatal hormones (Fine 2010; Jordan-Young 2010). However, as multiple feminist scholars have pointed out, brain development is a highly dynamic, interactive, and socially dependent process, via which "the social phenomenon of gender is literally incorporated, shaping the brain and the endocrine system," such that gendered world becomes part of and is reflected in our "cerebral biology" (Kaiser et al. 2009: 57; Rippon et al. 2014: 4). Thus, not only are neurological sex differences as much the result of one's development within a gendered environment as they are result of the proteins one's genes code for, they also exhibit a high-level of plasticity, calling into question the notion that such patterns are fixed or immutable. Spatial cognition – for example – is one of the few forms of cognition for which it is generally accepted that gendered differences exist. However, Feng, Spence, and Pratt (2007) have illustrated that playing an action video game for a mere ten hours can "virtually eliminate" gendered differences in spatial cognition.

The deep, dynamic entanglement between the natural and the environmental evidenced in neurological development, as well as the historically, theoretically, and materially co-constitutive nature of sex and gender more generally, has led many to argue for the use of a combined term such as "sex/gender" or "gender/sex" either in addition to or instead of sex and gender (e.g. Fausto-Sterling, Coll, and Lamarre 2012; Pitts-Taylor 2016). For example, van Anders and Dunn use the term gender/sex in their study of the influence of hormones on orgasm experience and sexual assertiveness in both men and women given that any differences found in the study of hormones in relation to gendered sexual differences (such as sexual aggression) cannot "knowingly be attributed to biology or gender socialization" (van Anders and Dunn 2009: 207). Rather than sex or gender, Van Anders (2015) uses the term gender/sex in her Sexual Configurations Theory (SCT) to refer to "whole people/identities and/or aspects of women, men, and people that relate to identity and/or cannot really be sourced specifically to sex or gender," while defining gender as referring to those socialized "aspects of masculinity, femininity, and gender-diversity" (as well as one's self, systems), and sex as "[a]spects of femaleness, maleness, and sex-related bodily features that are situated as biological, bodily, evolved, physical, and/or innate (e.g., vulvas, penises, breasts, body shape)" (van Anders 2015: 1181). Because this chapter is dedicated to exploring the social construction of sex (and along with it, the historical, theoretical, and material constitution of gender with which it is inextricably entangled), I shall treat them separately for the sake of clarity while recognizing that the aim of the chapter is, ultimately, to show that they cannot be separated. After exploring both the social construction of sex and its sociomaterial constitution, this chapter concludes by taking up two emerging threads in literature exploring the social construction of sex (historically and in the present) that promise to enrich – and complicate – already existing research by (1) emphasizing the temporal nature of sex (and gender) as specifically developmental objects of biomedical knowledge and intervention and (2) exploring the historical and contemporary co-construction of sex, race, and disability locally and transnationally.

The Social Construction of (Inter) Sex and Gender

Many of those feminists who began to argue for the social construction of sex in the late 1980s and throughout the 1990s did so by returning to the same place from which Oakley had initially adopted the sex/gender distinction and the notion of gender as socially constructed – the medical management of infants with intersex conditions under Dr. John Money's "Optimal Gender of Rearing" (OGR) treatment model. As multiple scholars have noted (Germon 2009; Hausman 1995; Rubin 2017), Oakley adopted her account of gender, and of the sex/gender distinction from psychoendocrinologist John Money (by way of psychoanalyst Robert Stoller), who developed it with colleagues Joan and John Hampson at Johns Hopkins Medical Center during the 1950s. It is in many ways unsurprising that feminist scholars across disciplines would turn to the management of intersex folks in order to reveal the social construction of sex as naturally dimorphic. Indeed, nowhere is this rendered clearer than in multitude of shifting –and, at times, conflicting – discourses, and practices regarding those bodies which challenge received definitions of male and female sex, both historically and today.

Feminist scholars like biologist Anne Fausto-Sterling (Fausto-Sterling 1985, 1993, 2000), sociologist Suzanne Kessler (1990), and critical theorist Judith Butler (1990) turned their critical attention of the categorization and management of intersex infants under the OGR treatment model, according to which infants with ambiguous genitalia were assigned sex via cosmetically normalizing genital surgeries and sterilizing gonadectomies. These feminist critics called into question the presumed naturalness of dimorphic sex that justified the pathologization of intersex bodies, and their management with invasive, experimental, and irreversible surgeries at the sake of reproductive capacity, and often later sexual sensation – and performed on individuals unable to provide informed consent. They argued that treatment recommendations were often guided by heterosexist norms regarding physical properties (e.g. penile length and vaginal capacity for penetration), as well as gendered behaviors (e.g. aggression and sexual potency for boys, and "reproductive/sexual-receptive potential for girls") (Dreger and Herndon 2009: 204). Further, clinical capacity to shape bodies and behaviors in coherence with these norms often played a greater role in decisions regarding sex assignment than biological indicators. These early feminist academic critics were quickly joined by intersex activists (Chase 1998, 1999) and bioethicists (Crouch 1998; Dreger 1998, 2000) who argued that under OGR the bioethical principles of autonomy, beneficence, and non-malfeasance were routinely violated for the sake of maintaining the fallacy of naturally binary sex. Historians of science and medicine situated contemporary biomedical discourse and practice with its conditions of possibility. These historical analyses outlined changes in the macro conceptualization of sexual difference in the West from antiquity until the eighteenth century (Laqueur 1990); tracked shifts in clinical practices (and the logics underwriting them) in response to technologically driven redefinitions of the "truth" of binary sexual difference during the nineteenth and twentieth centuries (Dreger 2000); and revealed the various sociocultural and material forces that served to shape the emergence and development of those disciplines central to the science and management of sex difference such as endocrinology (Oudshoorn 1994), cytology and chromosomal analysis (Richardson 2012, 2013), and neuroscientific studies of brain organization theory (Fine 2010; Jordan-Young 2010).

While this section shall devote itself to unfurling some of this narrative of the social construction of sex in humans that these (and many other) feminist scholars began articulating in the 1980s, such a summary is necessarily incomplete. This is because the discourses and practices it captures (as well as the discourses and practices about those discourses and practices) are themselves multiple and partial, contested and contestable – as all discourses and practices are. Recent feminist contributions have served to complicate and enrich this narrative, refining earlier analyses of historical and contemporary (re)definitions of both normal and abnormal sex (and gender) (Downing et al. 2015; Mak 2012; Reis 2009; Repo 2016; Rubin 2017). These interventions have deepened feminist understanding of sex as socioculturally and materially constructed, exposing the plethora of social norms, practices, knowledges, technologies, bureaucracies, institutions, and capacities implicated in its production and maintenance as binary and natural. Indeed, within biology, male and female sex is determined solely on the basis of gamete size – those members of a species who produce the smaller gametes ("sperm") are identified as males, whereas those who produce the larger gametes ("eggs") are the females (Roughgarden 2004: 24). However, even this distinction is acknowledged to be a convention, given that all the members of some species of algae, fungi, and protozoans produce the same size gametes. In these cases, the species is divided into genetic groups known as "mating types" (see Roughgarden 2004: 23–24).

In *Making Sex: Body and Gender from the Greeks to Freud*, Thomas Laqueur (1990) argued that Western understandings of sexual difference were radically reconstituted during the eighteenth century with the introduction of a "two-sex" model of sexual difference, and meaningfully revised histories of sex, and of the body in the process. Prior to this, Laqueur argued, Western thinking and practice was dominated by a "one-sex" model of sexual difference, according to which women were presumed to have the same sexual anatomy as men; however, that anatomy was inverted, remaining internal due to their lack of heat. Men, on the other hand, had sufficient heat to develop their genitalia outward, rendering them the more perfect sex. This "one-sex" was supplanted by a "two-sex" model, however, according to which the sexed bodies of women and men were considered to be two unique, oppositional, and incommensurate types. Laqueur identified the emergence (and persistence) of the "two-sex" model with modernity, which demanded "a single, consistent biology as the source and foundation of masculinity, and femininity" (Laqueur 1990: 61).

Laqueur's history of the emergence and eventual dominance of the "two-sex" model over the "one-sex" has been critiqued in terms of both its specifics and, more recently, with regard to his general thesis (King 2013). Nonetheless, the clear evidence Laqueur provided for the social construction of sex was – as King notes – immensely appealing (King 2013: 6). Further, Laqueur's account deeply impacted not only the history of sexual difference, but of the body more generally. King's (2013) argument that there was no clear moment in the eighteenth century when the "two-sex" model usurped the "one-sex," but rather that the two logics of sexual difference have co-existed since antiquity serves to complicate this initial history in a generative fashion. While the social construction of sex remains clear, these arguments highlight the contingent, and local nature of our sexual concepts, in their multiplicity. Conflicting logics of sexual difference, as well as interpretations of them, can (and often do) circulate simultaneously.

Dreger (1998) further elaborates this history, documenting shifts in biomedical definitions of sex (now conceptualized as naturally, and oppositional dimorphic) following the emergence of biology and clinical medicine as disciplines in the nineteenth century. Dreger tracks the evolution of biomedical definitions of the "truth" of sex difference (and practices regarding it) in response to technological advances in medicine, such as advancements in microscopy, and the increased availability of anesthetic. The introduction of the Klebsian, gonad-based system of sexual difference in 1876 ushered in the "Age of Gonads," argues Dreger, wherein the truth of sex could be found in gonadal tissue. It was already clear that both external genitalia and internal reproductive structures could be ambiguous. The Klebsian system preserved sex as naturally binary at a time when technology could not allow for the performance of living biopsies of gonadal tissue, such that "true hermaphrodite" could exist only in death. The capacity for increasingly specific elaboration of the body only served to reveal the instability of binary sex, as clinicians found themselves faced with living "true hermaphrodites," and gonadal tissue that was neither testicular nor ovarian. Dreger argues that the "Age of Gonads" gave way to the "Age of Conversion" with John Money's introduction of gender via the OGR treatment model in the 1950s, during which those with intersex conditions are surgically assigned a sex, and medically normalized in accordance with predicted future gender.

With the "Age of Conversion" then, the truth of sex becomes "optimal" gender – that is, the gender that can reliably be "constructed" (Dreger 2000; Kessler 1990, 1998), or "secured" (Clune-Taylor 2019) – by medical means. This required not only a theory of gender distinct from sex and irreducible to it, and of its acquisition, but also the development of particular biomedical disciplines, providing clinicians with the knowledge and technical capacity for normalizing the body, which Hausman (1995) argues is necessary for the emergence of gender as such. The development of those fields of biomedicine that came to facilitate the normalization of atypically sexed bodies (such as embryology, endocrinology, and genetics) only served to emphasize the instability of physical sex (Germon 2009: 33).[1] Further, advances in surgical techniques made during World War II meant that clinicians found themselves more capable physically normalizing intersex bodies than ever before, yet more uncertain of how to justify it. The solution to the "problem" of intersex management was provided by Money through the OGR treatment model, and the account of gender and of its development that model introduced.

Admittedly, Oakley's account of gender and of the sex/gender distinction is something of a reduction of Money's, who has been characterized as both the epitome of a social constructionist *and* as biological determinist who turned to environmental influences when it suited him (Rogers and Walsh 1982) – and much maligned for both views.[2] More recent feminist analyses of his work have provided a far more nuanced account of Money's model as interactionist regarding gender and its development in a revolutionarily complex manner for his time (Downing, Morland, and Sullivan 2015; Germon 2009; Hausman 1995; Karkazis 2008; Rubin 2017). However, this research has also shown his model to problematically presume both the natural dimorphism of gender (if not sex) and the "sedimentation" of gender within the brain after some critical period in development (Sullivan 2015). Further, Money has been accused by multiple scholars of holding an additive "biosocial" position rather than a truly interactionist one, insofar as it fails to consider how biological and social variables work in tandem, affecting the character of their respective influences beyond amount (Doell and Longino 1988; Jordan-Young 2010).

On the basis of his so-called "hermaphroditic" research begun during his doctoral program in clinical psychology at Harvard, Money "abandon[ed] the unitary definition of sex as male or female" as it was clear to him that the multiple variables of sex – chromosomes, gonads hormones, internal reproductive structures, and external genitalia – could develop independently of one another (Money 1995: 21). Not only was the assumption that sex could be "read" off the body via genitalia incorrect, but further, genitalia could itself be ambiguous. Moreover, none of these variables clearly correlated with gender which, as Money initially introduced it, encapsulated what we would today distinguish as gender identity, gender role or behavior and sexual orientation, though this would eventually be refined to Gender Identity/Role or G-I/R. Because gender corresponded to sex of rearing (or rather, the gender one is socialized as) for 95% of patients studied by Money and the Hampsons, the group identified sex of rearing as of primary importance in gender development, perhaps of even greater importance than biological variables given the stability of gender once "learned" (Clune-Taylor 2016; Karkazis 2008; Rubin 2017; Sullivan 2015). However, despite evidence to the contrary within his own research, he also posited sex-typical genitalia as necessary to "normal" psychosexual development as a man or a woman – ostensibly to reinforce gender socialization and identity for those doing the socializing and those being socialized (Feder 2014; Holmes 2008; Karkazis 2008). Subsequently, the treatment model developed by Money and his colleagues stressed early surgical assignment and unambiguous, binary gender socialization, boiling down his theoretically interactionist account to a social constructionist one in practice, undergirded by what Holmes (2008) refers to as Money's "genital determinism."

Money, then, is in many ways the first to clearly recognize and articulate the social construction of sex and the fallacy of its naturalization as dimorphic. Furthermore, with the introduction of gender, Money manages to – once again– renaturalize sex by presuming the naturalness of binary masculine and feminine genders with regard to which sex could be both indexed and created. In this sense, Money doesn't merely propose sex as "gender all along" 35 years prior to Butler (1990), but literally constitutes it as such, giving rise to the tripartite configuration of sex-gender-sexuality underwriting that grid of intelligibility known as the "heterosexual matrix." For this reason, gender has been multiply characterized as technology of power: as "a powerfully stabilizing factor at a time when technology was increasingly undermining" physiologically binary sex (Germon 2009: 62), as "a solution to the uncertainty of any absolute somatic sex" (Germon 2009: 35), as "a diagnostic category and treatment protocol" allowing clinicians "to predict and . . .literally fashion the sex [intersex patients] were 'supposed' to have all along"(Rubin 2017: 892), and as "the most recent historical apparatus to contain the body within a political economy of dimorphic sexual difference" (Germon 2009: 62). It also speaks to the character of sociocultural investment white supremacist, and gender normative, reproductive heterosexuality, reliant – as it is – on naturally dimorphic sex.

Over 30 years of feminist academic and intersex activist critique of the biomedical management of intersex infants under the OGR treatment model led to the introduction of the revised "Disorders of Sex Development" (DSD) treatment model by the American and European Pediatric Endocrine Associations in 2006, which was rapidly adopted internationally (Hughes 2010; Lee et al. 2006). After so many years of scholarship and activism specifically calling for the depathologization of intersex conditions, the DSD treatment model has proven to be quite controversial among intersex folks and their families, academics, and clinicians. Much of that controversy has centered on the pathologizing nature of the nomenclature, the processes out of which the treatment model and systems of nomenclature emerged, as well as the continued management of intersex infants under DSD with genital normalizing surgeries.[3] Empirical evidence indicates that not much has changed practically speaking under DSD, insofar as the frequency with which genital normalizing surgery is performed seems to be the same (if not higher) than it was under OGR (Creighton et al. 2014). However, insofar as the theory of gender development underwriting clinical recommendations made under DSD has changed, the treatment model achieves yet another reconstitution of our definitions of both normal and abnormal sex (and gender). I have argued that whereas OGR was underwritten – at least in practice – by a socially constructionist account of gender development treatment recommendations under DSD are based on biologically determinist account of gender development known as brain organization theory, which posits that gendered patterns of identity, behavior, cognition, and desires (including sexual ones) are the result of "organizing" effects of hormones on the brain in utero (Clune-Taylor 2019). Sex assignment is thus heavily influenced now by predictions regarding "brain gender" on the basis of assumptions regarding prenatal hormone exposure – something which cannot be tested directly.

Feminist scholars across disciplines have heavily critiqued brain organization theory for being both theoretically suspect and empirically inadequate (most notably Fine 2010; Jordan-Young 2010). Underwriting brain organization theory is the assumption that the well-established differentiating effects of prenatal hormones on genital development are mirrored within the brain, such that the same mechanism results in the sexual differentiation of "both sets" of organs required for reproduction – the brain and the genitalia (Jordan-Young 2010: 21). However, as Jordan-Young (2010) has noted, there are multiple, stark dissimilarities between these organs. First, brains are nowhere near as dichotomous as genitalia; while the latter can be reliably be sorted into male-typical and female-typical types by observers unaware of the sex of the individual they came from, this is not the case for the former (Jordan-Young 2010: 49). Indeed, the level of neuroanatomical dimorphism exhibited by human brains remains the subject of live debate, as does the question of whether or not "distinctively gendered patterns of brain function" exist at all (Jordan-Young 2010: 49; Rippon et al. 2014). For the majority of "social, cognitive, and personality variables," there is far more overlap than divergence between genders (Rippon et al. 2014: 4). Furthermore, brain development and genital development are deeply disanalogous. Human genitalia differentiate between the 7 and 17 weeks gestation and − barring intervention − those structures will develop in size, but remain generally fixed in form across the lifespan (Yiee and Baskin 2010). The human brain, on the other hand, is grossly underdeveloped at birth, and its developmental period is a markedly dynamic one, highly dependent on biological and social inputs, and characterized by atypically high levels of plasticity across the lifespan (Jordan-Young 2010; Rippon et al. 2014; Yiee and Baskin 2010). Indeed, the brain is − uniquely − "permanent[ly] plastic," undermining the notion that gendered neuroanatomical and functional differences are as fixed as brain organization theory posits (Rippon et al. 2014: 4). This leads Rippon et al. to argue that the principles that emerge from surveying gendered neurological research do not emphasize difference, but are overlap, mosaicism, contingency, and entanglement, and that these principles should guide and inform research design, analysis, and interpretation (Rippon et al. 2014: 2).

The relocation of the truth or − at the very least, the best biological indicator − of gender in gendered neuroanatomical and cognitive differences brings about an interesting collapse of the distinction between gender and sex at the level of the brain, revealing − yet again − the circular construction of both. If gendered differences in forms of cognition (such as language processing (Kaiser et al. 2009) or spatial rotation) are posited to be the result of sexed neuroanatomical differences, then the "black box" that is the brain becomes the site of both sex and gender. Further, the level of plasticity exhibited by the brain as pointed to in Rippon et al.'s survey or in Feng, Spence, and Pratt's (2007) research calls into question whether neuroanatomical differences are what determine functional differences, or whether they result from them. The complex, dynamic, interactive, and socioculturally dependent nature of neurological development via which the "the social phenomenon of gender is literally incorporated, shaping the brain and endocrine system," to become "part of our cerebral biology" is a clear example of the sociomaterial constitution of sex − the second sense in which sex is socially constructed (Kaiser et al. 2009: 57; Rippon et al. 2014: 4). Indeed, the fact that those bodily "sexed" (or is it "gendered"?) differences which we take to be indicative of gender (or is it "sex"?) are shaped by the world undermines whatever distinctions between sex/gender and nature/culture we might hope to maintain.

The Sociomaterial Constitution of Sex (and Race)

Since the 1980s, feminist scholars across disciplines have shown that physical and biologically gendered features are materially affected by social practices, shaping aspects of "sex, itself." The uniquely dynamic and socially dependent nature of neurological development renders the brain a privileged site for drawing the sociomaterial constitution of sex into view. It is often hard to imagine other seemingly more stable sexed characteristics might be so open to environmental inscription. However, feminists have been drawing connections between bodily sexed differences and social norms and practices since the early 1980s, and this research will only proliferate as science elucidates new mechanisms via which environmental influences become embodied (e.g. direct and indirect epigenetic effects), and as sociocultural investment in sexual difference persists. For example, Jagger (1983) argued that the naturalized difference in body size between men and women could be the result of the latter group receiving less food and resources due to a cultural devaluation of their gender/sex. Fausto-Sterling (2005) has explored the sociomaterial constitution of sex differences extensively, identifying the effects of socio-culturally and temporally specific gendered norms and practices on bone development, and arguing that sexed differences in muscularity so frequently identified with masculinity and femininity are not natural, and could be closed with shifts in sociocultural norms regarding activity and muscularity, as well as increased access to muscle building exercises (Fausto-Sterling 1985). She further applies her general claim – "that specific anatomies and physiologies are not fixed traits," but rather, "emerge over the lifecycle as a response to specific lived lives" – to race, undermining its use a typological category in medical research, and revealing its material co-constitution with sex/gender (Fausto-Sterling 2008: 658).

Motivated by concerns regarding increasing interest in genetic or genomic explanations for a range of problems ("from alcoholism to gender and racial health disparities") at the expense of insights from the social sciences, as well as the continued "use and abuse of concepts of race in medicine," Fausto-Sterling reviews hundreds of papers on race/gender and bone health (a field in which race and gender are inextricably entangled) to argue against the use of race as a typological kind (Fausto-Sterling 2008: 657–658). Not only is race, and hierarchies of it, socially constructed in medical research, through the persistent (and often inconsistent) use of racial and ethnic categories, but their use persists despite researchers' frequent inability to identify significant differences between racial groups. Moreover, the bare facts of the sociomaterial constitution of the (sexed, gendered, raced, abled) body – that is, that the "social *produces* the biological in a system of constant feedback between body and social experience"– render race a "poor object of study because it beckons us to structure problems in terms of nature *versus* nurture" (Fausto-Sterling 2008: 658, 683). Such a framing (in which, importantly, "time is held constant") is "ultimately futile," argues Fausto-Sterling, who advocates for the adoption of approaches from developmental and dynamic systems theories, emphasizing the dynamic interplay of the social and the biological over time (Fausto-Sterling 2005, 2008: 683). Thus, all sexed and racialized physiological differences – even our most stable, and most taken for granted – emerge out of, against, and through the environments in which they develop. As a result, the development of all physiological traits and their environment cannot – just like all other terms operationalized to map the nature/culture binary – be disentangled.

Emerging Engagements with the Social Construction of Sex

Feminists are still uncovering the myriad of ways in which sex is socially constructed across a variety of disciplines, and this work will continue to have import for feminist philosophers of science. One way to productively refine analyses of contemporary intersex management, and subsequently of sex and gender in themselves, is to reconfigure both as temporal (and temporally) constituted objects. Another way is to further elaborate the co-construction of sex with other axes of identity – particularly race and disability. Of course, even these two directions of analyses are themselves entangled for, as this work collectively reveals, individuals and populations are, necessarily – and with great local variation – sexed, racialized, and disabled both in and across time.

Sex and gender are natureculture in both the socially constructed and sociomaterially constituted senses. As a result, both our concepts of them and their instantiations in the world as embodied objects available for study and intervention are necessarily temporally specific. Further, recent research has highlighted the importance of their temporal dimension in their biomedical constitution as developmental objects. Fausto-Sterling's recent work on sex/race and bone health (Fausto-Sterling 2005, 2008), as well as gender development (Fausto-Sterling, Coll, and Lamarre 2012), emphasizes the need for specifically development approaches which examine health and disease patterns (including gendered and racialized ones) across the lifespan. As Sullivan (2015) points out, the temporality of gender development plays an important role in Money's account which unjustifiably presumes the "sedimentation" of binary brain gender during some critical developmental window. Similarly, brain organization theory posits a critical (prenatal) period of development as key to "sexing" the brain, and even critical feminist researchers have begun analyzing gender and sexual orientation developmentally. Moreover, I argue that the primary aim of intersex management is best understood as securing a specifically *cisgendered future* for an intersex individual, referring to a normalized trajectory of development over the lifespan across which a variety of sexed and gendered characteristics are maintained in "coherent" alignment (Clune-Taylor 2019). I use this term specifically to differentiate it from one's self-identification – though self-identification is one of the multiple sexed and gendered variables that make up cisgendered life – and to emphasize the social construction of this trajectory of development as cisgendered (and thus normal), discursively and materially. This account productively reconfigures cisgendered life as a developmental achievement, reliant on a variety of discursive, technological, and material achievements.

Kimberlé Crenshaw's introduction of intersectionality analyses rendered all subsequent work that failed to take seriously the co-constitution of gendered and racialized identity and the interlocking nature of oppression in some important sense of the term incomplete (Crenshaw 1989, 1991). Beyond this, Shelley Tremain's turn to intersex management in 2001s "On the Government of Disability" to reveal the social construction of both impairment *and* disability – another nature/culture binary dependent pair – not only gave rise to the field of critical disability studies but also highlighted the importance of exploring impairment, disability, and technologies of power implicated in the social constitution of bodies as sex/gendered and racialized. While some work – like that of Fausto-Sterling (2008) or, even earlier, Markowitz's (2001) "Pelvic Politics: Sexual Dimorphism and Sexual Difference" – has taken an intersectional approach to examining sexual difference, such analyses are rare in general with few seriously engaging disability, leaving many in feminist science studies to bemoan the relative lack of substantive intersectional analyses (Subramaniam 2009). However, scholarship that takes seriously both the contemporary and historical social co-construction of sex, race, and disability has begun to emerge, addressing important lacunae in the literature. This research has only served to reinforce the necessity of intersectional approaches to more capaciously and more accurately elaborating the co-constitution of sex, gender, race, and disability, both locally and transnationally. Further, it has highlighted the generative potential of a multiplicity of histories and disciplinary approaches to understanding the various biopolitical functions of sex/gender, race, and disability both historically and in the present. Indeed, as Markowitz noted "the ideology of sex/gender difference itself" does not "rest not on a simple binary of opposition between male and female but rather on a scale of racially coded degrees of sex/gender difference culminating in the manly European man and the feminine European woman" (Markowitz 2001: 391). That this "scale" was itself generally articulated in terms of pathology or abnormality only speaks to the historical emergence of sexed/gendered and racialized differences out of intimately entangled histories of colonialism, slavery, scientific racism, and clinical medicine, such that analyses of their social construction, in any sense of the term, must carefully be attended to and situate themselves in relation to these legacies.

C. Riley Snorton's *Black on Both Sides: A Racial History of Trans Identity* (2017) critically resituates the historical emergence of gynecology as a discipline within the make-shift hospital/slave quarters of Dr. J. Marion Simms. The inventor of multiple gynecological instruments and procedures (such as the cure for vesicovaginal fistulae), Simms developed and refined both via experiments on unanesthetized slaves (most notably Anarcha, Betsey, and Lucy) (Snorton 2017). Snorton's exploration of the co-constitutive emergence of blackness and transness begins from Spiller's claims regarding the "ungendering" of black folks by slavery (rendering heteronormative gender identity inaccessible to them), complicating and expanding upon traditional narratives regarding the birth of the gender (and the gender clinic) in the 1950s at Johns Hopkins

Medical Center (such as that outlined here) (Snorton 2017). Kyla Schuller's *The Biopolitics of Feeling: Race, Sex, and Science in the Nineteenth Century* similarly disrupts both hegemonic and feminist accounts of sex and racial difference, as well as "new" materialist hopes regarding the liberatory potential of analyses emphasizing the "plasticity, porosity, and vitality" of matter (Schuller 2018: 4). To counter traditional accounts regarding the dominance of biologically determinists notions of race during the nineteenth century, Schuller argues that sentimentalism played an unacknowledged biopolitical role in the nineteenth century, mingling with scientific discourse to consolidate political power in the "feeling body," and giving rise to a "palimpsestic model of race before genetics, in which racial status indexes the impressions that occur over the life span of individuals and the evolutionary time of races" (Schuller 2018: 12). This "sentimental biopolitics of life" distributed sex/gendered and racialized individuals along the great chain of being via their relative impressability – or the body's "energetic accumulation of sensory impressions and its capacity to regulate its engagement with the world outside the self" (Schuller 2018: 3). Subsequently, this dispositif "regulated the circulation of feeling through the population and delineated differential relational capacities of matter and, therefore the potential for evolutionary progress, as the modern concepts of race, sex, and species," such that race, sex, and species came to define a "body's relative claim s to life on the basis of perceived proportional vitality and inertia of the sensory and emotional faculties" (Schuller 2018: 5–6). Importantly, Schuller highlights that this "racial history of sexuality was a pathway for white women's political agency" citing the writings of Drs. Mary Blackwell (1821–1910) and Elizabeth Walker (1832–1919) – the first and second women to graduate with medical degrees in the Western world, respectively – who saw the bodies of "civilized" white women as "the culmination of racial potential" (Schuller 2018: 132).

As she acknowledges, Schuller's account disrupts "some of our most cherished scholarly and popular narratives," including binary distinctions between "social and organic processes; sentimental and scientific accounts of ontology and reason; biological and cultural interpretations of racial status; hegemonic and feminist versions of sex difference; and determinist and vitalist accounts of the capacities of matter" (Schuller 2018: 3). Moreover, in disrupting more traditional, biologically determinist accounts of race, sex, and species, Schuller complicates new materialist hopes for the political potential of "new" notions of "lively matter," identifying the ways in which new materialisms "generally animate racial thought" (Schuller 2018: 25–27). Rubin (2017) similarly disrupts traditional feminist engagements with sexual difference in the present, providing a much-needed transnational perspective to work on intersex, and underscoring the importance of diverse feminist perspectives and forms of analysis. In *Intersex Matters: Biomedical Embodiment, Gender Regulation, and Transnational Activism*, Rubin compares Western and transnational analyses of the controversy surrounding South African athlete Caster Semenya, whose sex was called into question after she won a gold medal in the women's 800 meter-dash at the International Association of Athletics Federations (IAAF) World Championship in August 2009 (Rubin 2017: 121–139). He reports that "[n]ot surprisingly" analyses from Western scholars "tended to agree that her treatment reaffirmed the need for intersectional approaches to subjectivity and power," while nonetheless failing to undertake such an approach (Rubin 2017: 134). Rather, the majority of Western analyses tended to frame Semenya's "case" as, centrally, about the social construction of sex and of intersex rights – despite the fact that Semenya herself has never identified as having an intersex condition nor as intersex. This response stood in stark contrast to popular and political reception to Semenya in South Africa following her 2009 win (and the eruption of international controversy regarding her "true sex"), that positioned the athlete as "not only a normatively sexed/gendered subject and proud symbol of South African athletic excellence and national pride, but also as an icon of postcolonial resistance to western domination" (Rubin 2017: 133). This reaction, argues Rubin, reproduced "sex/gender binarism as part of an effort to mobilize Black Nationalism as a challenge to Western imperialism," and must be read against "the intersecting eugenic and sexological legacies of colonialism" – legacies that feature, as Neville Hoad writes, the "shameful history of Sarah Baartman who was literally cut up and turned inside out for the world to see" (Hoad 2010: 402; Rubin 2017: 133).[4] However, Rubin notes that even transnational approaches, which quite rightly highlight the ways in which "national context, racialization, and imperial history overdetermine the visibility and subsequent management of bodies in doubt" – such as, allegedly, Semenya's – can nonetheless fail to consider the uneven globalization of Western models of medicalizing intersex bodies, and of rights-based intersex movements in their analyses (Rubin 2017: 135). Rubin concludes that there is "no one right way to read" the controversy over Caster Semenya – and indeed, given the myriad of transnational and local discourses, practices, investments, and institutions implicated in the social construction of bodies as of sexed/gendered, racialized, and (dis)abled, it is clear that more "capacious critical readings" are necessary if we hope to more accurately capture the life of folks like Semenya, as well as responses to it.

Meanwhile, emerging research on the sociomaterial co-constitution of gender/sex, race, and disability – such as work on the epigenetic effects of trauma on Black women, or on biological effects of trauma and/or "Adverse Childhood Experiences" (ACES) on women (and men) across races – promises to provide us with more representative accounts of the developmental sexualization, gendering, racialization, and disabling of humans, in all of its dynamic complexity (for example, Pear et al. 2017; Shannon Sullivan 2013). Jasbir Puar's (2017) *The Right to Maim: Debility, Capacity, Disability* lays out a critical feminist account of disability and strikes me as an articulation of the theoretical core of recent feminist work attempting "decolonize disability." In this project, Puar situates the constitution of impairment *and* disability within the functions of colonial power and transnational politics, as well as global capitalism and its racialized ecological effects. For example, Puar eerily echoes Tremain (2001) in her articulation of disability as existing "in relations to assemblages of capacity and debility, modulated across historical time, geopolitical space, institutional mandates and discursive regimes," rather than "a fixed state or attribute" (Puar 2017: xiv). Puar similarly attempts to productively disrupt traditional understandings of disability through the introduction of and an emphasis on the "biopolitics of debilitation" in order to foreground the "slow wearing down of populations instead of the event of becoming disabled" (Puar 2017: xiii–xiv). Puar links the "deliberate debilitation" of populations to the "racializing logics of security," noting that what counts as disability "is already overdetermined by white fragility on the one side, and the racialization of bodies that are expected to endure pain, suffering, and injury on the other" (Puar 2017: x, xiv).

Interventions like those of Puar (2017), Tremain (2001, 2017), Rubin (2017), Schuller (2018), and Snorton (2017) will only become more necessary for justice-oriented theory and practice as we are increasingly forced to recognize the co-constitutive nature of systems of oppression, both historically and in the present. As Puar notes, more than 50% of police shootings involve disabled folks, and as the twenty-first century continues its march forward, we shall increasingly be called on to wrestle with the disabling (and debilitating) effects of the always already sex/gendered and racialized system, that is, globalized capitalism in the form of climate change (Puar 2017: xii).

Related chapters: 7, 10, 15, 16, 18, 20.

Notes

1 For a critique of the timelines offered by Germon (2009) and Hausman (1995), see Mak (2012).
2 For an in-depth review of challenges to Money's position as a social constructionist, see Karkazis (2008: 63–80).
3 For more reaction to the DSD nomenclature/treatment model, and its development and adoption, see Davis (2015).
4 Sarah Baartman is perhaps better known by the derogatory "the Hottentot Venus."

References

Butler, J. (1990) *Gender Trouble: Feminism and the Subversion of Identity*. New York: Routledge.

Chase, C. (1999) "Rethinking Treatment for Ambiguous Genitalia [WWW Document]," *Pediatric Nursing*, http://link.galegroup.com/apps/doc/A55577832/AONE?sid=googlescholar (accessed 20.4.19).

Chase, C. (1998) "Surgical Progress Is Not the Answer to Intersexuality," *Journal of Clinical Ethics* 9, 385.

Clune-Taylor, C. (2019) "Securing Cisgendered Futures: Intersex Management under the 'Disorders of Sex Development' Treatment Model," *Hypatia* 34, 690–712. https://doi.org/10.1111/hypa.12494

Clune-Taylor, C.E. (2016) *From Intersex to Disorders of Sex Development: A Foucauldian Analysis of the Science, Ethics and Politics of the Medical Production of Cisgendered Lives* (PhD Thesis). University of Alberta.

Creighton, S.M., Michala, L., Mushtaq, I., and Yaron, M. (2014) "Childhood Surgery for Ambiguous Genitalia: Glimpses of Practice Changes or More of the Same?" *Psychology & Sexuality* 5, 34–43. https://doi.org/10.1080/19419899.2013.831214

Crenshaw, K. (1991) "Mapping the Margins: Intersectionality, Identity Politics, and Violence against Women of Color," *Stanford Law Review* 43, 1241. https://doi.org/10.2307/1229039

Crenshaw, K. (1989) "Demarginalizing the Intersection of Race and Sex: A Black Feminist Critique of Antidiscrimination Doctrine, Feminist Theory and Antiracist Politics," *University of Chicago Legal Forum* 1989, 139–168.

Crouch, R.A. (1998) "Betwixt and Between: The Past and Future of Intersexuality," *The Journal of Clinical Ethics* 9, 372.

Davis, G. (2015) *Contesting Intersex: The Dubious Diagnosis*. New York: New York University Press.

Doell, R.G. and Longino, H.E. (1988) "Sex Hormones and Human Behavior: A Critique of the Linear Model," *Journal of Homosexuality* 15, 55–78. https://doi.org/10.1300/J082v15n03_03

Downing, L., Morland, I., Sullivan, N. (eds.) (2015) *Fuckology : Critical Essays on John Money's Diagnostic Concepts* (First Edition). London: University of Chicago Press.

Dreger, A.D. (2000) *Hermaphrodites and the Medical Invention of Sex*. Cambridge, MA: Harvard University Press.

Dreger, A.D. (1998) " 'Ambiguous Sex': Or Ambivalent Medicine? Ethical Issues in the Treatment of Intersexuality," *The Hastings Center Report* 28, 24. https://doi.org/10.2307/3528648

Dreger, A.D. and Herndon, A.M. (2009) "Progress and Politics in the Intersex Rights Movement: Feminist Theory in Action," *GLQ: A Journal of Lesbian and Gay Studies* 15, 199–224. https://doi.org/10.1215/10642684-2008-134

Fausto-Sterling, A. (2008) "The Bare Bones of Race," *Social Studies of Science* 38, 657–694.

Fausto-Sterling, A. (2005) "The Bare Bones of Sex: Part 1—Sex and Gender," *Signs: Journal of Women in Culture and Society* 30, 1491–1527. https://doi.org/10.1086/424932

Fausto-Sterling, A. (2000) *Sexing the Body: Gender Politics and the Construction of Sexuality*. New York: Basic Books.

Fausto-Sterling, A. (1993) "The Five Sexes: Why Male and Female Are Not Enough," *Sciences* 33, 20. https://doi.org/10.1002/j.2326-1951.1993.tb03081.x

Fausto-Sterling, A. (1985) *Myths of Gender: Biological Theories about Women and Men* (Second Edition). New York: Basic Books.

Fausto-Sterling, A., Coll, C.G., Lamarre, M. (2012) "Sexing the Baby: Part 2 Applying Dynamic Systems Theory to the Emergences of Sex-related Differences in Infants and Toddlers," *Social Science & Medicine* 74, 1693–1702. https://doi.org/10.1016/j.socscimed.2011.06.027

Feder, E.K. (2014) *Making Sense of Intersex: Changing Ethical Perspectives in Biomedicine*. Bloomington: Indiana University Press.

Feng, J., Spence, I., and Pratt, J. (2007) "Playing an Action Video Game Reduces Gender Differences in Spatial Cognition," *Psychological Science* 18, 850–855. https://doi.org/10.1111/j.1467-9280.2007.01990.x

Fine, C. (2010) *Delusions of Gender: How Our Minds, Society, and Neurosexism Create Difference*. New York: W. W. Norton & Company.

Germon, J. (2009) *Gender: A Genealogy of an Idea*. New York: Palgrave Macmillan.

Haraway, D.J. (2003) *The Companion Species Manifesto: Dogs, People, and Significant Otherness*. Chicago, IL: University of Chicago Press. https://doi.org/10.5749/minnesota/9780816650477.001.0001

Hausman, B.L. (1995) *Changing Sex: Transsexualism, Technology, and the Idea of Gender*. Durham, NC: Duke University Press.

Hoad, N. (2010) " 'Run, Caster Semenya, Run!' Nativism and the Translations of Gender Variance," *Safundi* 11, 397–405. https://doi.org/10.1080/17533171.2010.511785

Holmes, M. (2008) *Intersex: A Perilous Difference*. Selinsgrove, PA: Susquehanna University Press.

Hughes, I.A. (2010) "The Quiet Revolution: Disorders of Sex Development," *Best Practice and Research. Clinical Endocrinology and Metabolism* 24, 159.

Jagger, A. (1983) "Human Biology in Feminist Theory: Sexual Equality Reconsidered," in Gould, C. (ed.) *Beyond Domination: New Perspectives on Women and Philosophy*. Lanham, MD: Rowman & Littlefield Publishers, Inc., 21–42.

Jordan-Young, R.M. (2010) *Brain Storm: The Flaws in the Science of Sex Differences*. Cambridge, MA: Harvard University Press.

Kaiser, A., Haller, S., Schmitz, S., and Nitsch, C. (2009) "On Sex/gender Related Similarities and Differences in fMRI Language Research," *Brain Research Reviews* 61, 49–59. https://doi.org/10.1016/j.brainresrev.2009.03.005

Karkazis, K. (2008) *Fixing Sex: Intersex, Medical Authority, and Lived Experience*. Durham, NC: Duke University Press.

Kessler, S.J. (1998) *Lessons from the Intersexed*. New Brunswick, NJ: Rutgers University Press.

Kessler, S.J. (1990) "The Medical Construction of Gender: Case Management of Intersexed Infants," *Signs: Journal of Women in Culture and Society* 16, 3–26. https://doi.org/10.1086/494643

King, H. (2013) *The One-Sex Body on Trial: The Classical and Early Modern Evidence*. Farnham, UK: Ashgate.

Laqueur, T. (1990) *Making Sex: Body and Gender from the Greeks to Freud*. Cambridge, MA: Harvard University Press.

Lee, P.A., Houk, C.P., Ahmed, S.F., and Hughes, I.A., in collaboration with the participants in the International Consensus Conference on Intersex organized by the Lawson Wilkins Pediatric Endocrine Society and the European Society for Paediatric Endocrinology (2006) "Consensus Statement on Management of Intersex Disorders," *Pediatrics* 118, e488–e500. https://doi.org/10.1542/peds.2006-0738

Mak, G. (2012) *Doubting Sex Inscriptions, Bodies and Selves in Nineteenth-Century Hermaphrodite Case Histories*. Manchester: Manchester University Press.

Markowitz, S. (2001) "Pelvic Politics: Sexual Dimorphism and Racial Difference," *Signs* 26, 389–414.

Money, J. (1995) *Gendermaps: Social Constructionism, Feminism and Sexosophical History*. London: Bloomsbury Academic.

Oudshoorn, N. (1994) *Beyond the Natural Body: An Archeology of Sex Hormones*. New York: Routledge.

Pear, V. A., Petito, L. C. and Abrams, B. (2017) "The Role of Maternal Adverse Childhood Experiences and Race in Intergenerational High-Risk Smoking Behaviors." *Nicotine & Tobacco Research* 19(5), 623–630. https://doi.org/10.1093/ntr/ntw295

Pitts-Taylor, V. (2016) *The Brain's Body: Neuroscience and Corporeal Politics*. Durham, NC: Duke University Press.

Puar, J.K. (2017) *The Right to Maim: Debility, Capacity, Disability*. Durham, NC: Duke University Press.

Reis, E. (2009) *Bodies in Doubt: An American History of Intersex*. Baltimore, MD: Johns Hopkins Press.

Repo, J. (2016) *The Biopolitics of Gender*. New York: Oxford University Press.

Richardson, S.S. (2013) *Sex Itself: The Search for Male and Female in the Human Genome*. Chicago, IL: University of Chicago Press.

Richardson, S.S. (2012) "Sexing the X: How the X Became the 'Female Chromosome'," *Signs: Journal of Women in Culture and Society* 37, 909–933. https://doi.org/10.1086/664477

Rippon, G., Jordan-Young, R., Kaiser, A., and Fine, C. (2014) "Recommendations for Sex/gender Neuroimaging Research: Key Principles and Implications for Research Design, Analysis, and Interpretation," *Frontiers in Human Neuroscience* 8, 1–13. https://doi.org/10.3389/fnhum.2014.00650

Rogers, L. and Walsh, J. (1982) "Shortcomings of the Psychomedical Research of John Money and Coworkers into Sex Differences in Behavior: Social and Political Implications," *Sex Roles* 8, 269–281. https://doi.org/10.1007/BF00287311

Roughgarden, J. (2004) *Evolution's Rainbow: Diversity, Gender, and Sexuality in Nature and People*. Berkeley: University of California Press.

Rubin, D.A. (2017) *Intersex Matters: Biomedical Embodiment, Gender Regulation, and Transnational Activism*. Albany: State University of New York Press.

Schuller, K. (2018) *The Biopolitics of Feeling: Race, Sex, and Science in the Nineteenth Century*. Durham, NC: Duke University Press.

Snorton, C.R. (2017) *Black on Both Sides: A Racial History of Trans Identity*. Minneapolis: University of Minnesota Press.

Subramaniam, B. (2009) "Moored Metamorphoses: A Retrospective Essay on Feminist Science Studies," *Signs: Journal of Women in Culture and Society* 34, 951–980. https://doi.org/10.1086/597147

Sullivan, N. (2015) "The Matter of Gender," in Downing, L., Morland, I., and Sullivan, N. (eds.) *Fuckology : Critical Essays on John Money's Diagnostic Concepts* (First Edition). London: University of Chicago Press, 19–40.

Sullivan, S. (2013) "Inheriting Racist Disparities in Health: Epigenetics and the Transgenerational Effects of White Racism," *Critical Philosophy of Race* 1(2), 190–218.

Tremain, S. (2017) *Foucault and Feminist Philosophy of Disability.* Ann Arbor: University of Michigan Press.

Tremain, S. (2001) "On the Government of Disability," *Social Theory and Practice* 27, 617–636. https://doi.org/10.5840/soctheorpract200127432

van Anders, S.M. (2015) "Beyond Sexual Orientation: Integrating Gender/Sex and Diverse Sexualities via Sexual Configurations Theory," *Archives of Sexual Behavior* 44, 1177–1213. https://doi.org/10.1007/s10508-015-0490-8

van Anders, S.M. and Dunn, E.J. (2009) "Are Gonadal Steroids Linked with Orgasm Perceptions and Sexual Assertiveness in Women and Men?" *Hormones and Behavior* 56, 206–213. https://doi.org/10.1016/j.yhbeh.2009.04.007

Yiee, J.H. and Baskin, L.S. (2010) "Penile Embryology and Anatomy," *The Scientific World Journal* 10, 1174–1179. https://doi.org/10.1100/tsw.2010.112

Further Reading

Gill-Peterson, J. (2018) *Histories of the Transgender Child.* Minneapolis: University of Minnesota Press.

Holmes, M. (ed.) (2016) *Critical Intersex.* New York: Routledge.

Latham, J. R. (2016) "Trans Men's Sexual Narrative-Practices: Introducing STS to Trans and Sexuality Studies," *Sexualities* 19(3), 347–368.

Latham, J. R. (2019) "Axiomatic: Constituting 'Transexuality' and Trans Sexualities in Medicine," *Sexualities* 22(1–2), 13–30.

Plemons, E. (2017) *The Look of a Woman: Facial Feminization Surgery and the Aims of Trans-Medicine.* Durham, NC: Duke University Press.

15

FEMINIST PERSPECTIVES ON VALUES IN SCIENCE

Kristen Intemann

Introduction

The history of science has shown us that values have influenced scientific decision-making in a variety of ways. Consider the case of hormone replacement therapy (HRT). Beginning the 1940s and 1950s, HRT became gradually accepted as the standard of care for treating symptoms of menopause in women such as hot flashes, night sweats, and sleep disturbances. But the number of physicians prescribing HRT and patients requesting it increased dramatically with the publication of the best-selling *Feminine Forever* in 1966, written by physician Robert A. Wilson (1966). The book argued that HRT not only could be used to treat certain symptoms but could "reverse" or "prevent" menopause and estrogen loss and "preserve" femininity by enhancing sexual drive, decreasing vaginal dryness, and preventing skin "aging and drooping." By the 1980s and 1990s, HRT was being widely claimed to not only be safe and effective for relieving menopausal symptoms but also as having a wide range of health benefits, including preventing cardiovascular disease, osteoporosis, and certain types of cancer (such as Lindsay et al. 1984; Stampfer and Colditz 1991; Grodstein et al. 1996; Gambacciani et al. 2000). In 2002, the Women's Health Initiative released its findings of one of the most comprehensive randomized clinical trials, which called into question the benefits of HRT and suggested that HRT (at least in some forms) actually *increased* risk of breast cancer, heart disease, stroke, and blot clots (at least in some women) (WHI 2002). Yet some have argued that this study should have come as no surprise, as there had been warnings that (a) there was little evidence supporting claims about the health benefits of HRT and (b) there had been evidence of potential risks for decades (Kirsh and Kreiger 2002). Subsequent findings suggest that the use of HRT in women is complex and both the risks and the benefits of HRT depend on a variety of factors, including the type of hormones delivered, the dose, the length of treatment, and a variety of patient-related factors.

While no doubt there are many reasons for this complicated history regarding the evidence for claims made about HRT, this is a case where a variety of value judgments operated in generating evidence for and against both scientific claims about HRT and clinical recommendations. Assumptions were made about what constitutes a "benefit" and a "risk" for women, what risks are acceptable, and how much evidence was needed to support particular claims. The idea that preventing "aging or drooping skin" constitutes a health benefit for women is a value judgment (and one clearly informed by feminine gender norms). Moreover, the assumption, often implicit, that the benefits of HRT outweigh potential risks depends not just on the probability of those risks, but how *good* the benefits are, as well as how *grave* the consequences of certain medical conditions would be. Finally, there were implicit value judgments made about how much evidence is needed in order to determine whether a treatment is safe or effective, as well as what kind of evidence (e.g. observational studies or randomized control trials) count as *good enough* evidence for assessing safety or effectiveness.

The idea that value judgments may influence science and evidentiary reasoning is not new and has been debated by a variety of philosophers of science over the years. But questions about the role of values in science arose in feminist scholarship with a particular set of concerns and were pursed with a unique set of aims. The goal of this chapter is to elucidate *why* feminist philosophers of science have focused on the role and nature of values in science (as compared to non-feminist philosophers interested in similar questions), to provide an overview of the ways that feminists have shown values to operate in science, and to distinguish two feminist approaches to thinking about *which* values or *whose* values might be endorsed by scientists or scientific communities. I will begin with an overview of the value-free ideal (VFI).

The Value-Free Ideal of Science and Its Critics

The view that science, ideally, ought to be "free of values" is challenging to characterize because its meaning has changed over time and has varied even among its advocates (Proctor 1991; Douglas 2009). It is *not* the claim that science and ethics are wholly distinct, or that values never have any role in science. Many recognize that *some* values play a role in judgments about what we take to be *good* evidence or in constituting the central aims of science (such as the value of understanding, prediction, or empirical adequacy of theories). The VFI of science is thus primarily concerned with limiting the role of "non-epistemic" values, or values thought to be distinct from the epistemic and cognitive aims of science. It is often assumed that moral, political, social, and economic values are non-epistemic in this sense. That is, they denote a *kind* of value that is distinct from (and possibly antithetical to) epistemic or cognitive aims. Of course, such values *can* be important to science in rather obvious ways. For example, such values may impact what sort of research is pursued or funded. Ethical values may also place certain constraints on scientists in, for example, the treatment of human subjects. The idea that science ought to be "value-free" has more been aimed at protecting what Heather Douglas (2009) has referred to as the "heart" of scientific reasoning. That is, it is the idea that non-epistemic values should play no role in the justification of scientific theories, or in determining *what evidence there is*, which theories are best supported by evidence, or ought to be accepted by scientists.

In the history of philosophy of science, the question of whether science ought to be value free was primarily concerned with the rationality of scientific reasoning and the objectivity of science. Thomas Kuhn (1962) has often been interpreted as suggesting that science is at least "partly subjective," in so far as assessing theories may depend on the values of scientists and how much weight they give to certain virtues of theory choice. Attempts to defend the VFI were often aimed at defending the rationality and objectivity of science. The concern about values in science was, to a large extent, a concern that the authority, rationality, and integrity of science might be compromised by "subjective" values, which are distinct from empirically verifiable scientific claims. Challenges to the VFI have typically been concerned with either showing that as an ideal it is unattainable, or, more recently, that the VFI might be inconsistent with the ethical obligations of scientists to think about how their decisions might impact the public (Douglas 2009; Elliott 2017).

Feminist philosophers of science, however, have had a distinct set of concerns in considering the relationship between values and science. In particular, the central worry is that the VFI stood as an obstacle that might reinforce rather than undermine social justice (and in particular, abolishing systems of oppression). The VFI was employed by both scientists and philosophers of science to dismiss several specific feminist concerns. First, feminist theorists were interested in the (arguably empirical) question about whether the historical exclusion of women and other underrepresented groups makes any epistemic difference to science. That is, does the gender, race, or class of scientists have any implications for what sorts of things are studied, how they are studied, or what evidence or knowledge is produced? If science is "value-free" and its methods are "objective" in the sense that they are independent of *who* is employing them, then it should not matter, from an epistemic perspective, *who* participates in science. Thus, the VFI was employed as a way to dismiss concerns about the lack of diversity and inclusivity in science. Of course, we might have *political* or *ethical* reasons for thinking that women should have access to education or careers in STEM. However, there was no recognized *epistemic* imperative for this on the grounds that it would make for better science insofar as science was thought to be ideally value-free.

Second, a growing number of feminist scholars and scientists in the 1970s and 1980s began to recognize instances where sexist and androcentric values and stereotypes seemed to influence science in negative ways. At best, these concerns were dismissed as isolated cases of "bad" science that had occurred precisely because some individual scientists went astray of the VFI. At worst, the concerns were met with the charge that feminists themselves were being "political" or introducing political considerations into judgments that should be "epistemic," "apolitical," and "objective." Letitia Meynell's chapter in this volume discusses how this occurred in evolutionary psychology and Robyn Blum's chapter demonstrates how this occurred in the history of neuroscience. Thus, feminists were motivated to consider whether sexist, racist, or other problematic biases were in fact isolated cases of "bad sciences" or whether even conscientious scientists, carefully adhering to the methods and norms of their disciplines, were still making value judgments (and value judgments that were themselves unjustified).

Finally, feminists have also noted that several celebrated successes of science have not actually resulted in benefits to women and other marginalized groups to the same extent as those with more resources or power. Despite advances in medicine, health inequalities have persisted. New technologies have tended to be accessible to those who have resources to afford them. And the sorts of things being studied failed, in some cases, to address the sorts of problems that marginalized groups take to be more pressing (See Kourany, this volume; Intemann and de-Melo-Martín 2010, 2014). Yet if the role of scientists is viewed, as per the VFI, as being narrowly concerned with evaluating whether data support particular theories, then this is not something scientists would appropriately worry about. Thus, a final motivation for feminist scholars to grapple with the VFI is to ensure that the knowledge, interventions, and new technologies generated by scientific inquiry can benefit everyone in society (as opposed to a select few). Feminist approaches to philosophy of science see science as an enterprise that plays an important role in society that like other social institutions can be examined through a lens of social justice.

Understanding the motivations for why feminists have been motivated to critically evaluate the VFI, I will now provide a brief and partial overview of the ways in which feminists have argued that values play an ineliminable role in scientific reasoning, often throughout the research process.

Feminist Challenges to the VFI

Many scholars have documented instances of sexist, androcentric, heterosexist, racist, and classist assumptions influencing science (Hubbard 1983; Fausto-Sterling 1985; Hrdy 1986; Martin 1991; Gould 1996; Wylie and Hankinson Nelson 2007; Longino 2013; Richardson 2013). Such bias is often unconscious and has occurred even among conscientious scientists aiming to adhere rigorously to scientific norms and methodologies. Feminist work revealed that the causes of biases were not values as such, but rather often involved implicit value judgments that were unacknowledged and, in many cases, unjustified (Longino 1990; Nelson 1990; Harding, 2008; Kourany 2010; Borgerson 2011; de Melo-Martín & Intemann 2011; Goldenberg 2015).

Correcting for such biases, then, was not stricter adherence to the VFI, but instead involved mechanisms for making value judgments more explicit, so that they could be critically evaluated. Scientists ought not aim to eliminate values from science, but ensure that those adopted are more justified and, in some sense, more "feminist." What follows will provide an overview of some of the ways that feminists have argued that values play necessary and important roles in scientific reasoning and practices in ways that may be inconsistent with the VFI.

One way that values play a role in scientific reasoning is in the selection and framing of research problems. As Kitcher argues (2001, 2011), science is not merely concerned with generating true beliefs about the world, but particularly interesting or significant truths. Public grant agencies evaluate grant proposals not only according to their intellectual merit but also in terms of their potential to have significant social impacts or address pressing social issues. Thus, ethical and social values are relevant to judgments about what should be studied or funded.

In biomedical research, the increasing presence of commercial interests has often directed research toward conditions that affect those with the resources to pay for medical interventions. It is well documented that 90% of biomedical research goes to diseases that affect 10% of the world's population, whereas diseases – particularly those widely affecting resource-poor countries, are neglected (Currat et al. 2004; Daniels 2008; Stevens 2008). Yet correcting this problem may require an explicit value judgment that research effort ought to be directed to those most in need, or those with the most pressing health problems. In such decisions, value–neutrality is not a viable option.

Yet values shape not only decisions about what to study broadly speaking but also how particular research problems are conceived or framed. For example, contraceptive research has largely focused on developing and testing interventions on women (Ringheim 1999; Doyal 2001). While there may be some good reasons for doing so, framing contraception solely as a problem for women means that the burdens of testing and taking contraceptive measures, as well as their possible side effects, fall largely on women. This is not to say that those conducting contraceptive research necessarily had bad intentions or thought that women should bear more risks in contraception. Indeed, many researchers may have even had the very good intention of wanting to empower women in making reproductive choices. However, the research problem was framed in a way that made an implicit judgment that women are those who should be responsible for contraception and that the risks associated with this could be justly placed on women. Making this value judgment explicit suggests that a better-grounded value judgment ought to frame research problems around contraception, for example the judgment that the burdens of contraception should be equally shared, or that we need contraceptive devices that equally empower the reproductive autonomy of males and females.

A second, more controversial, way that value judgments influence scientific reasoning is through observation and the description of data. As many have argued, observation is "theory laden" in that our perceptions of phenomena are mediated by background assumptions and our conceptual categories (Duhem 1906; Hanson 1958; Kuhn 1962). Our background assumptions and conceptual categories can also be influenced by social value judgments and gender norms. Emily Martin (1991), for example, has shown that accounts of human reproduction often employed language that ascribed gendered stereotypes to the sperm and egg, characterizing the egg as passively floating through the fallopian tubes and subsequently "penetrated" by the sperm that actively "burrows" into the egg. In this case, implicit gendered stereotypes obscured what was actually occurring. Subsequent research revealed that there is a complex chemical interaction between the surface of the egg and sperm. Yet, even in the newer accounts, language used to describe these processes remained gendered, attributing a femme fatale-like role to the egg in claiming that it "ensnares the sperm" with its sticky surface (Martin 1991).

Gendered stereotypes have also been reflected in observations of non-human animals. Early depictions of primates represented them as having many human-like qualities and social interactions

(Schiebinger 2004). Primatologists relied on the assumption that females are coy and choosy of potential mates, neglecting the fact that female primates often had multiple sexual partners (Hrdy 1986). This was neglected, not because male primatologists in the field at the time were particularly bad observers, but rather because their expectations and unconscious background beliefs directed them toward observing certain things and away from observing others, influencing their perception of primate behavior.

Yet encouraging researchers to be "neutral" toward all values may not be sufficient for correcting the sort of influence that stereotypes have. It may turn out that there are some metaphors or models by analogy that are in fact useful in science even if they are not strictly speaking literally true. What is important, however, is to be critically aware of when they are operating and the extent to which they are beneficial or harmful. In the earlier cases, the use of gendered stereotypes was not only empirically deficient but also potentially harmful insofar as it universalized and reinforced problematic gender norms that people are already inclined to mistakenly believe. Moreover, such stereotypes and norms have been used to justify social inequalities. Thus, scientists must make value judgments about whether their observations or descriptions may be informed by harmful stereotypes or not, or whether the language they use to describe data may have ethical as well as epistemic consequences. Once again, this is not compatible with the VFI, insofar as it may require making ethical value judgments.

Similarly, in archaeology, investigations into the evolution of human tool use reflect assumptions about, for example, what is a "tool." Historically, many archaeologists focused on tools used in hunting and killing animals (tools generally associated with men). Male archaeologists did not take baskets or reeds used for foraging to be "tools." Thus early accounts of the evolution of tool use in humans focused solely on the activities traditionally assigned to males. Again, archaeologists in the field at the time simply did not "see" baskets as tools and thus did not take them to be relevant data (Wylie and Nelson 2007). It wasn't until female and feminist archeologists started entering the field and began asking different questions – questions about the ways in which women may have contributed to the evolution of humans and tool use – that new data, such as baskets and reeds, was identified *as evidence*. In this case, values played an important and positive role in opening up new lines of evidence and correcting for the ways in which gendered norms and conceptions of labor may have limited or biased scientific theorizing.

A third way that values can play a role in scientific reasoning occurs in cases where the subject of science is itself normative or value-laden. Some areas of scientific research, particularly in the social sciences, may also involve concepts or objects of study that involve values. For example, measuring "harms" to children of divorce requires value judgments about what constitutes harm (Anderson 2004). Historically, harms to children were measured in terms of economic and psychological harms that result from not having two parents in the same home. Yet there are ways of conceiving of "harm" that might capture other impacts to children we are interested in, for example harms related to having a mother staying in an abusive relationship, or being economically dependent on a husband, or other inequalities that might exist by staying in a marriage. In this case, what constitutes a "harm" involves a value judgment about what constitute negative outcomes from which children ought to be protected. Insofar as some scientific concepts actually have normative content, then the VFI is misguided. Attempts to avoid "value-laden" concepts in science, even if possible, would surely not be desirable, as it would limit our ability to empirically understand certain features of our (normatively laden) world and human experience. Thus, the concern is not that concepts are value-laden, but that they may presuppose values that are obscured, partial, or unjustified.

Similarly, as we saw with HRT, there are a variety of areas of medical research that make assumptions about what constitutes a "risk" or a "health benefit" for patients. For example, studies on risks related to pregnancy and childbirth make implicit value judgments about what constitutes a risk, as well as what sorts of risks are acceptable or not (Lyerly et al. 2009; de Melo-Martín and Intemann 2012). Consider studies purporting to showing that home birth is less safe for low-risk women than hospital births (de Melo-Martín and Intemann 2012). Typically, such studies measure risks such as infant and maternal mortality rates (which in the United States are extremely low for low-risk women regardless of whether a birth occurs at a hospital). The assumption here is that even though the risks of death or serious complication are low in home births, hospital births will always be safer because of the staff and equipment available in case problems arise. However, those who opt for home birth typically do so because of other kinds of perceived risks with hospital births. Hospital births have much higher rates of c-section, episiotomies, fetal heart monitoring, and infection (Janssen et al. 2002, 2009; Johnson and Daviss 2005; Hutton et al. 2009). While many may judge these risks to be less grave than infant and maternal mortality, the probability that they will occur is far greater. Yet these kinds of risks have not been measured in many studies evaluating the "safety" of home birth. Thus, such studies make implicit value judgments about which adverse effects should be relevant to healthcare decisions.

Measuring safety and risks, whether it be for home birth, HRT, GMOs, the toxicity of a substance, or risks related to climate change, involves value judgments about what constitutes a "risk" and which risks are more acceptable than others. Insofar as risk assessment is an important aspect of scientific reasoning that necessarily relies on value judgments, the VFI is not tenable.

A fourth way that values enter into scientific reasoning is in the construction and use of particular models. Models are used in science to represent complex phenomena in ways that increase understanding or generate predictions. But models, by definition, are not literal representations of the way the world is. They select certain features to represent and neglect or simplify others. Models, of course, can be quite useful, but they necessarily rely on implicit assumptions about which variables are salient or important to understanding some phenomena and which are not. These judgments can also involve value judgments.

In biomedicine, researchers have often used exclusively male animals in preclinical research (Beery and Zucker 2011; Rice 2012; Kokras and Dalla 2014; Yoon et al. 2014). Rodent studies used to evaluate the effects of drugs on behavior use males nearly exclusively, in spite of the evidence showing sex differences in drug metabolism (Hughes 2007). The underrepresentation of females in animal models exists even in studies for diseases that predominantly affect women. For instance, although diagnoses for depressive and anxiety disorders, stroke, and thyroid diseases are significantly more common in women than in men, the majority of research with animals to study these disorders actually used male animals (Zucker and Beery 2010). Similarly, the majority of basic and preclinical studies conducted on experimental animals studying physiological, pharmacological, and endocrinological aspects of cardiovascular disease mostly use male animals (Beery and Zucker 2011).

This is not to say that the predominant use of males in animal models is because male researchers necessarily do not care about women's health or are sexist. The use of males in animal models has occurred in part because it was assumed that hormonal variability during the estrous cycle of females presents confounding variables that make it difficult to establish clear causal relationships necessary for understanding the mechanisms of disease and for testing the efficacy of interventions. This assumption is, however, dubious (Prendergast, Onishi, and Zucker 2014). Yet, more importantly, the decision to use male-only animal models for the sake of simplicity and predictive power relies on an implicit value judgment that these epistemic considerations are more important than the social consequences for female patients. That is, the dominance of such models may have the effect of generating guidelines for diagnosing diseases and treatment interventions that are less applicable for female patients (de Melo-Martín and Intemann 2016). Thus, there are concerns about whether this is a reasonable trade-off and who should make such decisions.

Similarly, there are concerns about the kinds of models that have been used in assessing the potential impacts of climate change (Agarwal 2002; Schneider and Kunz-Duriseti 2002). Many integrative assessment models measure the aggregate impacts of climate change on some phenomenon such as food production. Some have argued that although agriculture will decrease in places where droughts or rising sea-levels cause land to become unusable, this will be offset by increased food production in areas that will have milder climate conditions (Rosenzweig and Parry 1994; Parry et al. 2004). However, measuring the aggregative effects of climate change on food production may obscure the ways in which access to food may be affected in ways that reinforce or exacerbate existing social inequalities. As many have pointed out, developing just climate policies requires information not only about the aggregate impacts to be expected from climate change but also information about the distribution of those impacts to ensure that costs and benefits can be distributed equitably (Agarwal 2002; Schiermeier 2010). Thus, there is concern that current models fail to measure features of food production that fail to account for the interests of some groups, particularly those in the global south who are already arguably disadvantaged.

Hence, feminist theorists are concerned that choices made with respect to models are often accompanied by implicit and unexamined value judgments about what is important and whose interests ought to be protected. Particularly as members of these groups have been historically underrepresented within scientific research communities, there are concerns about which value judgments should guide modeling decisions and who should make these decisions (Intemann 2015). Yet, it is difficult to consider which value judgments should guide such decisions without first rejecting the VFI.

A final way that value judgments have been argued to enter into scientific reasoning is in the interpretation of data and consideration of alternative hypotheses. Assessing whether a particular hypothesis is justified, or ought to be accepted as the best explanation for the evidence, is not something that occurs in isolation. Scientists must consider not only whether evidence supports a hypothesis, but whether a hypothesis is better justified by the evidence than other hypotheses. That is, judgments about whether a hypothesis is justified are *comparative* judgments about how that hypothesis fares in relation to alternative explanations (Fausto-Sterling 1985; Longino 1990; Okruhlik 1994; Longino 2013). One consequence of this is that any values that influence which hypotheses are considered also play an indirect role in comparative judgments about which hypotheses are the "best" explanations (Okruhlik 1994).

Consider, for example, research on sex differences in visual-spatial abilities in the 1970s. One experiment was the rod and frame test, where subjects in a dark room were presented with a lighted rod bisecting a frame tilted at an angle and had to instruct the experimenter where to reposition the rod so that it would be perpendicular to the floor. Differences in performance on this test led researchers to conclude that males had superior visual-spatial abilities in virtue of differences in their biology (Fausto-Sterling 1985). According to Longino, one of the problems with these experiments was that the researchers conducting these experiments were overwhelmingly males with homogeneous values and interests (Longino 1990). As a result, they failed to see how the experimental design was biased so as to hinder the performance of females. As Fausto-Sterling (1985: 32) points out, female subjects were often tested in a dark room with male scientists, possibly making them uncomfortable and affecting performance. Scientists also neglected alternative hypotheses that would have accounted for the data equally well. For example, they never considered that because women are often socialized to be less assertive, they might have been less comfortable in having the researcher make continual minute adjustments to the rod (Fausto-Sterling 1985: 32). Indeed, when the experiment was subsequently altered to account for some of these potential variables, differences in performance disappeared (Fausto-Sterling 1985).

To summarize, this section has discussed five different ways that feminists have argued that values play a role in scientific reasoning. They can operate as implicit assumptions in (1) the framing research problems, (2) observing phenomena and describing data, (3) reasoning about value-laden concepts and assessing risks, (4) adopting particular models, and (5) collecting and interpreting evidence. While this list is not intended to be exhaustive, it illustrates a complexity and diversity of the ways that science and values can interact that is ignored by the VFI. Insofar as such value judgments are implicit or unnoticed, there are concerns that they may often uncritically reflect sexist, racist, or classes values and stereotypes, even when scientists are well-intentioned. In some cases, these value judgments lead to scientific conclusions that are unjustified, limited, or partial. But insofar as challenging these judgments often also requires making value judgments, feminists have not endorsed stricter adherence to the VFI as a way to address the problem. Rather, feminist work has shown that values may play important and necessary roles in generating new lines of evidence, correctly assessing risks and benefits, and directing science in ways that better meet the needs and interests of marginalized groups. Thus, the question is how might the negative of influence of values be avoided or minimized while the potential benefits of values in science be realized?

Feminist Alternatives to the VFI

Like non-feminist scholars who challenge the VFI, feminist science studies scholars have tended to agree that the VFI is both unattainable and misguided as an ideal for governing scientific reasoning or practice. Non-feminist philosophers of science have typically aimed to develop alternatives to the VFI that thus recognize the important role that social, ethical, or political values might play in science, while promoting mechanisms that protect the epistemic integrity of science (Douglas 2009). While feminists can happily endorse this aim, recall that feminists also have a unique set of concerns about the VFI. First, the VFI has been used to deny the epistemic value or diversity or inclusivity within scientific communities or among epistemic agents. Second, the VFI led scientists and some philosophers of science to dismiss cases of problematic biases as isolated instances of "bad science" and block feminist critiques as "politically motivated." Third, the VFI has created a culture of scientific practice and training where the products of science often fail to benefit marginalized groups or address the most pressing social needs. Indeed, the history of science is full of examples where the knowledge or innovations produced from science was knowledge that could be used to justify or reinforce existing inequalities. Thus, from a feminist perspective, a successful alternative to the VFI must address these additional concerns, as well as protect the epistemic integrity of science. That is, an alternative to the VFI must address:

1   Does the participation of historically excluded or underrepresented groups in science matter and, if so, how or why?
2   How might "feminist" values or aims (broadly construed) play a positive role in science?
3   How might we ensure that the products of science (knowledge, innovations, interventions) benefit marginalized groups in ways that reduce, rather than reinforce, systems of oppression?
4   Given the unavoidable (and often implicit) ways that values influence science, how can the epistemic integrity of science be protected?

In the following sections, I will consider two potential alternatives to the VFI, each of which can be seen as emerging from two feminist philosophies of science: critical contextual empiricism and feminist standpoint theory (discussed in chapters by Kirstin Borgerson and Catherine Hundleby in this volume).

The Value-Management Model

Rolin (2017) has articulated a "value-management" model, largely informed by Helen Longino's critical contextual empiricism (1990, 2002). According to this model, value judgments may play a variety of roles in science, so long as several mechanisms are adopted by scientific communities to "manage" the values that may influence individual scientists. By structuring science so that research communities are comprised of those with diverse values and criticism is encouraged, such communities as a whole can achieve a higher degree of objectivity. Following Longino's critical contextual empiricism, scientific knowledge is objective insofar as the organization of a diverse scientific community satisfies the following four conditions, which are also discussed in greater detail in Borgerson's chapter in this volume (Longino 1990: 76):

1 There must be recognized avenues for the criticism of evidence, of methods, and of assumptions and reasoning.
2 There must exist shared standards that critics can evoke.
3 The community as a whole must be responsive to such criticism.
4 Intellectual authority must be shared equally among qualified practitioners.

When scientific communities are structured so as to meet these criteria, then any value-laden assumptions inappropriately influencing scientific reasoning are more likely to be identified and critically evaluated (Longino 1990: 73–74, 80, 2002: 51). When values are different from one's own, it is easier to see when they are influencing scientific reasoning or description of data. Thus, a scientific community comprised of individuals with diverse ethical and political values will be able to identify and critically evaluate the ways that values influence the reasoning individual scientists.

This approach has several attractive features from a feminist perspective. First, it provides an ideal that does explain why diversity within a scientific community is epistemically valuable. Having diversity within scientific communities ensures that the community as a whole is likely to have a maximally representative pool of experiences, values, and interests to draw on in order to identify when stereotypes or problematic values or assumptions may be inappropriately influencing scientific reasoning in a particular case.

Second, on this view, feminist values can play important *positive* roles in science. Feminist values can play an appropriate role in scientific reasoning that requires value judgments, at least insofar as they are better-grounded or more relevant in a particular case than sexist or racist values. For example, if there are good reasons to believe that medical interventions ought to be as safe and effective for women as men, this provides good reason for thinking that male-only animal models should not be dominant in drug testing. If there is justification for taking the gathering activities of women to be valuable work, then baskets and reeds ought to be viewed as tools and data that should be accounted for in a theory of the evolution of tool use. If, after careful critical evaluation, there is agreement that c-section and episiotomies and infection are bad for pregnant women that ought to be minimized, then research on risks related to pregnancy ought to measure these outcomes. The idea here is that insofar as value judgments are sometimes *necessary and relevant* to scientific reasoning, then feminist values can play an appropriate role in making these decisions *so long as* they are justified or survive critical scrutiny by a diverse scientific community.

Third, there is reason to think that the value-management model might also have more success in producing research and innovations that would bring greater benefits to marginalized groups. If scientific communities include those with values and interests that have been historically underrepresented within science, then there is a greater chance that those with feminist, social justice, or decolonialist values and interests will pursue different research agendas and questions, alternatively frame research problems, and adopt new methodologies that will help direct research in ways that are more likely to benefit marginalized groups. Insofar as interests shape what is studied and how it is studied, incorporating diverse interests would broaden the range of phenomena studied and increase the creativity of scientific communities (Solomon 2006).

Finally, the value-management model can also meet the fourth criterion needed for any successful alternative to the VFI, that is, it allows for the unavoidable presence of values in science, but it does so in a way that builds in protections for the integrity of science. Scientists must still operate within a context of shared standards for evaluation (e.g. their theories must be empirically adequate) and the critical evaluation that is required by diverse participants ensures that those standards are adhered to and that individual biases will be caught or scrutinized.

While the value-management model certainly has significant strengths (and is clearly highly preferable to the VFI), there are some concerns that it may not go quite far enough. Specifically, the worry is that the value-management model appears to neglect the ways in which inequalities in power and resources can influence and limit both social and scientific practices in ways that may be of concern. The value-management approach to values appears to adopt something analogous to a liberal political philosophy. It suggests that any and all values are welcome in science, so long as they are critically evaluated on a level-playing field, where "players" are held to the same standards of evaluation.

There is some concern that this overlooks the social, cultural, and economic context in which science is currently practiced (Pinto 2014). That is, research occurs within a context where it is increasingly funded by private interests, and those with greater economic and social resources have greater power to determine what research programs are pursued, what sorts of methodologies might be required by regulatory standards, and what is (or is not) published. For example, clinical trials funded by pharmaceutical companies often retain the rights to withhold negative results (Elliott 2008; Intemann and de-Melo-Martín 2014). Also of increasing concern are cases of manufactured dissent, where privately funded scientists raise criticisms or cast doubt on research, such as research on the effects of smoking or environmental toxins, where commercial interests appear determined to generate dissenting research for the purpose of stalling public policy, confusing the public, and diverting the resources of other scientists (Michaels 2008; Oreskes and Conway 2010; Elliott 2011). It seems dubious that the ideal postulated by contextual empiricism is likely to be even approximated in this context.

Inequalities of power and resources also significantly still shape who participates in science. Access to resources that facilitate interest and training in STEM disciplines is not equitably distributed among populations (De Welde and Laursen 2011). For example, in the United States, one-third of K-12 schools are rural and one in five students attends a rural school; yet rural communities receive less funding for education than their urban counterparts, tend to have less access to technologies such as high-speed internet, and have a more difficult time recruiting and retaining qualified STEM teachers (Avery 2013). Many rural, poor, female, and indigenous STEM students also report a disconnect between academic STEM knowledge and their local knowledge practices, which they see as more relevant to the "real world" and things they care about (Avery and Kassam 2011). These factors make it more challenging for certain groups to pursue STEM education and careers.

In addition, the culture of STEM disciplines in the academy often fails to be inclusive, even when there is no explicit discrimination with respect to epistemic authority. For example, the work demands of scientists may be incompatible with the demands of parenting and other kinds of domestic duties in some contexts (Mason et al. 2013). Certain groups may experience harassment, microaggressions, or a chilly climate that make participation in scientific communities undesirable even if one's scientific work is not overtly discounted (Clark Blickenstaff 2005; De Welde and Laursen 2011; Wylie 2012).

The concern, then, is that without addressing these greater structural inequalities, the value-management approach will not really be able to assure that the needs and interests of underrepresented groups will genuinely be a part of scientific communities (or have a chance to challenge assumptions or value judgments that may be unjustified or have disproportionate influence). Of course, these concerns are largely pointing to concerns about whether the ideal proposed by this model is likely to be met and need not undermine the criteria the model offers. But given that we are looking for an alternative to the VFI that is also hopefully more realizable, perhaps more resources are needed.

The Social Justice Standpoint Model

I will identify a second approach offering an alternative to the VFI that draws on resources from standpoint feminist literature and decolonialist literature (e.g. Harding 2008; Mignolo 2009; Lugones 2010; Harding 2016). It might be seen as an approach that could add further requirements to the alternative provided to the value-management approach, rather than a whole-scale alternative. The social justice approach proceeds from a broad normative commitment to social justice (even though there may be multiple ways of understanding or promoting such a commitment). A social justice approach assumes that a central aim of publicly funded science (or at least a significant portion of publicly funded science that impacts citizens) ought to be achieving social justice. That is, science should aim to benefit all, but particularly those that are the least well-off. It views science as a vehicle for correcting past and current injustices and providing us with the knowledge and tools for ensuring that historically marginalized communities also have the opportunity to be healthy and flourish. In this sense, the social justice approach endorses and prioritizes some value over others (as opposed to embracing pluralism about *all* values). As the result of this value commitment, the social justice approach also calls on us to critically examine the ways that systems of power, oppression, and injustice influence the structure and practice of science, with an eye toward transforming them to be more just.

The social justice approach is largely informed by feminist standpoint theory. Feminist standpoint theory can be seen as offering a methodological resource for critically examining and understanding how hierarchical power structures limit and shape both *who* participates in knowledge production and the kind of knowledge that is produced. Standpoint feminists argue for adopting epistemic practices where researchers can achieve a feminist standpoint, or a critical examination of the ways in which power structures produce and maintain both epistemic and social inequalities. The aim here is to produce knowledge that is not only epistemically reliable but also that serves the needs of marginalized groups in ways that will reduce, rather than reinforce, social injustice.

This approach is also likely to be attractive from a feminist perspective for several reasons. First, it provides an account of why the participation of underrepresented groups in science is crucial. That is, members of marginalized groups are more likely to have access to experiences and evidence about how structures systematically disadvantage marginalized groups. Privilege is often easier to identify by those who do not have it. Moreover, this approach calls on scientific communities to identify, examine, and try to correct many of the barriers to participation in knowledge production that are created by inequalities.

Second, it advocates a prominent role for arguably feminist values. That is, it directs science toward the aim of social justice and to be attentive to social justice concerns throughout the research process. In this way, feminist values play a fairly robust role in guiding the sorts of value judgments that scientists should make in particular contexts.

Third, and relatedly, it is guided by an explicit political commitment to produce knowledge and innovations that are relevant and beneficial to marginalized groups. The aim is to employ methodologies that challenge, rather than reinforce, systems of oppression. Standpoint theorists proceed by "studying-up" or studying from the margins out, so as to understand how existing power structures shape and limit knowledge production, as well as the scientific phenomena under investigation (Crasnow 2007; Harding 2008; Intemann 2010; Intemann and de Melo-Martín 2011). Studying from the margins out requires researchers to work with members of marginalized groups to identify and frame scientific priorities problems, identify the full range of stakeholders affected by that problem, consider the ways in which the traditional conceptual frameworks relied upon may fail to serve the interests of different groups, investigate the ways the problem may manifest differently for different social groups, and examine how systems of oppression may play a role in or contribute to the problem.

Fourth, with respect to the integrity of science, the social justice approach does not deny the importance of science that is empirically adequate or justified. Social justice aims need not be taken to be at odds with the integrity of science. Indeed, presumably the achievement of knowledge that actually is effective in reducing inequalities or improving the lives and health of marginalized groups requires scientific information and interventions that are accurate and reliable. So, like the value-management model, the social justice approach is committed to some shared standards of evaluation (such as empirical adequacy) that can be used to protect the epistemic integrity of science.

The social justice approach may raise more questions than provide answers at this point. In particular: are there different (and potentially competing) conceptions of social justice and what do we do if they conflict? Should all research really be aimed at social justice, or is it possible that some research is valuable even if it does not advance these aims? How is social justice to be promoted in certain concrete cases of scientific decision-making? While I cannot address these important questions here, they are fruitful directions for future research.

Conclusions

The aim of this chapter has been to examine the ways in which feminist scholars have contributed to debates on values in science. Like many non-feminist philosophers of science, feminists have rejected the VFI as neither possible nor desirable, as evidenced by a large body of examples, only some of which were summarized here. But in examining and challenging the VFI, feminists have had a set of unique concerns. These concerns have also guided the alternative approaches that they have offered to values in science. The two approaches discussed here both offer important resources for correcting for sexist, androcentric, racist, and other biases that have influenced scientific reasoning, and promoting feminist aims. More research is needed to further develop these approaches and articulate a positive alternative to the VFI.

Related chapters: 6, 7, 16.

References

Agarwal, A. (2002) "A Southern Perspective on Curbing Global Climate Change," in Schneider, S., Rosencrantz, A., and Hiles, J.O. (eds.) *Climate Change Policy: A Survey*, Washington, DC: Island Press, pp. 375–392.

Anderson, E. (2004) "Uses of Value Judgments in Science: A General Argument, with Lessons from a Case Study of Feminist Research on Divorce," *Hypatia*, 19, pp. 1–24.

Avery, L. (2013) "Rural Science Education: Valuing Local Knowledge," *Theory into Practice* 52(1), pp. 28–35.

Avery, L.M. and Kassam, K.A. (2011) "Phronesis: Children's Local Rural Knowledge of Science and Engineering," *Journal of Research in Rural Education* 26, pp. 1–18.

Beery, A.K. and Zucker, I. (2011) "Sex Bias in Neuroscience and Biomedical Research," *Neuroscience and Biobehavioral Reviews* 35(3), pp. 565–572.

Borgerson, K. (2011) "Amending and Defending Critical Contextual Empiricism," *European Journal for Philosophy of Science* 1(3), pp. 435–449.

Clark Blickenstaff, J. (2005) "Women and Science Careers: Leaky Pipeline or Gender Filter?" *Gender and Education* 17(4), pp. 369–386.

Crasnow, S. (2007) "Feminist Anthropology and Sociology: Issues for Social Science," in Turner, S. and Risjord, M. (eds.) *Philosophy of Anthropology and Sociology*, North-Holland: Elsevier, pp. 755–789.

Currat, L., de Francisco, A., Al-Tuwaijri, S., Ghaffar, A., and Jupp, S. (2004) *10/90 Report on Health Research 2003–2004*, Geneva: Global Forum for Health Research.

Daniels, N. (2008) *Just Health: Meeting Health Needs Fairly*, New York: Cambridge University Press.

de Melo-Martín, I. and Intemann, K. (2011) "Feminist Resources for Biomedical Research: Lessons from the HPV Vaccines," *Hypatia* 26, pp. 79–101.

———. (2012) "Interpreting Evidence: Why Values Can Matter as Much as Science," *Perspectives in Biology and Medicine* 55(1), pp. 59–70.

———. (2016) "Gender in Medicine," in Solomon, M., Simon, J., and Kincaid, H (eds.) *The Routledge Companion to Philosophy of Medicine*, New York: Routledge, pp. 422–432.

De Welde, K. and Laursen, S. (2011) "The Glass Obstacle Course: Informal and Formal Barriers for Women Ph. D. Students in STEM Fields," *International Journal of Gender, Science and Technology* 3(3), pp. 571–595.

Douglas, H. (2009) *Science, Policy, and the Value-Free Ideal*, Pittsburgh: University of Pittsburgh Press.

Doyal, L. (2001) "Sex, Gender, and Heath: The Need for a New Approach," *British Medical Journal* 323(7320), pp. 1061–1063.

Duhem, P. ([1906] 1991) *The Aim and Structure of Physical Theory*, P. Wiener (trans.), Princeton, NJ: Princeton University Press.

Elliott, K. (2017) *A Tapestry of Values: An Introduction to Values in Science*, New York: Oxford University Press.

———. (2008) "Scientific Judgment and the Limit of Conflicts of Interest Policies," *Accountability in Research* 15(1), pp. 1–29.

———. (2011) *Is a Little Pollution Good for You?* New York: Oxford University Press.

Fausto-Sterling, A. (1985) *Myths of Gender: Biological Theories about Women and Men*, New York: Basic Books.

Gambacciani, M., Ciaponi, M., Cappagli, B., Benussi, C., and Genazzani, A.R. (2000) "Longitudinal Evaluation of Perimenopausal Femoral Bone Loss: Effects of a Low-Dose Oral Contraceptive Preparation on Bone Mineral Density and Metabolism," *Osteoporosis International* 11(6), pp. 544–548.

Goldenberg, M. (2015) "How Can Feminist Theories of Evidence Assist Clinical Reasoning and Decision-Making?" *Social Epistemology* 29(1), pp. 3–30.

Gould, S.J. (1996) *The Mismeasure of Man*, New York: Norton.

Grodstein, F., Stampfer, M.J., Manson, J.E., Colditz, G.A., Willett, W.C., Rosner, B., Speizer, F.E., and Hennekens, C.H. (1996) "Postmenopausal Estrogen and Progestin Use and the Risk of Cardiovascular Disease," *New England Journal of Medicine* 335(7), pp. 453–461.

Hanson, N.R. (1958) *Patterns of Discovery*, Cambridge, Cambridge University Press.

Harding, S. (2008) *Sciences from Below: Feminisms, Postcolonialities, and Modernities*, Durham, NC: Duke University Press.

———. (2016) "Latin American Decolonial Social Studies of Scientific Knowledge: Alliances and Tensions," *Science, Technology, & Human Values* 41(6), pp. 1063–1108.

Hrdy, S.B. (1986) "Empathy, Polyandry, and the Myth of the Coy Female," in Bleier, R. (ed.) *Feminist Approaches to Science*, New York: Pergamon Press, pp. 119–146.

Hubbard, R. (1983) "Have Only Men Evolved?" in Harding, S. and Hintika, M.B. (eds.) *Discovering Reality*, Dordecht: D. Reidel Publishing Company, pp. 45–59.

Hughes, R.N. (2007) "Sex Does Matter: Comments on the Prevalence of Male-Only Investigations of Drug Effects on Rodent Behaviour," *Behavioural Pharmacology* 18(7), pp. 583–589.

Hutton, E.K., Reitsma, A.H., and Kaufman, K. (2009) "Outcomes Associated with Planned Home and Planned Hospital Births in Low-Risk Women Attended by Midwives in Ontario, Canada, 2003–2006: A Retrospective Cohort Study," *Birth: Issues in Perinatal Care* 36(3), pp. 180–189.

Intemann, K. (2010) "25 Years of Feminist Empiricism and Standpoint Theory: Where Are We Now?" *Hypatia* 25(4), pp. 778–796.

———. (2015) "Distinguishing Between Legitimate and Illegitimate Values in Climate Modeling," *European Journal of Philosophy of Science* 5(2), pp. 217–232.

Intemann, K. and de-Melo-Martín, I. (2010) "Social Values and Scientific Evidence: The Case of the HPV Vaccines," *Biology & Philosophy* 25(2), pp. 203–213.

———. (2011) "Feminist Resources for Biomedical Research: Lessons from the HPV Vaccines," *Hypatia* 26(1), pp. 79–101.

———. (2014) "Addressing Problems in Profit-Driven Research: How can Feminist Conceptions of Objectivity Help?" *European Journal for Philosophy of Science* 4, pp. 135–151.

Janssen, P.A., Lee, S.K., Ryan, E.M., Etches, D.J., Farquharson, D.F., Peacock, D., and Klein, M.C. (2002) "Outcomes of Planned Home Births Versus Planned Hospital Births after Regulation of Midwifery in British Columbia," *CMAJ* 166(3), pp. 315–323.

Janssen, P.A., Saxell, L., Page, L.A., Klein, M.C., Liston, R.M., and Lee, S.K. (2009) "Outcomes of Planned Home Birth with Registered Midwife Versus Planned Hospital Birth with Midwife or Physician," *CMAJ* 181(6–7), pp. 377–383.

Johnson, K.C. and Daviss, B.A. (2005) "Outcomes of Planned Home Births with Certified Professional Midwives: Large Prospective Study in North America," *BMJ* 330(7505), pp. 1416–1419.

Kirsh, V. and Kreiger, N. (2002) "Estrogen and Estrogen–Progestin Replacement Therapy and Risk of Postmenopausal Breast Cancer in Canada," *Cancer Causes & Control* 13(6), pp. 583–590.

Kitcher, P. (2001) *Science, Truth, and Democracy*, Cambridge, MA: Cambridge University Press.

———. (2011) *Science in a Democratic Society*, Amherst: Prometheus Books.

Kokras, N., and Dalla, C. (2014) "Sex Differences in Animal Models of Psychiatric Disorders," *British Journal of Pharmacology* 171(20), pp. 4595–4619.

Kourany, J.A. (2010) *Philosophy of Science after Feminism*, New York: Oxford University Press.

Kuhn, T.S. (1962) *The Structure of Scientific Revolutions*, Chicago, IL: University of Chicago Press.

Lindsay, R., Hart, C.M., and Clark, D.M. (1984) "The Minimum Effective Dose of Estrogen for Prevention of Postmenopausal Bone Loss," *Obstetrics and Gynecology* 63, pp. 759–763.

Longino, H.E. (1990) *Science as Social Knowledge: Values and Objectivity in Scientific Inquiry*, Princeton, NJ: Princeton University Press.

———. (2002) *The Fate of Knowledge*, Princeton, NJ: Princeton University Press.

———. (2013) *Studying Human Behavior: How Scientists Investigate Aggression and Sexuality*, Chicago, IL: Chicago University Press.

Lugones, M. (2010) "Toward a Decolonial Feminism," *Hypatia* 25(4), pp. 742–759.

Lyerly A.D., Mitchel, L.M., Armstrong, E.M., Harris, L.H., Kukla, R., Kupperman, M., and Little, M.O. (2009) "Risk and the Pregnant Body," *Hastings Center Report* 39(6), pp. 34–42.

Martin, E. (1991) "The Egg and the Sperm: How Science Has Constructed a Romance Based on Stereotypical Male-Female Roles," *Signs* 16(3), pp. 485–501.

Mason, M.A., Wolfinger, N., and Goulden, M. (2013) *Do Babies Matter? Gender and Family in the Ivory Tower*, New Brunswick, NJ: Rutgers University Press.

Michaels, D. (2008) *Doubt Is Their Product*, Oxford: Oxford University Press.

Mignolo, W.D. (2009) "Epistemic Disobedience, Independent Thought and Decolonial Freedom," *Theory, Culture & Society*, 26(7–8), pp. 159–181.

Nelson, L.H. (1990) *Who Knows? From Quine to a Feminist Empiricism*, Philadelphia, PA: Temple University Press.

Okruhlik, K. (1994) "Gender and the Biological Sciences," *Canadian Journal of Philosophy* 24(suppl. 1), pp. 21–42.

Oreskes, N. and Conway, E. (2010) *Merchants of Doubt*, London: Bloomsbury.

Parry, M.L., Rosenzweig, C., Iglesias, A., Livemore, M., and Fischer, G. (2004) "Effects of Climate Change on Global Food Production under SRES Emissions and Socio-economic Scenarios," *Global Environmental Change* 14(1), pp. 53–67.

Pinto, M.F. (2014) "Philosophy of Science for Globalized Privatization: Uncovering Some Limitations of Contextual Empiricism," *Studies in the History and Philosophy of Science Part A* 47, pp. 10–17.

Prendergast, B.J., Onishi, K.G., and Zucker, I. (2014) "Female Mice Liberated for Inclusion in Neuroscience and Biomedical Research," *Neuroscience and Biobehavioral Reviews* 40, pp. 1–5.

Proctor, R. (1991) *Value-free Science? Purity and Power in Modern Knowledge*, Boston, MA: Harvard University Press.

Rice, J. (2012) "Animal Models: Not Close Enough," *Nature* 484(7393), p. S9.

Richardson, S.S. (2013) *Sex Itself: The Search for Male and Female in the Human Genome*, Chicago, IL: University of Chicago Press.

Ringheim, K. (1999) "Reversing the Downard Trend in Men's Share of Contraceptive Use," *Reproductive Health Matters* 7(14), pp. 83–96.

Rolin, K. (2017) "Can Social Diversity Be Best Incorporated into Science by Adopting the Social Value Management Ideal?" in Elliott, K. and Steele, D. (eds.) *Current Controversies in Values and Science*, New York, NY: Routledge, pp. 113–129.

Rosenzweig, C. and Parry, M.L. (1994) "Potential Impact of Climate Change on World Food Supply," *Nature* 367, pp. 133–138.

Schiebinger, L. (2004) *Nature's Body: Gender in the Making of Modern Science*, New Brunswick, NJ: Rutgers University Press.

Schiermeier, Q. (2010) "The Real Holes in Climate Science," *Nature* 463(7279), pp. 284–287.

Schneider, S.H. and Kunz-Duriseti, K. (2002) "Uncertainties and Climate Change Policy," in Schneider, S.G., Rosencranz, A., and Niles, J.O. (eds.) *Climate Change Policy: A Survey*, Washington, DC: Island Press, pp. 53–87.

Solomon, M. (2006) "Norms of Epistemic Diversity," *Episteme: A Journal of Social Epistemology* 3(1), pp. 23–36.

Stampfer, M.J. and Colditz, G.A. (1991) "Estrogen Replacement Therapy and Coronary Heart Disease: A Quantitative Assessment of the Epidemiologic Evidence," *Preventive Medicine* 20, pp. 47–63.

Stevens, P. (ed.) (2008) "Diseases of Poverty and the 10/90 Gap," in *Fighting the Diseases of Poverty*, New York: Routledge, pp. 126–140.

Wilson, R.A. (1966) *Feminine Forever*, New York: M. Evans and Company.

Women's Health Initiative (WHI). (2002) "Risks and Benefits of Estrogen Plus Progestin in Healthy Post-menopausal Women: Principal Results from the Women's Health Initiative Randomized Controlled Trial," *JAMA* 288, pp. 321–333.

Wylie, A. (2012) "Feminist Philosophy of Science: Standpoint Matters," *Proceedings and Addresses of the American Philosophical Association* 86(2), pp. 47–76.

Wylie, A. and Nelson, L.H. (2007) "Coming to Terms with the Values of Science: Insights from Feminist Science Studies Scholarship," in Kincaid, H., Dupre, J., and Wylie, A. (eds.), *Value-Free Science? Ideals and Illusions*, Oxford: Oxford University Press, pp. 58–86.

Yoon, D.Y., Mansukhani, N.A., Stubbs, V.C., Helenowski, I.B., Woodruff, T.K., and Kibbe, M.R. (2014) "Sex Bias Exists in Basic Science and Translational Surgical Research," *Surgery* 156(3), pp. 508–516.

Zucker, I. and Beery, A.K. (2010) "Males Still Dominate Animal Studies," *Nature* 465(7299), p. 690.

Additional Reading

Anderson, E. (2015). Feminist epistemology and philosophy of science. *Stanford Encyclopedia of Philosophy.* http://plato.stanford.edu/entries/feminism–epistemology/.

Crasnow, S., and Superson A. (eds.) (2012). *Out from the Shadows: Analytic Contributions to Traditional Philosophy*, Oxford: Oxford University Press.

Keller, E.F., and Longino, H. (eds.) (1996). *Feminism and Science*, Oxford: Oxford University Press.

16

SITUATED KNOWLEDGE AND OBJECTIVITY

Kristina Rolin

Introduction

Situated knowledge, broadly understood, is the view that the social location of the inquirer is of epistemic importance. However, not just any kind of social location is of interest to feminist philosophers of science. Feminist philosophers focus on those social locations that track systemic relations of power in the society. Systemic relations of power involve the ability of members of one social group, the dominant group, to constrain the choices available to members of another social group, the subordinate group. Systemic relations of power function as vehicles of domination when they constrain choices in ways that are harmful for members of the subordinate group. Like other social epistemologists, feminist philosophers of science draw attention to the social location of the inquirer, but they differ from others by focusing on the question of how relations of power and domination interact with knowledge.

Given the interest in power and domination, feminist philosophers of science have examined the epistemic significance of the inquirer's gender, ethnic identity, race, class, sexual identity, and (dis)ability. These attributes are morally and politically significant because in many national and cultural contexts they mark social locations divided by socio-economic inequalities or other ways in which social locations can be privileged or not privileged. As Alison Wylie explains, in feminist philosophy of science the situatedness of epistemic agents is construed in structural terms rather than as a matter of individual perspective or idiosyncratic skills and talents (2011: 162). Wylie emphasizes that the epistemically interesting features of social locations are not to be understood as essential properties of particular social groups. It is a matter of empirical inquiry to find out how the social location of the inquirer shapes her social experience in a particular context and how her social experience is relevant to specific research projects (Wylie 2003: 32).

In 1980s and 1990s, the idea of situated knowledge was advanced as a criticism of the myth-like understanding of objectivity as "the god-trick of seeing everything from nowhere" (Haraway 1991: 189). Feminist historian of science Donna Haraway insisted that all knowledge claims, including scientific knowledge claims, are situated, and their situatedness is a key to understanding who is accountable for them (1991: 191). She argued that knowledge claims provide merely a partial perspective on the object of inquiry, and therefore, the ideal of objectivity as impartiality is no longer plausible. In addition to questioning the "god-trick" version of objectivity, Haraway also questioned a version of relativism. According to Haraway, "Relativism is a way of being nowhere while claiming to be everywhere equally" (1991: 191). In her view, both objectivity as "god-trick" and relativism as its "mirror twin" obscure the epistemic importance of social location, the former by denying that knowledge claims are made in particular social locations, and the latter by denying that social locations can either impede or promote knowledge-seeking. While relativists grant equal epistemic standing to all social locations, Haraway, like many feminist philosophers, suggests that some social locations are better than others from an epistemic point of view.

In this chapter I discuss two questions that are motivated by the idea of situated knowledge. First, if the epistemic importance of social locations is contingent as many feminist philosophers suggest, how do social locations come to have epistemically interesting consequences, if not always, at least under some circumstances? Second, if relativism as epistemic equality of all social locations is problematic and if objectivity as freedom from social locations and partial perspectives is not a plausible alternative, how should the ideal of objectivity be redefined?

Local Knowledge and Social Experience

Social locations are thought to matter epistemically because they give rise to local knowledge and social experiences that are specific to the social location in question. Local knowledge is knowledge about a particular cultural, economic, or social practice and its circumstances, and it is best acquired by participating in the practice. Social experiences are social in two senses: they arise in particular social locations and they are shared with other people inhabiting similar social locations. The more complex societies are in terms of the division of labor and the more unequal citizens are in terms of their economic resources, education, and health, the more radically different the social experiences of citizens are likely to be. Moreover, the more pluralistic societies are in terms of political values and religious affiliations and the more multicultural they are in terms of ethnic identities and languages, the more likely it is that the social experiences of citizens will diverge. As Wylie explains, "social location systematically shapes and limits what we know, including tacit, experiential knowledge as well as explicit understanding, what we take knowledge to be as well as specific epistemic content" (2003: 31).

In scientific research, the social experience of the inquirer can be a source of criticism and creativity. Patricia Hill Collins argues that scientists and scholars who are "outsiders within" occupy an epistemically fruitful social location due to their first-hand experience of marginal or subordinate social locations in the society and access to the insider's perspective on academic knowledge production (2004: 103). Their unique social location enables them to assume a critical posture toward research that either ignores or distorts the reality of marginal or subordinate social locations. In virtue of the "creative tension of outsider within status" they are well positioned to identify anomalies in dominant scientific paradigms (2004: 122).

The idea of situated knowledge can highlight not only the social experience of the inquirer but also the social experiences of research participants and collaborators, that is, people who are studied and who can participate in the study in different ways, from agreeing to be observed and interviewed to contributing to the design of the study. With appropriate methods of participation, observation, and interview, scientists can do justice to local knowledge and the social experiences of research participants and collaborators. Sometimes scientists need to be engaged in social and political activism in order to earn the trust of research participants and collaborators. Relations of trust are of epistemic importance especially when the position of the scientist, on the one hand, and the positions of research participants and collaborators, on the other hand, are so unequal that they threaten to undermine the research process (Crasnow 2008: 1103).

Besides being a source of criticism and evidence, social experience can inspire new theoretical perspectives, that is, ways of conceptualizing the social phenomenon under study and the social mechanisms that are thought to generate and maintain the social phenomenon. By "social mechanism" I mean a constellation of human agents and activities that are organized so that they tend to bring about certain collective outcomes. For example, when a social scientist investigates gender-based discrimination in academia, she has to conceptualize discrimination so that she can distinguish it from non-discriminatory processes. Discrimination can be conceptualized narrowly by focusing merely on overt forms of discrimination such as not hiring women even when they have the best qualifications. A broader and more refined conceptualization of discrimination might include subtle forms of discrimination such as micro-inequities that have an impact on women scientists' motivation, confidence in their capabilities, opportunities for collaboration, and visibility (Rolin 2006; Wylie et al. 2007). A broader conceptualization of discrimination might include a more complex account of the social mechanisms that maintain discrimination. The social mechanisms can involve not merely a bias against women in hiring decisions but also social forces that undermine women's academic productivity. A social scientist who has herself experienced micro-inequities is likely to prefer the broader and more refined understanding of discrimination and its causes.

In this section, I have argued that social locations can matter epistemically by giving rise to local knowledge and social experiences which are sources of criticism, evidence, and theoretical perspectives. Moreover, the social location of the inquirer involves relations with research participants and collaborators. The ability of the inquirer to create and maintain mutual relations of trust may be crucial to her epistemic project. I have argued also that theoretical perspectives can grow out of social experiences because they highlight those aspects of natural or social reality the inquirer considers as significant. This insight is developed further in Helen Longino's critical contextual empiricism (CCE) which is the topic of the next section.

From Situation to Context

In CCE, situated knowledge is analyzed as contextual knowledge. CCE employs three different notions of context, an evidential, a specialty, and a social-cultural context. The first notion of context, an evidential context, figures in the argument that epistemic justification is relative to background assumptions because such assumptions are needed to establish the relevance of empirical evidence to a hypothesis or a theory (Longino 1990: 43). As Longino explains, "a state of affairs will only be taken to be evidence that something else is the case in light of some background belief or assumption asserting a connection between the two" (1990: 44). The second notion of context, a specialty context, plays a role in Longino's analysis of objectivity, in which she argues that objectivity is a function of a specialty community's practice rather than an individual scientist's observations and reasoning (1990: 74). The third notion of context, a social-cultural context, is employed in her analysis of the role of values in science, in which she argues that values belonging to the social and cultural context of science can enter into evidential context via background assumptions (1990: 83). Longino combines the three notions of context when she argues that we should adopt a community-based account of objectivity because values belonging to the social and cultural context of science can influence the evidential context of inquiry via background assumptions.

In CCE, the social experience of the inquirer is of epistemic interest especially when it functions as a source of value-laden background assumptions. While not all background assumptions "encode social values," many of them do (Longino 1990: 216). Value-laden background assumptions are difficult to identify when they are shared by all or most community members. In homogenous scientific communities, "they acquire an invisibility that renders them unavailable for criticism" (1990: 80). Longino argues that scientific communities benefit from heterogeneous social experiences and values because scientists are more likely to question value-laden background assumptions when the values in question differ from their own (2002: 131).

In CCE, the situatedness of scientific knowledge means also that scientific knowledge claims can *legitimately* be value-laden. Like many feminist philosophers of science, Longino (1990) rejects the value-free ideal, the view that contextual values are not allowed to play any role in the practices where scientific theories and hypotheses are justified and evaluated epistemically. As she explains, "contextual values, interests and value-laden assumptions *can* constrain scientific practice in such a way as to affect the results of inquiry and do so without violating constitutive rules of science" (1990: 83). Moreover, she does not believe that it is possible to eliminate the influence of moral and social values on epistemic justification "without seriously truncating the explanatory ambitions of the sciences" (1990: 223). In response to the worry that value-laden background assumptions lead to "unbridled relativism" (1990: 216), Longino introduces the ideal of "social value management" (2002: 50). The ideal recommends that the role of contextual values in scientific inquiry is analyzed, criticized, and judged as either acceptable or unacceptable by a specialty community that satisfied certain conditions (see the section on objectivity).

By emphasizing the role of specialty communities, Longino introduces yet another meaning to the idea of situated knowledge. Knowledge claims are situated in the sense that they are addressed to particular epistemic communities. In virtue of being members of epistemic communities, scientists have epistemic responsibilities toward other community members. This means that they have an obligation to engage criticism when it is appropriate, and to defend their knowledge claims by appealing to the standards of evidence and argumentation accepted by other community members. Knowledge claims can be situated also in the sense that they are addressed to particular non-academic audiences, for example, social groups who can use the results of research in their effort to solve pressing social, environmental, or health problems.

In this section, I have argued that the social location of the inquirer involves an epistemic community where her knowledge claims are accepted, rejected, or modified. In some cases, the social location of the inquirer involves also a non-academic audience interested in the application of research results. In sum, the thesis of situated knowledge, in a broad sense, is the idea that "knowledge is local in a profound way – knowledge is knowledge for and by a particular set of socially situated knowers" (Crasnow 2014: 147). In the next section, I discuss yet another way of theorizing the social location of the inquirer and its epistemic importance.

What Is a Standpoint?

Feminist standpoint theory (FST) advances the idea that the social location of the inquirer can be developed into a standpoint that is an epistemic resource in scientific inquiry. Thus, having a standpoint is not the same thing as occupying a particular social location or viewing the world from a particular perspective. A standpoint differs from a social location and a perspective in three ways. First, developing a standpoint requires that one is critically aware of the social conditions under which scientific knowledge is produced. As Wylie explains, "Standpoint theory concerns, then, not just the epistemic effects of *social location* but the effects and the emancipatory potential of a critical *standpoint on* knowledge production" (2012: 63).

Second, the formation of a standpoint is a collective project that involves shared values and interests, and sometimes also activism. Kristen Intemann argues that a standpoint involves a political commitment to producing scientific research that challenges systems of oppression (2010: 786). Research combined with activism is one way to generate knowledge of the ways relations of power function in society. As Sandra Harding explains, "[W]e can come to understand hidden aspects of social relations between the genders and the institutions that support these relations only through struggles to change them" (1991: 127). That scientists, scholars, and activists can share values and interests makes it possible to understand how otherwise differently located individuals come together to form a standpoint. Sharon Crasnow (2013) argues that a standpoint is properly attributed to an epistemic community that is also a political community. A political community is built on shared interests (2013: 420). As Crasnow explains, "Building such a community requires acknowledging diversity and discovering those shared interests" (2013: 420). This means that shared interests are not taken for granted but understood as an outcome of negotiation and coalition building (Crasnow 2014: 159).

The third feature of a standpoint follows from the second one. If the formation of a standpoint is a collective undertaking, then a standpoint is an achievement. As Harding explains, "A standpoint differs in this respect from a perspective, which anyone can have simply by 'opening one's eyes'" (1991: 127). This insight is called the "achievement thesis" (Crasnow 2013: 417). The term "achievement" describes both the process of building epistemic and political communities and the knowledge-generating struggles of such communities.

The concept of standpoint figures in the main thesis of FST, the thesis of epistemic advantage. This is the view that those who are unprivileged with respect to their social location are potentially privileged with respect to particular epistemic projects (Wylie 2003: 34). While subordinate and marginal social locations do not automatically give epistemic benefits to people who inhabit these locations, they can be a source of a standpoint that improves scientific research. According to Harding, the standpoint of feminist research is "less partial and less distorted than the picture of nature and social relations that emerges from conventional research" (1991: 121).

Not all feminist philosophers of science accept the thesis of epistemic advantage. The critics of FST argue that the thesis of epistemic advantage is undermined by the so called "bias paradox" (Antony 1993: 188–189; Longino 1999: 338). The bias paradox is the apparent contradiction between the thesis of epistemic advantage and the thesis of situated knowledge. Whereas the former states that some knowledge claims are less partial and less distorted than others, the latter states that all knowledge claims are situated and partial, thereby questioning the possibility of some claims being "less partial." The situated knowledge thesis threatens the epistemic advantage thesis because it suggests that there are no impartial standards that allow one to judge some situated knowledge claims as better than others. If FST wants to hold on to the thesis of epistemic advantage, the objection goes, it will have to explain what standards allow feminists to assess research conducted from a feminist standpoint as well as conventional research.

In response to the criticism, some feminist philosophers argue that the bias paradox can be dissolved by interpreting the thesis of epistemic advantage as an empirical hypothesis (Wylie 2003; Rolin 2006; Intemann 2010). While the thesis of epistemic advantage seems to rely on the assumption that there is a "view from nowhere," that is, an impartial standard that enables one to judge some situated knowledge claims as better than others, there is no need for such an assumption. In a particular context of inquiry, one can assess the relative merits of research conducted from a feminist standpoint in comparison to conventional research by applying standard epistemic values such as empirical adequacy and consistency. The key move is to specify an epistemic advantage by pinpointing a conceptual innovation, a novel body of evidence or another type of empirical success brought about by a feminist standpoint (Rolin 2006: 127).

One epistemic advantage is that a feminist standpoint can remedy epistemic injustices, that is, unfair treatment of persons in their capacity as an inquirer (Wylie 2011). The notion of epistemic injustice offers yet another way to understand how the social location of the inquirer can be epistemically consequential. One form of epistemic injustice is testimonial injustice which occurs when "prejudice causes a hearer to give a deflated level of credibility to a speaker's word" (Fricker 2007: 1). Another form of epistemic injustice is hermeneutical injustice which occurs when "a gap in collective interpretative resources puts someone at an unfair disadvantage when it comes to making sense of their social experience" (Fricker 2007: 1). In both cases a person is put into an epistemically unprivileged position because of social identity prejudice against her. However, according to FST, an epistemically unprivileged position is potentially a privileged one. A feminist standpoint can generate new knowledge by restoring credibility to victims of testimonial injustice and by correcting hermeneutical injustice with novel concepts such as "micro-inequities."

A feminist standpoint has the capacity to challenge relations of power and domination because it is akin to a social movement. A feminist scientific/intellectual movement provides feminist scientists, scholars, and activists with an opportunity to receive fruitful criticism for research which may be ignored in the larger scientific community (Rolin 2016: 17). Moreover, it enables them to generate evidence under social circumstances where relations of power tend to undermine their attempts to do so (2016: 16). This is because a social movement has the capacity to empower individuals, that is, to encourage them to act and speak in spite of or in response to the power wielded on them. Whereas an isolated individual is easily trapped in stereotypical images offered by prevailing relations of power, a feminist standpoint can empower her by transforming her self-definition and self-valuation (Collins 2004: 106). Scientists, scholars, and activists are also empowered by acquiring a sense of moral and political justification for speaking and acting in novel ways (Rolin 2016: 17). For example, the slogan "the personal is political" has provided many feminist scholars with a justification to use their own professional experiences, observations, and reflections as evidence in research (see e.g. Katila and Meriläinen 1999).

In this section, I have argued that knowledge can be situated in the sense that it is achieved from a standpoint. Whether an individual inquirer works in isolation or whether she participates in a feminist scientific/intellectual movement can make a huge difference to the epistemic outcome of her inquiry. As a participant in a feminist scientific/intellectual movement she can benefit from collective critical awareness of the social condition of knowledge production and feed-back from colleagues with whom she shares feminist values and interest. Moreover, by combining the generation of evidence with empowerment, she can generate novel evidence and theoretical perspectives under oppressive social conditions that could otherwise frustrate her efforts.

Intemann (2010) argues that contemporary versions of FST have so many features in common with feminist empiricism (e.g. CCE) that they are properly called feminist standpoint empiricism (FSE). Yet, FSE differs from CCE in its emphasis on the social experience of unprivileged people and the importance of building epistemic communities based on political commitments and shared interests. In the next section, I argue that these differences are also reflected in their respective conceptions of objectivity.

Feminist Approaches to Objectivity

Feminist philosophers of science are critical of the understanding of objectivity as freedom from social locations, partial perspectives, and values. Harding (1991) argues that the conventional understanding of objectivity as value-free science is too weak to identify sexist, racist, and homophobic assumptions in scientific research. Against the "weak" notion of objectivity, she advances the ideal of "strong" objectivity that requires systematic examination of background assumptions and methods through which knowledge is produced (1991: 149). According to Harding, strong objectivity is achieved from a feminist standpoint. As she explains, "starting thought from women's lives" increases the objectivity of research results because it challenges background assumptions that appear natural from the perspective of the lives of men in the dominant groups (1991: 150). More recently, Harding argues that "researchers should start research from outside the dominant conceptual framework – namely in the daily lives of oppressed groups such as women" (2015: 30). In her view, this increases objectivity because it enables scientists to detect the values, interests, and assumptions that serve the most powerful groups in the society, and might otherwise go unquestioned because the dominant groups are unlikely to challenge them (2015: 34).

For Harding, strong objectivity means also a commitment to cultural, sociological, and historical relativism when it comes to understanding socially situated knowledge claims and a rejection of epistemological relativism when it comes to comparing socially situated knowledge claims (1991: 156). As Harding explains, strong objectivity recognizes "the value of putting the subject or agent of knowledge in the same critical, causal plane as the object of her or his inquiry" (1991: 161). By this she means that inquirers should reflect on the epistemic effects of their own social locations as well as the processes through which they acquire knowledge.

While the advocates of FSE recognize the epistemic benefit of having a standpoint, they propose a different, more familiar conception of objectivity. According to Wylie, situated knowledge claims are objective when they satisfy widely accepted epistemic virtues such as empirical adequacy, internal and external consistency, and explanatory power (2003: 33). Intemann (2016) argues that in FSE the political and social aims of inquiry are partly constitutive of (as opposed to distinct from) the cognitive or epistemic aims of inquiry. This means that the interpretation of empirical adequacy depends on these other aims that define what type of evidence is relevant and how much evidence is needed. Crasnow (2014) proposes that "interest-based" objectivity is an appropriate ideal to both FST and FSE. Standpoints are interest-based in the sense that they are achieved by epistemic communities where scientists, scholars, and activists have common interests; yet, not only research results but also interests are subject to empirical constraints and such constraints are all that objectivity demands (2014: 157).

According to CCE, scientific knowledge is objective to the degree that a relevant community conforms to the four norms of "public venues," "uptake of criticism," "public standards," and "tempered equality of intellectual authority" (Longino 2002: 129–131, see also 1990: 76–81). Each of the four norms contributes to "transformative criticism" (1990: 76). The public venues norm requires that criticism of scientific research be given the same or nearly the same weight as original research (2002: 129). The uptake norm requires that each party to a critical exchange is ready to revise their views instead of merely "tolerating dissent" (2002: 129–130). The public standards norm requires that criticism appeals to at least some of the standards of evidence and argumentation recognized by the community (2002: 130–131). Finally, the tempered equality norm contributes to transformative criticism in two ways, by disqualifying those communities where certain perspectives dominate because of the political, social, or economic power of their adherents (1990: 78), and by making room for a diversity of perspectives which is likely to generate criticism (2002: 131).

Longino argues that a community practice constrained by the four norms advances objectivity because it forces scientists to examine critically the background assumptions that facilitate evidential reasoning as well as the moral and social values that may have motivated the choice of certain background assumptions (1990: 73). Without such a community practice, many ungrounded or even false assumptions may pass without criticism. As Longino explains, "As long as background beliefs can be articulated and subjected to criticism from the scientific community, they can be defended, modified, or abandoned in response to such criticism" (1990: 73–74). She adds that "As long as this kind of response is possible, the incorporation of hypotheses into the canon of scientific knowledge can be independent of any individual's subjective preferences" (1990: 74).

In sum, feminist philosophers of science argue that situated knowledge claims can be objective. In FST and FSE, it is a feminist standpoint that increases the objectivity of situated knowledge claims. By drawing on the social experience of unprivileged social groups in the society and the collective critical awareness of feminist scientists, scholars, and activist, a feminist standpoint is a good position to examine critically value-laden background assumptions and conventional methods of inquiry. In CCE, a larger scientific community is needed to increase the objectivity of situated knowledge claims. The scientific community should be open to criticism and inclusive of diverse perspectives, including the ones emerging from feminist standpoints.

Conclusion

The social location of the inquirer is an epistemic resource when it gives rise to local knowledge, social experience, criticism, evidence, and novel theoretical perspectives. It also situates the inquirer in a particular relation to other people, including research participants, collaborators, and potential users of knowledge. Most importantly, the inquirer is situated in particular epistemic communities. While all scientists and scholars are situated in disciplinary and specialty communities, feminist scientists and scholars are situated also in feminist research communities.

That knowledge claims are situated does not mean that they cannot be objective. Quite the contrary, feminist philosophers of science argue that the objectivity of scientific knowledge claims depends on scientific communities in which inquirers are situated. CCE holds the view that scientific knowledge is properly attributed to scientific communities and the objectivity of knowledge claims depends on how well these communities function epistemically. Ideally, an epistemically well-functioning community provides a platform for the criticism of sexist, racist, and heterosexist assumptions in research, and community members respond to criticism by revising such assumptions. The advocates of FST and FSE suggest that transformative criticism will not take place automatically as an effect of increased diversity and inclusion in scientific communities. While they agree that objectivity depends crucially on the ability of communities to detect and eliminate problematic assumptions, they emphasize the need to mobilize critical forces collectively. In their view, this involves the formation of a standpoint, an epistemic community with critical awareness, political commitments, and shared interests.

Related chapters: 6, 7, 15, 17.

References

Antony, L. (1993) "Quine as Feminist: The Radical Import of Naturalized Epistemology," in L. Antony and C. Witt (Eds.), *A Mind of One's Own: Feminist Essays on Reason and Objectivity*, pp. 185–225. Boulder: Westview Press.

Collins, P.H. (2004) "Learning from the Outsider Within: The Sociological Significance of Black Feminist Thought," in S. Harding (Ed.), *The Feminist Standpoint Theory Reader: Intellectual and Political Controversies*, pp. 103–126. New York and London: Routledge.

Crasnow, S. (2008) "Feminist Philosophy of Science: 'Standpoint' and Knowledge," *Science and Education* 17 (19), pp. 1089–1110.

Crasnow, S. (2013) "Feminist Philosophy of Science: Values and Objectivity," *Philosophy Compass* 8 (4), pp. 413–423.

Crasnow, S. (2014) "Feminist Standpoint Theory," in N. Cartwright and E. Montuschi (Eds.), *Philosophy of Social Science: A New Introduction*, pp. 145–161. Oxford and New York: Oxford University Press.

Fricker, M. (2007) *Epistemic Injustice: Power & the Ethics of Knowing*. Oxford and New York: Oxford University Press.

Haraway, D. (1991) "Situated Knowledges: The Science Question in Feminism and the Privilege of Partial Perspective," in D. Haraway (Ed.), *Simians, Cyborgs, and Women: The Reinvention of Nature*, pp. 183–201. London: Free Association Books.

Harding, S. (1991) *Whose Science? Whose Knowledge? Thinking from Women's Lives*. Ithaca: Cornell University Press.

Harding, S. (2015) *Objectivity and Diversity: Another Logic of Scientific Research*. Chicago: The University of Chicago Press.

Intemann, K. (2010) "25 Years of Feminist Empiricism and Standpoint Theory: Where Are We Now?" *Hypatia* 25 (4), pp. 778–796.

Intemann, K. (2016) "Feminist Standpoint," in L. Disch and M. Hawkesworth (Eds.), *The Oxford Handbook of Feminist Theory*, pp. 261–282. New York: Oxford University Press.

Katila, S., and Meriläinen, S. (1999) "A Serious Researcher or Just Another Nice Girl? Doing Gender in a Male Dominated Scientific Community," *Gender, Work, Organization* 6 (3): pp. 163–173.

Longino, H. (1990) *Science as Social Knowledge*. Princeton: Princeton University Press.

Longino, H. (1999) "Feminist Epistemology," in J. Greco and E. Sosa (Eds.), *The Blackwell Guide to Epistemology*, pp. 327–353. Oxford: Blackwell.

Longino, H. (2002) *The Fate of Knowledge*. Princeton: Princeton University Press.

Rolin, K. (2006) "The Bias Paradox in Feminist Standpoint Epistemology," *Episteme* 3 (1–2), pp. 125–136.

Rolin, K. (2016) "Values, Standpoints, and Scientific/intellectual Movements," *Studies in History and Philosophy of Science Part A* 56, pp. 11–19.

Wylie, A. (2003) "Why Standpoint Matters," in R. Figueroa and S. Harding (Eds.), *Science and Other Cultures: Issues in Philosophies of Science and Technology*, pp. 26–48. New York: Routledge.

Wylie, A. (2011) "What Knowers Know Well: Women, Work, and the Academy," in H. Grasswick (Ed.), *Feminist Epistemology and Philosophy of Science: Power in Knowledge*, pp. 157–79. Dordrecht: Springer.

Wylie, A. (2012) "Feminist Philosophy of Science: Standpoint Matters," *Proceedings and Addresses of the APA* 86 (2), pp. 47–76.

Wylie, A., Jakobsen, J., and Fosado, G. (2007) *Women, Work, and the Academy. Strategies for Responding to Post-Civil Rights Era' Gender Discrimination*. New York: Barnard Center for Research on Women.

Further Reading

Koskinen, I. (2019) "Defending a Risk Account of Scientific Objectivity," *British Journal for the Philosophy of Science*, https://doi.org/10.1093/bjps/axy053

Scheman, N. (2001) "Epistemology Resuscitated: Objectivity as Trustworthiness," in N. Tuana and S. Morgen (Eds.), *Engendering Rationalities*, pp. 23–52. Albany: State University of New York Press.

Wylie, A. (2015) "A Plurality of Pluralisms: Collaborative Practice in Archaeology," in F. Padovani, A. Richardson, and J.Y. Tsou, (Eds.), *Objectivity in Science: New Perspectives from Science and Technology Studies*, pp. 189–210. Boston Studies in the Philosophy and History of Science 130. Dordrecht and London: Springer.

17

IGNORANCE, SCIENCE, AND FEMINISM

Manuela Fernández Pinto

Introduction

Feminist philosophers of science have raised and endorsed some of the most profound critiques to traditional epistemology. First, most feminist philosophies of science are naturalized epistemologies insofar as they aim to be empirically adequate according to scientific standards, thus rejecting the foundationalist project of traditional epistemology (Longino 2002: 9; Potter 2006: 6). Second, feminist philosophers of science emphasize that scientists are socially and politically situated, following Haraway's claim that the knower is always embedded in a social and political context that inevitably influences their understanding of the world:

> The science question in feminism is about objectivity as positioned rationality. Its images are not the products of escape and transcendence of limits, i.e., the view from above, but the joining of partial views and halting voices into a collective subject position. . .
>
> *(Haraway 1988: 93)*

Third, feminist philosophers of science favor, to a lesser or greater extent, a social epistemology of science. Given their commitment to uncovering and combating androcentric and sexist values in research, feminist philosophers of science share the idea that science is value-laden, and emphasize the social location of knowledge production. Contrary to the Cartesian ideal, feminist philosophers of science favor a view of scientific knowledge in which the scientific community and society at large play a fundamental role in the production of knowledge (Potter 2006: 13).

More recently, feminist philosophers of science have contributed to developing a new and equally profound line of critique to traditional epistemology, which uncovers the crucial role that ignorance plays in our cognitive practices. Motivated, at least in part, by contemporary attacks to scientific knowledge, and in particular, by the manipulation of scientific research to favor partisan interests, such as the infamous case of the tobacco industry, the study of ignorance production has become an innovative approach to understanding the politics of science today. Challenging the traditional understanding of ignorance as a natural vacuum that needs to be filled with knowledge, the new epistemologies of ignorance, also known as agnotology or ignorance studies, have revealed the social and cultural dimensions of ignorance, and their overwhelming influence in knowledge production.

We can distinguish at least three approaches to the study of ignorance among science studies scholars: (i) agnotology (Proctor and Schiebinger 2008), (ii) epistemologies of ignorance

(Sullivan and Tuana 2006), and (iii) ignorance studies (Gross and McGoey 2015). While the epistemology of ignorance is more philosophical in character, ignorance studies are more descriptive and in line with the sociology of knowledge. In between these two, agnotology has been proposed as a tool for empirical inquiry, but with implicit and unclear normative aspirations (Fernández Pinto 2015). Although there are subtle differences among these approaches, for the purposes of this chapter I will not take such differences into account. Here I am interested in the study of ignorance in feminist philosophy of science, which I take to encompass all three approaches.

The aim of this chapter is to examine some of the key contributions of feminist philosophers of science to the study of ignorance. In order to do so, the second section provides a brief introduction to agnotology and its critical stance to traditional epistemology, presenting some of the conceptual clarifications that feminist philosophers of science and other scholars have contributed to this terrain. The third section illustrates how the study of ignorance can serve as a tool for feminist epistemology through an examination of case studies by feminist philosophers of science Nancy Tuana, Carla Fehr, and Janet Kourany. The fourth section examines the importance of ignorance studies for the feminist project in philosophy of science. Finally, I suggest some new directions worth pursuing in this area.

A Theoretical Framework for the Study of Ignorance

Just as a naturalized, situated, and social epistemology contributes to a more complex and empirically adequate understanding of knowledge, so does a more nuanced understanding of ignorance. In the words of Nancy Tuana:

> If we are to fully understand the complex practices of knowledge production and the variety of features that account for why something is known, we must also understand the practices that account for not knowing, that is, for our lack of knowledge about a phenomenon or, in some cases, an account of the practices that resulted in a group unlearning what was once a realm of knowledge. In other words, those who would strive to understand how we know must also develop epistemologies of ignorance.
>
> *(Tuana 2004: 194)*

Mainstream epistemology understands ignorance negatively as a mere lack of knowledge, where ignorance is overcome as knowledge is gained (Code 2006: 221). On the contrary, the new epistemologies of ignorance conceive ignorance as a complex phenomenon that can be socially constructed and maintained. In other words, ignorance can be the result of social practices and not just the default state of human understanding.

Political philosopher and race theorist Charles Mills gives a crucial account of how social practices can result in culturally sustained ignorance. Given a society structured by inequality and relations of power and domination, the world view of the dominant class will prevail, shaping what is considered acceptable and true, thus maintaining a distorted view of reality. Mills explains this point as part of his work on *The Racial Contract*:

> Thus in effect, on matters related to race, the Racial Contract prescribes for its signatories an inverted epistemology, an epistemology of ignorance, a particular pattern of localized and global cognitive dysfunctions (which are psychologically and socially functional), producing the ironic outcome that whites will in general be unable to understand the world they themselves have made.
>
> *(Mills 1997: 18)*

Mills later developed the concept of *white ignorance* to encompass this particular phenomenon of culturally sustained ignorance that results from racial inequality and white supremacy (2006: 15). As he clearly states, ignorance can very well be the result of social and political configurations that prescribe the way to understand the world around us. Furthermore, ignorance can be easily misunderstood and systematically mistaken for knowledge.

Historians of science Londa Schiebinger and Robert Proctor have also proposed an innovative approach to the study of ignorance, insisting on the importance of ignorance as a key concept for understanding scientific practice today. With this idea in mind, Proctor and Schiebinger co-edited the book *Agnotology: The Making and Unmaking of Ignorance* (2008), where they used the term *agnotology* to refer to this new terrain. In his introductory remarks, Proctor challenges the traditional conception of ignorance as a natural absence or vacuum that needs to be filled with knowledge:

> We need to think about the conscious, unconscious, and structural productions of ignorance, its diverse causes and conformations, whether brought about by neglect, forgetfulness, myopia, extinction, secrecy, or suppression. The point is to question the *naturalness* of ignorance, its causes and its distribution.
>
> *(Proctor 2008: 3)*

Contrary to the traditional conception of ignorance as a natural vacuum, Proctor suggests that ignorance can very well be the product of social practices. Accordingly, he distinguishes between ignorance as a passive construct or *lost realm*, the type of ignorance that inevitably grows from the social conditions in which science is made, and ignorance as an active construct or *strategic ploy*, the type of ignorance that can be deliberately made and maintained (2008: 3–6). He also acknowledges a positive type of ignorance, *virtous ignorance*, which prevents us from dangerous knowledge, for instance, when certain information is kept from juries to prevent a biased judgment (2008: 2).

Schiebinger illustrates the agnotological approach examining the case of West Indian abortifacients, of which much was known in the New World, but about which practically nothing traveled to the Old. The case helps us understand that knowledge is politically driven, and that studying the sources of our ignorance can help us understand the limits of our knowledge. In Schiebinger's words, "Agnotology traces the cultural politics of ignorance. It takes the measure of our ignorance, and analyzes why some knowledges are suppressed, lost, ignored, or abandoned, while others are embraced and come to shape our lives" (2008: 152).

A key feminist figure in the development of epistemologies of ignorance is philosopher of science Nancy Tuana. Together with Mills, Proctor, and Schiebinger, Tuana has emphasized the social and political dimensions of ignorance. In addition, Tuana has framed the study of ignorance as a core issue for feminist philosophy of science and feminist science studies more generally, bringing to our attention the importance of epistemologies of ignorance for uncovering and understanding sexism and androcentrism in science, as well as for enriching the science and values debate (2004: 226).

Tuana (2006) has also suggested a taxonomy of ignorance that takes into account its political dimensions, distinguishing: (i) what we know we don't know, but we don't care to know, such as in the case of male contraceptives, which were not pursued as a line of research because they were not deemed important by those in power to develop them (2006: 4–5), (ii) what we don't know we don't know, such as knowledge concerning clitoral physiology, which was precluded given the emphasis on female reproduction (2006: 6–8), (iii) what they don't want us to know, such as the dangers of estrogen, which pharmaceutical companies kept from women to protect their financial investment (2006: 9), (iv) what they don't know and don't want to know or *willful ignorance*, such as racial oppression, in which those in a position of privilege actively ignore the oppression of others (2006: 10–12), and (v) what we cannot know or *loving ignorance*, which is basically the recognition of what we cannot know given our cognitive limitations (2006: 15). The goal is to go beyond the traditional view of ignorance as epistemic oversight, uncovering the different other ways in which ignorance is created and sustained (Sullivan and Tuana 2006).

The scholars mentioned so far are interested in studying how ignorance is produced and maintained. However, there is also a lingering question of whether ignorance is always bad or not, or in other words, whether the epistemology of ignorance is a normative inquiry or not. Here it is important to notice that even though some of the proponents of agnotology (e.g. Robert Proctor and Naomi Oreskes) seem to endorse normative commitments, e.g. when studying the active production of ignorance in cases such as tobacco smoking and climate change, this has been identified as problematic, in so far as they do not provide a normative account of the distinction between knowledge and ignorance (for a detailed analysis of this problem, see Fernández Pinto 2015). On the contrary, and consistently with a naturalized epistemology, feminist philosopher of science Lorraine Code has made some important methodological remarks on the descriptive and genealogical character of the study of ignorance.

> It is because [an epistemology of ignorance] is not and cannot be a normative inquiry—at least not a priori and possibly not a posteriori—that it is best conceived as a genealogical inquiry into the power relations and structures of power that sustain, condone, or condemn ignorance. But in its diagnostic dimensions, it is an inquiry with a stronger descriptive-empirical and social-historical component that epistemology in an authorized sense would countenance. From my point of view this component is the source of its strength.
>
> *(Code 2006: 228)*

Even though from the point of view of mainstream epistemology the project of an epistemology of ignorance might seem paradoxical (given that an epistemology is literally a theory of knowledge and not of ignorance), from a naturalized stance the project is very much promising. In fact, the genealogical spirit of the epistemologies of ignorance has the potential to uncover mechanisms for ignorance production that remain occult for mainstream epistemology. In this way, the study of ignorance can serve as a tool for feminist philosophy, insofar as it contributes to unveiling sexist and androcentric practices in science and society.

Although ignorance has not been the main concern for feminist philosophers of science, one can find nevertheless insights into epistemologies of ignorance within some feminist approaches. For instance, Linda Martín Alcoff suggests that standpoint theory and in particular the work of Sandra Harding serve as a justification or motivation for an epistemology of ignorance.

> In terms of developing a "geography of the epistemic terrain," what follows most significantly from Harding's approach is that epistemic advantages and disadvantages accrue to social and group identifies per se rather than identities only in relation to a given context of inquiry (. . .) Thus ignorance is contextual, but there are patterns of ignorance associated with social and group identities.
>
> *(Martín-Alcoff 2006: 47)*

Insofar as members of oppressed groups have an epistemic advantage, members of dominant groups have an epistemic disadvantage. They have fewer reasons to question the status quo and more reasons to ignore its limitations (Harding 1991: 126). In this sense, standpoint theory provides also an explanation for the systematic production of ignorance.

Feminist Analyses of Ignorance Production

In addition to their theoretical contributions, feminist philosophers of science have also shown the methodological advantages of studying ignorance through a fruitful analysis of cases in the history of science, and more particularly in the history of medicine. From cases related to women's health, to cases regarding women's bodies, to issues of women's cognitive abilities, feminist analyses show that studying ignorance can help us understand practices of ignorance production related to women's oppression. Here we can draw attention to Nancy Tuana's research on female orgasm (2004) and the women's health movement (2006), and Carla Fehr's (2008) and Janet Kourany's (2015) studies on cognitive sex differences.

To begin, the study of ignorance production has shed light on our understanding (or lack thereof) of women's bodies. In her piece "Coming to Understand: Orgasm and the Epistemology of Ignorance" (2004), Tuana examines the history of scientific and common knowledge of female orgasm, focusing on the politics of ignorance surrounding female sexuality and women's genitalia (2004: 198). Tuana is interested in explaining why we seem to know more about male genitalia than about female genitalia, and why we seem to know about female genitalia when related to reproduction (uterus, ovaries, fallopian tubes), but not when related to pleasure (clitoris, labia, urethral and perineal sponges).

Through a careful examination of textbook images and theoretical assumptions about female reproduction and sexual pleasure, Tuana identifies an important gap in nonfeminist anatomies: the lack of interest and attention in mapping and understanding the clitoris and other pleasure-related parts of female genitalia (2004: 212), which she traces back to Freud and the conception of clitoral "excitability" as deviant (2004: 215). Ignorance related to female sexual pleasure, which by the way contrasts with women's multi-organismic capacities, reveals for Tuana the social values and politics related to sexual behavior: "Human women's orgasms are not denied, but they are carefully cultivated to avoid rupturing certain societal scripts" (2004: 224). Accordingly, an epistemology of ignorance contributes to a better understanding of what we know and why we know it: "Whose pleasures were enhanced by ignorance, and whose were suppressed by knowledge are complex questions that must be asked repeatedly in any study of the science of sexuality" (2004: 225).

A second example of ignorance production with a gender dimension is related to women's health issues. Tuana (2006) also uses the epistemologies of ignorance to examine the women's health movement in the United States. Her goal is to contribute to feminist accounts of science and situated knowledge by focusing on practices of ignorance production, especially ignorance about women's bodies and health (2006: 4). Using examples uncovered by the women's health movement, Tuana distinguishes between different types of ignorance, as I already mentioned in the previous section. The lack of male contraceptives illustrates the case knowing what we don't know, but really don't want to know (2006: 5). The lack of knowledge about clitoral physiology, the case examined in her 2004 paper, illustrates not knowing what we don't know: the emphasis on female reproduction obstructed the knowledge of the clitoris, not allowing scientists to understand what they did not know (2006: 6–8). To illustrate the type of ignorance resulting from cases of others not wanting us to know, Tuana uses the pharmaceutical industry's concealment of information regarding the dangers of estrogen in female oral contraceptives (2006: 9). Finally, willful ignorance is illustrated by the case of sterilization of women of color and disabled women, as well as by the case of incest, both in which historical facts have been systematically erased from public memory (2006: 11–12).

Thus, Tuana identifies different ways in which feminist health activists uncovered practices of ignorance production:

These feminist health activists were committed to uncovering the ways women's bodies had been ignored, to examining knowledge that had been withheld from women and certain

groups of men, to reclaiming knowledges that had been denied or suppressed and to developing new knowledge freed from the confines of traditional frameworks.

(2006: 2)

The epistemologies of ignorance serve as a tool for feminist inquiry.

A third and final example of feminist analysis of ignorance production addresses the contentious topic of cognitive sex differences. Carla Fehr (2008) also uses the epistemology of ignorance
as a theoretical framework for feminist research on the theories of cognitive sex differences and
their role in justifying the absence of women in higher education. As Tuana, Fehr is convinced
that paying attention to ignorance will bring fruitful results for feminist work:

> . . . we can begin to understand how a critical account of not knowing is essential in exam-
> ining persistent notions about inferiority and persistent inequalities— inequalities that are
> especially important in academe not only in terms of justice, but also in terms of the produc-
> tion of scientific knowledge.

(2008: 102–103)

Fehr examines in detail the history of the variability hypothesis, according to which there is
greater cognitive variability among men than among women resulting in more brilliant men than
women, including the mounting evidence against it, and concludes that:

> The resiliency of the variability hypothesis is a case of the epistemology of ignorance in ac-
> tion. False 'knowledge' of the variability of men's and women's intellectual abilities seems
> true because it coheres with the way that people expected the world to be.

(2008: 108)

An infamous case of such resilience was the use of the variability hypothesis by Lawrence Sum-
mers during a conference at Harvard in 2005. In her analysis of the case, Fehr brings to our
attention that our lack of knowledge regarding core critiques to the variability hypothesis, and
research on cognitive sex differences more generally, is related to gender inequalities and who is
considered a qualified knower:

> There are gaps in our cultural knowledge about women's intellectual abilities that empirical
> evidence seems unable to fill. Our culture's false 'knowledge' and ignorance about women's
> intellectual abilities persists in the minds of even the most educated people and in the face of
> vast bodies of evidence. One possible reason for this is that the false 'knowledge' seems true
> and retains currency because it coheres with the ways that both men and women expect the
> world to be.

(Fehr 2008: 112)

Accordingly, understanding why crucial evidence against the variability hypothesis is not known
becomes key to understanding why gender inequalities in the academy persist. Again, the epistemology of ignorance serves as a tool for feminist philosophy of science.

Janet Kourany (2015) also examines the case of cognitive sex differences from an agnotological perspective. For Kourany, science is an inevitable source of ignorance: deciding what lines of
research to pursue immediately creates ignorance about those lines that we decide not to pursue.
Organizing scientific communities in certain ways encourages the production of certain knowledge and discourages the production of other knowledge. In these and many other ways, science
inevitably contributes to producing ignorance at the same time that it produces knowledge. What

is important to notice is that this pattern of knowledge and ignorance production has different effects on different social groups. While some groups might benefit from the knowledge and ignorance produced, others might suffer. Epistemically, this pattern contributes to a partial and distorted view of the world; socially, it contributes to inequality, favoring some groups over others. Kourany illustrates how the knowledge and ignorance pattern impacts particular social group by examining the case of women.

Kourany identifies a series of examples in the history of science, where science has produced "knowledge" to serve the interests of men, while systematically producing ignorance about women (2015: 156). In archaeology, pivotal developments for human evolution, such as tools, fire, and hunting, were attributed to men. In medicine, women were systematically excluded from clinical trials, neglecting the production of knowledge on diseases that seriously affect women such as heart disease, AIDS, and breast cancer.

As Fehr, Kourany examines with more detail the case of cognitive sex differences. Women have been traditionally considered intellectually inferior to men and biological research has served as a tool trying to prove the stereotype. According to Kourany, it does not matter how many times the claim is contested, as it has frequently happened in the past century, research on cognitive sex differences continues, feeding the stereotype of women's intellectual inferiority. Kourany concludes:

> To be sure, scientific accounts that celebrate the past achievements of men while they make invisible—that is, promote ignorance of—the past achievements of women support, and are in turn supported by, scientific accounts that portray women as intellectually (emotionally, physically, . . .) inferior to men, and all this undermines women's self-esteem and the esteem of men. What's more, these accounts, because they suggest that women are worth less than men, also support a science that privileges the medical (economic, sexual, . . .) needs of men while it underserves and even promotes ignorance of the needs of women, and this further harms women. In short, the knowledge and ignorance produced by science work together to form a multiply distorted and multiply harmful picture of women.
>
> *(Kourany 2015: 160)*

In this way, understanding the pattern of knowledge and ignorance production that affects women helps us understand, in turn, how gender inequality, for instance with respect to mental ability, is created and sustained.

Through the analysis of gender-related cases in the history of science, feminist philosophers of science have shown that studying ignorance is useful for identifying and understanding the practices that have kept us from acquiring relevant knowledge regarding women's bodies, women's health, and women's cognitive abilities, thus making a significant contribution both to feminist and ignorance studies.

Ignorance as a Feminist Tool

Although the study of ignorance did not arise particularly or exclusively in feminist epistemology or philosophy of science, it has served as a fruitful methodological tool for feminist philosophy of science. As the last section tried to show, agnotological accounts of scientific research that impacts women can illuminate the different ways in which gender subordination and inequality are produced and maintained. Understanding why fundamental knowledge concerning women is not available—e.g. because it has been lost, because no one has paid attention, or because someone has deliberately kept it from us—is a necessary step toward countering ignorance and inequality, as well as promoting research toward a more balanced picture of reality. Insofar as the epistemologies of ignorance assist the fight against gender inequality, they are very much a tool for feminist inquiry. And insofar as they contribute to uncovering the patterns of ignorance production in science that impact women, they are very much a tool for feminist philosophy of science.

In addition, feminist philosophers of science have also made important contributions to clarifying the theoretical landscape for studying ignorance. As shown in the first section, feminist philosophers of science, such as Nancy Tuana, Lorraine Code, and Linda Martín-Alcoff, among others, have helped conceptualize what the study of ignorance amounts to: how to reframe the knowledge/ignorance dichotomy of traditional epistemology, how to include the social dimensions of ignorance production in epistemology, and how to map the different types of ignorance, and their political and social implications. This is a task for which feminist philosophers of science are particularly well-trained, insofar as feminist philosophy of science has constantly worked on the boundaries of the normative/descriptive distinction, looking for an epistemological account of scientific practice that is also empirically accurate. In this sense, the study of ignorance complements and is coherent with the naturalized, social, and situated epistemology found at the core of feminist philosophy of science.

Finally, the conceptual work that feminist philosophers of science have contributed to the epistemology of ignorance has served as a theoretical framework and methodological guide for other science studies scholars, social scientists, and gender scholars in their ignorance-related projects. Such influence can be found, for instance, in research about the role of ignorance in the study of women and drug addiction (Ettorre 2015), the access to information of young mothers (Greyson et al. 2017), education (Malewski and Jaramillo 2011), and the influence of evolutionary psychology in the humanities (Hall 2012). In this way, the contributions of feminist philosophers of science to the study of ignorance are having already an impact beyond philosophy.

New Directions

Before I finish, let me point out two lines of inquiry in a feminist philosophy of science concerned with practices of ignorance production that I consider worth pursuing. Feminist philosophy of science has given little attention to non-Western feminists and women of color, and to the issues that concern them such as intersectionality or South/North global relations. In this sense, the study of ignorance is a promising tool for uncovering patterns of exclusion, epistemic injustice, and ignorance production that relate to women of color in science, pretty much in the same way in which they are already contributing to uncover these patterns with respect to white women.

Feminist philosopher Mariana Ortega, for example, identifies a type of ignorance, which she names "loving, knowing ignorance," and which often characterizes the relation of white women toward women of color. In her own words:

> There are also those who suffer from what I have called loving, knowing ignorance; those who seem to have understood the need for a better way of perceiving but whose wanting leads them to continue to perceive arrogantly, to distort their objects of perception, all while thinking that they are loving perceivers.
>
> *(Ortega 2006: 60)*

The problem, according to Ortega, lies in a lack of engagement with the lives and experiences of women of color, which cannot be attained by simply reading non-white feminists: "There is a need to build relationships among white feminists and women of color; a need for a more active stance on the part of white feminists to learn about the experience of women who are not like them" (2006: 67). Otherwise, Ortega fears that "women of color will continue to be ignored, homogenized, and misunderstood" (2006: 67).

As Ortega suggests, the marginalization of and lack of engagement with women of color lead to practices of ignorance production that affect feminism itself. As the epistemology of ignorance has already established, whose knowledge is recovered and whose ignorance is fought are a matter of race and gender politics, in which women are not a homogeneous and equally marginalized group. In this sense, if the study of ignorance serves as a tool against gender inequality, it must also serve as a tool against the marginalization of women of color. In particular, agnotology can help uncover the practices of ignorance production that impact women of color and that go beyond the racial and gender ignorance suffered by their male and female counterparts.

Insofar as feminist philosophy of science is concerned with sexism and androcentrism in science, and insofar agnotology contributes to further understanding how sexism and androcentrism have epistemic impacts, feminist philosophy of science would gain much from using this tool for understanding other forms of ignorance production leading to gender inequality in science, particularly those concerning women of color.

Another area of inquiry where the study of ignorance is worth pursuing is related to the privatization and commercialization that science has experienced in recent decades. Within the current organization of science, where most research is both funded and developed by private companies (NSB 2018), new mechanisms for the production of ignorance have developed. Agnotology has been a fruitful approach to uncovering and documenting such mechanisms and how they operate (Fernández Pinto 2017), in cases ranging from the tobacco industry's support of cancer research (Proctor 2012), to the industry's undermining of climate science (Oreskes and Conway 2010), to the chemical industry's manipulations of toxicology studies (Michaels 2008). In addition, studies of ignorance production in the history of Big Pharma (Nik-Khah 2014) and the financial crisis (Mirowski 2013) show further contributions in this respect.

However, fewer contributions have been made regarding the ways in which the current commercialization of science foments ignorance that impacts women. Kourany and Fernández Pinto (2018) examine the case of breast cancer research, uncovering the ways in which private interests affect the production of health guidelines for breast cancer screening, obstructing the formation and dissemination of knowledge that is most relevant to women. In a similar vein, I have analyzed how practices of niche marketing, i.e. drug marketing targeted to specific social groups, broadly encouraged by the pharmaceutical industry, have led to patterns of knowledge and ignorance that deepen gender and race inequalities (Fernández Pinto 2018). More work however is needed if we seek to understand how commercial interests impact the knowledge and ignorance produced, and how they can contribute or not to gender inequality. Given the commercial framework under which scientific research develops nowadays, this is a pressing issue for feminist philosophy of science.

Conclusion

The main goal of this chapter was to examine the contributions of feminist philosophers of science to the recent field of agnotology or ignorance studies. After presenting the theoretical framework for the study of ignorance, highlighting the input of feminist philosophers, I then exemplified how feminist philosophers Nancy Tuana, Carla Fehr, and Janet Kourany have used this approach as a useful tool to analyze gender-related cases in the history of science. Accordingly, I argued that the study of ignorance presents an important opportunity for the feminist project in philosophy of science. Finally, I concluded by suggesting some directions for future research.

Acknowledgments

The author would like to thank Sharon Crasnow and Kristen Intemann for the invitation to participate in this volume, and to Kristen Intemann and Allison Wolf for their useful comments on previous drafts. A preliminary draft of this paper was presented in the Philosophy Colloquium at Universidad de los Andes.

Related chapters: 12, 15, 16, 18.

References

Code, L. (2006) "The Power of Ignorance," in Sullivan, S. and Tuana, N. (eds.) *Race and Epistemologies of Ignorance*, Albany, N.Y.: SUNY Press, pp. 213–230.

Ettorre, E. (2015) "Embodied Deviance, Gender, and Epistemologies of Ignorance: Re-visioning Drugs Use in a Neurochemical, Unjust World," *Substance Use & Misuse* 50(6), pp. 794–805.

Fehr, C. (2008) "Are Smart Men Smarter than Smart Women? The Epistemology of Ignorance, Women, and the Production of Knowledge," in May, A. (ed.) *The 'Woman Question' and Higher Education: Perspectives on Gender and Knowledge Production in America*, Cheltenham, UK: Edward Elgar, pp. 102–116.

Fernández Pinto, M. (2015) "Tensions in Agnotology: Normativity in the Studies of Commercially-Driven Ignorance," *Social Studies of Science* 45(2), pp. 294–315.

———. (2017) "To Know or Better Not to: Agnotology and the Social Construction of Ignorance in Commercially Driven Research," *Science & Technology Studies* 30(2), pp. 53–72.

———. (2018) "Democratizing Strategies for Industry-Funded Medical Research: A Cautionary Tale," *Philosophy of Science* 85, pp. 882–894.

Greyson, D., O'Brien, H., and Shoveller, J. (2017) "Constructing Knowledge and Ignorance in the Social Information Worlds of Young Mothers," *Proceedings of the Association for Information Science and Technology* 54(1), pp. 139–149.

Gross, M. and McGoey, L. (2015) *Routledge International Handbook of Ignorance Studies*, London and New York: Routledge.

Hall, K. (2012) " 'Not Much to Praise in Such Seeking and Finding': Evolutionary Psychology, the Biological Turn in the Humanities, and the Epistemology of Ignorance," *Hypatia* 27(1), pp. 28–49.

Haraway, D. (1988) "Situated Knowledges: The Science Question in Feminism and the Privilege of Partial Perspective," Reprinted in S. Harding (ed.). 2004. *The Feminist Standpoint Theory Reader: Intellectual and Political Controversies*, New York and London: Routledge, pp. 81–102.

Harding, S. (1991) *Whose Science? Whose Knowledge?: Thinking from Women's Lives*, Ithaca, N.Y.: Cornell University Press.

Kourany, J. (2015) "Science: For Better or Worse, A Source of Ignorance as Well as Knowledge," in Gross, M. and McGoey, L. (eds.) *Routledge International Handbook of Ignorance Studies*, London and New York: Routledge, pp. 155–164.

Kourany, J. and Fernández Pinto, M. (2018) "A Role for Science in Public Policy? The Obstacles, Illustrated by the Case of Breast Cancer Screening Policy," *Science, Technology, & Human Values* 43(5), pp. 917–943.

Longino, H. (2002) *The Fate of Knowledge*, Princeton, N.J.: Princeton University Press.

Malewski, E. and Jaramillo, N. (2011) *Epistemologies of Ignorance in Education*, Charlotte, N.C.: Information Age Pub.

Martín-Alcoff, L. (2006) "Epistemologies of Ignorance: Three Types," in Sullivan, S. and Tuana, N. (eds.) *Race and Epistemologies of Ignorance*, Albany, N.Y.: SUNY Press, pp. 39–58.

Michaels, D. (2008) *Doubt Is Their Product: How Industry's Assault on Science Threatens Your Health*, Oxford: Oxford University Press.

Mills, C. (1997) *The Racial Contract*, Ithaca, N.Y.: Cornell University Press.

———. (2006) "White Ignorance," in Sullivan, S. and Tuana, N. (eds.) *Race and Epistemologies of Ignorance*, Albany, N.Y.: SUNY Press, pp. 11–38.

Mirowski, P. (2013) *Never Let a Serious Crisis Go to Waste: How Neoliberalism Survived the Financial Meltdown*, London and New York: Verso.

National Science Board. (2018) *Science & Engineer Indicators 2018*, NSB-2018-1. Alexandria, VA: National Science Foundation. Available at https://www.nsf.gov/statistics/indicators/

Nik-Khah, E. (2014) "Neoliberal Pharmaceutical Science and the Chicago School of Economics," *Social Studies of Science* 44, pp. 489–517.

Oreskes, N. and Conway, E. (2010) *Merchants of Doubt: How a Handful of Scientists Obscured the Truth on Issues from Tobacco Smoke to Global Warming*, New York: Bloomsbury Press.

Ortega, M. (2006) "Being Lovingly, Knowingly Ignorant," *Hypatia* 21(3), pp. 56–74.

Potter, E. (2006) *Feminism and Philosophy of Science: An Introduction*, New York: Routledge.

Proctor, R. (2008). "Agnotology: A Misssing Term to Describe the Cultural Production of Ignorance (And Its Study)," in Proctor, R. and Schiebinger, L. (eds.) *Agnotology: The Making and Unmaking of Ignorance*, Stanford, CA: Stanford University Press, pp. 1–36.

———. (2012) *Golden Holocaust: Origins of the Cigarette Catastrophe and the Case for Abolition*, Berkeley: University of California Press.

Proctor, R. and Schiebinger, L. (2008) *Agnotology: The Making and Unmaking of Ignorance*, Stanford, CA: Stanford University Press.

Schiebinger, L. (2008). "West Indian Abortifacients and the Making of Ignorance," in Proctor, R. and Schiebinger, L. (eds.) *Agnotology: The Making and Unmaking of Ignorance*, Stanford, CA: Stanford University Press, pp. 149–162.

Sullivan, S. and Tuana, N. (2006) *Race and Epistemologies of Ignorance*, Albany, N.Y.: SUNY Press.

Tuana, N. (2004) "Coming to Understand: Orgasm and the Epistemology of Ignorance," *Hypatia* 19(1), pp. 194–232.

———. (2006) "The Speculum of Ignorance: The Women's Health Movement and Epistemologies of Ignorance," *Hypatia* 21, pp. 1–19.

18

HOW THE FACTS MIGHT GIVE US SOCIALLY RESPONSIBLE SCIENCE

Janet Kourany

Science is based on facts, not wishful thinking or revelation or speculation, facts that are systematically gathered by a community of enquirers through detailed observation and experiment. These facts are used to ground the rest of science, the concepts and laws and theories and so on; and it is this grounding in facts that makes science the most trusted source of knowledge we have, distinguishing it from all other enterprises that claim to produce knowledge. Of course, the success of science involves other factors besides its grounding in facts: a multitude of highly dedicated, imaginative contributors; a heady dose of genius now and then; a willingness to break with the ideas of the past; generous financial and social support; the availability of mathematics and other technological tools; and other factors as well. But its grounding in facts is the most important—the absolutely crucial and indispensable—ingredient of science's success, the ultimate source of science's authority.

So thought Francis Bacon and the other founders of modern science at the dawn of their great undertaking, and over the centuries "the facts" (or "data" or "evidence" or reports of what is observed in the world) have continued to enjoy the starring role in science. Oddly enough, however, the character of the star has been left very much in the dark even though the addition of further details may prove significant. After all, scientists don't gather just any facts and build science on their basis. Which facts they gather depends on a variety of factors: the areas of the world scientists choose to explore, the specific questions they raise about these areas, the methods they devise to answer the questions, the time and resources (even the number of scientists) they have at their disposal, and the concepts and assumptions they rely on for the job. As a result, the facts scientists gather at any time represent only a tiny fraction of all the different facts they could have gathered at that time—that is, all the different facts they would have gathered if they had chosen different areas to explore or raised different questions or had different resources to work with, or the like. And, of course, what shapes the areas scientists choose to explore, the questions they raise, the resources they have at their disposal, and the rest are values—scientists' epistemic (so-called truth productive) values, but also, and especially, their or their funders' social (ethical or political or economic or cultural) values. So, the star of the traditional account of science is not just "the facts" but a deliberately chosen and comparatively tiny selection of the facts. And it is *these* facts that ground the rest of science.

Bacon and all those illustrious scientists in the Royal Society and elsewhere he influenced would have found such additional details about the foundation of science alarming. For one thing, such details suggest that the knowledge produced by science is, and will forever remain, incomplete, while Bacon and the others expected that their new enterprise, if properly supported, might be completely wrapped up within their own lifetimes, or very soon thereafter. Of course, our expectations are now quite different, for we are regularly exposed to scientific and techno-logical breakthroughs that open up new vistas of research rather than close down old ones. The continuing incompleteness of science is thus very much acknowledged by us and even, in a way, welcomed, since it also makes possible the continuing progress of science. Still, these additional details about science's foundation might prove alarming to us as well as the founders, and for other reasons than the incompleteness of science. At least, many scientists, and especially feminist scientists, have suggested as much. So, the matter certainly bears looking into. This, at any rate, is what I shall attempt here.

What Feminist Scientists (and Would-Be Scientists) Have Taught Us about the Facts that Ground Science

If you are asked to picture a scientist, chances are you will picture a male scientist, especially a white, European or American, middle- or upper-class, male scientist. And if you are asked to name some well-known scientists, chances are you will have no difficulty naming any number of distinguished male scientists but will soon become stumped after naming just a few female sci-entists (or perhaps just one, Madame Curie). Our conception of scientists, in short, is very much a male affair. Yet, women have always engaged in science along with the men. Indeed, feminist historians of science have even discovered highly accomplished women scientists in the ancient world—such as the physician Merit-Ptah of ancient Egypt, the mathematician and astronomer Hypatia of ancient Greece, the chemist Tapputi-Belatekallim of ancient Mesopotamia (the world's first recorded chemist!), and the alchemist Fang of ancient China (see, for information about these and other celebrated women scientists, Alic 1986; Kass-Simon, Farnes, and Nash 1990; Ogilvie, Harvey, and Rossiter 2014). At various times and places, however, and especially during and after the Scientific Revolution, women were closed out of science, or at least pushed very much to the sidelines (Schiebinger 1989). And all this began to change only in the middle of the twentieth century, during second-wave feminism. But by then science had become a decidedly androcentric mode of enquiry: the areas of research pursued, the questions raised, the methods followed, the assumptions and concepts relied on all very much reflected what men desired and needed and had accomplished—in a word, men's values. And the facts were selected accordingly.

Consider, for example, the women who entered the sciences in increasing numbers during second-wave feminism. What these women found reported in the outcomes of social and natural science investigations was a torrent of facts relating to men together with a dearth of facts relating to women. They found, for example, facts in archaeology about men's contributions to the great turning points of human evolution but no facts about women's contributions; facts in medicine about men's problems with heart disease and stroke and, later, AIDS as well as other diseases but few facts about women's problems with these diseases; facts in economics and political science and sociology about men's rationality and agency and leadership styles and abilities but no facts about such characteristics in women; and so on. Even the titles of the works these women scien-tists eventually produced—such as biologist Ruth Hubbard's "Have Only Men Evolved?" (1979), or archaeologists Joan Gero and Margaret Conkey's *Engendering Archaeology: Women and Prehistory* (1991), or economists Marianne Ferber and Julie Nelson's *Beyond Economic Man: Feminist Theory and Economics* (1993), or health researcher Sue Rosser's *Women's Health—Missing from U.S. Medicine* (1994)—even the titles of the works these women scientists eventually produced bespoke the low visibility, indeed near invisibility, of women and the high visibility of men in the accumulated facts of their disciplines. True, these women scientists did find some facts relating to women in their disciplines—for example, a heady dose of facts regarding women and reproduction in medicine—but all too frequently the facts they did find, especially in psychology, reported on women's "deficiencies," where, of course, men were taken as the standard for what was not deficient.

Twentieth-century women scientists were not the first women to worry about this androcentric mode of fact gathering in the sciences, however. Nineteenth-century women, not permitted, like their twentieth-century sisters, to enter the sciences, could still diagnose its shortcomings. And some, such as Caroline Kennard and Eliza Burt Gamble, did, and with gusto. As they saw it, women had been thought inferior to men—intellectually, socially, physically, and even morally inferior—ever since ancient times, and so women were never expected to play any significant roles in the great exploits and achievements of humankind. Hence, no serious attention to them was ever considered warranted. Still, as Eliza Burt Gamble pointed out in her *The Evolution of Woman, an Inquiry into the Dogma of Her Inferiority to Man* (1894: vii–viii),

> With the dawn of scientific investigation it might have been hoped that the prejudices resulting from lower conditions of human society would disappear, and that in their stead would be set forth not only facts, but deductions from facts, better suited to the dawn of an intellectual age. . . . The ability, however, to collect facts, and the power to generalize and draw conclusions from them, avail little, when brought into direct opposition to deeply rooted prejudices.

So modern science simply followed the ancient tradition. The upshot was that the facts unearthed by modern scientific investigations, rather than undermining and displacing the old prejudicial picture of women, reinforced it instead.

What Feminist Scientists Are Still Teaching Us about the Facts that Ground Science

Of course, as already pointed out, change did begin in the middle of the twentieth century, the result of second-wave feminism. But change has been slow and far from complete even now. While some fields, such as cultural anthropology and primatology, have experienced something like a feminist makeover or are well on the way to it, many others have hardly changed at all. Consider, for instance, biomedical research (see, for what follows, Dusenbery 2018; Mazure and Jones 2015; Meinert 1995; Ovseiko et al. 2016; Rosser 1994; Schiebinger 1999, Schiebinger et al. 2020; Sherman, Temple and Merkatz 1995; Weisman and Cassard 1994). Until 1993—when Congress passed the National Institutes of Health Revitalization Act that mandated the inclusion of women and minority men as subjects in NIH-funded, US medical research—women tended to be neglected in both basic and clinical research. The neglect, at times, bordered on the absurd. Take hormone replacement therapy, the regime of estrogen or estrogen and progesterone regularly prescribed for postmenopausal women starting in the 1970s as a preventive for heart disease. Until the 1990s, the only randomized controlled trial testing the safety and efficacy of estrogen as a preventive for heart disease was conducted on men—8,341 men—and no women. Or take the NIH-supported Rockefeller University pilot study that looked at how obesity affected breast and uterine cancer that didn't enroll a single woman. Other examples are legion, and they showed the urgency of a measure like the Revitalization Act.[1]

Nonetheless, the neglect of women in biomedical research continued even after the passage of the Revitalization Act. For one thing, the Revitalization Act did not apply to early phase medical studies. In consequence, most basic research with animal models continued to focus on male animals and exclude females, most studies at the tissue and cellular levels either failed to report the donor sexes of the materials studied or reported that they were male, and early phase clinical trials (i.e. Phase I and II trials) didn't always include women. Even in the case of the medical studies to which the Revitalization Act did apply (i.e. Phase III clinical trials, the final stage after the earlier, smaller trials indicated safety and promise), women remained under-enrolled relative to their representation in the patient population, and the published results frequently did not include a breakdown by gender. What's more, the Revitalization Act applied only to NIH-funded research, not the sizeable amount of biomedical research funded by either private industry or foundations, and most of that research also failed to live up to the standards set by the Revitalization Act.

Ultimately, the problem was that many in the biomedical research community simply did not support the goals of the Revitalization Act. There were, of course, the old standby reasons—that including women of child-bearing age in drug studies could jeopardize the health and safety of any fetuses the women might be carrying, that men and women were so alike, anyway, that results obtained from studying men could always be validly applied to women, and finally (and quite inconsistently with the latter) that women, with their menstrual cycles, oral contraceptives, hormone therapies, pregnancies, and so on, would introduce too many complications to make "clean" results possible if they were included in the studies. But there was also the bottom line: that including women in clinical trials, or including female animals, tissues, or cells in earlier stage studies, would require much larger studies, more expense, and more work if done properly—that is, so as to ensure that an adequate analysis of results by sex and gender would be possible. And, in any case, there was the inertial resistance to changing old, entrenched procedures as well as the recognition that even the relatively weak requirements of the Revitalization Act were not being rigorously monitored and enforced even for NIH-funded research.

Small wonder that recent assessments of the progress brought about by the Revitalization Act have been quite negative. According to Carolyn Mazure, Director of Women's Health Research at Yale, for example:

> Steps have been taken in the United States to remedy the underrepresentation of women and the inadequate attention to sex and gender differences in research and regulatory approvals. However, progress has been painfully slow—stalling for long periods or sometimes reversing direction—and, consequently, not nearly enough progress has been made.
>
> *(see Mazure and Jones 2015: 2)*

Other researchers agree. Janice Werbinski, Founding President of the American College of Women's Health Physicians and Executive Director of the Sex and Gender Women's Health Collaborative (whose goal is to inform students and clinicians in sex-and gender-appropriate medicine), puts it this way: the expectation in the nineties was that just getting women enrolled in studies would "take care of the problem" of the lack of medical facts about women.

> But it's been twenty-five years and we now have a lot of research that includes women but women are still invisible. Researchers said, 'Okay, we included women,' but they weren't required to report their research by sex, so women's side effects and responses to medications and diseases were still invisible.
>
> *(quoted in Dusenbery 2018: 36)*

And the reports in other countries are the same. Concludes an international team of medical, gender, and policy researchers from the United Kingdom, Spain, Canada, Argentina, Qatar, Oman, Australia, Denmark, Germany, Sweden, Belgium, the United States, Brunei Darussalam, Zambia, Malawi, Italy, and South Africa: "Growing global investment in biomedical research is unlikely to result in outstanding science that benefits women and men equitably if current levels of conscious and unconscious gender bias in health research persist" (Ovseiko et al. 2016: 8).

Biomedical research is not the only field in which the facts gathered continue to deal mostly with men. Other fields, such as economics and political science, do precisely the same. Consider, first, economics (see, for what follows, Agenjo-Calderón and Gálvez-Muñoz 2019; Ferber and Nelson 1993, 2003; Nelson 1996a, 1996b, 2010; Tejani 2019; Waring 1992, 1997). Here the problem is not only that researchers use methods that privilege men (as is the case in biomedical research) but also that—in the words of feminist economist Julie Nelson—the conceptual framework designed to pursue this research "has been socially constructed to conform to a particular image of masculinity" (1996b: 24). What does this mean?

Take the conceptual framework of current mainstream economics ("neoclassical" economics). Its central concept is that of the market, a place where agents with stable preferences interact for the purposes of exchange. These agents may be individual persons or collectives of various kinds such as corporations, labor unions, and governments. The agents, in either case, exchange goods or services, with money facilitating the transactions. And the tool of choice for analyzing these transactions is mathematics. Indeed, high status is assigned in economics to formal mathematical models of these transactions.

What makes this conceptual framework "conform to a particular image of masculinity" is that the agents, even the collective agents such as corporations and labor unions, are assumed to act just like men: to be unfailingly rational, autonomous, and self-interested. That is, they are assumed to act the way men are expected to act by our current norms of masculinity, in contrast to the emotional, social, other-directed way women are expected to act by our current norms of femininity. Moreover, these agents' activities—their exchanges of goods and services—are understood to be part of the public sphere, facilitated, as they are, by money and described in the language of mathematics. And this sharply contrasts with all those exchanges of goods and services in the private sphere that women are expected to engage in, the ones facilitated by emotional attachments rather than money and for which a kind of description more subjective than the language of mathematics is thought to be appropriate. Finally, competition and the maximization of utility (or profit) under conditions of "scarcity" (unlimited desires coupled with limited resources for satisfying those desires) are central features of these agents' activities rather than the cooperation and social provisioning traditionally associated with women. Conceptualizing agents in this way is thus "an effective way to exclude women as political-economic subjects" (Agenjo-Calderón and Gálvez-Muñoz 2019: 144).

As a result, reports Nelson, "the discipline of economics has historically neglected subjects that particularly affect women, such as unpaid household labor, intra-family economic relations, labor market discrimination, and the social and emotional dimensions of occupations involving human relations and care" (2010: 1130).[2] Consider just women's activities in the family. Since the focus in mainstream economics is on the "public" sphere, "private" collectives such as the family naturally receive scant attention. And since the prototype for economic agents is individual persons, and masculine persons at that, when families *are* attended to, they are most commonly treated as if they were individuals themselves, with all their internal workings a "black box." Or they are treated as if they had a dominant "head" who makes all the decisions in accordance with "his" own (perhaps altruistic) preferences. Either way of treating the family, of course, leaves women invisible as agents in their own right in the family. More recently, however, families have been treated by some economists as (cooperative or noncooperative) collective decision-making partnerships. But since, here as elsewhere in mainstream economics, the focus is on simplified mathematical models portraying the interactions of rational, autonomous agents, these collective decision-making partnerships end by being models of marital *couples*. Children, not yet fully rational, certainly not autonomous, and threatening to the tractability of the models, are either conceptualized as "consumption goods" or not conceptualized at all. Left invisible, therefore, are women in the family as care-givers, as agents who historically have borne the bulk of the responsibility not only for domestic labor but also for the nurturance and education of children and the care of the sick and elderly. The upshot is that women's needs and priorities as well as contributions in families are left invisible.

The exclusive reliance on mathematical formalism and statistical analysis in mainstream economics is also problematic for women. For one thing, this approach eschews the more detailed kind of economic information about women (and men!) that can only be gathered by the use of focus groups, participant observation, more complex survey questions, and other innovative qualitative methods. For another, it leads to an overreliance on larger (for example, federal government) datasets that are already available and gathered for other purposes ("secondary datasets"), and these can have unexamined gender biases built into them. "For instance," explains feminist economist Sheba Tejani, "GDP typically measures market and not non-market production, household surveys do not investigate the distribution of resources within the household and labor-force surveys underestimate women's work because of gender biases in the way survey questions are framed." Unfortunately, the discipline's solution to these problems of inadequate data "has been to develop more sophisticated techniques of data manipulation" rather than a broadening of the methods used to gather data, especially data about women (Tejani 2019: 102).

The scene in political science is similar. Here the central concept—in both descriptive and normative enquiry—is power, "the abilities of social agents to affect the world in some way or other" (Isaac 2004: 54), though this concept is interpreted in different ways by different theorists. The four main modes of interpretation currently available are the voluntarist conception stemming from the social contract tradition, the hermeneutic conception stemming from German phenomenology, the structuralist conception stemming from the work of Marx and Durkheim, and the more recent poststructuralist conception stemming from the work of Michel Foucault. Each of these interpretative frameworks offers not only an elaboration of the concept of power but also a conception of human nature and the nature of social life. And yet, each, in the process, manages to produce ignorance of women.

Perhaps the most egregious of the four is the voluntarist conception. Initially put forward by Hobbes, this conception ties power to the voluntary intentions and strategies of individuals who seek to promote their own interests. And since the obstacles to be overcome frequently include the wills of other individuals, the voluntarist conception of power has been construed within political science as the capacity to get others to do what they would not otherwise do. What is involved in the modern conception, according to political scientist Jeffrey Isaac, is still well described by Hobbes and Hume, because their formulations make explicit what is only implicit in more recent accounts: that the voluntarist conception of power presupposes an atomistic view of social relations and a Humean account of causation. That is, individuals are envisioned rather like the atoms of a gas, and the scarce resources within their social world are envisioned rather like the confined space of the container that houses the gas. Hence, the conflicts experienced by individuals, that is, the collisions of the atoms. Power, here, is nothing more than empirical causation interpreted a la Hume: the ability simply to alter the path of another.

Understood in this way, the voluntarist conceptual framework within political science, no less than the neoclassical conceptual framework within economics, exemplifies Nelson's "image of masculinity." For again, individuals are portrayed as autonomous and self-interested, again they are portrayed as behaving in the ways men are expected to behave, and again the resources of the framework fail abysmally to provide information about women. Indeed, "one of the limitations of understanding power as domination or power over," as the voluntarist conception does, is that "it led to women being perpetually characterized as powerless" (Lloyd 2013: 120). By contrast, the work of feminist political scientists has provided a very different account of women. For example, investigations comparing the legislative and leadership styles of men and women, explains feminist political scientist Mary Hawkesworth, suggest that "women pursue cooperative legislative strategies while men prefer competitive, zero-sum tactics; and women are more oriented toward consensus, preferring less hierarchical, more participatory, and more collaborative approaches than their male counterparts" (2005: 145). What's more, women frequently seek to *empower* both themselves and other women, thereby enabling a (productive) *power to* rather than a (dominating) *power over* (see Lloyd 2013 as well as Kantola and Lombardo 2017); this, in fact, has been a hallmark of the women's movement. But none of this can be conceptualized within what Hawkesworth has called the "model of 'abstract masculinity'" offered by the voluntarist conceptual framework.

The upshot is that political science, too, along with economics and biomedical research, is filled with facts about men, not women, so that what emerges from all these fields continues to support the needs and interests of men while it fails to support—indeed, actually undercuts—the needs and interests of women (see Perez 2019 for many more examples).

What Other Scientists Are Teaching Us about the Facts that Ground Science

Feminist scientists are not the only ones concerned about the facts gathered and not gathered in science, and the conclusions drawn from them. Many other scientists share this same concern, though the immediate objects of their concern differ. Consider just two recent examples. The first regards Canadian scientists. In July 2012, 2,000 scientists from all over Canada marched in white lab coats through Ottawa carrying a coffin and tombstones. They then staged a mock funeral on Parliament Hill "to commemorate," as one speaker (then biology doctoral student Katie Gibbs) put it, "the untimely death of evidence in Canada" (Pedwell 2012; Smith 2012). Add to those 2,000 scientists the more than 800 scientists from 32 countries who wrote an open letter to Canada's Prime Minister Stephen Harper thereafter in support of the marchers (Chung 2014). Among the actions that precipitated the protest:

The Harper Administration had instituted sharp cutbacks in basic research and the overall funding of important research areas such as climate, energy, and environmental research. It had even tried to shut down world-class government research programs engaged with groundbreaking industrial pollution research and climate research such as the Experimental Lakes Area research station and the Polar Environment Atmospheric Research Laboratory (*Nature* Editorial 2012). And in place of all this government-run, environmentally relevant basic and applied research, the Harper Administration had pushed for government research partnered with industry and aimed at economic development (Hoag 2011).

The concrete results of these actions were jarring. Thousands of government research scientists were put out of work, and 200 scientific research institutions and more than a dozen federal science libraries were closed, all due to the cutbacks in funding. Scientific books and journals were literally thrown in dumpsters, invaluable data archives dating back a century were destroyed, and reams of publicly funded data and reports from government websites were deleted. In consequence, Canada dropped out of the world's top ten research and development performers, and it was said that Canada's basic climate and environmental science, in particular, had been set back for decades (Kingston 2015; Munro 2015).

As far as the protesting scientists were concerned, then, the actions of the Harper Administration meant that present and future evidence—facts!—that could be used to support a strong environmental and climate policy were simply being rubbed out by the Harper government, killed off. Hence the terminology that galvanized the protesters' movement: the "death of evidence." Indeed, according to the protesters, all these actions of the Harper Administration represented a "war on Canadian science," impeding its ability to continue to make important applied as well as basic contributions to science. Not surprisingly, therefore, the ten-year reign of Prime Minister

Stephen Harper and his Conservative Party ended with the Canadian election of October 2015, when the Liberal Party's Justin Trudeau was voted in as Canada's new Prime Minister.

Just a year later, however—in November 2016—the Republican Party's Donald Trump was voted in as President of the United States, and the kinds of actions that kill off scientific facts, such as major funding cuts for specific kinds of research and restrictions on communications from government science agencies, began yet again, though now in a different country. The protest *this* time, the so-called "March for Science" held on Earth Day in April 2017, was the largest science demonstration in history, taking place not only in Washington DC (where 100,000 people gathered) but also in more than 600 other cities all over the world (March for Science 2017; Smith-Spark and Hanna 2017). And many more protests were planned to take place (Kaplan 2017).

The current "replication crisis" in science is yet another example of scientists' concern with the facts gathered and not gathered (and/or not preserved) in science and the conclusions drawn from them. Replication, of course, is the successful reproducing of experimental results. Called the cornerstone of scientific method, it is an absolute requirement for the proper grounding of science. Yet, in recent years, given a scientific culture that consistently values and rewards new scientific work over replications of previous work, even attempts at replication in science have been relatively rare (see, for example, *Economist* 2013; Engber 2016b; Hastings 2017; Price 2011; Sheldrake 2015). So replication has been judged to be crucially important to science, but at the same time it has been treated as insignificant within the culture of science—a situation now viewed with alarm. As a result, serious efforts are currently underway to motivate the doing and publishing of replication studies. Funds have been allocated, large-scale replication studies have followed, and the results have been depressing. In every case, a surprisingly *low* percentage of the studies previously thought to be replicable *were* replicated (including studies done by the best scientists using the best methods and published in the best journals).

For example, in 2012 it was reported in *Nature* that scientists at the biotechnology company Amgen had attempted to replicate 53 "landmark" cancer studies, but only six of the 53 attempts were successful (11%) (Begley and Ellis 2012). In 2015 it was reported in *Science* that a collaboration of 270 researchers from all over the world had attempted to replicate 100 psychological studies that had been published in three top-tier psychology journals in 2008, but only 39 of those attempts were successful (Open Science Collaboration 2015). In 2016 a survey reported in *Nature* of 1,576 scientists from a variety of fields—chemistry, biology, physics and engineering, medicine, and earth and environmental science—found that more than 70% of those scientists had tried and failed to reproduce at least one other scientist's experiment and more than 50% had even failed to reproduce one of their own experiments (Baker 2016). In 2018 it was reported in *Nature Human Behaviour* that an attempt to replicate 21 social science experiments published between 2010 and 2015 in *Science* and *Nature* yielded only 13 successes, though even in the successful 13 the observed effect was on average only about 75% as large as in the original experiments (Nosek et al. 2018). And the list goes on. Meanwhile, these-top-of-the-line investigations that now failed in replication attempts had already exerted strong impacts in their fields, with citations sometimes running into the thousands and entire areas of study sometimes growing out of them (see, for example, Engber 2016a, 2016b; Ioannidis 2005).

So, all this has precipitated the current "replication crisis" across science, but especially in psychology and biomedical research. As scientists see it, they and their colleagues, under pressure to pursue, at an uncomfortably rapid pace, ever new and different—read "novel and original"—investigations rather than the more lackluster replication investigations science requires, end up shaping science's inventory of facts in intolerable ways. And what results are not only years of wasted research effort, wasted resources, and unwarranted conclusions but also years of inappropriate interventions—unhelpful medications, wrong diagnoses and treatments, missed opportunities.

Bacon's Promise (aka the Contract between Science and Society)

Many scientists have complained about science's inventory of facts, then, not just feminist scientists. But there is an important difference. There have been no marches, no irate letters from scientists all over the world, no lost elections, and no crisis across science in response to what the feminist scientists have disclosed, and this is unsettling. For the feminists' complaints regarding science's treatment of women concern *centuries* of missing facts in science's inventory of facts, with extensive social as well as epistemic repercussions, whereas, at worst, *decades*, not centuries, of missing facts are at issue in the other cases—the Harper case and the replication crisis. Shouldn't there be a more significant response than what has already occurred (such as the ill-fated Revitalization Act) in the feminist case? And if so, what kind of response should it be, and how should it be justified?

Think back to the dawn of modern science. It was then that a promise was made: if society would but support the new enterprise, society would be richly rewarded not only with unprecedented insights into the workings of the universe but also with all the benefits such insights would provide. Francis Bacon, one of the chief architects of the new science as well as one of its more exuberant press agents, promised in his *Novum Organum* that the knowledge science would offer would "establish and extend the power and dominion of the human race itself over the universe" for the benefit of all humankind ([1620] 1960): Book I, cxxix). As Bacon explained in his "Masculine Birth of Time," the human race had been thrust into "immeasurable helplessness and poverty" by the Fall from Eden, and needed to be rescued. And science would be the rescuer. Indeed, science would make humans once again the masters of nature as they had been in the Garden of Eden, and hence once again "peaceful, happy, prosperous and secure" ([1603] 1964). Bacon was more specific about the benefits he envisioned science would provide in his utopian *New Atlantis* ([1627] 2008). That work offered a blueprint for the new science, a blueprint that was later adopted by the Royal Society as well as other early scientific societies and that is still in effect today. In it he included as dividends from science such items as the curing of diseases and the preservation and prolongation of life, the control of plant and animal generation, the development of new materials and new modes of transportation ("through the air" and "under water"), and even new methods of defense—anticipating in the process many of the discoveries and inventions we all enjoy today.

Bacon's seventeenth-century promise was a precursor of what in more recent times has come to be known as the "contract between science and society." This so-called contract has been articulated by many, though the most famous articulation was offered by Vannevar Bush, the engineer and inventor who headed the Office of Scientific Research and Development (OSRD) during World War II. At the end of that war, Bush sent a report to US President Franklin D. Roosevelt that became the basis of US science policy for much of the twentieth century. In it, Bush promised that, if science is both supported by society and left free of societal control, its advances will bring

> more jobs, higher wages, shorter hours, more abundant crops, more leisure for recreation, for study, for learning how to live without the deadening drudgery which has been the burden of the common man for ages past. Advances in science will also bring higher standards of living, will lead to the prevention or cure of diseases, will promote conservation of our limited national resources, and will assure means of defense against aggression.
>
> *(1945: 10)*

Moreover, Bush added, such advances in science will be crucial for attaining these benefits: "Without scientific progress no amount of achievement in other directions can insure our health, prosperity, and security as a nation in the modern world" (1945: 11).

Bush, in the twentieth century, then, reiterated and further elaborated Bacon's seventeenth-century promise, and so, as I said, have many others. And it is this ongoing promise that has kept the majority of the public supporting science, depending on it for all sorts of policy decisions, trusting in it even when it has conflicted with other trusted sources of information such as religion. Science has been billed, right from the start, as a resource for all, to help all of us flourish. But, of course, women are more than half of this all. So, science should be providing the wherewithal for women to flourish along with the men. At a minimum, science should not be providing the wherewithal for women to be at a disadvantage relative to men, for this would not be consistent with the flourishing of women. But this is precisely what science has been doing up to the present, as has already been illustrated. For Bacon's promise, aka the contract between science and society, to be fulfilled, the wherewithal for gender equality has to be part of the package.

Taking Stock: Advice from Bacon as Well as Others

Let's retrace our steps. We noted at the outset that science is based on facts, just as everyone has always said. But we also noted that "the facts" in question are never all the available facts but only a tiny subset of them, and which facts these are is ultimately determined by values—social values as well as epistemic values, whether of scientists or their funders. As a result, we saw, the foundation of science has been heavily stocked, for centuries, with facts that serve the interests of men as well as, more recently, with facts that serve the interests of a not-always-environmentally sensitive government and industry and a not always responsible (either socially or epistemically) scientific community. Finally, we saw that many scientists—feminist scientists, Canadian scientists and their sympathizers, and (in the case of the replication crisis) nearly all scientists—have found all this alarming, and have sought change. And they have succeeded, at least to some extent. There has been a change of government in Canada leading to a change in the values that steer governmental fact finding there. There have been new sources of funding to do replication studies and changes in the culture and practices of science to encourage the doing of them. And there have been changes in the gender composition and politics of several scientific fields and even legislative changes that encourage the gathering of facts that serve the interests of women. None of these changes can reverse the failures of the past, of course, but they all make future research prospects much brighter. The problem is that the scope of past failures regarding the gathering of facts about women is so vast and the scope of present changes in the gender composition and politics of science (including legislative changes) is so modest that science will continue dramatically to privilege men at the expense of women unless a much more ambitious response is pursued. And Bacon's promise provides the rationale for such an ambitious response. But what kind of response should it be?

To answer this question, start with a thought experiment. Imagine a race in which half the runners have been made to carry heavy weights on their shoulders, and imagine that midway through the race there is a concern to make the race a fair one. What might be done to achieve this goal? One possibility would be to stop the race, take the weights off the shoulders of the runners who were carrying them, and then resume the race. This would hardly do the trick, however, for the disadvantage of the weights for the first half of the race would not have been overcome. A second possibility would be to stop the race, transfer the weights from the one group of runners to the other, and then resume the race. This would equalize the disadvantage of the weights for the two groups and thereby yield a fair race, but at the cost of treating the previously unweighted runners in the same cruel way the first group had been treated. By contrast, a third possibility would avoid that problem while still producing a fair race. It would be to give the previously weighted runners a head start for the second half of the race, providing an advantage to compensate for the previous disadvantage without harming the other runners in any way. This is the idea of affirmative action elaborated during the era of civil rights legislation in Martin Luther King's 1964 book *Why We Can't Wait* and Lyndon Johnson's 1965 graduation address at Howard University. Both men used a race metaphor to make the justification of their idea clear. King framed it this way:

> It is obvious that if a man is entered at the starting line of a race three hundred years after another man, the first would have to perform some impossible feat in order to catch up with his fellow runner.

"Something special" needs to be done "'for' him now to balance the equation and equip him to compete on a just and equal basis" (King 1964: 165). Johnson framed the metaphor slightly differently:

> You do not take a person who, for years, has been hobbled by chains and liberate him, bring him up to the starting line of a race and then say, 'You are free to compete with all the others,' and still justly believe that you have been completely fair. Thus it is not enough just to open the gates of opportunity. All our citizens must have the ability to walk through those gates.
>
> *(Johnson 1965)*

In other words, to make the race of our thought experiment fair the previously weighted runners have to be given "something special," some kind of head start after their weights are removed—enough of a head start so that they "all . . . have the ability" to win, that is, are all now as likely to win the race as the other runners.

The above thought experiment helps us consider how we might respond to the continuing paucity of facts serving the interests of women in science today. It suggests and at the same time offers an assessment of three possible responses. The first response is to enact something like the Revitalization Act but equipped with far more powerful provisions than that Act, applied far more widely to all the natural and social sciences, not just medical research, and actually enforced. Such a response would dramatically increase the gathering of facts serving the interests of women while still allowing the gathering of facts serving the interests of men. It would ensure that all publicly funded future research would always generate information about women as well as men. The problem is that, like the first possible fix in our thought experiment, it would do nothing to overcome the disadvantages of the past—the huge inventory of facts gathered over centuries that continue to serve the interests of men and that remain unbalanced by anything that comparably serves the interests of women. The situation portrayed in this first response, in other words, would exactly correspond to the man in King's metaphor who starts a race three centuries after his fellow runner, although the time would be closer to four centuries for the women.

But what if all public funding now and for the next few centuries only supported estrocentric research, that is, research focused on women. The facts gathered would then be solely about women's needs and experiences, exploits and accomplishments, with methods and concepts and assumptions and the rest supporting that aim. Like the second possible fix in our thought experiment, this *would* overcome the disadvantages of the past for women, for it would eventually yield equal inventories of facts serving the interests of men and women. But it would do this at the cost of treating men in the same unconscionable way women had been treated in the past (and, all too often, are still treated now). Such an estrocentric research program, in short, would be as unacceptable as the previous androcentric program had been.

This leaves the third possible response from our thought experiment, the affirmative action response. This response would amount to an *epistemic* affirmative action program for science. Such a program would: (1) privilege the gathering of facts about women, but only in areas in which that is what is currently needed to achieve gender equality in society; (2) de-privilege the gathering of facts about men, but only in areas in which that is what is currently needed to achieve gender equality in society. (These items would constitute the head start element of the program, the push for women within scientific research to compensate for the harm inflicted on women by the research of the past and, as we saw, even the present.) But this program would also (3) privilege egalitarian research, that is, the gathering of facts about both men and women that serve the interests of both men and women and also support gender equality. This item would be a further requirement of Bacon's promise, for it, along with the other two, would ensure the flourishing of both men and women, compliments of science.

With such a program in place, the hope of modern science that Eliza Burt Gamble expressed in her *The Evolution of Woman*—that "the prejudices resulting from lower conditions of human society would disappear, and that in their stead would be set forth not only facts, but deductions from facts, better suited to the dawn of an intellectual age"—might actually come to pass.

Related Chapters: 15, 16, 17.

Notes

1 In the congressional hearings that motivated passage of that legislation, Rep. Patricia Schroeder and Rep. Olympia Snowe, cochairs of the Congressional Caucus for Women's Issues,

> lambasted NIH leaders, and the medical research community as a whole, for compromising women's health. 'American women have been put at risk by medical practices that fail to include women in research studies,' said Rep. Patricia Schroeder. 'NIH's attitude has been to consider over half the population as some sort of special case,' Rep. Olympia Snowe charged.
>
> *(Dusenbery 2018: 25)*

2 "When this author was a student in the 1980s, economics students expressing a wish to study issues relating to women, families, or gender were frequently advised that these were *sociological*—not economic—topics. They likely still are so advised today, in many departments" (Nelson 2010: 1131).

References

Agenjo-Calderón, A. and Gálvez-Muñoz, L. (2019) "Feminist Economics: Theoretical and Political Dimensions," *American Journal of Economics and Sociology* 78(1), 137–166.

Alic, M. (1986) *Hypatia's Heritage: A History of Women in Science from Antiquity through the Nineteenth Century*, Boston: Beacon Press.

Bacon, F. ([1603] 1964) "The Masculine Birth of Time," in B. Farrington (ed. and trans.) *The Philosophy of Francis Bacon*, Liverpool: Liverpool University Press.

———. ([1620] 1960) *The New Organon and Related Writings* (ed. F. H. Anderson and trans. J. Spedding, R. L. Ellis, and D. D. Heath), Indianapolis and New York: Liberal Arts Press (Bobbs-Merrill).

———. ([1627] 2008) *The New Atlantis*, Project Gutenberg EBook #2434, https://www.gutenberg.org/files/2434/2434-h/2434-h.htm. Retrieved 1/16/20.

Baker, M. (2016, May 26) "1,500 Scientists Lift the Lid on Reproducibility: Survey Sheds Light on the 'Crisis' Rocking Research," *Nature* 533, 452–454.

Begley, C. G. and Ellis, L. M. (2012) "Drug Development: Raise Standards for Preclinical Cancer Research," *Nature* 483(7391), 531–533.

Bush, V. (1945) *Science, the Endless Frontier: A Report to the President*, Washington: U.S. Government Printing Office.

Chung, E. (2014) "Foreign Scientists Call on Stephen Harper to Restore Science Funding, Freedom," *CBC News* (October 20), https://www.cbc.ca/news/technology/foreign-scientists-call-on-stephen-harper-to-restore-science-funding-freedom-1.2806571. Retrieved 12/22/18.

Dusenbery, M. (2018) *Doing Harm: The Truth about How Bad Medicine and Lazy Science Leave Women Dismissed, Misdiagnosed, and Sick*, New York: HarperOne (HarperCollins).

Economist (2013) "How Science Goes Wrong," *The Economist* (October 19). *Expanded Academic ASAP*, http://link.galegroup.com/apps/doc/A345848248/EAIM?u=nd_ref&sid=EAIM&xid=5b5cb667. Accessed 12/24/18.

Engber, D. (2016a) "Everything Is Crumbling," *Slate* (March 6), http://www.slate.com/articles/health_ and_science/cover_story/2016/03/ego_depletion_an_influential_theory_in_psychology_may_have_ just_been_debunked.html

———. (2016b) "Cancer Research Is Broken," *Slate* (Future Tense) (April 19), http://www.slate.com/arti- cles/health_and_science/future_tense/2016/04/biomedicine_facing_a_worse_replication_crisis_than_ the_one_plaguing_psychology.html

Ferber, M. A. and Nelson, J. A. (eds.) (1993) *Beyond Economic Man: Feminist Theory and Economics*, Chicago and London: University of Chicago Press.

———. (eds.) (2003) *Feminist Economics Today: Beyond Economic Man*, Chicago and London: University of Chicago Press.

Gamble, E. B. (1894) *The Evolution of Woman, an Inquiry into the Dogma of Her Inferiority to Man*, London: G.P. Putnam's Sons.

Gero, J. and Conkey, M. (1991) *Engendering Archaeology: Women and Prehistory*, Oxford: Blackwell Publishers Ltd.

Hastings, C. (2017) "Are Replication Studies Unwelcome?" *Frontiers* (May 1), https://blog.frontiersin. org/2017/05/01/are-replication-studies-unwelcome/

Hawkesworth, M. (2005) "Engendering Political Science: An Immodest Proposal," *Politics & Gender* 1(1), 141–156.

Hoag, H. (2011, April 19) "Canadian Research Shift Makes Waves," *Nature* 472: 269, https://www.nature. com/news/2011/110419/full/472269a.html. Retrieved 12/22/18.

Hubbard, R. (1979) "Have Only Men Evolved?", in R. Hubbard, M. S. Henifin, and B. Fried (eds.) *Women Look at Biology Looking at Women: A Collection of Feminist Critiques*, Cambridge: Schenkman Publishing Co., 7–36.

Ioannidis, J. P. A. (2005) "Contradicted and Initially Stronger Effects in Highly Cited Clinical Research," *JAMA* 294(2), 218–228, https://jamanetwork.com/journals/jama/fullarticle/201218#note-joc50060-1 Retrieved 1/10/20.

Isaac, J. (2004) "Conceptions of Power," in M. Hawkesworth and M. Kogan (eds.) *Encyclopedia of Government and Politics*, Vol. 1, 2nd ed., London and New York: Routledge, 54–66.

Johnson, L. B. (1965) "To Fulfill These Rights," Commencement Address at Howard University, June 4, https://teachingamericanhistory.org/library/document/commencement-address-at-howard-university- to-fulfill-these-rights/

Kantola, J. and Lombardo, E. (2017) "Feminist Political Analysis: Exploring Strengths, Hegemonies and Limitations," *Feminist Theory* 18(3), 323–341.

Kaplan, S. (2017) "Six Months Later, the March for Science Tries to Build a Lasting Movement," *The Washington Post* (October 23), https://www.washingtonpost.com/news/speaking-of-science/wp/2017/ 10/23/six-months-later-the-march-for-science-tries-to-build-a-lasting-movement/?utm_term=.7f9 d9fb22413 Retrieved 12/22/18.

Kass-Simon, G., Farnes, P. and Nash, D. (eds.) (1990) *Women of Science: Righting the Record*, Bloomington: Indiana University Press.

King, M. L. Jr. (1964) *Why We Can't Wait*. New York: New American Library (Harper & Row).

Kingston, A. (2015) "Vanishing Canada: Why We're All Losers in Ottawa's War on Data," *Maclean's* (Septem- ber 18), https://www.macleans.ca/news/canada/vanishing-canada-why-were-all-losers-in-ottawas-war- on-data/ Retrieved 12/22/18.

Lloyd, M. (2013) "Power, Politics, Domination, and Oppression," in G. Waylen, K. Celis, J. Kantola, and S. L. Weldon (eds.) *The Oxford Handbook of Gender and Politics*, New York: Oxford University Press, doi: 10.1093/oxfordhb/9780199751457.013.0004

March for Science. (2017) "The Science Behind the March for Science Crowd Estimates," *Science Not Silence*, https://medium.com/marchforscience-blog/the-science-behind-the-march-for-science-crowd-estimates- f337adf2d665. Retrieved 12/22/18.

Mazure, C. and Jones, D. (2015) "Twenty Years and Still Counting: Including Women as Participants and Studying Sex and Gender in Biomedical Research," *BMC Women's Health* 15 (94). doi: 10.1186/s12905- 015-0251-9, https://www.ncbi.nlm.nih.gov/pmc/articles/PMC4624369/ Accessed July 30, 2017.

Meinert, C. L. (1995) "The Inclusion of Women in Clinical Trials," *Science* 269(5225): 795–796.

Munro, M. (2015, April 22) "Canadian Budget Pushes Applied Research," *Nature* 520(7549), https://www. nature.com/news/canadian-budget-pushes-applied-research-1.17305. Retrieved 12/22/18.

Nature Editorial (2012, July 19) "Death of Evidence: Changes to Canadian Science Raise Questions That the Government Must Answer," *Nature* 487, 271–272, https://www.nature.com/articles/487271b. Retrieved 12/22/18.

Nelson, J. A. (1996a) *Feminism, Objectivity and Economics*, London and New York: Routledge.

———. (1996b) "The Masculine Mindset of Economic Analysis," *Chronicle of Higher Education* 42(42), B3.

———. (2010) "Sociology, Economics, and Gender: Can Knowledge of the Past Contribute to a Better Future?" *American Journal of Economics and Sociology* 69(4), 1127–1154.

Nosek, B. A. et al. (2018, August 27) "Evaluating the Replicability of Social Science Experiments in *Nature* and *Science* Between 2010 and 2015," *Nature Human Behaviour* 2, 637–644.

Ogilvie, M., Harvey, J., and Rossiter, M. (2014) *The Biographical Dictionary of Women in Science: Pioneering Lives from Ancient Times to the Mid-20th Century*, New York: Routledge.

Open Science Collaboration. (2015, August 28) "Estimating the Reproducibility of Psychological Science," *Science* 349(6251), 943–951.

Ovseiko, P. V., Greenhalgh, T., Adam, P. et al. (2016) "A Global Call for Action to Include Gender in Research Impact Assessment," *Health Research Policy and Systems* 14(50), doi: 10.1186/s12961-016-0126-z

Pedwell, T. (2012) "Scientists Take Aim at Harper Cuts with 'Death of Evidence' Protest on Parliament Hill," *The Globe and Mail* (July 10), https://www.theglobeandmail.com/news/politics/scientists-take-aim-at-harper-cuts-with-death-of-evidence-protest-on-parliament-hill/article4403233/ Retrieved 12/22/18.

Perez, C. C. (2019) *Invisible Women: Data Bias in a World Designed for Men*, New York: Abrams Press.

Price, M. (2011) "To Replicate or Not to Replicate?" *Science* (December 2), https://www.sciencemag.org/careers/2011/12/replicate-or-not-replicate

Rosser, S. (1994) *Women's Health—Missing from U.S. Medicine*, Bloomington and Indianapolis: Indiana University Press.

Schiebinger, L. (1989) *The Mind Has No Sex?* Cambridge: Harvard University Press.

———. (1999) *Has Feminism Changed Science?* Cambridge: Harvard University Press.

Schiebinger, L. et al. (2020) "Designing Health and Biomedical Research," *Gendered Innovations in Science, Health and Medicine, Engineering, and Environment*, International Collaborative Project, http://genderedinnovations.stanford.edu/methods/health.html

Sheldrake, R. (2015) "The Replicability Crisis in Science," *Nature* (September 1), https://www.sheldrake.org/essays/the-replicability-crisis-in-science

Sherman, L. A., Temple, R. and Merkatz, R. B. (1995) "Women in Clinical Trials: An FDA Perspective," *Science* 269(5225): 793–795.

Smith, T. (2012) "Scientists Stage Mock Funeral to Protest Cuts to Research," *Canada.com* (July 11), http://www.canada.com/business/Scientists+stage+mock+funeral+protest+cuts+research/6913396/story.html. Retrieved 12/22/18.

Smith-Spark, L. and Hanna, J. (2017) "March for Science: Protesters Gather Worldwide to Support 'Evidence,' " *CNN*, https://www.cnn.com/2017/04/22/health/global-march-for-science/index.html. Retrieved 12/22/18.

Tejani, S. (2019) "What's Feminist about Feminist Economics?", *Journal of Economic Methodology* 26(2): 99–117.

Waring, M. J. (1992) "Economics," in C. Kramarae and D. Spender (eds.) *The Knowledge Explosion*, 303–309, New York and London: Teachers College Press.

———. (1997) *Three Masquerades: Essays on Equity, Work, and Hu(man) Rights*, Toronto: University of Toronto Press.

Weisman, C. S. and Cassard, S. D. (1994) "Health Consequences of Exclusion or Underrepresentation of Women in Clinical Studies (I)," in A. C. Mastroianni, R. Faden, and D. Federman (eds.) *Women and Health Research*, Vol. 2, 35–40, Washington: National Academy Press.

19

FEMINIST SCIENCE FOR THE PEOPLE

Feminist Approaches to Public Understanding of Science and Science Il/literacy

Sara Giordano

Western sciences have held epistemic authority to explain the world for several centuries now. To justify this status, there must be a continued belief and trust from not only those practicing science (scientists) but those who are outside of formal science laboratories (the public). Much effort has been put into the goal of maintaining a public mandate through pushes for increased science literacy and public understandings of science (Bauer 2009). An early assumption that has continued to drive many science literacy projects was that there is a "deficit" in public understanding of science and if this deficit is "filled," the public will buy-in to the epistemic authority of science. Critical scholars in science and technology studies (STS) have pushed against a deficit view of the public and instead argued for a more complex understanding of the relationships between sciences and societies (McNeil 2013). In part, this need to engage the public more than simply gain buy-in has led to calls for more democratic sciences (e.g. Roth & Barton 2004). In its newest form, democratic science has been proposed as a way to stimulate a better relationship between scientists and the public. Throughout this time, feminists have worked to challenge the monopoly on epistemic authority that Western sciences have enjoyed through various democratizing moves aimed at challenging (and expanding sometimes) what counts as legitimate knowledge and legitimate knowledge-seeking systems.

My work in the areas of democratic sciences has taken me in two directions. The first is developing feminist analyses of the more mainstream push for more democratic sciences over the last decade. The second is working to advance specifically feminist initiatives in democratic sciences. Integral to each direction is situating democratic science initiatives within a genealogy that includes the social movements of the 1960s and 70s and a history of Western science that traces its co-production from the fifteenth century to twenty-first century at the intersections of race, gender, sexuality, ability, and capitalism. In this chapter, I trace these genealogies to show the interconnectedness of these directions and to propose a way forward in developing feminist democratic sciences.

Science Literacy and Public Understandings of Science

To understand mainstream pushes for democratic sciences today, I begin by tracing the concept of "public understanding of science." The exact phrasing as such seems to have taken root in the mid-1980s in the United Kingdom and United States. However, the idea of needing to influence the public's understanding of science is traced back further than that in public understanding of science

(PUS) literature. Lewenstein (1992) argues in the first issue of the journal, *Public Understanding of Science* that there was an assumed demand for public understanding of science after World War II (WWII). This is an interesting point in history to consider for the United States and the history of science. Science literacy became part of a strong sense of nationalism. This makes sense since a general assumption that science was simply good for its own sake did not widely exist after WWII. Instead, who controlled science mattered. At the same time as scientific discovery was proposed as something that could benefit society, the scope of who was included in the society that mattered was clearly one with national boundaries. Germany could be reprimanded for its unethical scientific experimentation related to harming and killing millions of people, while the use of the atomic bomb by the United States in killing hundreds of thousands of people had to be defended as justified. Therefore, from at least the era post-WWII, the race to control science (and technology) and have a more scientifically literate public has been about developing a national pride around science and a general sense of nationalism strengthened through concerns about national security. The idea of science in the "wrong hands" that we see in today's bioethics discourses has roots in these cold war politics and has shifted through time to "war on terrorism" politics.

While the nationalist elements are not always identified in scholarship on science literacy, there appears to be consensus that much of the work of public understanding of science is actually about public *appreciation* of science (Bauer 2009; Lewenstein 1992). Further, there is an assumption throughout the framings of science literacy, public understanding of science, and even newer formations of "science and society" that the problem lies in some kind of public deficit around science (Bauer 2009). However, while we can read clear concerns about deficits into the discourses for science literacy and understanding, discourses based on the idea of democracy and inclusion around education circulate at the same time (Lewenstein 1992). That is, there is an assumption that the public's right to know about science and attempts to educate "all" can be seen throughout the history of such science education initiatives. In the latter part of the twentieth century (post 1960/70s social movements) the orientation to science literacy as one as furthering democracy not surprisingly draws on politics of inclusion, arguing for expanding science literacy to women, people of color, and poor people. Coupled with continued views of science literacy as aiming to correct a deficit and scientific knowledge as simple truths to be learned, this push for greater access to science for all does not include a participatory or agential approach for those subjects in need of understanding science. At times, this in fact further demeans these groups as ignorant as the cause of inclusion and access is touted.

We can uncover assumptions about who is "uneducated," whose "miseducation" or "ignorance" is most dangerous and what kinds of "misunderstandings" are most dangerous not only by reading policy documents and tracing popular news but also by reading literature within the academic subject of public understanding of science itself. In the first article of the journal *Public Understanding of Science*, two policy advocates with the Imperial Cancer Research Fund argue that more research is needed in understanding how to reach the public and also how to get scientists involved in helping the public to see them as normal people, trust them, and help with understandings of science (Bodmer & Wilkins 1992). The authors set the stage for the need for more scientist involvement by arguing that "unskilled working women" are the least likely to be able to be reached. The authors share their assumptions about how these women get their news and information – from TV game shows and tabloids (Bodmer & Wilkins 1992: 8). They suggest at one point that introducing a scientist as a soap opera character may help although it is not clear to them whether scientists would be open to this kind of representation or a television station would actually run such a program. The second fear that they present, beyond the uninformed working-class woman, is the negative image that animal rights activists had cultivated in the public. They do not name the activists but rather discuss polls and survey data that show that large segments of the public distrust scientists when discussing animal research and "a large, and apparently growing, section of the community seems to believe that scientists take pleasure in carrying out sadistic and pointless animal experiments in the name of medical research" (Bodmer & Wilkins 1992: 8). They present that information as connected to the concern that people might believe that "scientific curiosity . . . could lead to someone's child being born with yellow eyes and three noses" (Bodmer & Wilkins 1992: 8). The connection of these two seems to be spurious at best. It seems to act to discredit concerns about animal research by lumping in all distrust of science together. I focus on this article because it is the first article in the first journal dedicated to PUS research, thereby setting the stage for the field in many ways. Interestingly this article is in the same issue as other critical articles on the topic of public understandings of science. This reveals a close relationship between social scientists, scientists, and policy makers in the project of public understanding of science, highlighting a complicated and importantly non-innocent approach to the field.

Understanding scientific knowledge as historically situated has been a foundation of science studies for at least the last half-century (Harding 2001; Kuhn 1962). Therefore, we can analyze the article published in 1982 above within its historical context, both understanding better their argument and the historical moment by closely reading their argument. The concern for working-class women is something that makes sense within a post-1960s/70s multicultural inclusion-based neoliberalism. While in effect insulting the intelligence of these women, they are paternalistically offering to educate them – for their own good and for the good of society as a whole. The politics of including more people into the sphere of politics and economics was a response to widespread, international liberation movements of the 1960s and 1970s that called attention to the way political and economic systems were intertwined and dominated by European males. While these social movements called attention to the way that resources were distributed and controlled based on social categories of race, gender, sexuality, and ability, the end goal of these movements was not the simple inclusion of those marginalized into the existing systems of domination. Instead, many in these movements understood that these categories were not inevitable but were necessary for the building and maintenance of the current economic/political systems of domination. Therefore, the deeper goal was to challenge the racialized, gendered, capitalist system itself. Many activists knew and we can see now that the inclusion of some people from historically disenfranchised groups into a capitalist system allowed that system to thrive and instead created new lines around the categories of difference. For example, a politics of respectability has been used to justify the continued exclusion of poorer Black women from economic power by blaming it on individuals or cultural effects. Respectability in this case means following white, heterosexual, middle-class norms of how women should behave and the kinds of family structures they should live in. Political thought and legal action began to focus on how state support should be withheld from poor Black women because it only encouraged "undesireable" family and community structures (Ross & Solinger 2017). The focus on rights discourses has continued to create a system of those deserving and undeserving of rights. This shift has been critiqued in queer politics where the push for gay marriage has left the system of marriage itself unquestioned. This means that while healthcare benefits can be shared in same-sex couples with jobs that offer healthcare, jobless queers or those without health insurance options continue to remain outside of those deserving of such. So, in this case, certain queer people – who are disproportionately white from middle- and upper-class families – have been included while the vast amount of poor and working-class queers and non-queers of color have been further pushed to the margins. Changes to the boundaries of who was included and excluded made it more difficult to show the continued racialized and gendered violences continuing to happen, now often in the name of helping those become proper citizens (Hong 2015).

Arguments about animal research had made significant headway in cultural awareness – evidenced in mainstream Disney productions that focused on animal rights – from Dumbo to Bambi to a pointed critique of animal research itself in the *Secret of NIMH* (NIMH is the acronym for the National Institutes of Mental Health and the premise of this film is that rats who were subjected to experimentation became "intelligent" enough to escape from the laboratories). However, an attack on animal rights activists also meant attack on a large segment of feminist movements who were increasingly looking at the interconnection of environmental concern, animal abuses, and human oppression.

This one article then, importantly, sets the stage for engagement with the public in opposition with feminist challenges to scientific authority. That is, there were already agitations for democratic participation from feminist and other activist circles. Feminists had been creating their own medical knowledges and distributing this knowledge through projects such as *Our Bodies, Ourselves* that formed as part of a broad feminist women's health movement (Davis 2007). Also, the Black Panther Party made health justice a key part of their platform. They engaged in critiquing racist scientific data, providing health information and medical services in the face of a long history of medical neglect, and conducted public education and testing for sickle cell anemia (Nelson 2013). Following these earlier examples, in the 1980s racial justice movements in the United States called attention to environmental racism and began collecting data and building an environmental justice movement that continues to go up against corporate interests (Bullard 1993). During this same period, HIV/AIDS activists joined the ranks of citizen scientists by changing the way clinical trials were run and becoming directly involved in the changing definitions and scientific truths of the disease (Epstein 1996). These movements intersected, learned from each other, and at times directly collaborated.

So, why were those interested in public understanding of science not drawing on these examples and communities who were already directly engaging with knowledge production? One part of this answer is to go back again to understand that these initiatives were not open to a more participatory relationship between proper science and the public but rather sought to achieve a stronger affective public attachment to science – an appreciation and trust of science. Although this concern is broader than that of feminist distrust, we can assess from my close reading of the PUS article earlier that feminist disloyalty to science was a significant concern.

Some feminist science studies scholars have also lamented the lack of love for science that feminists have exhibited. I have argued elsewhere that we must carefully understand our affective attachments to science in the broader context of mainstream pushes for science literacy and feminist resistance to power/knowledge (Giordano 2017). Here I rehearse part of the argument I have made for a legitimate resistance through science illiteracy.

In Defense of Science Illiteracy and Unknowing Science

Feminist science studies scholars have documented the intentional illiteracy of feminists over the years as a resistance to the dominance of Western masculinist sciences. In Barr and Birke's (1998) work based on interviews with women about adult science education, they analyze what is commonly seen as ignorance of science as potentially resistance. Barr and Birke found that sometimes the women they interviewed knew the scientific answers but chose other non-scientific explanations to describe the natural world. They also point out that Emily Martin (2001) has similarly shown that what might at first appear as scientific illiteracy among working class women was actually a resistance to medical authority on the subject of menstruation. Therefore, Barr and Birke argue that women are not just "alienated" but there is also an "active resistance" to science proper.[1]

Independent artist, film maker, and activist, Lucía Egaña Rojas writes in "Notes on a Transfeminist Technology," that "A transfeminist technology will value illiteracy for its improductiveness for industry, as a way of finding paths unimagined by speed and productivity" (Rojas 2013: 1). Rojas further suggests a usefulness in creating new worlds by being gender illiterate as well as an acknowledgment of how the positive relationship between epistemic power/authority and literacy devalues the knowledge of many of the world's poorest inhabitants. I draw on Rojas' exploration of illiteracy to leave open space for the importance of science illiteracy as a failure that threatens the supremacy of scientific knowledges. What kinds of science illiteracy might we embrace as part of a project of destabilizing Western science and remaking knowledge production?

Another example comes from decolonial activists in South Africa who were part of the Fees Must Fall movement (that stemmed from Rhodes Must Fall) and came under intense fire when some suggested a total abolition of (Western) science before being able to remake new sciences. A video of one activist speaking at a university event about science went viral in October 2016 with the hashtag #ScienceMustFall. The commentaries on-line showed how attacking science was used to delegitimize the activists' movement. The suggestion that the activist made about science being a colonial project and needing to be abandoned before creating something new was not just rejected but taken as evidence of ignorance.

Is science illiteracy a worthwhile goal for creating a more participatory science practice? After all those who make decisions to reproduce our inequitable society everyday are those who are the best educated of our societies (Wynter 1994). Liberal education has been and continues to be an integral part of maintaining an unjust society. Therefore the question is not simply one of education or literacy. Instead many have argued over time for processes of unlearning (*popularly in terms of unlearning racism*), rewriting (Wynter 1994), unschooling (*a thread of student-focused homeschooling*), failure (e.g. Halberstam 2011 *in opposition to capitalist success*), or illiteracy (Rojas 2013). I argue that we must hold these possibilities open as work for a more just and democratic process of science. Next, I look at how social movements engagement with science may help us to understand new mainstream pushes for democratic sciences and the possibilities opened to further feminist democratic sciences.

New Democratic Sciences and Feminist Genealogies of Democratizing Science

While discourses of democracy (broader participation) have often circulated in conversations about science literacy, the result of who has been drawn into scientific knowledge production in each case has largely continued to be an elite class of intellectuals. Lewenstein (1992) documents this phenomenon in his analysis of the science literacy movement after WWII and its failure to appeal to a mass populace and instead settling for magazines aimed at intellectuals and policy makers. We might see the current push for more democratic sciences as a continuum with this kind of mismatch between rhetoric and discourse and reality.

Today, we see an emergence of more do-it-yourself, maker, and hacker communities. Within this larger context, there has been the development of DIY, community-based biology maker/hacker spaces. While not numerous, I have focused in my work on what the discourse of having accessible spaces for doing biology have done for the larger fields of biology, genetics, and synthetic biology (Giordano 2018). Through my research and others interested in these developments it has become clear that while the rhetoric of open science for all is used in the building of support for such endeavors, the practitioners in the actual spaces and the founders of such spaces are overwhelmingly those already scientifically trained in some way with many holding academic appointments and/or having held appointments in large biotech companies. As it turns out, most who are involved in this way seek to have more freedom to become biological entrepreneurs (Giordano 2018). That is, it is not necessarily about including more people but about challenging traditional modes of scientific discovery that the DIY biologists argue limit their creativity and ability to innovate. At first this was surprising to me because I assumed that community, DIY, democratic science would aim to not only create collective spaces for tinkering with biology but would be based on ideas of sharing both results and profits. Instead, I found that the successes that were highlighted by DIY biologists were those that earned a profit. What was particularly puzzling was that the DIY biologists did not seem to feel conflicted about this outcome.[2]

The feminist puzzlement I experienced stems from differing genealogies and definitions of democratic science and democracy itself. Sandra Harding argues that feminist science studies has made the case for democratizing science so "that those who bear the consequences of science and technology policy decisions have a proportionate share in making them – in short, that women's interests and values count" (Harding 2001: 302). The overall sentiment of this argument does represent a large strain in feminist and other critical science studies literature. However, the flattening of the category of "women" as universally oppressed and marginalized has been consistently challenged through the production of democratic sciences based on social movements and critical academic literature. Women of color feminists have long called attention to how the unmarked category of women stood in for the interests of white middle-class women (e.g. Moraga & Anzaldúa 1982; INCITE! Women of Color Against Violence 2006). Reproductive justice activists and scholars have also argued that the unit of concern for justice cannot be individuals or simply women but instead impact on whole communities and their resistance must be centered (Ross & Solinger 2017). Further, critical bioethicist, Harriet Washington (2006) convincingly argues that it is not simply a matter of including all women but that some women (primarily Black women) have been used for the benefit of other women's (primarily white women) health. The separation of different groups of women (and people of all genders) for dividing up the benefits (medicine and health) and harms (use and abuse) in medical research is not just unfortunate or coincidence but instead has been shown to be necessary and intentional under this system that she calls medical apartheid. That is, some people's bodies are undervalued so that they can justly be used for the benefit of others in our current systems. This is (hopefully) obviously not the only way to structure our health systems or larger societies. Therefore, to create the kind of democratic sciences that Harding (2001) argues for, those whose bodies have been subjected most to the medical gaze are who needs to be counted. We can see how this line of reasoning differs from the dominant calls for democratic sciences that do not take into account history or power. And also how simply starting from women's interests and values will not properly attend to the specific histories and power of the research being undertaken. That is, while sometimes we may find that the group most impacted by research is best described as women, many times this will be a subset of women and possibly a group that is not simply categorized by a universalized idea of a single gender.

This feminist call to democratize science is not simply for a greater diversity in STEM fields in the academy but a call for broadening the very definition of science, expanding what counts as scientific knowledge and whose knowledge is counted. Drawing on these traditions, feminist scholars have read reproductive justice and environmental justice activists as practicing democratic science in communities outside formal academic borders. These movements built on social movements of the 1960s and 1970s which as I argued before were not simply about gaining inclusion in unjust existing systems. Integral to these movements was a challenge to knowledge and truth itself. And within that, scientific and health knowledges were challenged and rewritten in the streets (in community health clinics, in meetings, in protests). Two prime examples of these challenges to scientific knowledge production, not just through critique but through the production of new knowledge, can be seen in the Black Panther Party's health activism (Nelson 2013) and the feminist women's health movements (Murphy 2012). By locating democratic sciences within these histories instead of more dominant narratives of science literacy based on calls for correcting public deficits in understanding or appreciation, we can both understand better the appropriation of democratic, inclusive science for entrepreneurialism and develop our own feminist democratic sciences as a continuation of our collective past engagement with science and particularly exploiting this current moment of interest in democratic science. What I mean by the first point is that we can learn from these movements how to spot co-optations of social justice-based democratic rhetoric in newer mainstream pushes for democratic science. We can see how, while likely not intentional on the part of the founders or participants, mainstream democratic science projects follow a neoliberal model of arguing for a colorblind and genderblind politics. The problem is that we have not achieved a raceless or genderless world. Previous social justice movements and current ones are focused on the reality of the systemic injustices we have based on the idea of racial and gendered difference. Assuming that opening up science to "everyone" will be possible without correcting deep injustice has meant that "everyone" only includes those who are economically successful. Economic success is limited based on race, gender, ability, and sexuality. And economic success must be limited to a select group under capitalism. The earlier movements which I draw on provided us with analyses that included the ways that exclusion based on difference had to do with the dominant economic system. A colorblind, genderblind politics does not make room to correct historic injustice thereby continuing the disenfranchisement of those who begin with a disadvantage. Below I look at some examples of how we can work within the current moment of interest in democratic science while challenging traditional modes of inquiry that refuse to analyze power and history.

Examples of Feminist Experiments in Democratic Sciences

So, what are examples of feminists creating democratic sciences? While I have argued that this work has been being done for a long time outside of the official label of democratic science or public science necessarily, it is worthwhile to use this current moment of increased interest in public/democratic sciences to put into practice our feminist science studies insights to call attention to the problems with universal, colorblind, and genderblind definitions of inclusion of all in science, and at the same time exploit this moment to further our experiments in testing out democratic sciences across academic borders that aim to challenge the status-quo power arrangements. Here I will briefly discuss three areas of democratic science that feminist science studies scholars have engaged in – (1) the science shop model, (2) community-based participatory research (CBPR) methods, and (3) public engagement. There are more examples of local projects across academic borders that feminist science studies scholars are participating in. Highlighting these is simply to provide some clear examples of how some of us have worked within more mainstream definitions of democratic science over the last several years. In each case, we were able to do so by basing our definition of democratic science and our goals on those provided by earlier social justice movements and decades of scholarship on science and health from feminist perspectives.

The science shop model comes originally from the Netherlands where student and other protests in the 1970s led to the implementation of science shops in universities. Science shops were set up to create a mechanism for community members to ask research questions that graduate students in the universities would then work to answer. This model, similar to the over general definition of democratic science used in today's movement to create community science labs, did not specify a social justice political orientation. Over time, these science shops have become a normal part of a neoliberal institution providing opportunities for businesses to interact directly with university researchers to meet their research interests for entrepreneurship and development. However, feminists (Barr & Birke 1998; Weasel 2001) have argued that the science shop could be used in feminist studies departments to change not only the kind of research questions asked but the way research is conducted – providing an interdisciplinary and an explicitly political commitment to social justice in its methodologies. Taking up this call for action, I have worked to experiment with the science shop model in my own universities, working with both undergraduate and graduate students to produce new kinds of knowledge on science questions brought to us by community partners (Cruz et al. 2019).

CBPR has gone through periods of popularity particularly in public health fields. While some argue for CBPR as necessarily being based on feminist and social justice traditions (Minkler & Wallerstein 2011), many obstacles to truly democratic work across the borders of the academy exist. In worst case scenarios, the language of CBPR is used to get free labor from communities and community buy-in for invasive research based on goals that remain set by funding agencies and university researchers. In its best attempts, researchers struggle with how to actually work on disrupting the power dynamics across the academic borders. Notably, Kim TallBear (2014) has suggested through her work that CBPR maintains boundaries that reinforce power imbalance between those inside and outside of the academy. She suggests "standing with" communities instead. This resonates with the work of others such as Michael Montoya (2013) who centers relationship building practices when he does community research. This represents therefore an intervention into the CBPR model that is in conversation with more mainstream uses of this method.

Some have used the language of public engagement to argue for more say in policy decisions and access to scientific results. Notable in this approach is the work of Natasha Myers and Max Liboiron with others in the Politics of Evidence working group and the *Write2Know* campaign in Canada (Downey & Zuiderent-Jerak 2016). The political starting place for this work highlights a challenge to traditional hierarchies of science rooted in colonial and racist histories. While recognizing the problems with overvaluing scientific knowledge, they find themselves in this moment concerned about the ways that certain scientific knowledges – particularly about environmental toxins – are kept from the public. They argue for access to this evidence, political power for those most impacted by environmental toxicity, and the ability to produce new kinds of scientific data in communities most impacted.

Conclusion

What I have laid out in this chapter is a way to evaluate our engagement with democratic sciences. I have challenged ideas of participation and democracy as simply good or as always moving in the "right" direction by complicating the motivations and impacts of different movements. I began by showing how the United States became interested in public understanding of science, science literacy, science appreciation, and how this became calls for public engagement and more democratic sciences. By presenting this history as intersecting with a political history of colonization, war, capitalism, exclusions and inclusions based on race, gender, and other categories of difference, I demonstrate that what may seem like simple, progressive moves to include more people may have the impact of maintaining long standing injustices. I offer social justice-based movements as an important starting point for producing feminist democratic sciences while I also warn of their potential appropriation. This chapter should allow those of us in the academy a way to interrogate our own motivations for democratic science, others' motivations, and the potential implications of each of these moves.

By drawing on social movements, I suggested here that we can broaden the concepts of science il/literacy. Therefore, I do not simply argue for more science appreciation or science engagement. Activists have demonstrated great skill at forming projects of knowledge production that hold critiques of dominant sciences while producing new forms of knowledge that might be considered science. That is, they practice an unlearning and a direct challenge to scientific knowledge at the same time as revaluing the knowledge that comes from their communities and movements. This brings attention to the question of what counts as science and whether our project should be to argue for other knowledge-seeking projects to count as sciences or to replace science's epistemic authority with knowledge-seeking projects by other names and with other histories. These approaches, I argue, are also not necessarily mutually exclusive. Those of us in the academy can and should learn from these practitioners of democratic, social justice science.

Related chapters: 11, 18, 32.

Notes

1 This is not a purist stance against anything that science and technology has produced. It is quite difficult in our current world system to imagine never using modern medicine, technologies such as phones or computers. The kind of resistance that I am highlighting is in specific instances where possible to offer counter-narratives and to ask questions instead of automatically assuming that scientific explanations are the most useful for their understanding of the world.

2 This does not mean that this was universally true since many people have entered these spaces with various political orientations and expectations; some may be disappointed and some of those who are disappointed cease to participate in the space and others have co-existed with these contradictions.

References

Barr, J., & Birke, L. I. A. (1998) *Common Science? Women, Science, and Knowledge*. Bloomington: Indiana University Press.

Bauer, M. W. (2009) "The Evolution of Public Understanding of Science—Discourse and Comparative Evidence," *Science, Technology and Society*, 14(2), pp. 221–240. https://doi.org/10.1177/097172180901400202

Bodmer, W., & Wilkins, J. (1992) "Research to Improve Public Understanding Programmes," *Public Understanding of Science*, 1(1), pp. 7–10.

Bullard, R. D. (1993) *Confronting Environmental Racism: Voices from the Grassroots*. Boston: South End Press.

Cruz, M., Jordan, J., Salinas, S. A. B., Jones, R., Thomas, S., Ney, A., & Giordano, S. (2019) "Using the Feminist Science Shop Model for Social Justice: A Case Study in Challenging the Nexus of Racist Policing and Medical Neglect," *Women's Studies*, 48(3), pp. 283–308.

Davis, K. (2007) *The Making of Our Bodies, Ourselves: How Feminism Travels across Borders*. Durham: Duke University Press.

Downey, G. L., & Zuiderent-Jerak, T. (2016) "Making and Doing: Engagement and Reflexive Learning in STS," in U. Felt, R. Fouché, C. A. Miller, & L. Smith-Doerr (eds.), *The Handbook of Science and Technology Studies*. Cambridge: MIT Press, pp. 223–250.

Epstein, S. (1996) *Impure Science: AIDS, Activism, and the Politics of Knowledge*. Berkeley: University of California Press.

Giordano, S. (2017) "Those Who Can't, Teach: Critical Science Literacy as a Queer Science of Failure," *Catalyst: Feminism, Theory, Technoscience*, 3(1), pp. 1–21.

Giordano, S. (2018) "New Democratic Sciences, Ethics, and Proper Publics," *Science, Technology, & Human Values*, 43(3), pp. 401–430.

Halberstam, J. (2011) *The Queer Art of Failure*. Durham: Duke University Press.

Harding, S. (2001) "After Absolute Neutrality: Expanding 'Science,' " in M. Mayberry, B. Subramaniam, & L. H. Weasel (eds.), *Feminist Science Studies: A New Generation*. Hove: Psychology Press Ltd, pp. 291–304.

Hong, G. K. (2015) "Neoliberalism," *Critical Ethnic Studies*, 1(1) 56–67. https://doi.org/10.5749/jcritethnstud.1.1.0056

INCITE! Women of Color Against Violence (2006) *Color of Violence: The INCITE! Anthology*. Cambridge: South End Press.

Kuhn, T. S. (1962) *The Structure of Scientific Revolutions*. Chicago: University of Chicago Press.

Lewenstein, B. V. (1992) "The Meaning of 'Public Understanding of Science' in the United States after World War II," *Public Understanding of Science*, 1(1), pp. 45–68.

Martin, E. (2001) *The Woman in the Body: A Cultural Analysis of Reproduction*. Boston: Beacon Press.

McNeil, M. (2013) "Between a Rock and a Hard Place: The Deficit Model, the Diffusion Model and Publics in STS," *Science as Culture*, 22(4), pp. 589–608.

Minkler, M., & Wallerstein, N. (2011) *Community-Based Participatory Research for Health: From Process to Outcomes*. Hoboken: John Wiley & Sons.

Montoya, M.J. (2013) "Potential Futures for a Healthy City: Community, Knowledge, and Hope for the Sciences of Life," *Current Anthropology*, 54(S7), pp. S45–S55.

Moraga, C., & Anzaldúa, G. (eds.) (1982) *This Bridge Called My Back: Writings by Radical Women of Color*. Watertown: Persephone Press.

Murphy, M. (2012) *Seizing the Means of Reproduction: Entanglements of Feminism, Health, and Technoscience*. Durham: Duke University Press.

Nelson, A. (2013) *Body and Soul: The Black Panther Party and the Fight against Medical Discrimination* (Reprint edition). Minneapolis: University of Minnesota Press.

Rojas, L. E. (2013, March 22) *Notes Towards a Transfeminist Technology.* Retrieved from http://www.lucysombra.org/TXT/03Notes-towards-a-transfeminist-technology.pdf

Ross, L., & Solinger, R. (2017) *Reproductive Justice: An Introduction, Volume 1.* Berkeley: University of California Press.

Roth, W.M. , & Barton, A.C. (2004) *Rethinking Scientific Literacy.* New York: Routledge.

TallBear, K. (2014) "Standing With and Speaking as Faith: A Feminist-Indigenous Approach to Inquiry," *Journal of Research Practice, 10*(2), p. 17.

Washington, H. A. (2006) *Medical Apartheid: The Dark History of Medical Experimentation on Black Americans from Colonial Times to the Present.* New York: Doubleday Books.

Weasel, L. H. (2001) "Laboratories Without Walls: The Science Shop as a Model for Feminist Community Science in Action," in M. Mayberry, B. Subramaniam, & L. H. Weasel (eds.), *Feminist Science Studies: A New Generation.* Hoven: Psychology Press Ltd., pp. 305–320.

Wynter, S. (1994) "A Black Studies Manifesto," *Forum NHI: Knowledge for the 21st Century,* 1(1), pp. 3–11.

PART IV

Feminist Philosophy of Science in Practice

Biology and Biomedical Sciences

20

FEMINIST PHILOSOPHY OF BIOLOGY

Lynn Hankinson Nelson

Introduction

Feminist philosophy of biology is a dynamic and multifaceted research tradition that is part of the research tradition of Feminist Science Studies that emerged over four decades ago. Feminist engagements with philosophy of biology share features with philosophy of biology more broadly, but also differ in their emphases and priorities in important ways. We begin with a brief discussion of some features they share.

Both feminist and non-feminist philosophy of biology involve work by biologists as well as philosophers and historians of science, who often collaborate. Both focus on topics of longstanding interest in philosophy of science – including the nature and strength of the evidence that supports scientific hypotheses and theories, scientific reasoning and methods, the nature of scientific explanations and theories, and the role of cognitive (i.e. epistemic) values in scientific practice.

Philosophers of biology have framed these issues in relation to one or more biological sciences, rather than physics – which was a common focus of mid-twentieth-century philosophy of science. This results in different emphases. Accounts of causal relationships in biology, as reflected in biological explanations and theories, often do not invoke "laws." This is a function of the complex causal relationships, including a role for environmental factors, in biological processes (see e.g. Sober 1993). In addition, some areas of biology draw on evolutionary theory, which highlights the historical, contingent, and multifaceted nature of evolutionary developments. However, some biological hypotheses about sex/gender, race, and sexuality are deterministic and have been the target of feminist criticism. We focus on those assuming alleged sex/gender and sex differences are based in biology. Conceptual issues are also a focus of philosophical engagements with biology. For example, there are many analyses of how the concept of "fitness" has been understood and debated in evolutionary theorizing. Feminists have asked whether the concept of "male aggression" has been clearly defined in hypotheses that maintain it is common in many species. Some philosophers of biology study relationships between biology and the broader scientific and/or social contexts in which it is undertaken (e.g. Ruse 2012). But a focus on relationships between social contexts, and the directions and content of biological research, is more common in feminist engagements (e.g. Bleier 1984; Fausto-Sterling 1985). As will become clear, most feminists who undertake such studies do not doubt that *evidence* has a central role in biological reasoning and practice. They view their analyses as contributing to fuller understandings of science than accounts of it that do not include such relationships.

There are other topics and issues of common interest in feminist and non-feminist philosophy of biology (Fehr 2011; Nelson 2017), but those noted are sufficient for the purposes of the present discussion. We now turn to general and specific focuses that distinguish feminist engagements from others in philosophy of biology. Most importantly, feminists analyze what they argue to be consequential issues involving *gender* and *biology*.

Gender and Biology

Feminist philosophers of biology (and, again, this includes biologists who were among the first to engage with the issues of interest in the tradition) focus on how, in various ways, the questions and hypotheses in many areas of biology are informed by unwarranted assumptions about sex and/or gender, and they often propose constructive alternatives they take to be more warranted by evidence. When discussing hypotheses proposing sex or gender differences among humans, we use the term "sex/gender" rather than "gender." This is because there is now substantial evidence that, at least in the human case, neither sex nor gender is stable or unaffected by the other. The editors of a recent collection of essays devoted to feminist engagements with neuroscience summarize the argument this way: "It is impossible to disentangle the effects of sex from those of gender" on women and men. It is clear, they argue, that our bodies, including our brains, change in response to environments and experiences (Bluhm et al. 2012: 4). See also Clune-Taylor in this volume.

We next describe some of the central areas of focus in feminist analyses of gender and biology. Often, feminists argue, the features of biological research listed below reflect "background assumptions" that are not recognized as requiring scrutiny (e.g. Longino and Doell 1983). The idea that background assumptions are features of scientific reasoning and, if they are incorrect, can lead to unwarranted conclusions did not originate in feminists' analyses of science. Philosopher Carl Hempel offered arguments for their role in the mid-twentieth century (Hempel 1966).

Some of the features in the following list are also a focus of feminists' analyses of other sciences. We emphasize feminists' arguments concerning their presence in and consequences for biological theorizing:

- Androcentrism (i.e. "male-centeredness"). Biological research and hypotheses that are androcentric take the activities, behaviors, and dispositions typically (or at least stereotypically) associated with men or males as their primary focus and fail to focus (or adequately focus) on those typically (or at least stereotypically) associated with women or females. Feminists argue that androcentrism has shaped the directions and content of research in areas of biology.

- Gender Stereotypes. A generalization is aptly described as "stereotypical" when it attributes a characteristic to all members of a group, when in fact it is not true of many (or even most) such individuals. Feminists point to the presence and role of gender stereotypes in various biological sciences. For example, women and females are often primarily or exclusively associated with reproductive activities (often viewed as "natural" and thus not requiring study) while males of the same species are associated with productive activities such as proving for and protecting females, and dominating social dynamics. Feminists maintain that these and other gender stereotypes have been consequently for biology, shaping research questions and hypotheses.

- Gendered Metaphors. Philosophers and others have devoted a lot of time to questions about the nature of metaphors, including their role in scientific reasoning. For our purposes, it is enough to note that metaphors attribute characteristics aptly attributed to some entities or phenomena, to other entities or phenomena to which they are not literally attributable. Many philosophers and scientists view metaphors as having a positive role in scientific reasoning – so long as metaphors are recognized *as metaphors* rather than literal, factual descriptions – as they can advance theorizing. Feminists argue that the gendered metaphors, which attribute gender characteristics to entities that are not sexed, that are used in some biological sciences are not of this kind. They reflect gender stereotypes and, as importantly, they are offered as if they are literal descriptions, rather than metaphors. For example, in the account of fertilization that prevailed into the 1980s, sperm were described as the primary "actors" in the process, competing with one another to capacitate a "passive" and "dormant" egg. In fact, although sperm capacitate eggs, eggs and features of the female reproductive tract capacitate sperm. The descriptions of sperm as active and of eggs as passive reflect longstanding gender stereotypes that contribute to the androcentrism (only some aspects of which have been mentioned here) of the account of fertilization in which they figure (see e.g. Martin 1991).

- Biological Determinism. Feminists have criticized hypotheses that explain what are alleged to be sex/gender and sex differences as based in biology. The differences proposed have included differences in behavior and temperament in many species, including humans, and differences in women's and men's psychology and cognitive abilities, among other characteristics. The specific biological determinants proposed to explain the alleged differences vary – they have been attributed to genetic differences and/or to assumed differences in the historical evolutionary selection pressures faced by males and females, including our ancestors and/or to pre and postnatal "male" hormones, among other factors. Feminists have criticized the accounts of many such differences on the grounds that many of the differences do not hold up under scrutiny, and that some can be explained on the basis of social factors of various kinds. Feminists also argue that biologically determinist hypotheses about sex/gender differences carry social and political implications.

- Evaluatively Thick Concepts. The idea that some concepts carry both empirical content *and* normative or evaluative content did not originate in feminist theorizing. But feminist philosopher Elizabeth Anderson argues that such concepts do have a role in scientific theorizing about the sexes and sex/genders (Anderson 2004). As suggested earlier, the gender stereotypes and gendered metaphors that are features of some biological theorizing are evaluatively thick. "Active" and "productive" typically associated with males have long been viewed as more valuable than the "passivity" and "reproductive activities" often attributed to females. And the evaluatively thick concepts just noted can be used and have been used to argue that men and women are suited for different roles, a view that carries significant normative implications.

- Equity Issues and Cognitive Values. Feminists have focused on equity issues involving women's participation in biology; and the cognitive values that, they argue, inform the practices of feminist biologists and often differ from those traditionally emphasized (Fehr 2011).

In the next two sections, we focus on case studies. They are representative of feminists' critical and constructive engagements with biology concerning the issues cited earlier. The engagements of which they are representative are extensive and include, in addition to the sciences we consider, analyses of research in developmental biology, biobehavioral sciences, neurobiology, and medical biology. We focus on only two cases in order to do justice to the research that feminists criticize, which is in many respects, *good science* in ways we later discuss; and to do justice to the evidence feminists cite in their critical and constructive arguments.

Sexual Selection Theory

Feminist engagements with evolutionary theorizing are extensive. We focus on feminists' analyses of Charles Darwin's hypothesis of sexual selection and briefly discuss the more recent theory of

Parental Investment argued to provide support for it. We consider feminists' critiques of the empirical adequacy of both and the alternatives they purpose.

In *On the Origin of Species*, Darwin proposed that "natural selection" is an important evolutionary mechanism – that is a mechanism that allows evolution to occur (Darwin 1859/2003). He also proposed sexual selection, the subject of this case study, as a secondary evolutionary mechanism. To understand Darwin's reasoning to sexual selection, we need to understand his arguments for how natural selection works.

Darwin's argument for natural selection was relatively straightforward and he described its premises as containing "no new facts," drawing as it did on then accepted hypotheses. Noting that there is variation in the traits of members of a species, and citing the hypothesis that "there is a struggle for existence" accepted by many of his scientific contemporaries, Darwin proposed the following: if a variation conveys an advantage, however small, in terms of survival, organisms with the variation will be more likely to survive than those without it, and will tend to pass on the variation to their offspring. Over time, Darwin argued, the accumulation of variations within a species would lead to the prevalence of a new trait, and eventually, if enough new traits emerged, to the emergence of a new species.

After presenting his argument for natural selection, Darwin went on to argue that any trait deleterious to an organism's survival would necessarily be selected *against*. In the *Origin* he states:

> It may metaphorically be said that natural selection is daily and hourly scrutinizing, throughout the world, the slightest variations; rejecting those that are bad, preserving and adding up all that are good; silently and insensibly working, whenever and wherever opportunity offers, at the improvement of each organic being in relation to its organic and inorganic conditions of life.
>
> *(Darwin 2003: 84)*

This understanding of natural selection led Darwin to pose sexual selection as a second evolutionary mechanism. He believed that some traits, found predominantly among males of various species, could not be explained by natural selection. The traits of concern seemed to him not only as not conducive to survival, but as working against it. The peacock's very large and colorful tail was one example he worried about. It is flamboyant enough to alert predators and so cumbersome that it hampers a peacock's ability to swiftly take flight or run to escape predators. In contrast, peahens have small brown tails. More generally, Darwin considered the brighter colors of the males of many species than their female counterparts, and the larger size of males relative to females in many species, as working against survival. How could natural selection allow for, rather than select against, such traits?

Darwin's solution was sexual selection. It explained the traits in question as promoting *reproductive success*, rather than survival. From the outset of the *Origin*, Darwin emphasized the importance of such success given that heritable variation is necessary to evolution and an organism's survival is necessary to but not sufficient for it. If traits such as larger size, bright plumage, and the like contributed to males' reproductive success, they pose no challenge to the role or efficacy of natural selection. In *The Descent of Man*, published in 1871 and arguing for human evolution, Darwin describes sexual selection as involving "the advantage which certain individuals have over other individuals of the same sex and species, in exclusive relation to reproduction" (1871, Vol. 1: 256).

In the *Origin* and the *Descent*, Darwin defined sexual selection in terms of two processes: members of one sex (typically males) compete with one another for access to the other sex (typically females); and, except in the human case, females choose mates. If traits such as large size help males compete with other males for access to females, and flamboyant traits such as bright colors are preferred by females, sexual selection explains the traits that initially concerned Darwin.

As noted earlier, Darwin did not attribute mate choice to human females, arguing that such choices were made by men. His reasoning was informed by the social context of Victorian England, as well as longstanding scientific views (dating as far back as Aristotle) about sex and sex/gender differences. And Darwin's account of sexual selection included more than sex and sex/gender differences in phenotypic traits, and these differences also reflected then-current scientific and social views. In the *Origin* and the *Descent*, Darwin argued for sex differences in what he called "secondary sex characteristics." In the *Origin*, he maintained that the behavior and psychology of males and females differ in significant ways. For example, he described males as "generally eager to mate with any female" and females as less eager (Darwin 2003: 156–158). In the *Descent*, Darwin described women as engaging in coy-like behavior – a description that, together with that of men as "eager to mate," took hold among some evolutionary theorists including some recent researchers in primatology, as we discuss in the second case study.

Darwin also argued that the differences in mating strategies and related behaviors he cited resulted in what he and many of his contemporaries took to be males' obvious superiority in valuable characteristics and abilities. Darwin attributed the differences to selection pressures facing males. In the *Origin*, he argued that males, given their need to compete for females and to provide for females and offspring, develop stronger passions, heightened intelligence, greater strength and sensory capacities, and heightened locomotive abilities. In the *Descent*, Darwin attributed the same and other characteristics to men in terms relevant to humans. He attributed men's superior powers of "observation, reason, invention, and imagination" to selection pressures that he thought were unique to them – competition for mates, protecting and providing for women and children, and "defense of community" (ibid.: 565). Darwin's assumptions and arguments about sex and sex/gender differences are obvious examples of androcentrism, gender stereotypes, and biological determinism. And although Darwin acknowledged that women enjoy some of the qualities just mentioned, including intelligence (but to a far less degree), he maintained that they do so only because they inherit them from men (Darwin 1871: 329).

Feminists have offered sustained critiques of the assumptions, claims, and arguments just summarized. Biologist Ruth Hubbard argued that Darwin's account of the greater selection pressures facing men, and of how women inherit traits such as intelligence from men, suggests that women have enjoyed "a free evolutionary ride" on the backs of their male conspecifics (Hubbard 1983). Other feminists have offered arguments and counter-examples challenging the androcentrism, gender stereotypes, and biological determinism characterizing Darwin's descriptions of females and males.

For example, feminists have criticized Darwin's assumption that a female's, including a woman's, survival and that of her offspring depend on the assistance of one or more males, as reflecting an acceptance of unwarranted gender stereotypes (e.g. Bleier 1984). Feminist biologists who study animal behavior in laboratories or in field research provide examples involving the species they observe (as we consider in the next section), to argue that Darwin's claim that in most species females are dependent on males is unwarranted (e.g. Hrdy 1986). Feminists also argue that Darwin's arguments for sex/gender differences in intelligence and reason are unwarranted and not nearly as related (if related at all) to his concerns about apparent counter-examples to natural selection, as they were related to then-current sociopolitical views about sex/gender (e.g. Bleier 1984; Hubbard 1983). As a further example, one we consider in some detail in relation to primatology, feminists cite numerous ways in which the females of a variety of primate species are far from coy or passive recipients of male sexual advances (e.g. Hrdy 1986).

Finally, feminists have focused on the social, political, and ethical implications of Darwin's account of sex/gender differences, including their impact on social beliefs, values, and policies (e.g. Hubbard 1983).

Other biologists and philosophers have been critical of sexual selection. Representative was the view of the prominent evolutionary biologist Ernst Mayr that its role was at most meager and that traits taken to be explained by it could be explained by natural selection (Mayr 1972). At the same time, as we discuss in our second case study, it appealed to by other biologists. And in the 1970s evolutionary biologist Robert L. Trivers introduced Parental Investment Theory and argued that it explained and supported Darwin's account of sexual selection (Trivers 1972). We briefly consider it here as it is influential and a focus of feminist criticism.

Citing experiments involving fruit-flies undertaken in the 1940s that indicated greater variance among males in terms of reproductive success than among females, in 1972 Trivers prosed that differences in "parental investment" explained the experimental results and Darwin's account of male and female mating strategies. Trivers proposed that the sex that "invests most" in offspring will be more selective in choosing a mate than the sex that invests less (ibid.: 139). He argued that in most species, "the male's only contribution to the survival of his offspring is his sex cells." In contrast, "female contributions clearly exceed male and by a large ration" (ibid.: 141). Gametic dimorphism – differences in the size and number of sperm and eggs – Trivers argued, contributes to sexually differentiated mating strategies. Female "investment" in gestation and lactation in many species, and things like "egg-sitting" in others, further contribute to the asymmetry in Parental Investment (ibid.: 146).

Parental Investment Theory has been the subject of sustained critique by feminists.

Several challenges to Trivers' core assumption about what Sociobiologist David Barash describes as the cheapness of sperm relative to eggs (Barash 1979) have been offered. They include that males of many species produce accessory secretions in addition to sperm, that high sperm production is generally required for successful insemination, and that in humans and some other species, males produce millions of sperm for each egg a female produces. An empirically adequate comparison of Parental Investment, feminists have argued, would take these factors into account (e.g. Tang-Martinez 2000).

Nor, feminists charge, does field research bear out the theory's predictions about sex-differentiated mating and parenting strategies, as we discuss when we focus on primatology. But the literature citing counter-examples is extensive. Feminists have also pointed to the metaphorical and evaluatively thick concept of "parental investment," as well as of the descriptions of sperm as "cheap" and eggs as "expensive" (e.g. in Barash 1979). Each applies economic concepts to biological phenomena (e.g. Nelson 2017).

Finally, feminist biologists have reminded other biologists of the criteria that explanations invoking sexual selection need to meet.

> Uncritical acceptance of [sexual selection's] ubiquitous occurrence (e.g., in Trivers 1972) does no service to evolutionary theory. . . heritable variance in reproductive success, *based on characters that are not favored by natural selection*, must be demonstrable before sexual selection may be invoked unequivocally.
>
> *(Spencer and Masters 1992: 301, emphasis added)*

Primatology

"Primate" is an *order* of the *class* mammals, and includes humans as well as gorillas, chimpanzees, orangutans, langurs, and extinct species. The name of the order derives from the Latin *primus* meaning "prime," "most important," or "the highest" order. The great apes, a sub-classification of primates that includes humans, are more dependent on learned behaviors for survival, and for this and other reasons are generally viewed as the most intelligent primates. Primatologists study nonhuman primates in laboratories and field studies, and their research takes various forms. Some study the behaviors and social dynamics of one or more specific primate species or compare one or more groups of the same species. Others work to identify relationships between the behaviors and social dynamics of one or more primate species, on the one hand, and evolutionary selection pressures, on the other hand. And among primatologists who focus on evolutionary issues, some seek to relate the behaviors and social structures of the primates they study to the evolution of human behaviors and social systems.

The rationale for this last focus is relatively straightforward. Several primate species – gorillas and two species of chimpanzee – are more closely related to us than is any other living species. It is believed that we and they split from a common ancestor some six million years ago, and this is supported by genetics. Individual humans differ from one another in about 0.1% of their DNA. Comparisons between us and the two species of chimpanzees – so-called ordinary chimpanzees and bonobos – that focus on the same aspects of their genomes indicate that the differences between chimpanzee and human DNA are about 1.2%. Differences in gorilla and human DNA are about 1.6%.

Our common ancestry and genetic as well as morphological similarities with gorillas and chimpanzees seem to warrant research into their evolutionary trajectory to learn about our own evolutionary history and its outcomes. For some primatologists and evolutionary theorists, it has also seemed reasonable to assume that these closely related primates can provide insights into "a basic human nature" that lies underneath the "overlay" of human culture (e.g. Wilson 1975). For perhaps obvious reasons, the extent to which biological rather than social/cultural factors contribute to or cause human behavior and ways of organizing social life is far from settled. For example, feminists and others point to problems such as anthropocentric and androcentric accounts of nonhuman primates' behavior that often characterize this area of research. The case study we next consider illuminates such problems.

In the 1960s and 1970s, groups of anthropologists undertook long-term field studies of savannah baboons in the Serengeti National Park in Tanzania. Feminist primatologist Susan Sperling notes that the behaviors and structure of social life that the anthropologists described as characteristic of savannah baboons were understood by them as likely characteristic of most primate species. They proposed a "primate pattern" that resulted from evolutionary selection pressures common to primates (Sperling 1991). They viewed the behaviors and social dynamics they observed in savannah baboons as "adaptations" that promote the survival and reproductive success of members of the species, and the stability of the social structure.

Influential anthropologists Irven DeVore and Sheldon Washburn were among those to argue that "male dominance" was one of, if not the most important, features of savannah baboon social structure. It functioned, they maintained, to organize and control the group in ways like the role of political leadership in human societies (DeVore and Washburn 1963). At the time, the hypothesis that male dominance is a core feature of primate social life was widely accepted and a central focus of studies of other primate groups.

There was also interest in relating the data about savannah baboons to human behavior and social dynamics. The hypothesis that early humans evolved in the African savannah under conditions like those of savannah baboons was understood to make it likely that there are strong similarities in behavior and social structures of the two species. Sperling and other feminists note a general interest in sex and sex/gender differences in behavior in the research undertaken. The anthropologists emphasized similarities between male dominance in baboon groups and men's dominance over women, and similarities in what they described as "divisions in labor by sex" in the two species. Males, they maintained, provide for and protect females and offspring, and maintain the social structure that benefits all; females take primary responsibility for the care of offspring (DeVore and Washburn 1963 7; Fedigan 1992, 2011). Field reports about the baboons also emphasized what anthropologists described as "male aggression," another "trait" they attributed to men, and related to evolutionary selection pressures (Fedigan 2001).

Feminist primatologists criticized the emphasis placed on the nature and role of male dominance in primate groups, the lack of attention to females' behavior and its impact on social dynamics, and the imposition of gender stereotypes on many primates. They cited examples that indicate that primate species exhibit a variety of behaviors and forms of social life that do not fit "the primate pattern" (e.g. Hrdy 1986). Empirically adequate accounts of nonhuman primate behaviors and social dynamics, they argued, would include attention to and knowledge about female primates, including their contributions to the social dynamics of their groups beyond those involving reproduction (e.g. Altmann 1980; Fedigan 1982; Hrdy 1977, 1986).

Jeanne Altmann was one of the primatologists who chose to study females, and "to take 'the female point of view' as part of [a] feminist approach to primatology" (Fedigan 2001: 49). For several years, Altmann worked in a long-term field study of the Amboseli savannah baboons in Kenya. Researchers collected data about social behavior, ecology, and demography; and they constructed mathematical models in relation to individuals and the relationships between them.

In an interview with feminist primatologist Donna Haraway, Altmann said that her own "self-identity – as a scientist, feminist, and mother" – led her to focus on female baboons. "Increasingly," according to Altmann, "it was screaming at me. These are the most interesting individuals; [interactions between baboon mothers and offspring] have the most evolutionary impact; this is where the ecological pressures are" (Haraway 1989: 312). In her field reports, Altmann described baboon mothers as engaging in what she described as doing several things at once, "juggling" between competing priorities (Altmann 1980). In her interview with Haraway, she noted that her reports about female baboons did not include the category "child care" because she could not find a way to separate females' activities in ways that would call for it (Haraway 1989: 313). In 1980, Altmann wrote of her observations of these females "budgeting" the time devoted to many activities, and that their interactions with juveniles as well as females and males in their troop were key factors in the group's social dynamics (Altmann 1980).

Altmann also reported significant interactions between adult males and infants, and that females developed special relationships with some males who would go on to protect them from other baboons (Altmann 1980). As feminist primatologist Sarah Blaffer Hrdy describes Altmann's findings, "Infants, then, are often the focal-point of elaborate male-female-infant relationships, relationships that are often initiated by the females themselves" (Hrdy 1986: 20). In contrast to the focuses of earlier studies of savannah baboons we discussed, Altmann argued that the "high drama" characterizing encounters and relationships between male baboons is not nearly as important to the social dynamics of baboon groups as are the "micro-practices" of females (Altmann 1980).

Hrdy has written articles and books based on her studies of hanuman langurs, rhesus monkeys, and other primates (e.g. Hrdy 1986, 2009). She has also focused on female primates. As with Altmann's observations of female savannah baboons, Hrdy's findings about primates she studies stand in stark contrast to those once thought to be representative of the "primate pattern." For example, in 1986 Hrdy described the results of her and others' studies of female primates as indicating that "a polyandrous component [the term refer to females who mate with more than one male] is at the core of the breeding systems of most troop-dwelling primates" (Hrdy 1986: 125). Females, she noted, mate with many males, each of whom may contribute a little bit toward the survival of offspring. Hrdy cited female savannah baboons, langurs, and female of other primate species as engaging in such behavior.

Altmann's and Hrdy's research priorities are representative of the research and its outcomes undertaken by feminist primatologists (see e.g. Fedigan 2001 for an overview).

Ongoing Issues

The case studies we have considered are representative of many other feminist engagements with biology. These engagements have engendered debates. One such debate concerns whether the research on which feminists focus are cases of "bad" or unsuccessful science, and thus do not provide valuable insights into biology itself. The case studies we considered, and they are representative, concern feminists' critiques of what is, or was, overall good science: research and hypotheses well in keeping with accepted general hypotheses and priorities, methods, and norms, of the sciences in which they were undertaken or proposed. Scientists gave the evidence they *recognized* its due, but feminists argued that there was more and/or different evidence to be considered.

Another debate concerns arguments that feminism has led to changes in some biological sciences' assumptions about sex and gender, approaches to them, and methods and hypotheses. The arguments that feminism has made a difference have been offered about evolutionary biology (Gowaty 2003), primatology (Fedigan 2001), and developmental biology (The Biology and Gender Study Group 1988). One alternative explanation, favored by some biologists and philosophers, is that such changes simply reflect sciences' tendency to "self-correct." Another is that was simply a coincidence that an influx of women (and some men) whose views about gender were affected by The Second Wave of The Women's Movement occurred as changes in biology concerning gender began to take place.

The second explanation has convinced few. Sarah Blaffer Hrdy's response, which focuses on changes in primatology, is representative of feminists' skeptical view of coincidence explanations.

> I seriously question whether it could have been just chance or historical sequence that caused a small group of primatologists in the 1960s, who happened to be mostly male, to focus on male-male competition and on the numbers of matings a male achieved, while a subsequent group of researchers, including many women (beginning in the 1970s) started to shift the focus to female behaviors having long-term consequences for the fates of infants.
>
> *(Hrdy 1986: 136)*

The issues involved in the explanation crediting scientists with an ability and tendency to "self-correct" are more complicated. For one thing, scientific hypotheses and theories do change, and are even abandoned, in response to new evidence. But in the cases we have considered and there are many others, it was not scientists who developed androcentric hypotheses who eventually recognized them as problematic, as the passage from Hrdy addresses.

In addition, we have considered examples in which biologists credit feminism with enabling them to recognize androcentrism and other problems related to gender characterizing research in their fields. Altmann's and Hrdy's descriptions of feminism's impact on their research are paralleled in the accounts of other feminist biologists. But there are women working in fields such as evolutionary biology, developmental biology, and primatology in which it is generally acknowledged that approaches to gender have undergone significant change, who reject explanations of such changes that link them to feminism (as noted, respectively, in Gilbert and Rader 2001; Fedigan 2001).

Primatologist Linda M. Fedigan describes the views that underly the arguments women in primatology, who do not recognize a role for feminism in the field. Fedigan suggests that, despite the contributions of researchers who self-identify as feminists, feminism is perceived by many primatologists, including some women, as "outside" science, not "within" it. Relatedly, many continue to understand feminism as a purely political and value-laden perspective that is incompatible with good science (Fedigan 2001: 66).

These debates are likely to continue, but the kinds of changes that have led to them continue as well (Schiebinger 1999).

Looking Forward: Interest in Socially Relevant Philosophy of Science

Interest among some philosophers of science in making their discipline "more socially relevant" is reflected in a special issue of the philosophy journal *Synthese* devoted to the topic. (*Synthese* 2010). The editors of the issue as well as many contributors are feminist scientists and philosophers of science. A core goal of those interested in making philosophy of science more socially relevant is to find ways to foster and promote socially responsible science – science that is reflexive, recognizing that it may reflect unwarranted socially relevant assumptions, and impact social beliefs and policies in ways that are harmful to some (see Kourany in this volume for detailed discussion of the notion). The introduction to the special issue called for exploring ways in which philosophy of science "can provide social benefits, as well as benefits to scientific practice and philosophy itself" (Plaisance and Fehr 2010: 301).

Such "expansion" was not seen as involving the abandonment of some of the discipline's emphases, but reflected interest in how philosophers can broaden their analyses of epistemological issues, for example, in ways that are relevant to the impact of scientific research on the lay public and in ways that can contribute to scientists' abilities to positively engage with the public affected by their research. The editors of the issue argued, and contributors offered similar arguments, that socially relevant philosophy of science is pluralistic; "it includes philosophical engagement with scientific research on socially relevant topics, [and] philosophical activities that attend to the interactions among scientists and various communities that contribute to and are affected by scientific research" (ibid.: 302).

Contributors emphasized the need to identify ways in which interactions between scientists and those who their research will impact can foster trust, knowledge on the part of scientists of the perspectives of such publics about potential impacts of their research, and knowledge among the public of the rationale and assumptions of the research in question.

More recently, The Consortium for Socially Relevant Philosophy of/in Science and Engineering (SRPoiSE) was founded. Its members include feminists who were instrumental in introducing the topic; but the word "in" in the name of the consortium reflects that its members include scientists and engineers who engage these issues in their own research and fields that are socially relevant. The mission statement of SRPoiSE reflects these interests and efforts.

> This consortium supports, advances, and conducts philosophical work that is related to science and engineering and that contributes to public welfare and collective wellbeing. We aim to improve the capacities of philosophers of all specializations to collaborate and engage with scientists, engineers, policy makers, and a wide range of publics to foster epistemically and ethically responsible scientific and technological research.
>
> We are particularly interested in addressing complex social and environmental problems and in fostering the ability of researchers in science and engineering to do so as well. We seek to understand and ameliorate conceptual and institutional barriers to collaborative research across these groups.
>
> (*SrPo/iSE, http://srpoise.org/, accessed December 11, 2019*)

The consortium has hosted conferences attended by scientists, engineers, philosophers, and members of other academic disciplines.

These are just two of the developments that reflect interest in collaborative projects to address the social impact and social implications of scientific research. Increasingly feminists' interests in relationships between ethics and science align with those of many other scientists and philosophers. Included in these concerns are values in science, scientific fallibility, and the consequences of scientific research for human well-being and public policy. Although it is not possible to predict what directions scientific and philosophical attention to these issues will take in the future, it seems likely that scientists and science scholars, feminists and non-feminists, will find ways to work collaboratively in their efforts to bring about socially responsible science.

This is a positive development in light of the concerns feminists raise, and we have considered, about research in areas of biology.

Related Chapters: 5, 15, 17, 18.

References

Altmann, J. (1980) *Baboon Mothers and Infant*, Cambridge, MA: Harvard University Press.

Anderson, E. (2004) "Uses of Value Judgments in Science: A General Argument with Lessons from a Case Study of Feminist Research on Divorce," in Nelson, L. H. and Wylie, A. (eds.) *Hypatia* 19 (1): Special Issue on Feminist Science Studies, pp. 1–24.

Barash, D. (1979) *The Whisperings Within*, New York: Harper & Row.

The Biology & Gender Study Group. (1988) "The Importance of Feminist Critique for Contemporary Cell Biology," *Hypatia* 3 (1): 61–76.

Bleier, R. (1984) *Science and Gender: A Critique of Biology and Its Theories on Women*, Oxford: Pergamon Press.

Bluhm, R., Jacobson, A. J. and Maibom, H. L. (eds.) (2012) *Neurofeminism: Essays at the Intersection of Feminist Theory and Cognitive Science*, New York: Palgrave Macmillan.

Darwin, C. (1859) *On the Origin of Species*, London: John Murray. Reprinted Wildside Press (2003).

———. (1871) *The Descent of Man and Selection in Relation to Sex*, London: John Murray.

DeVore, I. and S. L. Washburn (1963) "Baboon Ecology and Human Evolution," in Howell, F. C. and Bourliere, F. (eds.) *African Ecology and Human Evolution*, Chicago, IL: Aldine Press, pp. 355–367.

Fausto-Sterling, A. (1985) *Myths of Gender: Biological Theories about Women and Men*, New York: Basic Books.

Fedigan, L. M. (1882) *Primate Paradigms: Sex Roles and Social Bonds*. Fountain Valley, CA: Eden Press.

——— (1992) *Primate Paradigms: Sex Roles and Social Bonds*. 2nd Edition. Chicago, IL: University of Chicago Press.

——— (2001) "The Paradox of Feminist Primatology: The Goddess Discipline," in Creager, A. N., Lunbeck, E. and Schiebinger, L. (eds.) *Feminism in Twentieth Century Science, Technology and Medicine*, Chicago, IL: University of Chicago Press.

Fehr, C. (2011) "Feminist Philosophy of Biology," in Zalta, E. N. (ed.) *The Stanford Encyclopedia of Philosophy* (Fall 2011 Edition), http://plato.stanford.edu/archives/fall2011/entries/feminist-philosophy-biology/

Gilbert, S. and Rader, K. A. (2001) "Revisiting Women, Gender, and Feminism in Developmental Biology," in Creager, A. N., Lunbeck, E. and Schiebinger, L. (eds.) *Feminism in Twentieth Century Science, Technology and Medicine*, Chicago, IL: University of Chicago Press, pp. 73–97

Gowaty, P. A. (2003) "Sexual Natures: How Feminism Changed Evolutionary Biology," *Signs* 28 (3), pp. 901–921.

Haraway, D. (1989) *Primate Visions: Gender, Race, and Nature in the World of Modern Science*, Abingdon: Routledge.

Hempel, C. (1966) *Philosophy of Natural Science*, New York: Prentice Hall.

Hrdy, S. B. (1977) *The Langurs of Abu: Female and Male Strategies of Reproduction*, Cambridge, MA: Harvard University Press.

———. (1986) "Empathy, Polyandry, and the Myth of the Coy Female," in Bleier, R. (ed.) *Feminist Approaches to Science*, New York: Pergamon Press, pp. 139–152.

——— (2009) *The Woman That Never Evolved*, Cambridge, MA: Harvard University Press.

Hubbard, R. (1983) "Have Only Men Evolved?" in Harding, S. and Hintikka, M. (eds.) *Discovering Reality: Feminist Perspectives on Epistemology, Metaphysics, Methodology, and Philosophy of Science*, Dordrecht: D. Reidel, pp. 45–70.

Martin, E. (1991) "The Egg and The Sperm: How Science Has Constructed a Romance Based on Stereotypical Male-Female Roles," *Signs* 16 (3), pp. 485–501.

Mayr, E. (1972) "Sexual Selection and Natural Selection," in Campbell, B. (ed.) *Sexual Selection and the Descent of Man*, London: Heinemann, pp. 87–104.

Nelson, L.H. (2017) "Feminist and Non-Feminist Philosophy of Biology," in Amoretti, M. C. and Vassallo, N. (eds.) *Meta-Philosophical Reflections on Feminist Philosophy of Science*, New York, Dordrecht, and London: Springer, pp. 55–74.

Plaisance, K. and Fehr, C. (2010) "Socially Relevant Philosophy of Science: An Introduction," *Synthese. Special Issue: Making Philosophy of Science More Socially Relevant* 177 (3), pp. 301–316.

Ruse, M. (2012) *The Philosophy of Human Evolution*, Cambridge, UK: Cambridge University Press.

Schiebinger, L. (1999) *Has Feminism Changed Science?* Cambridge, MA: Harvard University Press.

Sober, E. (1993) *Philosophy of Biology*, Boulder, CO and San Francisco, CA: Westview Press.

Spencer, H. and Masters, J. (1992) "Sexual Selection: Contemporary Debates," in Keller, E. F. and Lloyd, E. A. (eds.) *Keywords in Evolutionary Biology*, Cambridge, MA: Harvard University Press, pp. 288–304.

Sperling, S. (1991) "Baboons with Briefcases: Feminism, Functionalism, and Sociobiology in the Evolution of Primate Gender," *Signs* 17 (1), pp. 1–27.

Synthese. (2010) *Special Issue: Making Philosophy of Science More Socially Relevant* 177 (3), K. Plaisance and C. Fehr (eds.). New York: Springer International.

Tang-Martinez, Z. (2000) "Paradigms and Primates: Bateman's Principle, Passive Females, and Perspectives from Other Taxa," in Strum, S. C. and Fedigan, L. M. (eds.) *Primate Encounters: Models of Science, Gender, and Society*, Chicago, IL: University of Chicago Press, pp. 261–274.

Trivers, R. (1972) "Parental Investment and Sexual Selection," in B. Campbell (ed.) *Sexual Selection and the Descent of Man*, Chicago, IL: Aldine-Atherton, pp. 136–179.

Wilson, E. O. (1975) *Sociobiology: The New Synthesis*, Cambridge, MA: Harvard University Press.

Suggested Further Readings

Lloyd, E. A. (1993) "Pre-Theoretical Assumptions in Evolutionary Explanations of Female Sexuality," *Philosophical Studies* 69 (2/3), pp. 139–153. An influential article arguing that the assumption that women's sexuality is fundamentally linked to reproduction is bases on social beliefs and agendas, rather than evidence.

Longino, H. and Doell, R. (1983) "Body, Bias, and Behavior: A Comparative Analysis of Reasoning in Two Areas of Biological Science," *Signs* 9 (2), pp. 106–127. An accessible and influential feminist analysis of reasoning about gender in biological science.

Nelson, L. H. (2017) *Biology and Feminism: A Philosophical Introduction*, Cambridge, UK: Cambridge University Press. An accessible and balanced account of philosophical issues raised in and by feminists' engagements with areas of biology.

21

OBSERVING PRIMATES

Gender, Power, and Knowledge in Primatology

Maria Botero

While observing a group of chimpanzees in the wild, at Gombe National Park, I was attacked by Frodo, one of the most famous chimpanzees of the Kasakela community. Frodo's long list of violent deeds included assaults on primatologist Jane Goodall, "Far Side" cartoonist Gary Larson, and several filmmakers and researchers. Almost everyone who did research at the Kasakela community during Frodo's lifetime has a Frodo-attack story. In my case I was observing a group of chimpanzees and Frodo went out of his way to push me to the ground. I got to see his majestic and terrifying face from a few inches away. Then, as if nothing had happened, he moved away without harming me. After I got up from the ground, several of the local researchers (who are permanent data collectors on the site) said "pole" (pole can be translated roughly as "sorry" and is frequently used to express compassion) and looked at me with smiles and concern. I felt I gained stature in their eyes, a kind of a "badge of honor" after the attack.

I was told by other researchers that, given that chimpanzees are a dimorphic species (e.g. males and females can be distinguished based on external physiological characteristics), chimpanzees also are adept at spotting sex differences in humans. What is even more interesting, they told me, is that chimpanzees use these differences in including human researchers in their social interactions. It has been argued that male chimpanzees tend to engage in a series of aggressive displays and encounters to gain higher ranking in their groups and social status. From the beginnings of primatology (see for example Goodall 1986), it has been argued that this social hierarchy strategy frequently starts with aggressive attacks on low ranking female chimpanzees and juveniles, then, moving upward, the males will start attacking higher ranked female and male chimpanzees in their group. In some cases, as was described by Goodall, it has been observed that human females are used as "pawns" in these social interactions.

This experience was interesting from a feminist perspective. First, it made me think about sex differences and about the role these differences play in power relationships and about how these differences transverse human and nonhuman primates. Second, it illustrated how these sex differences can translate into gender and race issues in the way researchers interact with each other, and how pre-conceived ideas of sex and gender may play a role in the way we understand nonhuman primates. These kinds of observations are not uncommon in the field of animal behavior. In this chapter, I show how some of the main feminist critiques are applicable to the observation of nonhuman animals. I focus on one of the issues described in feminist critiques of science: when scientists represent themselves as neutral, they fail to recognize the ways their values have shaped their inquiry. This, in turn, prevents a critical analysis of these values and an understanding of new aspects of the world and new perspectives in their area of study. I focus on examples of observations of primates in the wild as a way of illustrating these ideas.[1] I argue that, in the history of primatology, it is possible to observe that openness to diverse approaches has allowed researchers to understand primates in new and more complex ways. This chapter does not attempt to provide an exhaustive exploration of feminist critiques; rather, the aim is to explore some of these ideas and describe how they account for the ways nonhuman primates have been observed in primatology.

In what continues I will examine the relationship between primatology and feminist critiques of science; primatology and feminist critiques have a rich history in which each discipline has had an impact on the other. Several of the main points of this history can be summarized in two main areas: the relationship between primatology and different conceptions of human nature and how primatology has often been described as a "feminist science."

First, I will focus on fundamental role that primatology has played in our understanding the evolution of humans, in particular, in creating a notion of human nature. Much feminist criticism of science consists in exposing the androcentric and sexist biases in scientific research, especially in theories about women, sexuality, and gender differences. Since the beginnings of primatology, the history of the study of human evolution has been permeated by accounts of sex, gender, and race. For example, Darwin argued that "the average standard of mental power in man must be above that of women" for that reason he concludes that "man has ultimately become superior to woman" (Darwin 1871: 327–328). Primatology has played an essential role in this discussion, examining the categories used to define human and nonhuman nature. As Haraway has argued, primatology "is a discourse on first principles; and it is a major practice for western 20th century people to construct and negotiate the boundaries of human and animal, gender and sex, west and other, culture and nature, whole and part" (1984: 493).

Early on, research in primatology seemed to support ideas similar to the ideas argued by Darwin through descriptions of sex differences in nonhuman primates. For example, many of the primate studies conducted in the 1960s and 1970s identified a more-or-less linear dominance hierarchy, consisting of males at the top and females in lower ranking positions (Lancaster 1973). These issues continued to be perpetuated in primatological accounts of sex differences and reproduction. For example, according to Robert L. Trivers (1972), exaggerated characteristics in apes are most often seen in males. He argues that the reasons are that males have little to do with offspring rearing and that their efforts are directed toward obtaining sexually receptive females. Meanwhile, we do not find this radical dimorphism in females because they are predisposed to make a greater investment in the production of offspring and the rearing of that offspring. Thus, Trivers concludes that competition for these receptive females may have influenced sexual dimorphism. This perspective began to be challenged in the late 1970s and 1980s when more women entered the field of primatology. For example, one of the most famous examples is the compilation edited by Meredith Small, *Female Primates: Studies by Women Primatologists* (1984); Small argues that

> When scientists first decided to study the social organization of monkeys and apes in their natural habitats, the emphasis was primarily on the behavior of males. It was widely assumed that males determined the social structure of the group. Missing from most of the early studies was any serious consideration of the activities of females. *Female Primates: Studies by Women Primatologists* presents original data on a wide variety of female primates, with articles ranging from parenting behavior to old age. This volume is also an acknowledgment of the large number of women who have focused their research on primates.
>
> *(Small 1984, dust jacket)*

Sarah Blaffer Hrdy (1981) argued in a similar way that continuous reports of "competitive males and mothering females" are currently "entrenched" in academia; however, beginning in the 1970s, academics became increasingly aware that evolutionary forces worked not only on males but on females as well. As Hrdy describes, primatologists began to focus on cooperation and competition among and between both sexes. Fedigan (1983) was also instrumental in denying the notion that nonhuman female primates are passive "objects" and that a female's best reproductive strategy is to choose a more "dominant" male. Fedigan notes that these assumptions were developed in the absence of systematic research on the actual choice of sexual partners despite the fact that female choice is central to Darwin's original model of sexual selection.

In short, primatology can be used as an example of how the research of female researchers (e.g. the assumptions they made visible, the different questions they raised and pursued, the evidence they generated) resulted in support for a different conception of both human and nonhuman primates. The research of these female researchers can be characterized as an example of feminist empiricism (Anderson 2004; Longino 1990; Nelson 1990). First, the kind of scientific knowledge generated by these researchers is *contextually* situated − it demonstrates how the aims, cognitive values, and methods used by primatologists can vary depending on the kind of research they are interested in pursuing. Second, the kind of research resulted by these researchers is *socially* situated −it demonstrates how cognitive values depend on social ideas such as assumptions about the behavioral characteristics of females and males. It is important to notice that there are several definitions of feminist empiricism (for an analysis of key differences and overlapping ideas on the role that empiricism plays in feminist empiricism and standpoint feminism, see Intemann 2010). I will argue, in what continues, that primatology has adopted a variety of perspectives. I will focus on diverse empiricist and contextualist feminist epistemological views to show that including women in primatology creates new areas of research based on women's experiences, thus improving the empirical methods used in primatology.

I will now focus in a second aspect of the relationship between primatology and feminist critiques of science, how primatology has been described a "feminist science." Primatology has been a locus of attention for feminist critiques that focus on the roles that gender, politics, and race play in the observation of nonhuman primates. Several influential books including *The Woman that Never Evolved* (1981) by Sarah Blaffer Hrdy and *Primate Paradigms* by Linda Marie Fedigan (1992) changed the way primates are studied. As a result several authors have used primatology as an example of a feminist science or, at least, as the focus of feminist critiques. One of the best-known examples is the critical examination of the observation of primates presented in *Primate Visions* by Donna Haraway (1989). However, the relationship between these kinds of critiques and primatology is a complex one. This complex history can be illustrated through the example of the reception that *Primate Visions* received when it was published; several feminist scholars praised Haraway's attempt to illustrate how gender, politics, and race play pivotal roles in the creation of knowledge in primatology. For example, Fausto-Sterling (1990) argued that *Primate Visions* is one of the most important books to come along in the last twenty years; meanwhile the reception of primatologists was generally not positive one; as Fedigan (2005) argues, "infuriating" is an adjective frequently found in reviews of the work of animal behavior researchers.

Moreover, many of the leading figures in primatology have been female. Perhaps the most famous of these are Jane Goodall, Diane Fossey, and Birute Galdikas, whose research focused, respectively, on the three great apes: chimpanzees, gorillas, and orangutans. These researchers put primatology in the spotlight as a science that was led by female researchers. It is important to acknowledge that there are many other important female primatologists who led significant changes in the field of primatology; focusing on these leading figures is not meant to ignore the role of many other primatologists in the field. The reason I focused on these three figures is because of their impact on the general public at the time and because of the way they embodied the challenges faced by female scientists in the 1960s and 1970s.

The gender of these leading figures has created a sense of hope; primatology has been described as a model of an "equal-opportunity" discipline because of the common perception that women are better represented in primatology than in other similar fields. For example, Schiebinger (2001) estimated that women made up 80% of graduate students pursuing PhD in primatology, up from 50% in the 1970s. Because of the large number of women, Schiebinger has even asserted that "Primatology is widely celebrated as a feminist science." However, this might be a case of false hope that has clouded the story. As Addessi, Borgi, and Palagi (2012) found that even though the proportion of women occupying academic positions in biological sciences has increased in the past few decades (in the past 15 years, the proportion of female primatologists increased from 38% of the early 1990s to 57% of 2008), women are still under-represented in senior academic ranks, compared with their male colleagues. Arguing that primatology suffers the phenomenon of "glass ceiling," the authors show that even if primatology does attract more female students than male students, at the full professor level, male faculty members significantly outnumber female faculty. Addessi, Borgi, and Palagi also showed that male members of international primatological societies publish significantly more than their female colleagues, and that the scientific achievements of female primatologists (in terms of number and type of publications) do not always match their professional achievements (in terms of academic position). However, Addessi, Borgi, and Palagi also found that the fewer articles female primatologists' have had greater impact than those by their male colleagues.

As described, the relationship between primatology and feminist critiques of science can be examined from multiple perspectives. In this chapter I focus on one of these aspects, the work of some of the pioneers of primatology, and show how their methods are in line with some of the ideas of feminist epistemology and the feminist critiques of objectivity.

In what follows I focus on some of the aspects of these primatologists lives as researchers and on the impact they had on the study of primates, and I use these examples as illustrations of some of the ideas brought forward by feminist critiques of science. This account is not meant to be exhaustive but rather to give the reader an overview of some of these critiques and to illustrate these critiques with the work of these primatologists.

The first example I will examine is the role that Louis Leakey played in the development of primatology when he deliberately chose to support three female researchers to conduct studies in great apes, the *Trimates*, a term coined by Leakey to refer to Jane Goodall, Diane Fossey, and Birute Galdikas. This decision impacted what we know about apes and how we study apes, and, as described earlier, it brought hope to generations of women who wanted to become primatologists. However, as Haraway (1989) warned us, the way this narrative is presented can radically affect the way we understand the role of women in science.

Leakey choose to support women in their research of apes because women are "more sensitive to mother-infant relations, are less likely to arouse aggression in males," and, unlike trained scientists, tend not to "see too much" (Hayes 1990: 117). As argued earlier, Leakey's decision advanced the field of primatology by opening the doors to women; however, his characterization of female primatologists is problematic from a feminist standpoint. First, as Crasnow (2008) argues, women are not automatically epistemically privileged. To illustrate this point, let's take a look at one of the earliest critiques of evolutionary accounts of gender. Antoinette Brown Blackwell argued in 1875 that Spencer and Darwin were wrong in considering that evolution had selected men as an organism superior to females. She argued that, as a woman, she was able to provide a perspective that differed from theirs: "Only a woman can approach the subject from a feminine standpoint, and there are none but beginners among us in this class of investigations" (Brown Blackwell 1875: 22). It is important to notice that what provides this characteristic point of view for Brown Blackwell is not being a woman per se but the experiences that she has as a woman in that particular period: "Experience must have more weight than any amount of outside observation" (Brown Blackwell 1875: 7). Moreover, this perspective must be based on observation; as she argues: "It is to the most rigid scientific methods of investigation that we must undoubtedly look for a final and authoritive decision as to a woman's legitimate nature and functions" (Brown Blackwell 1875: 231–232). Thus, Brown Blackwell embodies what I am trying to demonstrate with these three primatologists. As will be shown in what continues, there is something about the experience of being a female researcher that allows these researchers to adapt methods, questions, and ways of approaching the nonhuman subjects that will add to the complexity of nonhuman animal science. However, what these primatologists added to our knowledge of nonhuman primates cannot be reduced to their gender; being a woman may have contributed to the ways in which each changed the way we understand nonhuman primates, but that is not a sufficient condition to explain the advances brought forward by their epistemic privilege.

Second, Leakey's characterization of the advantages of including women in primatological research is an abstraction that makes uniform the different academic backgrounds of each of these primatologists (before they engaged in animal observation[2]) and ignores how particular backgrounds contributed to their research. Each of their individual backgrounds was a fundamental piece of their epistemic privilege; for example, it is possible to conclude that Fossey's work as an occupational therapist (OT) with children with language delays influenced the way in which she approached and imitated gorillas and facilitated her observations. OTs working with children frequently provide movement activities such as swinging, crashing onto huge bean bags, and jumping on trampolines. They may use wide chewable pencil tops and jewelry to provide calming input that helps children sit and focus.[3] Fossey's strategy for habituating gorillas included a similar approach; she was able to get quite close when she "knuckle-walked." She would also chew on celery when she was near the groups, drawing them closer to her.

Third, several accounts recall Leakey as "choosing" these primatologists as researchers; however, upon close examination of these accounts, we can observe that each of these women made with great sacrifices to pursue these positions. For example, Fossey's determination to have this research position included having an unnecessary appendectomy before her research (this was just a test, designed by Leakey, that, to his surprise, Fossey took seriously). Finally, by defending his choice through this characterization of women, he reinforced traditional female gender images: passivity, sensitivity, and lack of professional training. Thus, Leakey's characterization of these primatologists is a reductive view that ignores how the different characteristics and methodological choices of these researchers would influence their approaches to apes.

However, it is important to acknowledge that there were also particular power dynamics that occurred in their histories as researchers, dynamics that came about because they were women. For that reason, in what continues I will focus on how the personal journeys of these researchers can serve as illustrations of power dynamics.

The personal journeys of Jane Goodall, Dian Fossey, and Birute Galdikas provide historical accounts of the role that power and oppression played for female researchers at that time. For example, their personal lives were scrutinized in great detail in ways that the personal lives of male scientists were not scrutinized. For example, Jane Goodall's divorce and marriage was the subject of a People magazine article.[4] She was described in the local newspapers in the 1970s as "pert scientist" and a "comely Miss," and as Goodall recalls, at the time it was often claimed that her fame was due to her legs.[5] Diane Fossey's personal life, in particular her romantic life, was heavily scrutinized; for example, in Harold Hayes' biography of her, *The Dark Romance of Dian Fossey*, it is possible to find several unprofessional references to Fossey's appearance (Hayes 1990: 57, 169, 217, 246) and an index entry of "romantic affairs of" (with a total of 56 page references). This is in accordance with the way female researchers were portrayed at the time; for example, Bruno Bettelheim claimed that "as much as women want to be good scientists . . . they want first and foremost to be womanly companions of men and to be mothers" (Bettelheim 1965: 15).

Their personal journeys are also an illustration of the history of violence to women. For example, Dian Fossey (De la Bédoyère and Fossey 2005; Fossey 1983) fought valiantly for the conservation of gorillas, and as a result, she was subjected to extreme violence in response for her conservation efforts; she was raped and finally violently murdered.[6] Birute Galdikas, concerned by the rapid clearing of orangutan habitat, has also engaged in conservationist projects. She has lobbied the Indonesian government to set aside parks and to curb illegal logging and orangutan trading, but as a result, she has been threatened, harassed, and even kidnapped by those who oppose her work.[7]

In what continues I narrate this history in a different way, focusing on how some of the practices of these primatologists changed the ways of the science of animal behavior, in particular, our understanding of the demarcation of human and nonhuman animals. One of the radical changes that Fossey and Goodall introduced to primatology was the adoption of revolutionary methods in the observation of animal behavior. These new methods were criticized for being anthropomorphic and lacking detachment. However, as I will argue, these very characteristics contributed to a process conducive to better understanding of nonhuman primates.

The way Fossey observed gorillas was radically different than the way previous researchers had observed the behavior of gorillas. First, as Fossey (1983) described, she observed the animals from a distance, silently, hidden, but after a few months, contacts between Fossey and the gorillas helped her "win the animals' acceptance." Fossey did not announce her presence to the gorillas; she adopted the gorillas' movements and habits. She "imitated their contentment vocalizations . . . [and] crunched wild celery stalks. She crouched, eyes averted, scratching herself loud and long, as gorillas do" (Montgomery 2009: 53). Even during the rare times she used binoculars, she "always wrapped vines around [them] in an attempt to disguise the potentially threatening glass eyes from the shy animals" (Fossey 1983: 11). She realized that standing upright "increased the animals' apprehension." That discovery marked the beginning of "[her] knuckle-walking days" (Fossey 1983: 13). Fossey never wanted to challenge the gorillas with her presence; she claimed that "any observer is an intruder . . . and must remember that the rights of that animal supersede human interests" (Fossey 1983: 14). Goodall adopted a similar methodology, observing from a distance and approaching them slowly, in gentle ways, until they tolerated her presence. Part of this methodology included giving the chimpanzees names instead of the numbers that were customary at the time.

These novel methodological approaches were received as anthropomorphic biases. Fossey was viewed as "victim of anthropomorphism, the attribution of human characteristics to animals" (Hayes 1990: 140). This anthropomorphic stance is frowned upon because such an "unscientific attitude" results in human reason undermined by human feeling (Hayes 1990: 140). Similarly, in the 1960s the Royal Society rejected of Jane Goodall's early papers on the grounds that they referred to chimpanzees as 'he' and 'she' rather than 'it' (Montgomery 2009).

To understand the significance of this change in methods and of the charges of anthropomorphism, it is necessary to understand the general role that anthropomorphism plays in the observation of nonhuman animals. Since the beginnings of the study of animal behavior, researchers have been preoccupied with identifying the ideal conditions under which they can provide a proof (or at least verify) that animals possess cognitive capacities. One of the most influential ways in which comparative psychologists have attempted to maintain rigor when ascribing mental capacities to animals is through what is known as Morgan's Canon.[8] This canon originated with a statement made by Conwy Lloyd Morgan, who argued that

> [i]n no case may we interpret an action as the outcome of the exercise of a higher psychical faculty, if it can be interpreted as the outcome of the exercise of one which stands lower in the psychological scale.
>
> *(Morgan 1894: 53)*

There are several interpretations of this statement (see Fitzpatrick 2008). In the most prevalent, it is argued that following Morgan's Canon means that, when confronted with several explanations for animal behavior, researchers must adopt the explanation that uses "less complex processes" (such as associative learning) as opposed to explanations that involve "more complex cognitive capacities" (such as symbolic or rule-based reasoning and declarative knowledge) (see for example Crystal and Foote 2009; Dwyer and Burgess 2011). In other words, when researchers adopt Morgan's canon, they use it as a normative claim that where presented with two competing explanations of an observed animal behavior that are *equally plausible*, researchers in *all cases* must adopt the *simplest* lower-process explanation. The argument is that this is a way to achieve objective knowledge in the study of animal behavior.

In line with feminist critiques (Harding 1991, 1998; Longino 1990, 2001; Potter 2001; Wylie 1997) when scientists represent themselves as neutral in their research (denying that their own values could influence their research), this denial blocks their recognition of the ways their values have shaped their inquiry. In other words, denying the role that values play in our acquisition of knowledge prevents the exposure and critical scrutiny of these values. This kind of denial can be observed in the application of Morgan's Canon: researchers follow it as a normative stance, and there is no scrutiny of the reasons that the simplest explanation must always be chosen. As Mikhalevich (2018) argues, what lies underneath this Canon is a kind of bias, a "simplicity heuristic." That is, scientists following this canon are advised, for the sake of simplicity, against developing alternative, more complex, explanations of cognitive models. However, as Mikhalevich argues, no clear arguments are given to justify why simplicity is preferable.

Morgan's Canon gains power as a normative stance for the following reason: it is a defense against a perceived threat of anthropomorphism. Several authors (see for example Burghardt 1985; Sober 1998) have argued that Morgan viewed his fellow researchers' interpretations of animal behavior as biased toward anthropomorphism and thought that this bias needed to be counterbalanced. This fear of anthropomorphism is adopted as a value that stems from an unexamined presupposition that human and nonhuman animals must be different and that humans are cognitively superior to nonhuman animals. Indeed, it seems that it is not a matter of preferring "simpler explanations" but a matter of how we are defining "simpler." Associated behavior is only "simpler" **if** one assumes that the alternative explanation would require us to revise other beliefs (e.g. human superiority). Moreover, upon close examination of the use of this canon, we realize that behind this use of Morgan's canon as a normative stance, there are a series of values and interests including the justification of not attributing cognitive capacities to nonhuman animals, so that these nonhuman animals can be used in animal experimentation (see Botero 2019).

In short, Morgan's Canon is positioned as a way of obtaining a "neutral" and "objective" view of animal behavior. This is part of a long tradition in the study of animal behavior, in particular, comparative cognition, where universal categories such as human and nonhuman animals and "lower" and higher" cognitive capacities are used to enforce normative accounts that guarantee "objective," "real," or "non-biased" observation of animals. However, adopting these standards of "neutrality" and "objectivity" through Morgan's Canon has contributed to blurring the recognition of the ways the researchers' values such as the fear of anthropomorphism or the advance of biomedical research have shaped researchers' inquiry, and thereby preventing the exposure of these values to critical scrutiny.

A similar attitude can be observed in terms of attachment to the subjects observed. Fossey argued that her "*intentions* to remain a detached scientific observer dissolved" (Fossey 1983). Most other scientists who worked alongside Fossey (at her Karisoke Research Center in the Virungas or back in the United States) objected to her research methods. Several of her critics argue that "Dian failed to learn the most important rule of male-dominated empirical science: the rule of separation, of distance from her study subjects" (Montgomery 2009: 144). Other critiques focused on the fact that "she treated [the gorillas] as friends. . . . She had personal likes and dislikes of the individual animals . . . [which] affected her interpretation of their behavior" (Hayes 1990: 292). These critics are starting from the perspective of detachment as an ideal form of detachment as objectivity, according to which good scientists should adopt an emotionally distanced, controlling stance toward their objects of study.

However, despite these kinds of criticism, the attitude adopted by researchers like Fossey and Goodall toward the subjects of research has been accepted by other disciplines. In ways akin to anthropological data gathered through *participant observation* (e.g. researchers live with and participate in the activities of the members of the group studied, while observing and recording these activities), the research methods proposed by Fossey and Goodall are closer to ethnographic methods. Adopting this kind of ethnographic methods entails the importance of including both the observed behavior of the nonhuman primates and the primatologist's personal reflections while conducting these observations. Within this perspective, it is acknowledged that everything a researcher observes is filtered through their personal and academic background, and, as such, it is important to be aware of how these elements play roles in shaping the methods used for observation.

This approach is in line with what several feminist authors have argued on the critical epistemic role that emotions serve because they help to guide observers to evaluate relevant features of the world that otherwise would not be noticed (Anderson 2004; Jaggar 1989; Little 1995). For example, according to Anderson, "emotional experiences" (e.g. experiences of persons, things, events, or states of the world through affective states) can serve as evidence as value judgments whose cognitive components can be tested for representational adequacy. Following Anderson, the emotional experience of a researcher who observes nonhuman animals, or is choosing among different concepts or hypotheses, can be used as part of the evidence. These kinds of emotional experiences have epistemic value that accepts individuality and differences in the subjects observed; the inclusion of this diversity can potentially help researchers discover a wider range of evidence. This does not mean that factual and value judgments play the same role in research. These kinds of emotional value judgments can guide a researchers' choice of hypothesis or research methods; however, ultimately, only empirical research can help researchers evaluate the cognitive and representational elements embedded in these emotional experiences. In other words, what Anderson and other authors are showing us is that a "feeling for the [individual] organism" may sensitize a scientist to critical data (Keller 1983; Ruetsche 2004). This is true for both Fossey's and Goodall's innovative methodologies. Fossey's discoveries about the mountain gorilla were many, but one of the most important was the gentleness of the animals' treatment of each other (and to Fossey), a discovery that differs from previous accounts. In addition, Fossey realized that her subjects had "strong bonds of kinship [which] contributed toward the cohesiveness of a gorilla family unit" (Fossey1983: 105). Goodall contributed to our understanding of chimpanzees in observing for the first time their use of tools and of hunting behavior (Goodall 2010).

Similar approaches can be found in several primatologists. Rees (2007) conducted a textual analysis of 11 popular accounts published from 1964 to 2001 and found that primatologists produced accounts of their research, for non-specialist audiences, that treated their research subjects in "anthropomorphic" ways, attributing to them individual personalities, with their own histories, motives, and agendas, and conveying a sense of inter-subjectivity between researcher and subject. Rees found that these tendencies appeared to deepen over time: as the primatologists gained more experience observing nonhuman primates, their description of nonhuman primates seemed to become more "person-like." Moreover, according to Rees, researchers portray themselves not as simple observers of their primate subjects but as active participants in relationships with them: several of these researchers engage in speculation about the mental lives of non-verbal primates; others describe their irritation at watching a focal animal subject running into a thicket where it cannot be followed. Rees noticed that several of these primatologists argue that the professional journals do not convey a full sense of what it is like to live and work with primates in the field, and that "speculative" anthropomorphism remains a legitimate and useful aspect of the research process. Rees's account shows how the use of "anthropomorphic" accounts and "attached" perspectives continues to permeate primatology to these days.

In short, what Goodall and Fossey brought to the field of primatology is different than bringing a "woman's perspective" as Leakey had hoped. Rather, Fossey and Goodall brought a unique perspective that allows us to understand the complexity of nonhuman primates. Understanding an "other" who is not human requires a specific perspective, and these primatologists showed us that we cannot think of apes in terms of an organism that can be "objectively" known, independently of those who observe them. This feminist critique targets traditional views of objectivity that assumes that to achieve knowledge of the way things "objectively" or "really" are, independent of knowers, one's beliefs must be guided by the nature of the object, not by the presuppositions and biases of the knower. Feminists (Longino 1990; Nelson 1990) argue instead that researchers play a fundamental role: they conceive of and represent the object of knowledge, decide what aspects of it to study, chose how to interpret evidence concerning the object, and determine how to represent the conclusions, all of these choices have an effect in the resulting knowledge. When we deny the researcher's influence and pretend that sound scientific theories are the products solely of external guidance, we are unable to see how the choices made by researchers affect knowledge. Moreover, in line with what feminist epistemologists urge, what these primatologists added to the study of primates is restructuring of scientific practices to be *open to different* social influences.

This does not mean that there is no objectivity or that methodologies cannot be evaluated as more adequate than others in their aim of helping us understand nonhuman animals. In fact, the exact opposite is true, in line with Harding's (2004) concept of "strong objectivity"; instead of adopting neutrality as a standard for objectivity, we must strive for clarifying the aim and specifying how the chosen method contributes to this aim makes for a better science. One of the most cited papers in the study of animal behavior is the perfect illustration of this idea, "Observational study of behavior: Sampling methods" (Altmann 1974). Despite being written in 1974 this paper continues to be used as a guide to the observation of animal behavior. Altmann evaluated a range of sampling methods and in doing so developed a method, focal-animal sampling, that undermined previous research (which had generated sexist accounts of social interactions among nonhuman primates) and that enabled research on female primates and on novel topics such as mothering. As Haraway (1989) argues, Altmann, troubled by the androcentric focus in primatology and the less than rigorous methodologies, brought her perspective as a woman, a feminist, a mother, and a mathematically trained scientist to bear on improving research methods. In line with Harding's (2004) notion of strong objectivity, Altmann provided a way of bringing to light the way we study primates; she provided a way of scrutinizing how this knowledge is produced, how problems are selected, how each research hypothesis requires a different study design. The collection of data is now a point of discussion open for critical evaluation. Altmann's approach recognizes that the interests and values of inquirers vary and that researchers will select background assumptions in part because the assumptions fit with the researcher's interests. For example, if a primatologist adopts the value of understanding behavior in terms of individual strategies of investment in time and energy, they will focus on behaviors such as male hunting and aggression that fit this background model. Others, who adopt values such as cooperation, will focus instead on cooperation in females as it better fits this model. This innovation in the scrutiny of the choices in methods was a response to research which was conducted through what is called "opportunistic sampling," where bias in the researcher's may affect the resulting data. For example, since, in many primate societies, males are larger than females, males inevitably caught the researchers' eye and were recorded more frequently. Females were thus unintentionally discriminated against

(Fedigan & Fedigan 1989). In line with that feminist epistemology (Longino 2001), Altmann, in her 1974 paper, is adopting a perspective where she accepts that there is no one-neutral approach to studying primates; she demonstrated that no single theory captures who a nonhuman primate is, that different ways of classifying phenomena will reveal different patterns useful to different practical interests.

Feminist epistemologists (Haraway 1991; Harding 1998; Longino 2001) have argued that this kind of pluralism of theories and research programs should be accepted as a normal feature of science because it helps researchers discover and understand new aspects of the world and see these aspects from new perspectives. However, it is important to notice that there is no agreement on whether it is the perspectives of primatologists who are taking these different approaches (and who are women) that produces better knowledge (see for example Harding 1998) or the mere fact of the diversity of approaches that produces better knowledge (see for example Longino 2001). In the history of primatology, it is possible to observe both approaches. Altmann's analysis can be interpreted as acknowledging that there is a diversity of methods that can be used in research and a recognition that this diversity results in better knowledge. However, when we consider other contributions discussed earlier, such as the ones produced by Fossey, Goodall, and Galdikas among others, we can also argue that in the history of primatology we can also observe researchers are adopting feminist values such as adopting the notion of strong objectivity and accepting alternative forms of evidence, such as emotions, as ways of obtaining better knowledge about primates.

To summarize, in this chapter I have attempted to show, through the examples of work done by two of the most well-known primatologists, how feminist critiques relate to primatology. I would like to finish this chapter by briefly describing an avenue of future research. Some have argued that most of the papers discussed in this area are part of the Western view of primatology and that non-Western approaches would bring different elements to the discussion. For example, Haraway and others (see Fuentes 2011; Rees 2001) call attention to the fact that in Western primatology, there is an acknowledgment of the shared ancestry of primates and humans, and yet "anthropocentric perceptions have situated humans at one extreme of the human–animal continuum and therefore apart from the hundreds of other primate species studied by primatologists" (Strier 2011). The pervasive distance between human and nonhuman primates is a fundamental part of Western primatology (Fouts 2000); however, this distance is not necessarily part of all primatologies. Unlike Western primatology, establishing the human–animal divide is of little importance in non-Western primatologies, as can be observed in Japanese primatology research (Imanishi 1960). Instead, the general aim in Japanese primatology is to acknowledge "human-like characteristics in animals" (de Waal 2003). As a result, concepts such as cultural traditions in primates were put forward by Japanese primatologists before they were first described in the West (Asquith 1986; Takasaki 2000). Thus, it is important to keep in mind that much of what we know about primates is situated within the background of Western beliefs confirming what some feminist critiques (Longino 1990) have already argued: that data (observations, measurements) are evaluated among background assumptions of social, cultural, and political values.

This entails that even the feminist perspectives explored in this chapter are situated within this specific Western background and that it is important not to forget that this perspective is just one among many different perspectives. For example, when considering feminist critiques in order to understand the identity of female primatologists, it is important to consider how much of this work has focused on white female primatologists from the United States, Canada, or England and that these feminist critiques may not be the same as feminist critiques done from different perspectives. For example, the challenges experienced by Latina researchers within a context of masculine pride (machismo) may be particular to Latin American women researchers (Bernal et al. 2019).

From a personal perspective, I am a Latina primatologist, so my identity is complex (to say the least). When I was conducting research at Gombe National Park, we would frequently be walking in Kigoma, a nearby town, for provisions. One afternoon a group of kids pointed at me and called me "muzungo" (white person). I have never been called white before, but I realized that in that specific context, I was white. Following Gloria Anzaldúa (1987), many of the feminist texts do not really capture the experience of the Latina, whose identity goes beyond the black/white binary. Anzaldúa describes her queer Chicana identity as *un choque*. The term *choque* refers to a crash, intermeshing, or collision between incompatible frames of reference for identity, particularly when the identity is derived from dominant paradigms. This kind of *choque* influenced and opened me to a different way of conducting research. In that same field season, I remember looking at the women in town carrying their babies on their backs. This reminded me of my childhood and the women back in Silvia, Colombia, who also carried their babies in their backs. In the field, I also observed that chimpanzee mothers carry their infants close on their backs, on their ventral area. At first glance, the experience of being a female primatologist born in a Latin-American country and educated in Canada should not have an effect on my methodological design or my hypothesis. Moreover, from a classical perspective on "objectivity," my emotional reaction of familiarity when observing these commonalities is irrelevant to my research project because it is a subjective experience. However, this experience made me think of the prevalence of gaze for the study of the development of the social mind in both human and nonhuman primates and how these studies are mostly conducted in Western, industrialized countries where gaze is a common form of interaction. It made me think how this pervasive use of gaze caused us to ignore other modes of interaction that might be used in non-industrialized cultures and in non-Western cultures. Finally, it made me think of how this pervasive use of gaze may be descriptive of humans but will not include the modes of interaction preferred by chimpanzees (for an overview of these positions, see Botero 2016, 2018). Thus, part of future research must focus on alternative ways of understanding these feminist critiques, ways that go beyond Western-industrialized frameworks. This conclusion is in accordance with the argument presented in this chapter, primatology as an example of the application of diverse (and sometimes conflicting) feminists ideas on science and scientific methods. However, I would like to take this idea a step further and argue for an even more pluralistic approach to the ways that we examine the influence of experience and context on scientific practices. I am arguing that we consider the experiences of non-Western, non-industrialized primatologists as another way in which primatology can be contextually situated and socially situated science.

In conclusion, primatology and feminist critiques of epistemology have a long history together. Is primatology a feminist science? My tentative answer is "no"; rather it is a science that illustrates how diverse approaches to feminist philosophy of science can be applied in science and how this diversity contributes to improving our understanding of primates. The history of the study of apes in the wild provides us with a window through which we can understand how, in the history of the study of animal behavior, it is possible to observe that openness to a diversity of approaches and feminist values, such as accepting modes of observation that have traditionally been considered anthropomorphic or not objective, has an important place in the study of animal behavior and has allowed researchers to understand primates in new and more complex ways.

Related chapters: 14, 15, 16, 20.

Notes

1 There are many examples of researchers and research studies that could potentially illustrate these and many other feminist critiques such as studies in field and laboratory of monkeys such as the ones done by Harry Harlow, study of apes in humans' homes such as in the case of Kellogg and laboratories such as Yerkes Primate Centers. These examples can provide another set of feminist critiques to science, in many cases opposed to the examples presented through the studies examined here. However, because

of space constraints I will focus only on the example of observations of primates in the wild to illustrate the ideas discussed.

2 Jane Goodall had experience in administrative positions, Dian Fossey had a degree in Occupational Therapy, and Birute Galdikas had an undergraduate degree in Psychology and Zoology and a master's degree in Anthropology.

3 Retrieved from https://childmind.org/article/occupational-therapists-what-do-they-do/ February 27, 2019.

4 "Jane Goodall & Derek Bryceson Share a Marriage and a Love of the African Wilds," by Peter Kovler and Judy Lansing, October 24, 1977, https://people.com/archive/jane-goodall-derek-bryceson-share-a-marriage-and-a-love-of-the-african-wilds-vol-8-no-17/

5 Jane Goodall: Bride of Gombe, by Parker Bauer, https://www.weeklystandard.com/parker-bauer/jane-goodall-bride-of-gombe. Accessed February 23, 2019.

6 It is important to acknowledge that Fossey also engaged in violent acts toward poachers. See for example this entry blog based on Hayes' problematic biography of Fossey, https://www.ladyscience.com/ideas/time-to-stop-lionizing-dian-fossey-conservation. I believe that this is part of the narrative I am trying to demonstrate; war often results in unthinkable actions and a women who enter this domain will have a specific perspective of the experience. My aim is not to excuse Fossey's actions but rather to focus on the fact that she was engaged in a violent war that resulted in extreme violence to the women, men, and gorillas involved in the conflict; I believe that is important to understand the effects that this war had on her perspective as a researcher.

7 Pioneering Primatologist, Nature, PBS, http://www.pbs.org/wnet/nature/orphan-king-pioneering-primatologist/11410/ accessed February 23, 2019.

8 Despite how recently this particular use of Morgan's Canon been scrutinized and criticized as a valid tool for understanding the animal mind (Andrews and Huss 2014; Buckner 2011; de Waal 1999; Fitzpatrick 2008; Sober 2005; Starzak 2017), this idea continues to be highly influential among comparative cognition researchers. To this day, the canon continues to be used as a warning against anthropomorphism (see for example Kennedy 1992).

References

Addessi, E., Borgi, M., and Palagi, E. (2012) "Is Primatology an Equal-opportunity Discipline?" *PloS One* 7(1), p. e30458. doi:10.1371/journal.pone.0030458.

Altmann, J. (1974) "Observational Study of Behavior: Sampling Methods," *Behaviour* 49(3–4), pp. 227–266.

Anderson, E. (2004) "Uses of Value Judgments in Science: A General Argument, With Lessons From a Case Study of Feminist Research on Divorce," *Hypatia: A Journal of Feminist Philosophy* 19(1), pp. 1–24.

Andrews, K., and Huss, B. (2014) "Anthropomorphism, Anthropectomy, and the Null Hypothesis," *Biology & Philosophy* 29(5), pp. 711–729.

Anzaldúa, G. (1987) *Borderlands/La Frontera: The New Mestiza*. San Francisco: Aunt Lute Books.

Asquith, P. J. (1986) "Anthropomorphism and the Japanese and Western Traditions in Primatology," in Else, J. G. and Lee, P. C. (eds) *Primate Ontogeny, Cognition, and Social Behavior*, Cambridge: Cambridge University Press, pp. 61–71.

Bernal, X., Rojas, B., Pinto-E. M. A., Mendoza-Henao, A., Herrera-Montes, A., Herrera-Montes, M. I. and Cáceres Franco, A. P. (2019) "Empowering Latina Scientists," *Science* 22, pp. 825–826.

Bettelheim, B. (1965) "Commitment Required of a Woman," in Mattfeld, J. A. and Van Aken, C. G. (eds.) *Women and the Scientific Professions: The MIT Symposium on American Women in Science and Engineering*, Cambridge: The MIT Press, pp. 1–19.

Blackwell, A. L. B. (1875) *The Sexes Throughout Nature*, New York: GP Putnam.

Botero, M. (2016) "Tactless Scientists, Ignoring Touch in the Study of Joint Attention," *Philosophical Psychology* 29, pp. 1200–1214.

Botero, M. (2018) "Bringing Touch Back to the Study of Emotions in Human and Non-Human Primates: A Theoretical Exploration," *International Journal of Comparative Psychology* 30(10), pp. 1–17.

Botero, M. (2019) "Understanding the Moral Implications of Morgan's Canon," in Fischer, B. (ed.) *The Routledge Handbook of Animal Ethics*, London and New York: Routledge, pp. 35–42.

Buckner, C. (2011) "Two Approaches to the Distinction between Cognition and 'Mere Association'," *International Journal of Comparative Psychology* 24(4), pp. 314–348.

Burghardt, G. (1985) "Animal Awareness-current Perceptions and Historical Perspective," *American Psychologist* 40, pp. 905–919.

Crasnow, S. (2008) "Feminist philosophy of science: 'standpoint' and knowledge," *Science & Education* 17(10), pp. 1089–1110.

Crystal, J. D., and Foote, A. L. (2009) "Metacognition in Animals," *Comparative Cognition & Behavior Reviews* 4(1), pp. 1–16.

De la Bédoyère, C., and Fossey, D. (2005) *No One Loved Gorillas More: Dian Fossey, Letters from the Mist*, Bath: Palazzo Editions Limited.

Darwin, C. (1871/2004) *The Descent of Man*. Reprinted in 2004. New York: Penguin Classics Series.

de Waal F. (1999) "Anthropomorphism and A: Consistency in Our Thinking about Humans and Other Animals," *Philos Top* 27, pp. 225–280.

de Waal F. (2003) "Silent Invasion: Imanishi's Primatology and Cultural Bias in Science," *Animal Cognition* 6, pp. 293–299.

Dwyer, D. M., and Burgess, K. V. (2011) "Rational Accounts of Animal Behaviour? Lessons from C. Lloyd Morgan's Canon," *International Journal of Comparative Psychology* 24(4), pp. 349–364.

Fausto-Sterling, A. (1990) "Essay Review: Primate Visions, a Model for Historians of Science?" *Journal of the History of Biology* 23(2), pp. 329–333.

Fedigan, L. M. (1983) "Dominance and Reproductive Success in Primates," *American Journal of Physical Anthropology* 26(S1), pp. 91–129.

Fedigan, L. M. (1992) *Primate Paradigms: Sex Roles and Social Bonds*, Chicago: University of Chicago Press.

Fedigan, L. M. (2005) "Is Primatology a Feminist Science?" in Hager, L.D (ed.) *Women in Human Evolution*, New York: Routledge, pp. 69–88.

Fedigan, L. M., and Fedigan, L. (1989) "Gender and the Study of Primates," *Critical Reviews of Gender and Anthropology*, pp. 41–64.

Fitzpatrick, S. (2008) "Doing Away with Morgan's Canon," *Mind and Language* 23(2), pp. 224–246.

Fossey, D. (1983) *Gorillas in the Mist*, Boston: Houghton Mifflin.

Fouts R. (2000) "One-on-One With Our Closest Cousins: My Best Friend Is a Chimp," *Psychology Today* 32(4), pp. 68–73.

Fuentes, A. (2011) "Being Human and Doing Primatology: National, Socioeconomic, and Ethnic Influences on Primatological Practice," *American Journal of Primatology* 73(3), pp. 233–237.

Goodall, J. (1986) *The Chimpanzees of Gombe: Patterns of Behavior*, Cambridge: Belknap Harvard.

Goodall, J. (2010) *Through a Window: My Thirty Years with the Chimpanzees of Gombe*, New York: Houghton Mifflin Harcourt.

Haraway, D. J. (1984) "Primatology is Politics by Other Means, *PSA: Proceedings of the Biennial Meeting of the Philosophy of Science Association* 2, pp. 489–524.

Haraway, D. J. (1989) *Primate Visions : Gender, Race, and Nature in the World of Modern Science*, New York : Routledge

Haraway, D. (1991) "Situated Knowledges: The Science Question in Feminism and the Privilege of Partial Perspective". In Haraway, D., *Simians, Cyborgs and Women: The Reinvention of Nature*, New York: Routledge, pp. 183–201.

Harding, S. (1991) *Whose Science? Whose Knowledge? Thinking from Women's Lives*, Ithaca: Cornell University Press.

Harding, S. (1998) *Is Science Multicultural? Postcolonialisms, Feminisms, and Epistemologies*. Bloomington: Indiana University Press.

Hayes, H. (1990) *The Dark Romance of Dian Fossey*, New York: Simon and Schuster.

Hrdy, S. B. (1981/2009) *The Woman that Never Evolved*, Cambridge: Harvard University Press.

Imanishi, K. (1960) "Social Organization of Subhuman Primates in their Natural Habitat," *Current Anthropology* 1, pp. 393–407.

Intemann, K. (2010) "25 Years of Feminist Empiricism and Standpoint Theory: Where Are We Now?," *Hypatia* 25(4), pp. 778–796.

Jaggar, A. M. (1989) "Love and Knowledge: Emotion in Feminist Epistemology," *Inquiry* 32(2), pp. 151–176.

Keller, E. F. (1983) *A Feeling for the Organism*, San Francisco: W.H. Freeman

Kennedy, J. S. (1992) *The New Anthropomorphism*, Cambridge: Cambridge University Press.

Lancaster, J. B. (1973) "Stimulus Response-Praise of Achieving Female Monkey," *Psychology Today*, 7(4), p. 30.

Little, M. (1995) "Seeing and Caring: The Role of Affect in Feminist Moral Epistemology," *Hypatia* 10, pp. 117–137.

Longino, H. (1990) *Science as Social Knowledge*, Princeton: Princeton University Press.

Longino, H. (2001) *The Fate of Knowledge*, Princeton: Princeton University Press.

Mikhalevich, I. (2018) "Simplicity and Cognitive Models: Avoiding Old Mistakes in New Experimental Contexts," in Andrews, K. and Beck, J. (eds.) *The Routledge Handbook of Animal Minds*, London and New York: Routledge, pp. 427–437.

Montgomery, S. (2009) *Walking with the Great Apes: Jane Goodall, Dian Fossey, Biruté Galdikas*, White River Junction: Chelsea Green Publishing.

Morgan, C. (1894) *An Introduction to Comparative Psychology*, London: Walter Scott Publishing Company.

Nelson, L. H. (1990) *Who Knows: From Quine to a Feminist Empiricism*, Philadelphia: Temple University Press.

Potter, E. (2001) *Gender and Boyle's Law of Gases*, Bloomington: Indiana University Press.

Rees, A. (2001) "Anthropomorphism, Anthropocentrism, and Anecdote: Primatologists on Primatology," *Science, Technology, & Human Values* 26(2), pp. 227–247.

Rees, A. (2007) Reflections on the field: Primatology, popular science and the politics of personhood. *Social Studies of Science* 37(6), 881–907.

Ruetsche, L. (2004) "Virtue and Contingent History: Possibilities for Feminist Epistemology," *Hypatia* 19(1), pp. 73–101.

Schiebinger, L. (2001) *Has Feminism Changed Science?* Cambridge: Harvard University Press.

Small, M. F. (1984) *Female Primates: Studies by Women Primatologists* (Vol. 4). New York: Alan R. Liss

Sober, E. (1998) "Morgan's Canon," in Cummins, D. and Allen, C. (eds.) *The Evolution of Mind*, Oxford: Oxford University Press, pp. 224–242.

Sober, E. (2005) "Comparative Psychology Meets Evolutionary Biology: Morgan's Canon and Cladistic Parsimony," in Mitman, G. and Datson, L. (eds.) *Thinking with Animals: New Perspectives on Anthropomorphism*, New York: Columbia University Press, pp. 85–99.

Starzak, T. (2017) "Interpretations without Justification: A General Argument Against Morgan's Canon," *Synthese* 194(5), pp. 1681–1701.

Strier, K. B. (2011) "Why Anthropology Needs Primatology," *General Anthropology* 18(1), pp. 1–8.

Takasaki, H. (2000) "Traditions of the Kyoto School of Field Primatology in Japan," in Strum, S. and Fedigan, L. M. (eds.) *Primate Encounters: Models of Science, Gender, and Society*, Chicago: University of Chicago Press, pp. 151–164.

Trivers, R. L. (1972) "Parental Investment and Sexual Selection," in B. Campbell (ed.) *Sexual Selection and the Descent of Man, 1871–1971* , Chicago: Aldine, , pp. 136–179.

Wylie, A. (1997) "Good Science, Bad Science, or Science as Usual? Feminist Critiques of Science," in Hager, L. D. (ed.) *Women in Human Evolution*, New York: Routledge, pp. 29–54

22

THE GENDERED NATURE OF REPROGENETIC TECHNOLOGIES

Inmaculada de Melo-Martín

Introduction

Since the birth of Louise Brown in 1978 by in vitro fertilization (IVF), reprogenetic technologies have become clinical routine. More than 7 million babies have been born worldwide with the help of IVF and associated techniques (IVF for short hereafter) (ESHRE 2018). In some European countries over 5% of children are born through these technologies (De Geyter et al. 2018), and IVF accounts for 1.7% of all infants born in the United States (CDC 2018). Although initially developed to address infertility caused by blocked fallopian tubes, these technologies are now a common treatment for a variety of other fertility problems, including inability to produce eggs, poor sperm quality, endometriosis, and unexplained infertility, as well as used to select embryos with or without particular genetic variants or traits (Kuliev and Rechitsky 2017). And according to some reports, these technologies have now been used to bring into the world children whose genomes have been edited with the goal of preventing HIV infection (Cyranoski and Ledford 2018).

These technologies are important not only because of the technical feats they have accomplished. Reprogenetic technologies, which combine the power of reproductive technologies with the tools of genetic science and technology, affect some of the most important aspects of human existence: our desire to reproduce, to form families, and to ensure the health and well-being of our offspring. They potentially give us an unprecedented and sophisticated level of control over whether and when to have children and over who can and cannot be born.

Unsurprisingly, the development and use of these technologies raise a broad host of ethical and epistemic issues. Because they make embryos available for experimentation and generate surplus embryos, they raise questions regarding the appropriate ways to treat these entities and about how we come to conceptualize them. Semen and eggs can now be sold and bought and women can agree to gestate babies for others; consequently these technologies bring about concerns about the commercialization of reproductive material and women's exploitation. Because they permit the involvement of third parties in the creation of a child, reprogenetic technologies also challenge kinship notions, problematize the relation between biological contributions and parental rights and responsibilities, and bring forth the threat of the commodification of children. In addition, because these technologies enable the selection of embryos with or without particular genetic variants and they might be used to manipulate embryos' genomes in various ways, their use raises concerns about designer babies, personal identity, eugenic control, risks to future generations, and changes to notions of health and disease. Of course, the ability to select and manipulate future offspring can also affect normative notions of parenting, judgments about risks and potential benefits, and considerations about the moral status of manipulated human embryos.

Space limitations prevent a detailed discussion of all the philosophical issues associated with reprogenetics. An appropriate response to these ethical issues requires, however, an adequate understanding of what they involve. Such understanding will be flawed unless we attend to the gendered nature of reprogenetic technologies and to the social context in which these technologies are developed and implemented. My goal in this chapter is to provide a brief overview of the ways in which recognizing the gendered nature of reprogenetics leads to a more sound analysis of these technologies, one that gives us not only a correct description of such technologies but also reveals the ways in which they might either exacerbate or undermine gender oppression. Feminist work in this area is particularly relevant given both the attention to gender that feminism calls for and the aim of equity that it promotes. I began with a brief description of reprogenetic technologies.

Reprogenetic Technologies

The term "reprogenetics" generally refers to practices that combine reproductive technologies and genetic tools (Knowles and Kaebnick 2007). They allow the creation, storage, and genetic manipulation of gametes and embryos with the aim of reproduction.[1] Importantly, differently from reproductive technologies, which enable individuals to have a child, reprogenetic technologies aim at allowing prospective parents to have a *particular* kind of child, that is, a child with or without certain traits.

Clinicians have been using some of these technologies, such as IVF, for four decades. Researchers are however constantly updating many of the procedures and techniques involved in IVF: new fertility drugs, cryopreservation protocols for gametes and embryos, and fertilization methods (Niederberger et al. 2018). In its most basic form, that is, when the woman undergoing IVF provides her own eggs and her husband or partner supplies the sperm, IVF involves several steps (Niederberger et al. 2018). The first one, ovarian stimulation, aims to produce multiple oocytes in a cycle and requires that women inject a number of fertility drugs in order to increase the chance of collecting multiple eggs (Polat et al. 2014). The second step, oocyte retrieval, involves sedation and insertion of a thin needle, connected to a suction device, through the woman's vagina and into the ovary and follicles containing the eggs to retrieve them. In the fertilization stage, assessed eggs are fertilized by either combining them with prepared sperm and cultured or by injecting a selected sperm directly into the egg using a technique called intracytoplasmic sperm injection (ICSI) (Babayev et al. 2014). Once fertilization occurs, professionals pass the embryos to a laboratory incubator containing special growth medium and monitor them for appropriate development. It is at this stage of embryo culture that preimplantation genetic diagnosis (PGD), which involves the removal of one or more cells from an embryo in order to test for chromosomal abnormalities or genetic mutations in the genome (Kuliev and Rechitsky 2017), can be used to select for particular embryos. Finally, clinicians transfer the embryos – usually more than one – into the woman's body by inserting a catheter loaded with them through the cervix and into the uterus. If not all embryos are transferred, they can be cryopreserved for future use.

Other technologies, such as mitochondrial replacement techniques (MTRs), are much newer, with reports of only a handful of children born through their application (Zhang et al. 2016, 2017) and clinical trials currently under way in some countries like the United Kingdom (Lyon 2017). In the United States, the FDA oversees these technologies, and has not yet approved them for clinical use. MRTs give some women at risk of transmitting mitochondrial diseases the opportunity to have unaffected and genetically related children. These techniques involve the transplantation of pronuclei, meiotic spindle, or polar bodies from the oocytes of a woman who has mutant mitochondrial DNA into the cytoplasm of enucleated donor oocytes that have no identified mitochondrial mutations (Reznichenko et al. 2016). Because mitochondria have their own genome, the resulting embryos contain DNA from three different individuals. Significantly, the offspring of women – though not of men[2] – who use these techniques will inherit the mitochondrial DNA from the donor of the eggs, and so will future generations. This makes MRTs germline modification techniques.

The majority of the genetic and molecular tools used in reprogenetics are also all relatively recent developments. PGD, for example, which as mentioned earlier involves the removal of one or more cells from an embryo to assess the existence of chromosomal abnormalities or genetic mutations in the genome (Kuliev and Rechitsky 2017), was introduced only in the early 1990s (Handyside et al. 1990). Until recently, clinicians used PGD mainly to test for disorders caused by chromosomal abnormalities (e.g. Down syndrome), X-linked diseases (e.g. Duchenne muscular dystrophy and hemophilia), and single-gene disorders (e.g. Huntington's disease and sickle cell anemia) (Collins 2013). However, the applications for PGD have expanded rapidly and clinicians can now test for approximately 400 different conditions, including late-onset, lower-penetrance mutations such as BRCA mutations associated with hereditary breast and ovarian cancer (Kuliev and Rechitsky 2017). They are also using PGD to allow sex selection for the purposes of "family balancing," and for human leukocyte antigen (HLA) matching to ensure the birth of a baby who can become a tissue donor for an existing sibling who has some disease (Kuliev and Rechitsky 2017).

Some of the molecular technologies that researchers use today to manipulate the genetic make-up of cells have become ubiquitous in biomedical investigations (Bak et al. 2018). Current targeted genome-editing technologies using restriction endonucleases, for instance, provide the ability to insert, remove, or replace DNA in precise ways. Researchers can use these tools not only to study gene function, biological mechanisms, and disease pathology but also to treat or cure particular diseases (Bak et al. 2018; Hussain et al. 2019). These systems are significantly more efficient and more accurate than older technologies and have dramatically increased scientists' power to manipulate genomes. Indeed, researchers have already used one of these systems, CRISPR/-Cas9, to create genetically modified macaque monkeys (Niu et al. 2014), and the system promises to be useful in various other ways (Hussain et al. 2019).

Concerns raised by the use of this technique in human embryos led the scientific community to call for a moratorium on this type of research (Baltimore et al. 2015). Nonetheless, several groups in jurisdictions where these interventions are lawful have reported on the use of genome-editing tools to modify the genomes of human embryos for research purposes (Fogarty et al. 2017; Liang et al. 2017; Ma et al. 2017). Furthermore, in November 2018, a Chinese researcher named He Jiankui claimed that his team had transferred human embryos edited to disable a genetic pathway that HIV uses to infect cells and that twin baby girls had been born (Cyranoski and Ledford 2018). Significant questions remain about He's claims because the data have not been published in a peer-reviewed journal as of yet. If true however, the birth of these babies has significant implications for the use of genome-editing tools in reproduction.

In general, a rapid transfer from the laboratory to the clinic to routine care has characterized reprogenetic technologies, in many cases with scant evidence of safety and efficacy. Moreover, although the initial indications for many reprogenetic techniques were limited (i.e. directed toward particular infertility problems or particular genetic mutations), their application has steadily broadened. Clinicians now use these technologies in cases in which neither infertility nor the risk of transmitting some genetic mutation is present. In fact, some companies such as 23andMe and GenePeeks are going further and now offer services to the general public that aim at the prediction of disease risks and other phenotypic traits including height, eye color, sex, and a variety of personality characteristics not of existing embryos, but of hypothetical ones (Couzin-Frankel 2012; DeFrancesco 2014).

Reprogenetic Technologies: Attending to Gender

Determining the permissibility of technological developments is an important philosophical task. Doing so involves an appropriate evaluation of their risks and potential benefits. Advocates of reprogenetic technologies present us with an array of potential benefits that individuals and society can derive (DeGrazia 2012; Harris 2007, 2016; Savulescu 2005, 2006; Savulescu et al. 2015; Silver 1997; Smith et al. 2012). Such potential benefits include increased reproductive choice, as well as the possibility that our offspring will have longer lives, be free from severe diseases and disabilities, more intelligent, and better able to enjoy life more fully. All of these improvements are said to benefit not only the individuals who have been selected or enhanced but also society in general, as ultimately these improvements are likely to result in an increase in productivity and in more just societies (Buchanan 2008). Many who defend the development and implementation of reprogenetics also acknowledge possible harms that can result from their use including risks to human health. Nonetheless, advocates often argue that appropriate research, regulatory practices, and the consent of those using these technologies are sufficient to manage such risks.

Mainstream risk evaluations of reprogenetics, however, presuppose a value-free and gender-neutral conception of science and technology (Harris 2007, 2016; Savulescu 2005; Savulescu et al. 2015). Advocates see these technologies as simple tools that users can employ in appropriate or inappropriate ways. Also betraying their conceptions of these techniques as value-free is proponents' insistence that their immoral or potentially faulty application is not an argument against the technologies themselves but against their users or against flawed regulations. An —striking – absence of any significant discussion about the role of women in the development and implementation of reprogenetics and about the risks to women's bodies and lives also characterize mainstream risk evaluations of reprogenetics (DeGrazia 2012; Harris 2007; Robertson 2003; Savulescu 2005).

Nonetheless, a variety of studies have shown that the value-neutrality conception of science and technology is both untenable and undesirable. Feminist work has been foundational and crucial to these developments (see, for instance, Fausto-Sterling 1992; Haraway 1991; Harding 1986; Intemann 2001; Keller 1985; Longino 1990; MacKenzie and Wajcman 1985; Rolin 2004; Wajcman 1991; Wylie 2001). Although there is not a unified feminist position regarding reprogenetic technologies, feminist scholars have been critical also in calling attention to the gendered nature of these technological developments (see, for instance, Arditti et al. 1984; Baylis 2014; Callahan 1995; Corea 1987; de Melo-Martín 2017; Dickenson 2007; Donchin 1993; Duden 1993; Franklin 2013; Mahowald 2000; Overall 1987; Parens and Asch 2000; Purdy 1996; Roberts 1997; Rothman 1989; Scully 2008; Sherwin 1992; Thompson 2005).

But if science and technology in general, and reprogenetics in particular, are gendered, then any appropriate assessment of these technologies must take into account their gendered nature. This is so for several reasons. First, attending to the gendered nature of reprogenetics calls attention to the fact that the development and use of these technologies impose disproportionate amount of risks and burdens on women. Although often couples make decisions about whether, when, and what type of technologies to use, only women's bodies are directly involved in the use of IVF, and all the genetic interventions necessary for selecting or manipulating embryos require the use of IVF. These risks are not simply those usually involved in reproduction. Women's bodies and women's reproductive materials are essential for the development and implementation of reprogenetic technologies. They provide the eggs, whether for themselves, for other women, or for research, necessary. They receive the hormonal injections, undergo the surgeries, and suffer the physical and psychological side effects associated with the use of these procedures. They gestate, give birth to, and usually rear the babies selected or manipulated with the help of such technologies. Insofar as the use of these technologies often results in multiple pregnancies (Kissin et al. 2015), women alone also run the risks associated with multiple births, including miscarriages, pregnancy-related high blood pressure, gestational diabetes, and delivery by cesarean section (Qin et al. 2015). Due mainly to prematurity, multiple births also have negative effects on the babies, who can suffer a variety of health complications (Murray and Norman 2014). Given that often women are children's primary caretakers, the birth of multiple babies also has a significant effect on women's physical and psychosocial well-being. Clearly, an analysis of these technologies that treats them as gender-neutral simply conceals the differential burdens of these technologies on men and women's bodies.

Second, attention to the gendered nature of reprogenetics reveals the differential effects that reproductive decisions have on men and women's lives. Proponents of these technologies often give as a primary reason for the existence of these technologies that they expand women's reproductive choices. But increases in choices come not only with benefits but also costs (Dworkin 1982; Schwartz 2004; Velleman 1992). One such cost is that of managing information in order to choose appropriately. The development and introduction of whole genome sequencing technologies make these costs more significant as these techniques can generate unparalleled amounts of genetic information about an individual (Reuter et al. 2015). The more information that can be gathered about the genomic endowment of embryos the more difficult, and thus more costly, will be making a decision.

Moreover, increased choices in reproduction also come with increased scrutiny, particularly the decisions of women who do not fit the normative understanding of the "good" mother such as poor women, women of color, and women considered to have some disability (Asch 1999; Kleege 2006; Roberts 1997). Women's decisions about whether to get pregnant, when to do so, and how to manage their pregnancy are open to inspection. Pregnant women must control what they put into their bodies, sacrifice their pleasures and desires in order to limit even the slightest of risks to their fetuses, and submit to expert medical knowledge (Duden 1993; Ettorre 2002; Kukla 2005). Indeed, as evidence shows, many women feel they have little choice to say "no" when prenatal genetic testing is recommended by health care professionals (d'Agincourt-Canning 2006; Mahowald 2000). Of course, these consequences of increased choices can affect everyone and not just women. Nonetheless, because reproductive decisions are thought to be the responsibility of women, constraints on some choices, e.g. deciding not to have genetic testing, and the opening of others, e.g. deciding not to have a child who will be thought disabled, will be of particular relevance to women.

Similarly, given the relationships between choice, responsibility, and blame, increased reproductive choices are associated also with these costs. Once one is aware that a particular choice is available, e.g. to use PGD to select against embryos with undesirable characteristics or gene-editing techniques to improve a particular trait, the failure to choose counts against one, as one is now responsible, and can be held responsible, for the choice in question (Dworkin 1982). Women can thus both blame themselves and be blamed by others for the choices they make regarding whether to use or not these technologies and how to use them.

Of course, that introducing new reproductive choices can have differential negative costs for women does not mean that it is always better to have fewer choices. In many contexts, having increased choices promotes well-being and thus choices are valuable. However, that these costs of increased choice exist is reason not simply to attend to them but to do so in a way that acknowledges the gendered nature of reproductive decisions.

Third, awareness of the gendered nature of reprogenetics calls for a recognition of the ways in which the development and use of reprogenetics risk furthering injustices against women. Because the primary aim of reprogenetic technologies is to create "better" babies, reprogenetic technologies are likely to expand the already extensive surveillance and control of women's choices regarding reproduction – all in the service of ensuring the well-being of fetuses and future children. Women will likely be expected to make particular kinds of choices, i.e. to use these technologies to select embryos with desirable traits or improve some characteristics, to discard those with traits that are thought to fall outside accepted norms. In a context where the health and well-being of fetuses are a priority to women themselves and to others, women will likely suffer moral condemnation when their choices do not fit what is expected of them.

Importantly, reprogenetic technologies also have disparate effects in differently situated women. In the United States, for instance, white, economically well-off women constitute the main users of these technologies, in part because access to reprogenetic technologies depends on ability to pay (Spar 2006). Women with lower socioeconomic status are however often the ones providing eggs and serving as gestational carriers, as the growing market in cross border reproductive care shows (Dickenson 2011; Donchin 2010; Twine 2015). Similarly, white, middle class women are encouraged to use these technologies, but a variety of laws and institutional practices discourage women of color from having children (Roberts 1997). Moreover, given that the main goal of these technologies is to prevent the birth of children with particular diseases and disabilities, women who are thought to be at higher risk of having children considered disabled are a target group (Parens and Asch 1999; Scully 2008; Shakespeare 2006; Tremain 2001). Hence, attention to the ways in which the intersectionality of gender with other social categories such as race, disability status, class, nationality can lead to more robust and appropriate assessments of reprogenetics.

Fourth, a gender-attentive understanding of reprogenetics can challenge the status quo and offer constructive solutions to problems that make the use of these technologies problematic. For instance, the increasing use of reprogenetic technologies has resulted in a growing use of egg providers both for reproductive and research purposes. Nonetheless, little attention has been given to the long-term health risks of fertility drugs (Woodriff et al. 2014). Development of gene-editing techniques aimed at enhancement of embryos is likely to increase the need for women to provide eggs (Dickenson 2013). Attention to the gendered nature of these technologies thus emphasizes the necessity of careful collection and analysis of data on egg providers, the importance of national egg and embryo donor registries, and the need for long-term studies and long-term follow-up.

Likewise, the use of these technologies, with the possibility of selling reproductive materials and gestating babies to give to others, raises the specter of women's exploitation. The lack of consistency in national regulations, the differences in economic incentives between various countries, pronatalist ideologies, all contribute to a context in which egg providers and surrogate mothers are more likely to be harmed by exploitation. Recognizing the gendered nature of reprogenetics can result in assessments that demand that appropriate mechanisms be put in place to eliminate or reduce these dangers.

Evaluations that attend to the gendered nature of reprogenetic technologies can also insist on the need to ensure that women are appropriately informed about the risks of ovarian stimulating drugs and other physical and psychological consequences of providing eggs. Indeed, given that increased choice is accompanied by increased responsibility –that one will choose wisely— the need to ensure that women have adequate information about their choices imposes obligations on the part of clinicians and researchers to produce and make such information available and accessible.

Moreover, it is well known that access to suitable prenatal care contributes to the well-being of women and their children. Nonetheless, in 2009–2010 in the United States, more than 17% of recent mothers reported that they were not able to access prenatal care as early as they had wanted (DHHS 2013). Women of color report being particularly affected by such lack of access. Unsurprisingly one of the most common barriers to getting prenatal care at all or as early as desired includes limited resources. Worldwide, inadequate attention to pregnancy and childbirth claimed the lives in 2013 of 289,000, and most could have been prevented (WHO 2014b). Furthermore, almost 3 million babies die every year in their first month of life, at least in part because their mothers, while pregnant, did not have access to adequate health care and nutrition (WHO 2014a). Recognition of the gendered nature of reprogenetics technologies and the ways in which women's reproductive decisions can be constrained can lead to forcefully and heartily defend the need for adequate access to integrated and flexible prenatal services.

Fifth, attention to the gendered nature of reprogenetic technologies provides us with insights into how these technologies can reinforce oppressive gender norms. Take, for instance, ideals of motherhood. There is little doubt that many women find motherhood desirable and valuable in multiple ways. Nonetheless, prevalent notions of motherhood and family, which sanction inequalities between men and women, have worked as prescriptions in the service of gender oppression (Badinter 1981; de Beauvoir [1949]1993; De la Concha and Osborne 2004; Ehrenreich and English 2005; Firestone 1970; LaChance Adams and Lundquist 2013; Okin 1989; Rich 1976). Such notions minimize the physical and emotional work that mothers do, impose norms about "good" and "bad" mothers that create difficult-to-fulfill expectations, and contribute to marginalizing those mothers, such as poor women, women of color, or those with disabilities, who do not conform to such norms (Badinter2011; Ehrenreich and English 2005; Gillies 2006; Kukla 2005; O'Reilly 2006; Roberts 1997).

Reprogenetic technologies might seem ideal candidates to dismantle rather than reinforce normative notions of motherhood. These technologies secure the separation between sexuality and reproduction initiated by contraceptives and thus can disrupt the identification between woman and mother and challenge notions of biological imperatives by allowing women to have children later. Furthermore, reprogenetic techniques allow the separation between genetic, gestational, and social maternity. They thus disturb the natural and continuous process that was motherhood and substitute it for one that is highly technologized and fractioned. The division of the different aspects of the reproductive process – gametes, gestation, parenting – that is a consequence of the use of these techniques also has the power to contribute to the demystification of the nuclear family, an institution that has been fundamental in fostering gender oppression (Okin 1989). This is because these technologies now allow women and men alone and homosexual couples to create families that contravene those norms (Mamo 2007). Moreover, donation and cryopreservation techniques can involve several women in the creation of a baby further contesting ideal notions of motherhood.

Nevertheless, attending to the gendered nature of reprogenetics can lead to a more careful assessment of their disruptive potentials. Indeed, from their initial introduction in the clinic, feminist scholars have called attention to the ways in which these technologies reinforce oppressive notions of motherhood and family (Arditti et al. 1984; Corea 1985, 1987; Rothman 1989; Rowland 1992). For instance, prior to the routine use of reprogenetics, women who could not have children could appeal to having exhausted their options in trying to become mothers. The introduction of reprogenetic technologies however opens alternatives that need to be tried. And because often the causal mechanisms of infertility are unknown, it is difficult to determine when one should stop trying after multiple failures (Thompson 2005). Hence, far from minimizing essentialist and reductionist connections between women and motherhood, these technologies can buttress them. Moreover, reprogenetic technologies embody the asymmetry of reproductive relationships by making women's bodies alone the sites of treatment even when infertility affects both men and women. Reprogenetic techniques thus secure the patriarchal control over women's bodies and strengthen believes about women's primary responsibility for reproductive matters.

Similarly, cryopreservation techniques directed to fertile women, which permit them to delay motherhood, have the ability to undermine normative notions of motherhood by liberating women from the biological imperative. As we have seen, however, obtaining eggs still involves considerable physical, emotional, and financial costs for women. More importantly, the development of these techniques necessarily presupposes women's desire to become mothers at some point or another in their lives. Indeed, such desire seems so obvious that women are expected to take whatever measures exist – including risky and financially costly ones – to ensure that they can satisfy it. The fact that, at the moment when women undergo procedures of cryopreservation of their eggs, they have no wish to become mothers is not an obstacle to predicting that at some point they will desire to be so.

To call attention to the ways in which reprogenetic technologies reinforce – in both subtle and patent ways – problematic notions of motherhood and family is not to say that technological development necessarily determines human values. It might well be the case that the use of re-progenetics could contribute to undermine oppressive gender norms. However, for that to be the case, it is necessary to attend to the gendered nature of these technologies.

Conclusion

Reprogenetic technologies present us with wondrous possibilities but also with many ethical challenges. Dealing with such challenges requires a correct analysis of these technologies. A disregard of the gendered nature of reprogenetics can only provide us with manifestly inadequate evaluations. Such evaluations fail to take into account the crucial role of women in the development and implementation of these technologies as well as the impacts that they have on women's lives. That women are the ones who must undergo IVF and the associated risks, the ones who get pregnant, the ones who are often held responsible for reproductive decisions, should be of central concern to any appropriate assessment of reprogenetics. A gender attentive analysis also calls attention to how these technologies can contribute to furthering injustices against women and thus draw attention to the structural problems that must be transformed to avoid such injustices. Whether the development and implementation of reprogenetic technologies ultimately contribute to more or less oppressive societies depends, at least in part, on ensuring that our assessments do not neglect the gendered nature of these technologies.

Related chapters: 11, 15, 33.

Notes

1 These technologies can also be used for research purposes but the focus of the chapter will be on reproduction.
2 Mitochondria are inherited maternally.

References

Arditti, R., Klein, R. and Minden, S. (1984) *Test-Tube Women: What Future for Motherhood?* London: Pandora Press.

Asch, A. (1999) "Prenatal Diagnosis and Selective Abortion: A Challenge to Practice and Policy," *American Journal of Public Health*, 89(11), 1649–1657.

Babayev, S. N., Park, C. W. and Bukulmez, O. (2014) "Intracytoplasmic Sperm Injection Indications: How Rigorous?," *Seminars in Reproductive Medicine*, 32(4), 283–290.

Badinter, E. (1981) *Mother Love: Myth and Reality*. New York: Macmillan.

Badinter, E. (2011) *The Conflict: How Modern Motherhood Undermines the Status of Women*. Translated by A. Hunter. New York: Metropolitan Books/Henry Holt and Co.

Bak, R. O., Gomez-Ospina, N. and Porteus, M. H. (2018) "Gene Editing on Center Stage," *Trends in Genetics*, 34(8), 600–611.

Baltimore, D., Berg, P., Botchan, M., et al. (2015) "Biotechnology. A Prudent Path Forward for Genomic Engineering and Germline Gene Modification," *Science*, 348(6230), 36–38.

Baylis, F. (2014) *Family-Making: Contemporary Ethical Challenges*. New York: Oxford University Press.

Buchanan, A. (2008) "Enhancement and the Ethics of Development," *Kennedy Institute of Ethics Journal*, 18(1), 1–34.

Callahan, J. C. (1995) *Reproduction, Ethics, and the Law: Feminist Perspectives*. Bloomington: Indiana University Press.

Centers for Disease Control and Prevention (CDC). (2018) *Assisted Reproductive Technologies*, https://www.cdc.gov/art/artdata/index.html.

Collins, S. C. (2013) "Preimplantation Genetic Diagnosis: Technical Advances and Expanding Applications," *Current Opinion in Obstetrics & Gynecology*, 25(3), 201–206.

Corea, G. (1985) *The Mother Machine: Reproductive Technologies from Artificial Insemination To Artificial Wombs*. New York: Harper & Row.

———. (1987) *Man-Made Women: How New Reproductive Technologies Affect Women*. 1st Midland book edn. Bloomington: Indiana University Press.

Couzin-Frankel, J. (2012) "Genetics. New Company Pushes the Envelope on Pre-Conception Testing," *Science*, 338(6105), 315–316.

Cyranoski, D. and Ledford, H. (2018) " 'International Outcry over Genome-Edited Baby Claim," *Nature*, 563(7733), 607–608.

d'Agincourt-Canning, L. (2006) "Genetic Testing for Hereditary Breast and Ovarian Cancer: Responsibility and Choice," *Qualitative Health Research*, 16(1), 97–118.

de Beauvoir, S. (1993) *The Second Sex*. Translated by H. Parshley. New York: Alfred A. Knopf.

De Geyter, C., Calhaz-Jorge, C., Kupka, M. S., et al. (2018) "ART in Europe, 2014: Results Generated from European Registries by ESHRE: The European IVF-monitoring Consortium (EIM) for the European Society of Human Reproduction and Embryology (ESHRE)," *Human Reproduction*, 33(9), pp. 1586–1601.

De la Concha, A. and Osborne, R. (eds.) (2004) *Las Mujeres Y Los Niños Primero: Discursos De La Maternidad*. Barcelona: Icaria.

de Melo-Martín, I. (2017) *Rethinking Reprogenetics: Enhancing Ethical Analyses of Reprogenetic Technologies*. New York: Oxford University Press.

DeFrancesco, L. (2014) "23andMe's Designer Baby Patent," *Nature Biotechnology*, 32(1), 8.

DeGrazia, D. (2012) *Creation Ethics: Reproduction, Genetics, and Quality of Life*. New York: Oxford University Press.

Dickenson, D. L. (2007) *Property in the Body: Feminist Perspectives*. Cambridge: Cambridge University Press.

——— (2011) "Regulating (or Not) Reproductive Medicine: An Alternative to Letting the Market Decide," *Indian Journal of Medical Ethics*, 8(3), 175–179.

———. (2013) "The Commercialization of Human Eggs In Mitochondrial Replacement Research," *The New Bioethics: A Multidisciplinary Journal of Biotechnology and the Body*, 19(1), 18–29.

Donchin, A. (1993) *Procreation, Power and Subjectivity: Feminist Approaches to New Reproductive Technologies*. Wellesley: Center for Research on Women.

Donchin, A. (2010) "Reproductive Tourism and the Quest for Global Gender Justice," *Bioethics*, 24(7), 323–332.

Duden, B. (1993) *Disembodying Women: Perspectives on Pregnancy and the Unborn*. Cambridge: Harvard University Press.

Dworkin, G. (1982) "Is More Choice Better than Less?," *Midwest Studies in Philosophy*, 7(1), 47–61.

Ehrenreich, B. and English, D. (2005) *For Her Own Good: Two Centuries of the Experts' Advice to Women*. New York: Anchor Books.

Ettorre, E. (2002) *Reproductive Genetics, Gender, and the Body*. New York: Routledge.

European Society of Human Reproduction and Embryology (ESHRE). (2018). "ART Fact Sheet." Available at: https://www.eshre.eu/~/media/sitecore-files/.../ART-fact-sheet_vFebr18_VG.pdf?la=en

Fausto-Sterling, A. (1992) *Myths of Gender: Biological Theories about Women and Men*, 2nd edn. New York: Basic Books.

Firestone, S. (1970) *The Dialectic of Sex: The Case for Feminist Revolution*. New York: Morrow.

Fogarty, N. M. E., McCarthy, A., Snijders, K. E., et al. (2017) "Genome Editing Reveals a Role for Oct4 in Human Embryogenesis," *Nature*, 550(7674), pp. 67–73.

Franklin, S. (2013) *Biological Relatives: IVF, Stem Cells, and the Future of Kinship*. Durham: Duke University Press.

Gillies, V. (2006) *Marginalised Mothers: Exploring Working Class Experiences of Parenting*. New York: Routledge.

Handyside, A. H., Kontogianni, E. H., Hardy, K. and Winston, R. M. (1990) "Pregnancies from Biopsied Human Preimplantation Embryos Sexed by Y-Specific DNA Amplification," *Nature*, 344(6268), 768–770.

Haraway, D. J. (1991) *Simians, Cyborgs, and Women: The Reinvention of Nature*. New York: Routledge.

Harding, S. G. (1986) *The Science Question in Feminism*. Ithaca: Cornell University Press.

Harris, J. (2007) *Enhancing Evolution: The Ethical Case for Making Better People*. Princeton: Princeton University Press.

———. (2016) "Germline Modification and the Burden of Human Existence," *Cambridge Quarterly Healthcare Ethics*, 25(1), pp. 6–18 .

Hussain, W., Mahmood, T., Hussain, J., Ali, N., Shah, T., Qayyum, S. and Khan, I. (2019) "CRISPR/Cas system: A Game Changing Genome Editing Technology, to Treat Human Genetic Diseases," *Gene*, 685, 70–75.

Intemann, K. (2001) "Science and Values: Are Value Judgments Always Irrelevant to the Justification of Scientific Claims?," *Philosophy of Science*, 68(3), pp. S506–S518.

Keller, E. F. (1985) *Reflections on Gender and Science*. New Haven: Yale University Press.

Kissin, D. M., Kulkarni, A. D., Mneimneh, A., et al. (2015) "Embryo Transfer Practices and Multiple Births Resulting from Assisted Reproductive Technology: An Opportunity for Prevention," *Fertility and Sterility*, 103(4), 954–961.

Kleege, G. (2006) *Blind Rage: Letters to Helen Keller*. Washington: Gallaudet University Press.

Knowles, L. P. and Kaebnick, G. E. (2007) *Reprogenetics: Law, Policy, and Ethical Issues*. Baltimore: Johns Hopkins University Press.

Kukla, R. (2005) *Mass Hysteria: Medicine, Culture, and Mothers' Bodies*. Lanham: Rowman & Littlefield.

Kuliev, A. and Rechitsky, S. (2017) "Preimplantation Genetic Testing: Current Challenges and Future Prospects," *Expert Review of Molecular Diagnostics*, 17(12), 1071–1088.

LaChance Adams, S. and Lundquist, C. R. (2013) *Coming to Life: Philosophies of Pregnancy, Childbirth, and Mothering*. New York: Fordham University Press.

Liang, P., Ding, C., Sun, H., Xie, X., et al. (2017) "Correction of Beta-Thalassemia Mutant by Base Editor in Human Embryos," *Protein & Cell*, 8(11), pp. 811–822.

Longino, H. E. (1990) *Science as Social Knowledge: Values and Objectivity in Scientific Inquiry*. Princeton: Princeton University Press.

Lyon, J. (2017) "Sanctioned UK Trial of Mitochondrial Transfer Nears," *JAMA*, 317(5), 462–464.

Ma, H., Marti-Gutierrez, N., Park, S.-W., et al. (2017) "Correction of a Pathogenic Gene Mutation in Human Embryos," *Nature*, 548(7668), 413–419.

MacKenzie, D. A. and Wajcman, J. (1985) *The Social Shaping of Technology: How the Refrigerator Got Its Hum*. Philadelphia: Open University Press.

Mahowald, M. B. (2000) *Genes, Women, Equality*. New York: Oxford University Press.

Mamo, L. (2007) *Queering Reproduction: Achieving Pregnancy in the Age of Technoscience*. Durham: Duke University Press.

Murray, S. R. and Norman, J. E. (2014) "Multiple Pregnancies Following Assisted Reproductive Technologies – A Happy Consequence or Double Trouble?," *Seminars in Fetal & Neonatal Medicine*, 19(4), 222–227.

Niederberger, C., Pellicer, A., Cohen, J., et al. (2018) "Forty Years of IVF," *Fertility and Sterility*, 110(2), 185–324.

Niu, Y., Shen, B., Cui, Y., et al. (2014) "Generation of Gene-Modified Cynomolgus Monkey via Cas9/RNA-Mediated Gene Targeting in One-Cell Embryos." *Cell*, 156(4), 836–843.

Okin, S. M. (1989) *Justice, Gender, and the Family*. New York: Basic Books.

O'Reilly, A. (2006) *Rocking the Cradle: Thoughts on Feminism, Motherhood, and the Possibility of Empowered Mothering*. Toronto: Demeter Press.Overall, C. (1987) *Ethics and Human Reproduction: A Feminist Analysis*. Boston: Allen & Unwin.

Parens, E. and Asch, A. (1999) "The Disability Rights Critique of Prenatal Genetic Testing. Reflections and Recommendations," *The Hastings Center Report*, 29(5), S1–22.

Parens, E. and Asch, A. (2000) *Prenatal Testing and Disability Rights*. Washington: Georgetown University Press.

Polat, M., Bozdag, G. and Yarali, H. (2014) "Best Protocol for Controlled Ovarian Hyperstimulation in Assisted Reproductive Technologies: Fact or Opinion?," *Seminars in Reproductive Medicine*, 32(4), 262–271.

Purdy, L. M. (1996) *Reproducing Persons: Issues in Feminist Bioethics*. Ithaca: Cornell University Press.

Qin, J., Wang, H., Sheng, X., Liang, D., Tan, H. and Xia, J. (2015) "Pregnancy-Related Complications and Adverse Pregnancy Outcomes in Multiple Pregnancies Resulting From Assisted Reproductive Technology: A Meta-Analysis of Cohort Studies," *Fertility and Sterility*, 103(6), 1492–508.e1–7.

Reuter, J. A., Spacek, D. V. and Snyder, M. P. (2015) "High-Throughput Sequencing Technologies," *Molecular Cell*, 58(4), 586–597.

Reznichenko, A. S., Huyser, C. and Pepper, M. S. (2016) "Mitochondrial Transfer: Implications for Assisted Reproductive Technologies," *Applied and Translational Genomics*, 11, 40–47.

Rich, A. (1976) *Of Woman Born: Motherhood as Experience and Institution*. New York: Norton.

Roberts, D. E. (1997) *Killing the Black Body: Race, Reproduction, and the Meaning of Liberty*. New York: Pantheon Books.

Robertson, J. A. (2003) "Procreative Liberty in the Era of Genomics," *American Journal of Law and Medicine*, 29(4), 439–487.

Rolin, K. (2004) "Why Gender Is a Relevant Factor in the Social Epistemology of Scientific Inquiry," *Philosophy of Science*, 71(5), 880–891.

Rothman, B. K. (1989) *Recreating Motherhood: Ideology and Technology in a Patriarchal Society*. New York: Norton.

Rowland, R. (1992) *Living Laboratories: Women and Reproductive Technologies*. Bloomington: Indiana University Press.

Savulescu, J. (2005) "New Breeds of Humans: The Moral Obligation to Enhance," *Reproductive Biomedicine Online*, 10, 36–39.

———. (2006) "Justice, Fairness, and Enhancement," *Annals of NY Academy of Science*, 1093, 321–338.

Savulescu, J., Pugh, J., Douglas, T. and Gyngell, C. (2015) "The Moral Imperative to Continue Gene Editing Research on Human Embryos," *Protein & Cell*, 6(7), 476–479.

Schwartz, B. (2004) *The Paradox of Choice: Why More Is Less*. New York: Ecco.

Scully, J. L. (2008) *Disability Bioethics: Moral Bodies, Moral Difference*. Lanham: Rowman & Littlefield.

Shakespeare, T. (2006) *Disability Rights and Wrongs*. New York: Routledge.

Sherwin, S. (1992) *No Longer Patient: Feminist Ethics and Health Care*. Philadelphia: Temple University Press.

Silver, L. M. (1997) *Remaking Eden: Cloning and Beyond in a Brave New World*. New York: Avon Books.

Smith, K. R., Chan, S. and Harris, J. (2012) "Human Germline Genetic Modification: Scientific and Bioethical Perspectives," *Archives of Medical Research*, 43(7), 491–513.

Spar, D. L. (2006) *The Baby Business: How Money, Science, and Politics Drive the Commerce of Conception*. Boston: Harvard Business School Press.

Thompson, C. (2005) *Making Parents: The Ontological Choreography of Reproductive Technologies*. Cambridge: MIT Press.

Tremain, S. (2001) "On the Government of Disability," *Social Theory and Practice*, 27, 617–636.

Twine, F. W. (2015) *Outsourcing the Womb: Race, Class and Gestational Surrogacy in a Global Market*, 2nd ed. New York: Routledge, Taylor & Francis Group.

U.S. Department of Health and Human Services Health Resources and Services Administration (DHHS) (2013) *Child Health USA 2013*. Rockville: U.S. Department of Health and Human Services.

Velleman, J. D. (1992) "Against the Right to Die," *Journal of Medical Philosophy*, 17(6), 665–681.

Wajcman, J. (1991) *Feminism Confronts Technology*. Cambridge: Polity Press.

Woodriff, M., Sauer, M. V. and Klitzman, R. (2014) "Advocating for Longitudinal Follow-up of the Health and Welfare of Egg Donors," *Fertility and Sterility*, 102(3), 662–666.

World Health Organization (WHO). (2014a) *Children: Reducing Mortality*. Available at http://www.who.int/mediacentre/factsheets/fs178/en/.

World Health Organization (WHO). (2014b) *Maternal Mortality*. Available at http://www.who.int/mediacentre/factsheets/fs348/en/.

Wylie, A. (2001) "Doing Social Science as a Feminist: The Engendering of Archaeology," in Creager, A. N. H., Lunbeck, E. and Schiebinger, L. (eds.) *Feminism in Twentieth Century Science, Technology, and Medicine*. Chicago: University of Chicago Press, pp. 23–45.

Zhang, J., Liu, H., Luo, S., et al. (2017) "Live Birth Derived from Oocyte Spindle Transfer to Prevent Mitochondrial Disease," *Reproductive Biomedicine Online*, 34(4), 361–368.

Zhang, J., Zhuang, G. L., Zeng, Y., Grifo, J., Acosta, C., Shu, Y. M. and Liu, H. (2016) "Pregnancy Derived from Human Zygote Pronuclear Transfer in a Patient Who Had Arrested Embryos after IVF," *Reproductive Biomedicine Online*, 33(4), 529–533.

Psychology, Cognitive Science, and Neuroscience

23

WHAT'S WRONG WITH (NARROW) EVOLUTIONARY PSYCHOLOGY

Letitia Meynell

Introduction

In order to understand the many and varied criticisms of evolutionary psychology one must first understand what exactly is meant by the term itself. At first glance, it appears self-explanatory. What else could evolutionary psychology be other than the study of the evolution of psychological states, cognition, and behavior? So construed, evolutionary psychology scopes over a huge array of projects in disciplines ranging from classical ethology, to behavioral ecology, comparative psychology, primatology, and more. However, the vast majority of critics—feminist and otherwise—who have taken aim at evolutionary psychology (e.g. Buller 2005; Fausto-Sterling et al. 1997; Griffiths 2008; Lloyd 2015; Meynell 2012; Nelson 2017; Smith 2020; Stotz and Griffiths 2003; Weaver 2017) direct their critical gaze toward a very specific research program that started around 1990. This narrow research program, often called narrow evolutionary psychology (NEP), is unusually easy to define, at least *in theory*, as key proponents have done their critics the courtesy of writing manifestos that clearly articulate their central commitments (e.g. Tooby and Cosmides 1992, 2005). In what follows I rely heavily on one of these—John Tooby and Leda Cosmides' "Conceptual Foundations of Evolutionary Psychology" from *The Handbook of Evolutionary Psychology* (2005) (which defends their account of an evolved universal human nature, comprising mental modules). However, *in practice*, delineating the scope of this research program is rather more challenging, as NEP's results and methods have seeped into other evolutionary and psychological projects.

Of course, even if NEP has fundamental flaws this doesn't entail anything for evolutionary psychology *broadly* speaking. This is a good thing too as, at the risk of stating the obvious, various animals, including humans, have psychological states and, unless we are willing to reject the natural historical basis of contemporary biology, the cognitive and affective capacities associated with these states must have evolved. Clearly, there is a plausible scientific research program here, albeit one that suffers all the usual epistemic limitations of the historical sciences. The problems with NEP go far beyond these challenges. As countless critical assessments have shown, NEP is inconsistent with our best current understanding of evolution and neuroscience; it is poorly motivated, and methodologically broken. As if these scientific failings weren't enough, NEP is also politically pernicious. It is awash in number of sexist assumptions and purports to discover a number of gendered psychological traits that repeat various sexist stereotypes (though, sadly, this does not distinguish it from a number of other research programs in associated areas). Furthermore, the singular focus on defining a human nature that, while supposedly universal, explicitly leaves out a number of humans is clearly ableist.

In this chapter, I will put NEP in context, review some of its key commitments, and rehearse some of the most important criticisms of the program. I will then consider a case study that exemplifies both these problems and the inherent circularity of a key research method employed by narrow evolutionary psychologists. I conclude the chapter with a few remarks about a new research program, feminist (narrow) evolutionary psychology, and some reflections on the rhetorical positioning of NEP.

NEP in Context[1]

While the evolutionary study of behavior might reasonably be thought to go back to Charles Darwin, robust organized research programs really took hold in the mid-twentieth century with the work of Konrad Lorenz and Niko Tinbergen (Griffiths 2008: 393). Classical ethology contrasted with behaviorism, the dominant research program for animal behavior in the North America at the time, which tended to eschew evolutionary questions and study general learning mechanisms under laboratory conditions. Though certainly ethologists postulated some general mechanisms, they focused primarily on species-typical behaviors of animals in their natural habitats (Griffiths 2008: 394–395).

As we will find it useful for criticizing NEP below, it is worth spelling out the basic commitments of classical ethology as stated in Tinbergen's four questions (1963). Really, these are four different ways of answering one question, namely, "Why do these animals behave as they do?" (1963: 411). If I ask, for instance, why does this bird build a particular kind of nest I might be asking about (i) efficient causation—what prompts the behavior and the mechanics of its execution, say the perception of lengthening days in the spring, or the internal neurological mechanisms that perceive and respond to this stimulus; or (ii) ontogeny—the developmental history of the individual, perhaps for this species an individual must learn to build the nest by copying conspecifics; or (iii) current survival value—a nest helps the bird to incubate their eggs thus promoting their reproductive success, or perhaps it attracts mates (or both). These first three questions are amenable to direct empirical investigation. Evolution, the fourth question, is significantly less empirically tractable. As Tinbergen notes, there are no behavioral fossils (1963: 427), and, as is widely recognized, you cannot simply infer evolutionary function from current function.

The idea that a trait that evolved for one function might be coopted for a whole new role has even been given a name—exaptation (Gould and Vrba 1982). This concept was introduced to discourage lazy inferences from current function to historical function and, indeed, to remind evolutionary theorists that some traits have no function at all. Inevitably, answering the question of how a particular behavior evolved depends on comparison with other species or (and it's an inclusive or) Tinbergen's other three questions. That nests are ubiquitous among birds and were built by some dinosaurs tells us something about the deep evolutionary history of nest building and suggests cross-species comparisons might be a useful way of answering the evolutionary question regarding why this particular bird builds that kind of nest. Not only will it help us articulate the many possible roles that nests might play for any species, but if we find that those species most closely related to the bird in question share the same nest building behaviors this is *prima facie* evidence for thinking that their last common ancestor behaved that way too.

As we can see from Tinbergen's four whys, classical ethology offers an empirically tractable, rigorous approach to the study of the evolution of behavior.[2] However, it was eclipsed by sociobiology and behavioral ecology in the 1970s (Griffiths 2008: 398 ff.). In contrast to classical ethology, these newer approaches tended to depend more on the methods of population biology and other mathematical models (like kin selection models). Also, they adopted various simplifying assumptions. Perhaps most notably, sociobiologists tended to eschew concerns with development and neurology (efficient causation in ethology) and often collapsed current survival value into evolution. This produced a troubling adaptationist tendency to assume that all behavioral traits were produced by natural selection, ignoring other evolutionary processes. Sociobiology gained particular attention through the publication of E.O. Wilson's book on the topic (1975), in part because it extended its discussion of the evolution of animal behavior to include humans.

Wilson's book and sociobiology more generally came under considerable attack. Feminists and other progressives questioned many of the putative findings of sociobiology, such as the biological programming of women to care for men and children (e.g. Hubbard 1988) or the adaptive value of rape (e.g. Dusek 1984), and generally bemoaned the sexist presuppositions woven throughout the program (e.g. Gould et al. 1979; Rose and Rose 1986). Stephen Jay Gould and Richard Lewontin had sociobiology as one of their targets in their famous paper, "The Spandrels of San Marco and the Panglossian Paradigm" (1979). They challenged the adaptationism of many evolutionary theorists, as well as their peculiar tendency of conceptually dissecting the whole organism into a collection of atomic traits. Despite the cogency of their critique, adaptationism, in particular, has been difficult to dislodge. In a remarkable paper, Ernst Mayr articulated what seems to be a common position, conceding the correctness of Gould and Lewontin's critique while maintaining adaptationist assumptions for purely methodological reasons (1983). That methodological adaptationism inevitably distorts research findings and is, in fact, quite unnecessary for a successful evolutionary research program was exactly Gould and Lewontin's point, and has recently been reprised by Elisabeth Lloyd (2015).

NEP self-consciously emerged as a successor to sociobiology, albeit with a considerably narrowed focus on humans alone. NEP retains sociobiology's fondness for formalisms and a tendency to view evolution through the lens of replicator dynamics as well as what Sara Weaver calls the Bateman-Trivers-Parker paradigm, which emphasizes sex dimorphism, driven by gamete size and parental investment, creating ardent competitive males and coy but nurturing females (Weaver 2017: §2.4.1). Where sociobiology falls short, from a NEP perspective, is in failing to recognize the promise of the "cognitive revolution" (Tooby and Cosmides 2005: 15) and the concomitant insight that the brain *is* a computer (2005: 16). Moreover, while sociobiologists attempt to explain the evolution of behaviors, NEP attempts to investigate the cognitive mechanisms underlying them.

Their main foil, however, is what Tooby and Cosmides call the Standard Social Science Model—basically all social science research except NEP—which they identify as profoundly defective and in need of complete replacement (2005: 6–7). The principle sin of the Standard Social Science Model, we are told, is the assumption that the mind is a general purpose information processing machine—"a blank slate" that simply acquires information (2005: 6, 15). On the contrary, they assert, NEP shows that the human mind comes "factory equipped" with a set of distinct content-specific modules that have been designed by natural selection to solve the evolutionary problems faced by our recent ancestors (2005: 15). NEP's critics, particularly from the social sciences, are characterized as scientifically illiterate (at least when it comes to evolutionary theory), defending degenerate research programs simply for their own self-interest (2005: 2–3). This rhetorical stance may seem rather ironic in light of what follows.

NEP's Theoretical Commitments Assessed

The ultimate goal of NEP is to construct "a set of empirically validated, high-resolution models of the evolved mechanisms that collectively constitute *universal human nature*" (Tooby and Cosmides 2005: 5, emphasis mine). This "human nature," we are told, comprises the set of mental modules that evolved and ultimately became fixed in the Pleistocene (roughly 2.5 million to 12,000 years ago). As we shall see, neither the commitment to human essences—often referred to as the monomorphic mind thesis—the modular view of mind—called the massive modularity thesis—nor the singular focus on the Pleistocene are well-defended or, indeed, plausible. Although much more could be said about other assumptions of NEP, I will focus on these three as they are commonly criticized.

The massive modularity thesis goes hand-in-hand with the idea that the mind/brain *is* a computer. Much the same way that your own personal computer can have an indefinite number of possible outputs, despite the highly specific functions of its components, so too highly specific mental modules can give rise to the remarkable diversity of human behavior (Tooby and Cosmides 2005: 18). However, some variants are, according to Tooby and Cosmides, dysfunctional—a result of broken programs. One of their examples is the putative module that produces the link between following another's gaze and inferring that they want what they are looking at. "There is an inference circuit," they claim, "a reasoning instinct—that produces this inference. When the circuit is broken, or fails to develop, the inference cannot be made. Those with autism fail [the gaze-following] task because they lack this reasoning instinct" (2005: 19). Such deficits, they claim, prove the massive modularity thesis: "If the mind consisted of a domain-general knowledge acquisition system, narrow impairments of this kind would not be possible" (2005: 19).[3]

There are a number of problems with this reasoning but let me start by noting that it rests on a false dichotomy. Just because the mind/brain is not a blank slate it does not follow that it must be a set of highly content-specific modules. Even if there were evidence for some highly modular content-specific cognitive mechanisms—perhaps even a gaze-following module—this does not entail or even suggest that all cognitive processing is highly modular. This logical point is buoyed by empirical results from cognitive neuroscience. In "The Seven Sins of Evolutionary Psychology" (2000) neuroscientists Jaak and Jules Panksepp note that despite there being "brain/mind modules for all of our basic sensory/perceptual and motor processes. . ., we can now be equally confident that there is also a great amount of general-purpose computational space" (2000: 115). Moreover, significant parts of the brain's architecture are "not straitforwardly compatible with any highly resolved, genetically-governed modular point of view" (2000: 116). Indeed, the idea of fixed mental modules seems straightforwardly incompatible with the remarkable plasticity of the brain, its much-celebrated capacity to rewire itself, and the diversity of brain morphology. Consider cognitively typical individuals that are morphologically anomalous such as lacking a complete corpus callosum (Tovar-Moll et al. 2014) or even a whole hemisphere (Battro 2000).

The supposedly extreme modularity of the brain/mind grounds the argument that Tooby and Cosmides proffer for the monomorphic mind. The argument is a priori, which is unsurprising given, as Paul Griffiths concisely puts it, "genetics and developmental biology provide no reason to accept [it]" (2008: 408). Complex adaptations are intricate machines, Tooby and Cosmides claim, which require "complex specification at the genetic level" (2005: 36). Moreover, "[l]ike any other intricate machine, the parts of a complex adaptation must all be present and fit together precisely if the adaptation is to work properly" (2005: 37). (In a related literature this is called "irreducible complexity" [Meyer 2000: 154–158].) They continue,

> If in a given generation, different individuals had different complex adaptations each of which was coded by a different suite of genes, then during the formation of the gametes for the next generation the random sampling of subsets of the parental genes would break apart each suite. . ..[T]he only way that each generation can be supplied with the genetic specification for complex adaptations is if the entire suite of genes necessary for coding for each complex adaptation is effectively universal and hence reliably supplied by each parent regardless of which genes are sampled.
>
> *(2005: 37)*

Thus the massive modularity of the mind must be maintained by its being monomorphic in the population.

Any appeal this argument might seem to enjoy disintegrates in the face of the biological evidence. For most traits there is significant redundancy at both developmental and genetic levels, which allows for the same genes to be expressed in a variety of ways and different genes to manifest the same phenotypic trait in different individuals (Jablonka and Lamb 2007: 356–357). As has been noted in a rather different context, the kind of irreducible complexity and complex specificity advanced by Tooby and Cosmides is manifestly bad design precisely because of its fragility (Sarkar 2011: 299–300). Moreover, one would expect that a trial and error haphazard "designer" like natural selection would typically produce phenotypes that are robust in the face of significant genetic variation.

Indeed, it is difficult to imagine how mental modules could evolve at all were the genetics underlying behavior as fragile as Tooby and Cosmides suggest. After all, natural selection—the differential survival and reproduction of individuals with different versions of a trait in a population—is impossible without there being variants of the trait under selection. It takes many generations, to produce a complex adaptation through the gradual piecemeal tinkering of natural selection. The more complex an adaptation the more likely that its evolutionary path was indirect, being selected initially for one role and then being coopted for another—a sequence of secondary adaptations on primary exaptations (Gould and Vrba 1982: 11–12). This requires that incipient stages on the way to the final module be either neutral or have some other function, which is difficult to jibe with an irreducibly complex machine. Equally puzzling is how extant modules could adapt to the addition of new ones given their extraordinary genetic fragility. According to the picture of the monomorphic mind offered by Tooby and Cosmides, the very genetic variations required for the evolution of such adaptations threaten the functional architecture of the mind/brain as a whole.

As much as anything else, this point suggests the impropriety of the "Evolutionary" moniker for NEP. It is vanishingly unlikely that evolutionary processes like natural selection would produce a perfectly engineered set of traits that are irreducibly complex, highly specified (without redundancy), perfectly integrated, and *universal* in a population. Fiat creation by an intelligent designer is the only process that could produce the large set of complex, content-specific, integrated, extremely genetically fragile, and evolutionarily fixed (though sometimes maladaptive) mental modules that are presupposed in Tooby and Cosmides' defense of the monomorphic mind.

Despite NEP's commitment to the idea of a universal human nature, the picture of humanity offered is, in fact, dimorphic. This is not, strictly speaking, irreconcilable with an irreducibly complex, universal suite of genes (though there is no reason to think it is true). Tooby and Cosmides postulate that differences between the sexes (and other life stage differences) "must be based on a genetic architecture that is largely universal and simply activated by an environmental trigger or a simple genetic switch such as a single locus (e.g., the unrecombining regions of the Y chromosome [which produces androgens])" (2005: 38). The idea seems to be that behavioral differences between the sexes are determined by cascading developmental effects that shape the mind as much as the reproductive organs.

At first glance, this appears to unite NEP theoretically with brain organization theory—another research program that looks to explain the biological basis for human gender dimorphism and has undergone significant feminist criticism (Fine 2010; Jordan-Young 2010). However, the unity is superficial. Even allowing the significant conceptual and methodological problems with the construction of dimorphism in brain organization theory, NEP is, in important ways, worse. As Blackless et al. (2000) have pointed out, dimorphism can be either moderate or absolute. There is moderate sexual dimorphism when differences are only found on average, with substantial overlap between the sexes (as, for instance, height). Absolute dimorphism is when there is no overlap or continuity between the two morphs (there isn't any physiological sex difference that will serve here, but if we weaken "absolute" to "relatively absolute," genitalia or gametes serve as adequate examples). Unlike brain organization theory (and most other research programs committed to behavioral sex differences), NEP's mono(di)morphic mind thesis seems to entail two absolutely dimorphic sets of complex, content-specific modules. Just as it is difficult to see how the monomorphic mind could evolve, it is difficult to see how these modules could develop. Given that sex-specific modules are supposedly hormonally regulated, one would expect a range of variants as hormones fluctuate in different people, given different environments, to different degrees throughout their development (Fausto-Sterling 1997: 250–254). This means that even if genetic monomorphism were plausible (and it isn't), the idea that the Y chromosome could initiate a hormonal cascade producing two absolutely dimorphic minds seems pretty implausible.

The third oft-criticized feature of NEP is the idea of the Environment of Evolutionary Adaptedness (EEA). Narrow evolutionary psychologists emphasize that the human mind evolved in the Pleistocene and maintain that the relevant EEA was determined by "the hunter-gatherer lifestyle," which is the imagined backdrop for the evolutionary narratives that ground their hypotheses. Importantly, this environment is not an actual environment but instead "refers jointly to the problems hunter-gatherers had to solve and the conditions under which they solved them" (Tooby and Cosmides 2005: 22).

By stipulating a distinct environment for our hominid ancestors that has nothing to do with the physical environment, NEP simply assumes human uniqueness. This is achieved by some conceptual sleight of hand, treating an evolutionary population as a cultural group—hunter gatherers—rather than a biological one—say, social omnivores. This at once subtly denies that culture could be relevant for the evolution of other primates (despite evidence to the contrary [Caldwell and Whiten 2011]), while simultaneously denying that anything other than culture could have affected human cognitive evolution or that there might be continuities in cognitive, affective, and social capacities and behavior between humans and other animals.

Equally bizarre is the idea that the evolution of human behavior stopped at the end of the Pleistocene. There is, after all, evidence of physiological evolution in the last 10,000 years—for instance, the evolution of lactose digestion in agricultural groups who raise cattle (Gerbault et al. 2011). More peculiar yet is the idea that the evolution of the human mind began in the Pleistocene. The implausibility of this claim is increased by the fact that most of the so-called "problems" that our ancestors had to evolve to solve are shared by other animals. Tooby and Cosmides mention "finding a mate, cooperating with others, hunting, gathering, protecting children, navigating, avoiding predators, avoiding exploitation" (2005: 19); predator vigilance, prey stalking, mate selection, childbirth, parental care, coalition formation, disease avoidance (2005: 20), while David Buss (1995) emphasizes the sex specificity of adaptive problems noting that men face one set—e.g. "Paternity uncertainty," "Identifying reproductively valuable women," "Gaining sexual access to women"(1995: 165)—while women face another—e.g. "Identifying men who are able to invest" and "Identifying men who are willing to invest" (1995: 165). Even if these were the problems driving evolutionary change, many would, presumably, be shared by sexual animals generally. It makes little sense to speculate about our hominid ancestors when we can just investigate closely related species in the present, which provides at least prima facie evidence for the psychological traits of our shared ancestors. Of course, when one looks at the other currently extant great apes—Chimpanzees, Bonobos, Gorillas, and Orangutans—one doesn't find a single dominant way of life or reproductive strategy, but rather a remarkable diversity (Campbell et al. 2011). This suggests that, at least when it comes to adaptive cognitive traits, including those relevant to reproduction, there are a wide variety of different social structures with a variety of different "problems" and possible "solutions" that might have driven the psychological evolution of our ancestors.

Ultimately, there are no good reasons for adopting any of these central theses of NEP. Certainly, not everyone working in NEP explicitly adopts each of these commitments (and even Tooby and Cosmides vacillate on the degree of modularity of the mind, the universality of human nature [2005: 30–40] and whether it's just the Pleistocene that matters [43]). However, because of the cumulative character of science, the massive modularity thesis, the monomorphic mind thesis, and the presupposition of human uniqueness and concomitant focus on the Pleistocene seep into many related research programs and studies, and the results of NEP are often uncritically accepted. One of the most powerful reasons given for accepting NEP's findings is that their hypotheses are almost always confirmed by experiment. However, as we shall see in the next two sections, these apparently impressive results are due to a fundamentally flawed method and actually provide no grounds for accepting NEP's evolutionary hypotheses.

A Case Study—NEP's Methods Assessed

Again Tooby and Cosmides serve us by not only articulating NEP's "systematic method for using theories of adaptive function and principles of good design for discovering new programs" (Tooby and Cosmides 2005: 28) but also offering an exemplary case study. First the method: one starts by imagining the EEA and coming up with various theories about the adaptive problems faced by our ancestors—remembering, of course, that many of these problems will be sex-specific. Then,

> From the model of an adaptive problem, the researcher develops a task analysis of the kinds of computations necessary for solving that problem, concentrating on what would count as a well-designed program given the adaptive function under consideration. Based on this task analysis, hypotheses can be formulated about what kinds of programs might actually have evolved.
>
> *(2005: 28)*

Next, one operationalizes the hypothesized program, specifying a test hypothesis, i.e. deciding what observable contemporary behaviors will count in support of or against its existence. Then one tests the test hypothesis. If it is confirmed in one population one then tests it against alternatives and cross-culturally (2005: 28).

Philosophers of science will notice how far away this is from the ideal of coming up with a hypothesis and trying to refute it. As Karl Popper noted long ago, confirmations of a theory are easily found; genuine tests are ones that attempt to falsify hypotheses (Popper 1963: 36). (Narrow evolutionary psychologists cannot even test their hypotheses against competitors [Fausto-Sterling 1997: 244] because, unless these competing hypotheses follow the tenets of NEP, they belong to the Standard Social Science Model, which, by stipulation, is a nonstarter.) One might demur that, as this ideal is rarely met, it is unfair to hold NEP to this standard. However, the problems with NEP's hypotheses and methods are particularly acute. Because the evolutionary hypotheses are not about behaviors, which are only signals of the underlying programs under study, any given module may have "different expressions triggered by different environmental, or social conditions, or local calibration by specific circumstances" (Tooby and Cosmides 2005: 28). This neatly ensures that a negative test result need never be taken to falsify or even provide significant evidence against a given evolutionary hypothesis, let alone the underlying theoretical commitments. Ad hoc hypotheses designed to save theories from falsification, which were derided by Popper (1963: 37), are built right into NEP's method.

We now turn to a case study to see NEP's method in action. In "Formidability and the Logic of Human Anger" (or as I call it, "Why Jocks are Jerks and Pretty Girls are Bitches") Aaron Sell, Tooby, and Cosmides (STC) test their recalibration theory of anger. "This theory proposes that anger is produced by a neurocognitive program engineered by natural selection to use bargaining tactics to resolve conflicts of interest in favor of the angry individual" (2009: 15073). They lay out the problem faced by our ancestors: "For a given choice set involving self and other, how much weight should be placed on the welfare of the other compared with the self?" (2009: 15073). Their hypothesized *program* "integrate[s] the welfare-relevant inputs (e.g., cues of kinship, formidability)" and weighs the "welfare tradeoff ratio" (WTR) in these encounters so that we can judge when the optimal solution in some encounter is to put another's welfare before our own (2009: 15073). Each of us has a pre-set ratio in our "motivational architecture" that skews either toward being self or other favoring, depending on what we need and what we have to offer (2009: 15074).

Now, assuming the existence of WTRs, they "could have selected for adaptations whose function is to reach out into the brain of a target, so to speak, and adjust upwards that target's WTR toward oneself" (Sell et al. 2009: 15074). This is the function of anger—"to recalibrate the WTR_{ji} [where j is the target and i is the actor] in the target's brain, increasing its magnitude so that the target subsequently places more weight on the welfare of the angry individual" (2009: 15074). Basically, when one detects that one is not being treated as well as one deserves, one gets angry to persuade the other to treat one better. But what can possibly ground these judgments? There are "two interpersonal negotiating tactics available to organisms: inflicting costs (aggression); or withdrawing or downregulating expected benefits" (2009: 15074). When the inflicted costs or denied benefits for the target of anger are greater than the costs of increasing the WTR in favor of the angry individual, it is rational for the target to recalibrate the ratio. Importantly, "this threshold also defines the conditions in which anger should be triggered in the actor" (2009: 15074). STC explain:

> Because these factors give the actor greater leverage over the target, individuals who are more formidable and individuals who are better able to confer benefits should feel 'entitled to' a higher WTR from others (that is, they should expect better treatment), should get angry when they do not receive it, and should (other things being equal) prevail more in conflicts of interest.
>
> *(Sell et al. 2009: 15074)*

While STC readily admit that in a "noisy, uncertain world" there are many other ways to "inflict costs" and "confer benefits," "for simplicity of operationalization" they "selected two for an empirical test of the model: strength and attractiveness" (2009: 15074). STC then hypothesize about the ways in which the EEA would have shaped the sexes differently. They take a moment to reflect on the evolutionary importance of strength in men for inflicting costs (i.e. thumping people) or threatening to inflict them, and then explain that, because of the supposedly vast strength differences between men and women, strength is unlikely to matter for women (2009: 15075). However, although attractiveness is important for both sexes, so that attractive people "register a higher conferral index, and implicitly expect a higher WTR from others" (2009: 15075), attractiveness grounds a weighting strategy that is typically deployed by women. Because access to female reproductive capacity "was a far greater limiting factor for male fitness than access to male sexuality was for females. . ., even small changes in the probability of a woman's granting sexual access constitute a powerful benefit" (2009: 15075).

It is worth reflecting on the special pleading that goes into these narratives. Even were we to grant that there are vast differences in strength between the sexes, this would not lead to a switching off of women's strength-based anger program as the WTR module ranks comparative capacities to "inflict costs." Relatedly, the benefit of reproductive access is not one that women can confer on other women and while doubtless reproductive technologies will correct this lacuna soon, this still wouldn't make it relevant for the EEA. However, women, attractive or not, can still "inflict costs" on each other; indeed one imagines that this might be a good strategy for strong women to improve their comparative attractiveness to males with good genes. This suggests the kind of risky prediction of which Popper would have approved: according to the EEA described by STC, attractive women should *only* get angry with men—the only possible target of their benefit conferral—and they should tend *not* to get angry with women who are stronger than they are. Unfortunately, neither these hypotheses, nor any other alternative hypotheses that might refute either the existence of the WTR or the use of anger in calibrating it, were tested.

Instead, STC tested 11 predictions: that strength and attractiveness lead to "(i) [and (iv)] greater success in resolving conflicts in one's favor, (ii) [and (v)] greater sense of entitlement. . ., and (iii) [and (vi)] greater anger-proneness"; that strength has a greater effect for men (vii) and attractiveness for women (viii); and that stronger men have a greater history of fighting (ix) and approve the use of force in personal (x) and state (xi) conflicts (2009: 15074). Various measures of anger proneness and aggression were operationalized through a survey asking participants to assess their own tendencies and attitudes; strength was operationalized through three measures—self and other assessment, lifting strength, and bicep circumference—while attractiveness was based purely on self-assessment in comparison to others. Two separate studies were done; one with "62 men (mean age, 21) . . .recruited from a gym at the University of California, Santa Barbara (UCSB)" and the other with "125 men (mean age, 20) and 156 women (mean age, 19) . . .recruited from the UCSB student center" (2009: 15078).

As is so often the case for NEP, their results are impressive. STC found statistically significant correlations "between physical strength and the seven anger-relevant instruments" (2009: 15075) in men but not in women (with the exception of "success in conflict" [2009: 15076]). However, "attractive women look like strong men on all measures [in their pattern of responses] except (as expected) their history of fighting" (2009: 15076). (The case of attractive men was thought to be confounded by the fact that strength was also a feature of male attractiveness and results were mixed.)

STC summarize their results: "Eleven predictions were derived from the recalibrational theory of anger, and all of them were empirically supported. No other theory predicts this complex, subtle, and precise pattern of results" (2009: 15077). Contra the self-aggrandizing rhetoric, in fact, any theory that predicts that people tend to conform to social type would predict these results. After all, it is a persistent theme in North American teen popular culture that jocks are jerks and pretty girls are bitches. That social pressures and self-fulfilling prophesies would produce conformity to this trope among undergraduates and gym members at a California university is hardly surprising and seems, minimally, a reasonable alternative hypothesis for which one does not require any evolutionary speculations. STC might respond that there are always alternative hypotheses and, until they are tested and shown to be better, their results confirm all 11 test hypotheses and thus provide strong evidence in favor of the recalibration theory of anger.

While a Popperian would give such a response short shrift given the apparent lack of riskiness of STC's predictions, in fact, the methodological flaws are even more profound. The method of NEP described at the beginning of this section inevitably produces predictions (or test hypotheses) that are so unrisky that their confirmation provides no reason at all for thinking that their evolutionary hypotheses are true. To see this, we need to go back to Tinbergen's four questions. We start with his insight about the speculative character of claims about evolution and his recognition that the only way to ground a hypothesis about the evolution of behavior is by comparison with closely related species or by inference from the other three causes—efficient causation, development, and current survival value. NEP rules out the relevance of comparisons with our closest kin by stipulating that the EEA is limited to the Pleistocene (which significantly postdates our lineage's divergence from that of chimpanzees). The only alternative is to acquire one's hypotheses from the other three causes, i.e. contemporary behavior. There simply isn't anywhere else to go. Thus, when imagining an "adaptive problem encountered by human ancestors, including what information would potentially have been present in past environments for solving that problem" (Tooby and Cosmides 2005: 28), all a narrow evolutionary psychologist can do is project what they believe about contemporary behavior and imagine familiar scenarios dressed up in an EEA setting.

Once we recognize that NEP's hypotheses cannot but rest on their beliefs about contemporary human behavior we see that their method is hopelessly circular.[4] Informed by folk observations of human behavior they project these into a fictional Pleistocene past, constructing narratives in terms of adaptive problems, mental modules, and computations. Then they operationalize their evolutionary hypotheses coming up with test hypotheses about contemporary behaviors and, lo and behold, their hypotheses are confirmed. To suggest that their speculative evolutionary hypotheses are "no more likely to be falsified than the hypotheses advanced by nonevolutionary researchers" (2005: 25) surely understates the case; it is highly unlikely that any of their hypotheses will be ever be falsified as the hypotheses themselves rest on the folk observations of the populations that are ultimately used to test them.

It is a mistake, however, to think that the evolutionary narrative is irrelevant. NEP's detour through evolutionary scenarios is what justifies their claims that they are discovering the mental modules that comprise universal human nature. This has the unnerving implication that we are stuck with them.[5]

Values, Science, and Rhetoric

Perhaps it is a fatalistic acceptance of innate cognitive and affective sex dimorphism that has inspired some researchers to form a new subdiscipline, "Feminist Evolutionary Psychology" (FNEP) (Weaver 2017). The appropriateness of the feminist moniker is certainly up for debate, as Sara Weaver has shown in her careful analysis of this research program (2017). While Weaver has found real efforts in FNEP studies to treat women as active evolutionary agents and address women's issues, such as gender inequality, she also notes a failure to carefully engage feminist criticisms of NEP (or feminist scholarship more generally), as well as a promotion of "social roles that are conducive to a patriarchal society" (2017: 116). Of course, insofar as they adopt the theoretical assumptions and methods of NEP—massive modularity, the mono(di)morphic mind, the EEA, adaptationism, and a circular method (and they typically do [Weaver 2017])—FNEP will be profoundly flawed, regardless of any feminist scruples.

As I have argued in this chapter, the massive modularity thesis is inconsistent with contemporary neuroscience; and, when considered in concert with the monomorphic mind thesis, it is difficult to see how this hypothesized set of fragile, highly specific mental modules could possibly evolve. The absolute sexual dimorphism that appears to be assumed by NEP is equally implausible, given the fluctuating character of hormones and the contingency of development. The EEA surreptitiously presupposes a human uniqueness that is highly unlikely given that the adaptive problems supposedly faced by hominids are mostly shared not only with our primate relatives but by many other sexual animals (e.g. mate recognition). Insofar as narrow evolutionary psychologists engage in weak tests of hypotheses drawn from imaginary evolutionary scenarios that are unavoidably dependent on folk conceptions of sex differences and behaviors, their apparent empirical successes provide no reason to accept their evolutionary hypotheses. Alternative methods that don't assume human uniqueness—for instance, using cross-species comparisons—promise a more robust foundation for the study of the evolution of human behavior, though they too should be expected to be consistent with contemporary neuroscience, developmental biology, and evolutionary theory and be able to withstand feminist scrutiny.

This brings me back to the point with which I began. A rejection of NEP entails nothing about the legitimacy of evolutionary psychology broadly construed. Certainly, research that depends on the results of NEP or relies on one (or more) of its problematic theoretical commitments will be seriously flawed; but nothing in evolutionary theory or psychology requires one to do this

Of course, critics have taken exception to more than just the scientific failings of NEP. Feminists, in particular, point to moral and political problems with this research, and these are considerable. There is, for instance, the ableism implicit a view of *universal* human nature that diagnoses neurodiverse people as missing various mental modules, effectively denying that they have a fully human nature. And, of course, the presupposition of gender essentialism—assuming ardent, competitive males and coy but nurturing females—is unlikely to prove conducive to a post-patriarchal society where women, queer, and transgender people are truly equal.

Interestingly, the political commitments of feminist critics are often cited by advocates of NEP as grounds to dismiss them. Some accuse feminists of committing the Naturalistic Fallacy—confusing findings of fact with statements of value (Nelson 2017)—while simultaneously maintaining that NEP has no political agenda and that their scientific results have no normative implications. As Lynn Hankinson Nelson notes, this fails to appreciate the force of the feminist critique. NEP's hypotheses are "evaluatively thick," making claims that "have both empirical and normative content" (2017: 259)—an insight that STC's study of "Formidability and the Logic of Anger" exemplifies.

One is left wondering if these dismissals of feminism are a rhetorical tactic more than anything else. Indeed, on inspection, the rhetorical strategies of NEP mirror those one finds in the contemporary radical right. They accuse their detractors of their own central failings, identifying themselves as the plucky defenders of truth, common sense, and justice—standing up in the face of an ideologically driven foe. Then they use the moral licensing that they attain from this positioning to simply ignore the many careful, well-evidenced criticisms leveled against them. Because they evince exactly the flaws that they lay at their detractors' feet, this means that we can summarize the failings of NEP roughly in their own words simply by swapping "NEP" in for "the Standard Social Science Model":

> For almost [three decades] adherence to [NEP] has been strongly moralized within [parts of] the scholarly world, immunizing key aspects from criticism and reform. . ..As a result, in the international scholarly community, criteria for belief fixation have often strayed disturbingly far from the scientific merits of the issues involved, whenever research trajectories produce results that threaten to undermine the credibility of [NEP]. Nevertheless, in recent decades, the strain of ignoring, exceptionalizing, or explaining away the growing weight of evidence contradicting [NEP] has become severe. Equally, reexaminations of the arguments advanced in favor of the moral necessity of [NEP] suggest that they—at best—result from misplaced fears. . ..Indeed, we may all have been complicit in vast tides of human suffering—suffering that might have been prevented if the scientific community had not chosen to postpone or forgo a more veridical social and behavioral science.
>
> *(2005: 7, additions mine)*

Acknowledgments

I thank audiences at CSU Fresno, Mount Allison University, Dalhousie University, and the University of King's College for helping me work through these ideas.

Related chapters: 6, 10, 11, 14, 24, 25, 26.

Notes

1 In this section and the next I draw heavily from Paul Griffiths' (2008) excellent overview (see also Meynell 2012).
2 It's tempting to romanticize mid-twentieth-century ethology. However, Lorenz was a member of the Nazi Party who flourished under the Third Reich and a number of his theories reflected and supported

Nazi eugenic ideology (Klopfer 1994). In using classical ethology to provide useful correctives to NEP I do not mean to suggest that it is not also fraught with various pernicious political attitudes and problems.

3 There are reasons to think that this supposedly narrow impairment may be more complex and variable than Tooby and Cosmides suggest (see e.g. Thorup et al. 2017).

4 Stotz and Griffiths (2003) make a similar point, with rather more nuance and detail than I can offer here; I owe the substance of this analysis to Griffiths' presentation on the topic at the *Idea of Evolution* Workshop, Dalhousie University, October 2009.

5 See Morton et al. (2009) for some troubling empirical confirmation of this worry.

References

Battro, A. (2000) *Half a Brain Is Enough: The Story of Nico*. Cambridge: Cambridge University Press.

Blackless, M., et al. (2000) "How Sexually Dimorphic Are We?" *Review and Synthesis, American Journal of Human Biology* 12(2): 151–166.

Buller, D. (2005) *Adapting Minds: Evolutionary Psychology and the Persistent Quest for Human Nature*. Cambridge: MIT Press.

Buss, D. (1995) "Psychological Sex Differences: Origins Through Sexual Selection," *American Psychologist* 50(3): 164–168.

Caldwell, C. and A. Whiten (2011) "Social Learning in Monkeys and Apes," in C. Campbell, et al. (eds.), *Primates in Perspective*. Oxford: Oxford University Press, pp. 652–662.

Campbell, C., et al. (eds.) (2011) *Primates in Perspective*. Oxford: Oxford University Press.

Dusek, V. (1984) "Sociobiology and Rape," *Science for the People* 16(1): 10–16.

Fausto-Sterling, A. (1997) "Beyond Difference: A Biologist's Perspective," *Journal of Social Issues* 53(2): 233–258.

Fausto-Sterling, A., P. Gowaty and M. Zuk (1997) "Review: Evolutionary Psychology and Darwinian Feminism," *Feminist Studies* 23(2): 402–417.

Fine, C. (2010) *Delusions of Gender: How Our Minds, Society and Neurosexism Create Difference*. New York: W.W. Norton and Company.

Gerbault, P., et al. (2011) "Evolution of Lactase Persistence: An Example of Human Niche Construction," *Philosophical Transactions of the Royal Society B Biological Sciences* 366(1566): 863–877.

Gould, S.J., et al. (1979) "The Politics of Sociobiology," *The New York Review of Books* 26(9). https://www.nybooks.com/articles/1979/05/31/the-politics-of-sociobiology/. Accessed Sept 6, 2020.

Gould, S.J. and R. Lewontin (1979) "The Spandrels of San Marco and the Planglossian Paradigm: A Critique of the Adaptationist Programme," *Proceedings of the Royal Society of London, Series B* 205(1161): 581–598.

Gould, S.J. and E. Vrba (1982) "Exaptation—A Missing Term in the Science of Form," *Paleobiology* 8(1), 4–15.

Griffiths, P. (2008) "Ethology, Sociobiology and Evolutionary Psychology," in S. Sarkar and A. Plutyinski (eds.) *A Companion to Philosophy of Biology*. Oxford: Blackwell.

Hubbard, R. (1988) "Science, Facts, and Feminism," *Hypatia*, 3(1): 5–17.

Jablonka, E. and M.J. Lamb (2007) "Précis of *Evolution in Four Dimensions*," *Behavioral and Brain Sciences* 30(4): 353–365.

Jordan-Young, R. (2010) *Brain Storm: The Flaws in the Science of Sex Differences*. Cambridge: Harvard University.

Klopfer, P. (1994) "Konrad Lorenz and the National Socialists: On the Politics of Ethology," *International Journal of Comparative Psychology* 7(4): 202–208.

Lloyd, E. (2015) "Adaptationism and the Logic of Research Questions: How to Think Clearly About Evolutionary Causes," *Biological Theory* 10(4): 343–362.

Mayr, E. (1983) "How to Carry Out the Adaptationist Program?" *The American Naturalist* 121(3): 324–334.

Meyer, S., (2000) "Qualified Agreement: Modern Science and the Return of the 'God Hypothesis,' " in R. Carlson (ed.) *Science and Christianity: Four Views*. Downers Grove: InterVarsity Press.

Meynell, L. (2012) "Evolutionary Psychology, Ethology, and Essentialism (Because What They Don't Know Can Hurt Us)," *Hypatia* 27(1): 3–27.

Morton, T., et al. (2009) "Theorizing Gender in the Face of Social Change: Is There Anything Essential about Essentialism?" *Journal of Personality and Social Psychology* 96(3): 653–664.

Nelson, L.H. (2017) "Evolutionary Psychology, Feminist Critiques Thereof, and the Naturalistic Fallacy," in M. Ruse and R. Richards (eds.) *The Cambridge Handbook of Evolutionary Ethics*. Cambridge: Cambridge University Press.

Panksepp, J. and J. Panksepp (2000) "The Seven Sins of Evolutionary Psychology," *Evolution and Cognition* 6(2): 108–131.

Popper, K. (1963) *Conjectures and Refutations*. New York: Basic Books.

Rose, S. and H. Rose (1986) "Less than Human Nature: Biology and the New Right," *Race and Class* XX-VII(3): 47–66.

Sarkar, S. (2011) "The Science Question in Intelligent Design," *Synthese* 178(2): 291–305.

Sell, A., J. Tooby and L. Cosmides (2009) "Formidability and the Logic of Human Anger," *Proceedings of the National Academy of Sciences* 106(35): 15073–15078.

Smith, S. (2020) "Is Evolutionary Psychology Possible?" *Biological Theory*, 15: 39–49.

Stotz, K. and P. Griffiths (2003) "Dancing in the Dark: Evolutionary Psychology and the Argument from Design," in S. Scher and R. Rauscher (eds.) *Evolutionary Psychology: Alternative Approaches*. Dordrecht: Kluwer, pp. 135–160.

Thorup, E., J. Kleberg and T. Falck-Ytter (2017) "Gaze Following in Children with Autism: Do High Interest Objects Boost Performance?" *Journal of Autism and Developmental Disorders* 47(3): 626–635.

Tinbergen, N. (1963) "On Aims and Methods of Ethology," *Zeitschrift für Tierpsychologie* 20: 410–433.

Tooby, J. and L. Cosmides (1992) "The Psychological Foundations of Culture," in J. Barkow, L. Cosmides and J. Tooby (eds.) *The Adapted Mind: Evolutionary Psychology and the Generation of Culture*. New York: Oxford University Press, pp. 19–136.

Tooby, J. and L. Cosmides (2005) "Conceptual Foundations of Evolutionary Psychology," in D. Buss (ed.), *The Handbook of Evolutionary Psychology*. Hoboken: John Wiley and Sons.

Tovar-Moll, F., et al. (2014) "Structural and Functional Brain Rewiring Clarifies Preserved Interhemispheric Transfer in Humans Born Without the Corpus Callosum. *PNAS* 111(21): 7843–7848.

Weaver, S. (2017) *A Constructive Critical Assessment of Feminist Evolutionary Psychology* (Doctoral dissertation). Retrieved from UWSpace (https://uwspace.uwaterloo.ca/bitstream/handle/10012/12772/Weaver_Sara.pdf).

Wilson, E.O. (1975) *Sociobiology: The New Synthesis*. Cambridge: Harvard University Press.

Further Reading

Buller, D. (2005) *Adapting Minds: Evolutionary Psychology and the Persistent Quest for Human Nature*. Cambridge: MIT Press. (A general, accessible overview of NEP and the problems with it.)

Gowaty, Patricia (ed.) (1997) *Feminism and Evolutionary Biology: Boundaries, Intersections, and Frontiers*. New York: Chapman and Hall. (A collection of essays, mostly by scientists, offering a variety of perspectives on feminist engagements with evolutionary biology.)

Lloyd, E. (2005) *The Case of the Female Orgasm: Bias in the Science of Evolution*. Cambridge: Harvard University Press. (An excellent case study of poor evolutionary reasoning about a sex-specific trait.)

Prum, R. (2017) *The Evolution of Beauty: How Darwin's Forgotten Theory of Mate Choice Shapes the Animal World—and Us*. New York: Doubleday. (An interesting [if controversial] account of sexual selection that challenges many of the presuppositions of NEP.)

Rose, H. and S. Rose (eds.) (2001) *Alas Poor Darwin: Arguments against Evolutionary Psychology*. London: Vintage. (A useful interdisciplinary collection of essays criticizing NEP.)

24

NEUROSEXISM AND OUR UNDERSTANDING OF SEX DIFFERENCES IN THE BRAIN

Robyn Bluhm

A recent special issue of the *Journal of Neuroscience Research* proclaims that research on sex differences in the brain is "an issue whose time has come." In his introduction to the issue, guest editor Larry Cahill describes his foray into this area of research in the early twenty-first century: a senior colleague warned him to be careful, because the study of sex differences was the "third rail." Cahill agreed that this area was so controversial that "exploring sex influences was indeed a terrific way for a brain scientist not studying reproductive functions to lose credibility at best, and at worst, become a pariah in the eyes of the neuroscience mainstream" (2017: 12). In part because of this, he suggested, very few neuroscientists studied sex differences.

Feminist analyses of research in neuroscience and psychology, however, tell a different story about research on sex differences. Scholars from various disciplinary backgrounds, including neuroscience, history, psychology, and philosophy, have documented a long history of scientists searching for sex differences in the brain, as well as sex differences in behavior, personality, and cognitive abilities that were, in turn, attributed to as-yet-undiscovered differences in the brain. They have also argued that this research relies on stereotypes about women and men and on a framework that views sex differences as inevitable, because they are rooted in "biology" (as opposed to, for example, differences in the socialization of girls and boys). Cordelia Fine (2008) has described such research as exhibiting "neurosexism," which has unsurprisingly been poorly received by scientists who study sex differences.

My aim in this chapter is to show where and how contemporary neuroscience research on sex differences is characterized by neurosexism. I will begin by using a historical case study, Elizabeth Fee's research on the nineteenth-century science of craniology, that provides a paradigm case of neurosexism. This case, together with Fine's explanations of neurosexism, shows that neurosexism has both scientific and social aspects, which can be further divided into five distinct forms of neurosexism. Next, I describe the basic framework that guides contemporary sex difference research in neuroscience, which traces sex differences in brain structure and function to differences in exposure to prenatal sex hormones and to differences in the adaptive challenges faced by our female and male ancestors. With this background in place, I turn to feminist criticisms of sex difference research, as well as the response of researchers in this area, to show that although much has changed since the days of craniology, there is still significant neurosexism in contemporary discussions of the neuroscience of sex differences.

Craniology and Neurosexism

During the second half of the nineteenth century, social reformers began to question the "natural" social order, which placed wealthy white men at the apex of society. These social movements informed first-wave feminism and led to the demands of women for education and for the right to vote. At the same time, a corresponding wave of biological research on sex difference aimed to prove women's weakness, intellectual limitations, and inability to participate in public life. Elizabeth Fee has analyzed the anthropological field of craniology, which was deployed during that period to show that women's brains (or, more properly, their skulls) were different from men's – and different in ways that demonstrated feminine inferiority.

In its early days, craniology made some explicit assumptions that served as a theoretical justification for using measurements of the skull to draw conclusions about the brain. Scientists believed, first, that brains were analogous to muscles, in that both kinds of tissue increased in size with increasing "strength." Second, the skull closely reflected the shape and size of the brain, so that it could be used as an adequate proxy for the brain itself (Fee 1979: 420). Third, the brain functioned as a unitary whole (rather than having different mental functions localized to different areas of the brain), so that a bigger brain was a better one. Together, these assumptions made it reasonable to believe that skull size reflected brain function. Drawing on these background assumptions, craniologists demonstrated that, on average, women's skulls were smaller than men's, providing an explanation for their inferior intelligence (421).

Leaving aside the fact that, on average, women were smaller than men, so that differences the size of their skulls were proportionate to the differences in the size of the rest of the skeleton, craniologists soon realized that they faced a problem. If the absolute size of the skull was an accurate reflection of intelligence, "then the elephant and the whale must be the lords of creation" (ibid.). At this point, craniologists began to engage in a series of intellectual contortions to identify the true way to measure women's intellectual inferiority. They considered the relative size of the skull to the rest of the body, though this was also problematic, as some studies showed that by this measure women's skulls were larger than men's. They then began to consider differences *within* the skull such as the proportion of cranial bones to facial bones, the proportion of parietal to frontal bones, the ratio of the width and length of the skull, the angle between different parts of the skull, or the angle between the skull and the vertebral column (421–426). Over the course of several decades,

> [t]he number of cranial angles and indices multiplied at a prodigious rate…in what may be called the Baroque period of craniology. Each person involved in craniological research employed his own favorite measurements and many contributed new ones to the growing number already in existence.
>
> *(426)*

Note, too, that these changes moved craniological measures away from their foundational assumptions, as many of these new measurements focused on aspects of the skull that were not relevant to beliefs about the function of the brain. By the early twentieth century, Fee concludes, "craniology as a scientific specialty had collapsed under its own weight" (432).[1]

Despite the many methods craniologists used to try to explain women's inferiority, one thing that never wavered during the rise and fall of craniology was the belief that women were less intelligent than men. Moreover, social reformers who pushed for women to receive an education – or even to be allowed to vote – were ignoring the natural order of things. In other words, the craniologists' motivations were explicitly political:

> In the guise of defending true womanhood, the English anthropologists wanted to protect women from the false doctrines of reformers by presenting the true facts and natural laws

which governed woman's role…It was obvious to them that the social reformers were anti-scientific. Either the reformers were ignorant of, or afraid of, the facts of anatomy and physiology, for these were seldom mentioned in feminist tracts. Instead, their arguments were couched in abstract terms, such as "justice", "freedom", and "humanity"…In contrast, the anthropologists said they wanted simply to assign woman her true place in nature, so that she could live in accordance with her biological destiny.

(416–417)

Craniology, as analyzed by Fee, is a paradigm case of neurosexism. The term "neurosexism" was coined by Cordelia Fine to refer to "[t]he ugly rush to cloak old-fashioned sexism in the respectable and authoritative language of neuroscience" (2008: 69), which she observed in a number of popular neuroscience-based self-help books. She later extended the concept to assess neuroscience research itself, suggesting that research exhibits neurosexism when it "reinforces and legitimates traditional gender roles and stereotypes in ways that are not scientifically justified" or when it involves "bias in the way that…research on sex differences is conducted and interpreted, detrimental effects for understanding the complex phenomenon of gender, as well as harmful social and psychological effects from the reification of gender roles" (2013: 370). There are therefore two main aspects or manifestations of "neurosexism": first, an explicitly social one that uses neuroscience research on sex differences to draw conclusions about gender differences and gender roles, and second, one that involves the use of scientific practices that tend to enable the former kind of neurosexism.

The case of craniology clearly involves both kinds of neurosexism. With regard to the "scientific neurosexism" involved in craniological research, the dubious methods used by craniologists meant that their conclusions owed more to gender stereotypes than to their data. There was also "social neurosexism," as the craniologists began with the explicit aim of "explaining" women's inferiority through cutting-edge scientific methods, and they drew sweeping conclusions about appropriate gender roles based on their conclusions. Drawing on Fee's analysis, I would also add an additional form of neurosexism that was not discussed by Fine: the dismissal of critics as being unscientific, or unwilling to face the biological facts. As I will show later in the chapter, this kind of neurosexism has both social and scientific aspects.

In the case of contemporary neuroscience research on sex differences, however, things are more complicated. Later in the chapter, I will argue that although researchers today are not overtly sexist, both scientific and social neurosexism are still problems in contemporary neuroscience research. In order to show this, however, I will first describe the scientific framework that informs this research.

A Framework for the Neuroscience of Sex Differences

Just as craniology flourished during a time in which women's rights advocates and other social reformers were demanding changes that threatened the existing (and purportedly "natural") social order, the 1960s and 1970s saw another period of both social change and research on sex differences. It also saw the emergence of explicitly feminist criticisms of research on sex differences in neuroscience and in psychology. In many ways, both the guiding assumptions and the feminist criticisms of sex difference research remain the same now as they were in the 1970s. The framework that emerged during this period for studying sex differences has remained largely unchanged and feminist critics therefore continue to raise similar questions and criticisms.

Sex difference research in neuroscience is guided by the idea that prenatal exposure to sex hormones (androgens and estrogens) permanently "organizes" the brain to take a basic masculine or feminine form. This idea dates back to 1959, with the publication of a paper that examined the effects of prenatal exposure to testosterone on female guinea pigs (Phoenix et al. 1959). The animals were born with "masculinized" genitalia, which was consistent with the so-called Jost paradigm, in which the chromosomal sex of an animal determined its gonadal sex, and the gonads, in turn, determined the development of the animal's reproductive organs. Phoenix et al. extended this idea to investigate reproductive behaviors and found that, when injected with more testosterone as adults, female guinea pigs that had been exposed to high levels of testosterone during fetal development displayed more frequent male-typical sexual behavior (e.g. mounting) than animals that had not been exposed to prenatal testosterone. When injected with the "female" sex hormones, estrogen and progesterone, as adults, the animals were less likely than unexposed females to display female-typical sexual behavior (e.g. lordosis).

Subsequent studies further developed this line of research, ultimately leading researchers to conclude that early exposure to sex hormones had a permanent "organizing" effect on the brain, influencing neural structures and thus also neural function. When an animal reached sexual maturity, the increased production of sex hormones (mimicked by the injections given in the Phoenix et al. study) had "activating" effects on sexual behavior. In other words, normal sexual behavior depended on both the prenatal organizing effect of sex hormones on the brain, and their activating effects once sexual maturity was reached.[2]

It is difficult to overstate the influence of brain organization theory. In a review article published to mark the 50th anniversary of the original Phoenix et al. paper, Arthur Arnold notes that these researchers "provided a conceptual framework that has been repeatedly tested and improved since 1959, but has not been substantially undermined by experimental findings in the intervening half century" (2009: 570). Note, however, that the research I have described was conducted on rodents, not on human beings. Moreover, it focused primarily on the influence of sex hormones on reproductive physiology and behavior. But it did not take long for researchers interested in human sexuality and sexual development to begin to draw on brain organization theory. From there, it was a short step to investigating the potential influence of prenatal hormones on a wider range of gendered behaviors and attributes.

Initially, research on brain organization in humans was conducted in individuals with intersex conditions, on the grounds that these conditions involved exposure to the "wrong" hormone levels for the individual's sex. For example, congenital adrenal hyperplasia (CAH) is a condition in which female fetuses are exposed to higher-than-normal levels of androgens during development, with the result that they are born with genitalia that look masculine, or intermediate between female structures. John Money and Anke Ehrhardt conducted a number of studies involving girls and women with CAH, to see whether they were more likely than their sisters without CAH to have masculine tendencies such as (in younger girls) interest in boys' toys and in rough-and-tumble play or (in older girls and women) an increased interest in a career, decreased interest in motherhood, or sexual attraction to women (Money and Ehrdardt 1972).

To summarize, by the early 1970s, brain organization theory was well established in research on animal behavior and had been extended to an explanation of "masculine" behaviors in women, and "feminine" behaviors in men, with intersex conditions. The next question was whether and how this framework could be used to study sex differences in women and men without intersex conditions. Unlike animal researchers, who could inject hormones into their research subjects and kill them to study their brains, researchers studying human behavior were comparatively limited. Psychologists could study behavioral differences between women and men and, for individuals with intersex conditions, could link them to known alterations in sex hormone levels. But for a long time, the only way to investigate the neural basis of those behavioral differences was to measure differences in the size and shape of particular brain structures in donated cadaveric brains.[3]

Things began to change in the 1990s, for two reasons. First, the development of neuroimaging techniques made it possible to study sex differences in the structure or function of the brains of living humans. (I will return to neuroimaging research below, in discussing feminist criticisms of this area of research.) Second, the increasing popularity of evolutionary psychology provided a theoretical account that specified the kinds of behavioral sex differences that were likely to exist and that linked those behaviors to biological (specifically genetic) factors. While neuroimaging contributed a new method of investigation for sex difference researchers, evolutionary psychology buttressed the theoretical framework for this research.[4]

Evolutionary psychology aims to provide an account of sex differences by explaining them as a product of differences in the adaptive pressures faced by our female and our male ancestors. This basic idea derives from Darwin, who hypothesized two mechanisms of evolution: natural selection explains the evolution of traits related to survival, while sexual selection explained the evolution of traits related to reproductive success.[5] Evolutionary *biology* draws on the concept of sexual selection to explain the evolution of physical traits like the male peacock's tail; evolutionary *psychology* extends this analytic framework to explain the evolution of specific psychological mechanisms. On this account, sex differences in behavior are the result of sex differences in these psychological mechanisms, which, in turn, are due to the distinct "information processing problems" related to reproduction faced by our male, versus our female, ancestors. For example, the prominent evolutionary psychologist David Buss explains, "men have faced the adaptive problem of uncertainty of paternity in putative offspring" (1995: 164), while women needed to "secure a reliable or replenishable supply of resources to carry them through pregnancy and lactation," which they did in part "by preferring mates who showed the ability to accrue resources and the willingness to provide them for particular women" (ibid.).

As with brain organization theory, researchers have used evolutionary psychology to explain sex differences in behaviors that are not obviously related to reproduction. Perhaps the most elaborate such account is Simon Baron-Cohen's theory that the "female" brain evolved primarily for empathizing with others, while the "male" brain evolved primarily for systematizing. These two specialties, he hypothesizes, conferred evolutionary advantages on our ancestors. For females, these included mothering, making friends, gossiping, and "decoding their male partner's next move," which would have led to "greater success in avoiding spousal aggression" (129). Masculine advantages, conferred by the systematizing brain, included the ability to make and use tools, to engage in both hunting and trading, and to understand the social dominance hierarchy (see Baron-Cohen, Chapter 9, see also Buss 1995: 65).

Baron-Cohen also makes explicit the link between evolutionary psychology and brain organization theory; systematizing ability is the result of the influence of prenatal testosterone on the brain development (2003: Ch. 8). Thus, while brain organization theory provides a developmental account of the causes of sex differences in individuals, evolutionary psychology provides an account of sex differences at the species level, resulting in a consistent explanation for the existence of sex differences in the brain and in behavior. Neuroscience researchers discussing sex differences may therefore cite the findings of evolutionary psychologists and vice versa. Together, these developmental and evolutionary approaches provide a way of understanding sex differences that link genetics, hormonal, neural, and behavioral research.

(Where) Is There Neurosexism in Contemporary Sex Difference Research?

Having outlined the framework that guides contemporary neuroscience research on sex differences, I will now turn to a discussion of feminist criticisms of this research. In understanding the argument that sex difference research is associated with neurosexism, it is important to be clear about both the scope and the nature of these claims. With regard to scope, there are two important points to note. First, feminist critics are interested in assessing research that aims to explain (purported) sex differences in behavior, ability, or psychological characteristics such as personality traits. There is no inherent worry about neurosexism when scientists discuss, for example, differences in neuroreceptor density or in cortical thickness. Problems arise only when too-quick leaps are made from these biological differences to behavioral differences. Related to this, it is worth mentioning that feminist critics of neuroscience do not simply deny that there are areas of the brain that, on average, are different in structure or function in women and men. (As Rebecca Jordan-Young has emphatically said "I do not reject the idea of sex differences in the brain as either 'dangerous' or implausible – to be honest, I'm somewhat amazed that there are not *more* sex differences in the brain" (2010: 10).) Second, feminist critics focus on claims about sex differences in human beings. There is no inherent worry about neurosexism in research on, for example hormonal effects on rodent behavior. Rather, neurosexism becomes a worry when results in animals are too quickly extrapolated to explain or to posit human differences.

With regard to the nature of neurosexism, I argued earlier that craniology can serve as a paradigm case of both "scientific" and "social" neurosexism. Using Fine's description of neurosexism and Fee's case study, I identified five different aspects of neurosexism, or ways that it can manifest. Scientific neurosexism occurs when gender stereotypes influence the way that scientific research itself is conducted, in ways that raise epistemological problems. It may manifest in two different ways: (1) bias in research on sex differences, including the framework guiding such research, and the methods and inferences used in particular studies or, more broadly, (2) detrimental effects for understanding gender. Social neurosexism occurs when widely held gender stereotypes inform, or are reinforced by, neuroscience research on sex differences. There are also two aspects of social neurosexism: (3) the sexist motivation of explaining women's inferiority using science; (4) harmful social effects of sex difference research, often via its coverage in popular works and in the media. Finally, (5) the dismissal of critics of sex difference research as unscientific and motivated solely by political factors is social in that it takes place largely in the public sphere, but also epistemic in that it reflects an attitude toward feminist criticism that directly affects the way research on sex differences is conducted.

In the following sections, I address the question of whether and how contemporary research on sex differences in neuroscience meets each of these five aspects of neurosexism. Where applicable, I draw not only on sex difference research itself but also on the responses of sex difference researchers to feminist criticism. Ultimately, I argue, while contemporary research is certainly not sexist in the way that craniology was, a more complex form of neurosexism still characterizes some sex difference research and much of the discussion of sex difference research in the public sphere.

Scientific Neurosexism

Identification of bias in sex difference research is a central theme in feminist criticisms of neuroscience, with critics showing when this research relies on problematic methodological choices and/or when scientists draw conclusions about gender differences that are unsupported by the data. This line of critique originated in the 1970s and 1980s, with what Sandra Harding (1986) called "spontaneous feminist empiricism"; feminist biologists drew on the standards for research held by their scientific community to argue that, in many cases, sex difference research did not measure up to these standards (e.g. Bleier 1984; Fausto-Sterling 1985, Birke 1986).

In the past decade, this kind of criticism has seen a resurgence, with contemporary feminist critics engaging in the same kind of close analysis that was being conducted 30 or 40 years ago – and often identifying similar problems. Much of this contemporary criticism focuses on functional neuroimaging research, which has made it possible to relate activity in the brains of living human beings to the performance of specific cognitive tasks such as language processing (Kaiser et al. 2009), emotion processing (Bluhm 2013a, b), or visuospatial reasoning (Bentley et al., 2019a, b).

A smaller number of works have taken a broader scope, notably Rebecca Jordan-Young's comprehensive analysis of brain organization theory (2010) and Cordelia Fine's books on problems with the science of sex differences, which include extended discussions of brain organization theory and evolutionary psychology (2010, 2017).[6]

These critical works use several strategies to assess sex difference research. For example, they may examine the methods used by researchers, or make explicit the influence of gender stereotypes and beliefs about sex differences on how scientists interpret their data. Unsurprisingly, the methodological problems identified by feminist critics tend to make it more likely that a study will find differences between women and men. For example, Kaiser et al. (2009) have shown that studies examining sex differences in language processing may present neuroimaging results separately for women and for men to show that some areas of the brain are active in one group but not the other. Without a direct statistical comparison, however, it cannot be concluded that a brain area is significantly *more* active in that group than in the other (see also Bluhm 2013a). Yet because the within-group images "have their own expressive impact," they "quickly lead to the assumption of differences" (Kaiser et al. 2009: 54). Moreover, even when the data obtained in a study are less than unequivocal, scientists may draw on gender stereotypes to interpret their results as supporting the hypothesis that a sex difference has been identified. For example, studies that look at cognitive control over emotion processing conclude that "women are emotional and men are more likely to process emotional stimuli cognitively," whether their neuroimaging results show more, less, or similar levels of activity in relevant areas of the brain (Bluhm 2013b: 882; see also Fine 2010).

Brain organization theory itself was originally viewed as a hypothesis about the causes of sex differences in behavior, but, as Rebecca Jordan-Young has pointed out, it has now become so entrenched that it "is now only rarely identified as a theory and is usually incorporated into research as a 'background fact' of development" (2010: 37). As a "fact," it can be used to explain *any* sex differences observed – the difference is ultimately due to the influence of prenatal exposure to sex hormones. I suggest that the same is also true for evolutionary psychology, which attributes behavioral differences to different adaptive pressures related to reproduction. The move from hypothesis to background fact makes these accounts of the causes of difference effectively unfalsifiable. Moreover, taking brain organization theory, or the account of sex selection underlying evolutionary psychology, to explain observed sex differences also contributes to the second kind of scientific neurosexism: detrimental effects for understanding gender. This is because these explanations say that any observed sex differences are ultimately due to factors that act well before birth (prenatal hormones, or adaptive challenges faced by our distant ancestors), rather than adequately addressing the role of gendered experience in shaping sex differences in brains and in behavior.

This claim needs a bit more clarification. While researchers working within the framework provided by evolutionary psychology and brain organization theory do think that differences in adaptive environments and in exposure to prenatal hormones cause observed sex differences, they are not complete biological determinists. Rather, they agree with their feminist critics that environmental and social influences shape both brain and behavior – and some people on both sides of the issue have even argued that it is not even helpful to think about these as distinct processes whose specific influence can be analyzed (Fausto-Sterling 1985, 2000; Buss 1995; Keller 2010). Yet this framework also biases researchers toward emphasizing innate factors and viewing social and environmental factors as acting on a neural foundation that is inherently sexed. When researchers say things like "[s]ex differences…are due to both biological and social factors" (Baron-Cohen 2003: 117) or talk about "the biological basis of human brain differences" (McCarthy and Arnold 2011: 1), it is clear that they are emphasizing the importance of innate, nonsocial factors. The question to be addressed regarding social factors, then, is as David Buss puts it with regard to evolutionary psychology, "[w]hich social, cultural, and contextual inputs moderate the magnitude of expressed sex differences" (1995: 164). Note that this means that researchers are much less likely to consider the possibility that observed differences are the *result* of social, cultural, and contextual inputs, as women and men are viewed as being "naturally" different.

Helen Longino has described the idea that women and men are fundamentally different as "the assumption of a thoroughgoing dimorphism or sexual essentialism" (1990: 129) and has argued that it is a third patriarchal value that, together with androcentrism and sexism, has influenced science. Longino links this essentialism to heterosexism, but I want to emphasize that, in the context of neuroscience research, it also underwrites sexism. While sex difference researchers may, and do, explicitly deny that one sex is superior, or that this is implied by sex difference research (Buss 1995: 197; Cahill 2014), the idea that women and men – and their brains – are fundamentally and "naturally" different leaves room for the nature of those differences to be specified by social forms of neurosexism that appear to be supported by science.

Social Neurosexism

While biological forms of neurosexism have not changed much since feminist criticisms of sex difference research first emerged, or even since the days of craniology, contemporary forms of social neurosexism are in many ways very different than the neurosexism identified by Fee. Recall the anthropologists who developed craniology as a means of proving, scientifically, women's intellectual inferiority. Contemporary scientists are clearly *not* sexist in this sense: to my knowledge, no contemporary sex difference researcher has claimed that one sex is superior to the other. In fact, they may explicitly say that women and men are equal (Cahill 2014), or that their research does not imply that one sex is superior (Buss 1995). Given this, it is understandable that researchers bristle at the perceived suggestion that they are (neuro)sexist.[7]

Yet the fact that researchers themselves may not draw sexist conclusions from their research does not mean that their research cannot be used to support sexist claims. As I noted in the previous section, the belief that "female" and "male" brains are essentially different is itself neutral with respect to the nature of those differences. But the claim of difference does not exist independent of the social context within which it is made – a context that is rife with gender stereotypes. In some cases, particularly in cognitive neuroimaging studies, the influence of deeply ingrained beliefs about gender is clear. Studies of sex differences in emotion processing, for example, begin from the assumption that women are more emotional than men (Bluhm 2013a, b), while studies of visuospatial reasoning start from the assumption that men have stronger visuospatial skills (Bentley et al., 2019a, b). Even here, though, scientists are generally cautious about drawing conclusions about the relationship between neural differences and stereotypical gender differences in behavior, personality, and social roles.

By contrast, popular discussions of sex difference research in neuroscience are often incautious, sometimes to the point of complete inaccuracy. For example, *The Guardian*'s review of Simon Baron-Cohen's book begins by asking

> Why do most men use the phone to exchange information rather than have a chat? Why do so many women love talking about feelings and relationships? We may suspect profound differences between the sexes but do they stand up to scientific scrutiny?
>
> *(Adam 2003)*

There is also a whole genre of "pop neuroscience" that uses and abuses sex difference research to "explain" gender differences such as "Why men don't listen and women can't read maps" (Pease and Pease 2000).[8]

This kind of irresponsible discussion of sex differences in popular neuroscience books and in the media is the form of neurosexism that originally led Fine to coin the term. It is also one that many sex difference researchers are very aware, and very critical, of. In their review of Rebecca Jordan-Young's *Brainstorm* (2010) and Cordelia Fine's *Delusions of Gender* (2010), two prominent advocates of sex difference research acknowledge the force of these criticisms (McCarthy and Arnold 2011). Specifically, they point to the problems raised by popular neuroscience books about sex differences that overstate the conclusions reached by scientists and they acknowledge that, in reality, interpreting the meaning and the cause of neural sex differences is challenging.[9] Yet they also criticize both Jordan-Young's and Fine's books in ways that suggest that they have misunderstood the scope of their arguments. As I noted earlier, concerns about neurosexism are relevant only when they relate to human beings, but McCarthy and Arnold suggest that it is a shortcoming of Jordan-Young's book that it is "humancentric." Moreover, they say that because both authors limit themselves to criticizing problematic studies, they fail to take account of the large numbers of well-conducted studies that show sex differences in the brain. The well-conducted studies, however, are not the problem identified by feminist researchers, who, like McCarthy and Arnold, worry only when these well-conducted studies are used to support problematic claims about the differences between (human) women and men. Combatting this misuse of neuroscientific findings about sex difference is an area where both the proponents and the critics of this area of research can, and do, agree.

At their best, responses to feminist criticisms are thoughtful, yet critical, engagements with the issues raised; McCarthy and Arnold's book review is an example of this. Too often, though, feminist criticisms of sex difference research are poorly received; at their worst, these responses are reminiscent of the craniologists' responses to their critics as politically motivated and un- or antiscientific. This is the fifth form of neurosexism I identified earlier, that is, the unwarranted dismissal of feminist criticisms, and of the critics themselves.

The characterization of feminist criticisms of sex difference research as being (purely) politically motivated is a recurring theme. I began this chapter by quoting Larry Cahill's description of his experiences when his research began to turn to the study of sex differences: he and his mentors viewed this research as a potential threat to a scientist's credibility. While Cahill does not specify the nature of this threat, others are not as circumspect. In his blurb of Simon Baron-Cohen's book, Steven Pinker lauds the author for writing a book that is "unlike many books on this vexed subject…neither politically correct nor politically oblivious." Baron-Cohen himself opens his review of Fine's book *Delusions of Gender* by saying that the author has a "barely veiled agenda" of showing that all sex differences are caused by social forces, resulting in a "mistaken blurring of science with politics." Elsewhere, he refers to her "strident, extreme denial of the role that biology might play in giving rise to any sex differences in the brain" (Baron-Cohen 2010: 904).[10] Similarly, the evolutionary biologist Jerry Coyne titles a blog post about an op-ed written by Daphna Joel and Cordelia Fine "Ideology trumps science once again" (Coyne 2018) The impression one gets is of a brave cadre of beleaguered scientists, searching for the neutral truth about sex differences in the face of fierce, politically motivated opposition. Yet, as I have shown throughout the chapter, sex difference research is itself politically influenced (even when it is not explicitly politically motivated); moreover, much feminist criticism of this research is scientific in nature.

The scientific criticisms, however, may also be met with a neurosexist response, specifically, the suggestion that feminist critics are ignorant of science. In a blog post for the Dana Foundation, for example, Larry Cahill repeatedly refers to them as "non-neuroscientists." This is clearly inaccurate; as I noted at the beginning of the chapter, many feminist critics of sex difference research, including ones whose work Cahill targets in this post, are neuroscientists or cognitive neuropsychologists. He also chides his critics for "misunderstanding…some key facts of brain biology" and "finds it amusing" that they insist on the importance of brain plasticity, as if the idea of neuroplasticity had never occurred to him. Both his tone and his refusal to engage with the substance of their criticisms show that, like the craniologists, Cahill simply dismisses his critics as unscientific.

As with the second form of social neurosexism, much of the pushback against feminist criticisms occurs in the media and other public venues such as blog posts. Moreover, it addresses the political nature of debates over sex differences. Yet in addition to its social aspects, this fifth form of social neurosexism is also epistemic. Helen Longino (1990) has emphasized the importance of both the uptake of criticism by a scientific community and the (tempered) equality of participants in the community. By failing to engage appropriately with critics, at least some sex difference researchers are losing the opportunity to critically reflect on their own practices, methods, and assumptions.

Conclusion

Both sex difference research and society have made great progress since the days of craniology. Yet critics are right to worry about neurosexism in contemporary research. While it is less overt than it used to be, neurosexism affects scientific research directly both by shaping the methods used and the inferences drawn in sex difference research and by biasing explanations of differences toward innate factors. It also slows scientific progress when scientists refuse to engage with criticism because they dismiss their critics as unscientific and politically motivated. Moreover, these forms of neurosexism enable discussions in the public sphere that (mis)use sex difference research to support gender stereotypes, a problem that should concern both critics and proponents of this research.

Related chapters: 6, 14, 23, 25.

Notes

1 The collapse of craniology did not mean the end of the scientific investigation of mental differences. Fee notes that craniology was simply replaced by emerging methods of intelligence testing in psychology.
2 For a long time, testosterone was the main hormone of interest; it was believed to "masculinize" the brain, which would otherwise develop along a "default" female pathway. More recently, researchers have become more interested in the organizing effects of "female" sex hormones, particularly estrogen, and recognized their importance for normal development in females, though it is still common to refer to the female brain as a default (see for example Azvolinsky 2015).
3 Proxy measures for the influence of testosterone on the brain have also been used. The most common of these is the ratio of the second and fourth fingers, which tends to be different in women and men and which has therefore been hypothesized to reflect exposure to prenatal testosterone. See Voracek (2011) and Valla and Ceci (2011) for discussion.
4 To clarify, evolutionary psychology is interested in the evolution of a wide variety of cognitive and behavioral traits, not just in sex differences (see e.g. Cosmides and Tooby 1997).
5 It is also a "descendant" of sociobiology, which aims to explain social behaviors in biological, primarily evolutionary terms (Driscoll 2018).
6 Because these books, particularly the two earlier ones, have been the topic of book reviews and media coverage, I will be discussing them in greater detail in the section on social neurosexism.
7 Relatedly, Jeff Lockhart (2020) has pointed out that many sex difference researchers deny that the history of their field has any relevance to contemporary work.
8 The answer, apparently, is that *Men are Like Waffles—Women are Like Spaghetti* (Farrel and Farrel 2017).
9 They also acknowledge that it is controversial whether brain organization theory can even be applied to the human brain (2011: 1).
10 Note, too, Baron-Cohen's opposition of "biological" with "social," as discussed earlier.

References

Adam, D. (2003) "His 'n' Hers," *The Guardian*, May 17. https://www.theguardian.com/books/2003/may/17/featuresreviews.guardianreview6

Arnold, A. (2009) "The Organizational-Activational Hypothesis as the Foundation for a Unified Theory of Sexual Differentiation of all Mammalian Tissues," *Hormones and Behavior* 55: 570–578.

Azvolinsky, A. (2015) "Female Brain Maintained By Methylation," *The Scientist*, March 30. https://www. the-scientist.com/daily-news/female-brain-maintained-by-methylation-35724

Baron-Cohen, S. (2003) *The Essential Difference: Men, Women, and the Extreme Male Brain*. New York: Basic Books.

Baron-Cohen, S. (2010) "Delusions of Gender – 'Neurosexism', Biology, and Politics," *The Psychologist* 23(11): 904–905. https://thepsychologist.bps.org.uk/volume-23/edition-11/book-reviews

Bentley, V., Kleinherenbrink, A., Rippon, G., Schellenberg, D., Schmitz, S. (2019a) "Improving Practices for Investigating Spatial "Stuff": Part I: Critical Gender Perspectives on Current Research Practices," *Scholar & Feminist Online* 15(2): 2019.

Bentley, V., Kleinherenbrink, A., Rippon, G., Schellenberg, D., Schmitz, S. (2019b) "Improving Practices for Investigating Spatial "Stuff": Part II: Considerations from Critical Neurogenderings Perspectives," *Scholar & Feminist Online* 15(2): 2019.

Birke, L. (1986) *Women, Feminism and Biology: The Feminist Challenge*. Brighton: Harvester Press.

Bleier, R. (1984). *Science and Gender: A Critique of Biology and Its Theories on Women*. New York: Pergamon Press.

Bluhm, R. (2013a) "Self-Fulfilling Prophecies: The Influence of Gender Stereotypes on Functional Neuro-imaging Research on Emotion," *Hypatia* 28(4): 870–886.

Bluhm, R. (2013b) "New Research, Old Problems: Methodological and Ethical Issues in fMRI Research Examining Sex/Gender Differences in Emotional Processing," *Neuroethics* 6(2): 319–330.

Buss, D. (1995) "Psychological Sex Differences: Origins through Natural Selection," *American Psychologist* 50(3): 164–168.

Cahill, L. (2014) "Equal ≠ The Same: Sex Differences in the Human Brain," *Cerebrum*. https://www.ncbi. nlm.nih.gov/pmc/articles/PMC4087190/.

Cahill L. (2017) "An Issue Whose Time Has Come," *Journal of Neuroscience Research* 95(1–2): 12–13.

Cosmides L., Tooby, J. (1997) "Evolutionary Psychology: A Primer." https://www.cep.ucsb.edu/primer. html

Coyne, J. (2018) "Ideology Trumps Science Once Again: Daphna Joel and Cordelia Fine Deny the Notion of 'Male Vs. Female Brains'," *Why Evolution Is True*, December 17. https://whyevolutionistrue.word-press.com/2018/12/07/ideology-trumps-science-once-again-daphna-joel-and-cordelia-fine-deny-the-notion-of-male-vs-female-brains/

Driscoll, C. (2018) "Sociobiology," in *The Stanford Encyclopedia of Philosophy*. Edward N. Zalta (Ed.) https:// plato.stanford.edu/archives/spr2018/entries/sociobiology/

Farrel B., Farrel, P. (2017) *Men are Like Waffles – Women are Like Spaghetti: Understanding and Delighting in Your Differences*. Eugene, OR: Harvest House Publishers. Revised edition.

Fausto-Sterling, A. (1985) *Myths of Gender: Biological Theories about Women and Men*. New York: Basic Books. 1st edition.

Fausto-Sterling, A. (2000) *Sexing the Body: Gender Politics and the Construction of Sexuality*. New York: Basic Books.

Fee, E. (1979) "Nineteenth-Century Craniology: The Study of the Female Skull," *Bulletin of the History of Medicine* 53(3): 415–433.

Fine, C. (2008) "Will Working Mothers' Brains Explode? The Popular New Genre of Neurosexism," *Neuroethics* 1(1): 69–72.

Fine, C. (2010) *Delusions of Gender: How Our Minds, Society, and Neurosexism Create Difference*. New York: Norton.

Fine, C. (2013) "Is There Sexism in Functional Neuroimaging Investigations of Sex Differences?," *Neuroethics* 6(2): 369–409.

Fine, C. (2017) *Testosterone Rex*. New York: W.W. Norton & Company.

Harding, S. (1986) *The Science Question in Feminism*. Ithaca: Cornell University Press.

Jordan-Young, R. (2010) *Brain Storm: The Flaws in the Science of Sex Differences*. Cambridge, MA: Harvard University Press.

Kaiser A, Haller, S., Schmitz, S., Nitsch. C. (2009) "On Sex/Gender Related Similarities and Differences in fMRI Language Research," *Brain Research Review* 61(2): 49–59.

Keller, E.F. (2010) *The Mirage of a Space Between Nature and Nurture*. Durham NC: Duke University Press.

Lockhart, J.W. (2020) "'A Large and Longstanding Body': Historical Authority in the Science of Sex." In L.D. Valencia-Garcia (Ed.) *Far-Right Revisionism and the End of History: Alt/Histories*. New York: Routledge. pp. 359–386.

Longino, H. (1990) *Science as Social Knowledge: Values and Objectivity in Scientific Inquiry*. Princeton, NJ: Princeton University Press.

McCarthy, M., Arnold. A. P. (2011) "Tempests and Tales: Challenges to the Study of Sex Differences in the Brain," *Biology of Sex Differences* 2(4): 1–4.

Money, J., Ehrhardt, A.A. (1972) *Man and Woman, Boy and Girl: Gender Identity from Conception to Maturity.* Baltimore, MD: Johns Hopkins University Press.

Pease A., Pease B. (2000) *Why Men Don't Listen and Women Can't Read Maps: How We're Different and What to Do About It.* New York: Welcome Rain Press.

Phoenix, C.H., Goy, R.W., Gerall, A. A., Young, W.C. (1959) "Organizing Action of Prenatally Administered Testosterone Propionate on the Tissues Mediating Mating Behavior in the Female Guinea Pig," *Endocrinology* 65: 369–382.

Valla J.M., Ceci, S.J. (2011) "Can Sex Differences in Science Be Tied to the Long Reach of Prenatal Hormones: Brain Organization Theory, Digit Ratio (2D/4D), and Sex Differences in Preferences and Cognition," *Perspectives on Psychological Science* 6(2): 143–146.

Voracek, M. (2011) "Special Issue Preamble: Digit Ratio (2D:4D) and Individual Differences Research," *Personality and Individual Differences* 51: 367–370.

25

FEMINISM AND COGNITIVE NEUROSCIENCE

Vanessa Bentley

Feminist neuroscience has been primarily concerned with "sex differences" research, that is, how are male and female brains different? (For historical accounts of brain-related sex/gender difference research, see Bluhm this volume; Fee 1979; Fine 2010; Bluhm 2012). Despite contributions in other areas using insights from feminist theory, the recalcitrance of the neuroscientific community to take a more nuanced approach to sex and gender has left those working on feminist neuroscience to continue to focus on sex/gender differences.

One of the earliest and most significant (yet still overlooked) contributions of feminist science was connecting with theory in gender studies to query the ways that "sex" and "gender" are used. Whereas sex is generally assumed to deal with genes, hormones, and reproduction, such that there are two sexes (male and female), and humans share an evolutionary bond to other animal species, gender is associated with social roles as constructed within a society. Fausto-Sterling (2000), Kaiser et al. (2009), van Anders (2015), Clune-Taylor (this volume), and others have shown that the clear distinction between sex and gender cannot be made, for example, culturally created definitions of sex have changed (external genitalia, internal sex organs, chromosomes, hormones) showing that sex, along with gender, is socially constructed. Many feminist and gender studies researchers prefer the terms "sex/gender" or "gender/sex" for this reason, as well as to clarify that whatever difference is found, it is often hard to trace the cause of the difference – whether it be genes and hormones (innate, nature, biology) or social experience in a gendered society (learned, nurture, social experience).

This chapter provides an overview of some of the feminist and queer responses to research in cognitive neuroscience. Cognitive neuroscience investigates the neural substrates of behavior and cognition. One of the most important methods in cognitive neuroscience is cognitive brain imaging or neuroimaging. Although there are a number of neuroimaging tools, one of the most widely used is magnetic resonance imaging (MRI), which uses the magnetic qualities of water and the different densities of brain structures (white matter, gray matter, blood, cerebrospinal fluid) to construct a three-dimensional image of the brain. Functional MRI (fMRI) allows for the imaging and analysis of changes in blood flow in the brain. It is assumed that blood flow responds to meet the resource needs of brain areas that are active in a given task. Thus, fMRI is thought to show what brain areas are involved in a cognitive task.

Feminist work in cognitive neuroscience is interdisciplinary in nature, with epistemological, methodological, ontological, and ethical critiques coming from philosophers, historians, gender studies scholars, discourse analysts, journalists, neuroscientists, psychologists, neuroendocrinologists, and probably more. Most of the material reviewed here is part of a network of feminist brain scientists, the NeuroGenderings network, a transdisciplinary network of 'neurofeminist' scholars who aim to critically examine neuro-scientific knowledge production and to develop differentiated approaches for a more gender adequate neuroscientific research. Feminist neuroscientists generally seek to elaborate the relation between gender and the brain beyond biological determinism but still engaging with the materiality of the brain.

(https://neurogenderings.wordpress.com/)

I've organized the chapter into two main parts. In the first half the chapter focuses on feminist critiques of cognitive neuroscience practice – the theoretical and background assumptions, the methods, the analysis and reporting of results, and the research question. In the second half the chapter reviews feminist approaches to address the problems critiqued in the first half.

Critiquing Assumptions

In the previous chapter, Bluhm discussed brain organization theory (or the hard-wiring paradigm) as a research assumption. Another assumption is sex essentialism, which is the view that men and women are essentially different due to biological sex. Sex essentialism assumes a sex-binary where everyone is cisgender, which leaves out an array of individuals who are transgender, nonbinary, gender-fluid, or intersex. One way that sex essentialism occurs is through gender stereotypes finding their way into research (Bluhm 2013b, this volume). Gender stereotypes may enter through reverse inference, in which researchers infer what cognitive state or process a subject (or group of subjects) was engaged in from looking at what areas of the brain were active. Whereas the problem of reverse inference is well known within cognitive neuroscience, Fine (2013) explores it in cases of sex/gender differences (see also Fine et al. 2019). Fine (2013) surveyed all of the fMRI studies published in 2009–2010 that referred to sex differences in the title of the article, finding 39 studies. Of these, 14 studies suggested reverse inferences in addition to containing behavioral data that could be brought to bear on the inference. In 11 out of the 14, "the relevant available behavioral data were inconsistent with, or unsupportive of, a reverse inference made" (Fine 2013: 380). In other words, researchers ignored their behavioral data to offer an explanation based on gender stereotypes. For example, in one case, researchers suggested that observed activation differences between men and women may indicate that women "need to partake in a higher degree of mental consideration before making a risky response, and render women more risk averse" (Lee et al: 1308, as cited in Fine 2013: 396), but they did not find sex/gender differences in response time or in number of risky responses. In other words, women weren't actually making less risky responses, nor were they deliberating longer than men before making their choices, but the researchers assumed women were "risk-averse."

Bluhm (2013b) writes: "In this, the research continues practices long criticized by feminist scholars for adding an aura of scientific legitimacy to gender stereotypes" (882). The assumption of sex essentialism leads to gender stereotypes (such as greater emotionality in women), which are "supported" by research, and which further reinforce sex essentialism (such as emotionality as an essential characteristic of women).

Joel (2011, 2012) argues that sex essentialism requires the fulfillment of two parts: (1) brain regions are dimorphic (with clear "male" and "female" forms); and (2) brains are internally consistent (a "male" brain has all "male-typed" brain forms and no "female-typed"). She reviews animal data showing how environmental experiences (such as stress) affect the brain, resulting in exaggerating, reversing, abolishing, and creating sex differences. Thus, she argues, brain regions are not dimorphic; the first requirement fails.

Regarding the second requirement that brains are internally consistent in form, Joel et al. (2015) analyzed four MRI and three behavioral datasets. By demarcating "male" and "female" forms (defined as the most extreme 33%), they show that the vast majority of brains are not 100% male or 100% female. They used multiple datasets and multiple measures (voxel-based morphometry, surface-based analysis, diffusion tensor imaging, and behavioral data such as personality traits, attitudes, interests, activities, and gender-stereotypical behaviors). Across all brain-based measures, brains demonstrating "internal consistency" (only male-type, female-type, or intermediate-type) ranged from 0.7% to 10.4%. That means that only 0.7–10.4% of individuals had a wholly "male," wholly "female," or wholly "intermediate" brain. Around 90% of brains (individuals) displayed some mix of masculine and feminine attributes. Of that 90%, 23–53% of brains displayed "substantial variability," meaning that they contained some "male-typed" brain regions and some "female-typed," whereas the rest of the brains were male-intermediate mixes or female-intermediate mixes. As for behavioral measures, individuals scoring consistently "male," "female," or intermediate" ranged from 0.1% to 1.8%, with 55–70% of individuals demonstrating substantial variability, again meaning that 55–70% of individuals have both "masculine-" and "feminine-"coded behaviors and interests. What this means, then, is that if we take the first part of sex essentialism seriously (sexual dimorphism), the vast majority of brains are mosaics of male and female forms. They state:

> The lack of internal consistency in human brain and gender characteristics undermines the dimorphic view of human brain and behavior and calls for a shift in our conceptualization of the relations between sex and the brain. Specifically, we should shift from thinking about brains as falling into two classes, one typical of males and one typical of females, to appreciating the variability of the human brain mosaic.
>
> *(Joel et al. 2015: 15472)*

According to Joel et al. (2015), one cannot predict an individual's brain mosaic based on knowing an individual's sex/gender. Nor can one visually inspect a human brain and classify it as male or female.[1]

Another essentialist concern has to do with the study of sexuality. The sex-essentialist assumption of separate male and female essences relating to biological sex is often paired with the assumption of heteronormativity, which is the view that individuals are naturally sexually attracted to individuals of the opposite sex. Due to space constraints, I cannot review this research in depth here, but see Fausto-Sterling (2000), Jordan-Young (2010), van Anders (2015), Dussauge and Kaiser (2012), and Kaiser and Dussauge (2015) for concerns related to the study of sexuality based on a sex essentialist framework.

Animal models for understanding human biology and behavior are another assumption critiqued by feminists (Fine 2010; Jordan-Young 2010; Eliot and Richardson 2016; Fine et al. 2019; Gungor et al. 2019). Given that biological mechanisms are often conserved across species, which serves as a rationale for biomedical research on animals to understand human biology and to develop pharmaceuticals for human use, a lot of research on sex differences is conducted using animals and generalizing to humans. However, sex differences differ across animal species, and human society is much more complicated than animal society, in addition to the fact that there are many different human societies. For example, female rats appear to be less anxious than male rats, which is the opposite of what is observed in humans (Gungor et al. 2019). As such, it is not an easy, direct route from sex differences in animals to sex/gender differences in humans.

Critiquing Methods

Bluhm (2013a) and Kaiser et al. (2009) criticize the within-group analyses some fMRI studies employ (see discussion in Bluhm this volume). Another analysis concern is the use of region of interest (ROI) analyses (Bluhm 2013a; Fine 2013). In ROI analyses, rather than looking at the activity of the whole brain, the study focuses only on those areas that are thought to be important for the task. Looking at fewer brain areas means that researchers have to correct for only a small number of comparisons, so smaller statistical differences may be found. Thus, it reduces the problem of false negative results, which is a statistical error in which a "real" result was erroneously missed (such as a correction for multiple comparisons "correcting away" a real result). According to Fine (2013), by focusing on ROIs and not whole-brain analyses, researchers miss out on seeing how the whole brain is involved: "Thus the decision to use this analysis technique implicitly prioritizes the finding of a sex difference over theoretical advance and the avoidance of false-positive errors" (400).

Another critique feminists have raised involves failing to correct for brain size (Rippon et al. 2014; Fine et al. 2019). Head and brain size are correlated with body size. In addition, male humans are generally larger than female humans. Thus, when comparing male and female brain areas, researchers must take into account that the males may be larger in size and may have larger brains.

Feminist critics have also expressed concern with the small sample sizes in many studies of sex/gender difference (Fine 2013; Fine et al. 2019). According to Fine (2013), an adequate sample size for between-group comparisons is at least 20 to avoid false positive findings. Fine investigated group sizes in two years of fMRI data (2009 and 2010) where sex differences were referred to in the title of the article. Focusing on the 22 studies that made sex comparisons only, and not those that made sex-by-group comparisons, she found that the mean number of males was 13.5 (±6.4) and females was 13.8 (±6.5). Only 3 out of the 22 met the 20 subjects-per-group suggestion. Meynell (2012) comments: "Arguably, the only way you could justify using such small numbers statistically, is if you are already committed to a strong sex dimorphism and you are confident of capturing male and female natures in your study" (25).

The last methodological critique regards nature vs. nurture. A research team's commitment to a nature-based vs. nurture-based explanation for sex/gender differences is properly considered a background assumption or explicit framework. However, I discuss it as a methodological critique because the commitment a research team has translates to how the study is designed. Researchers know that the brain can be shaped by both genetic or hormonal factors and the experiences the individual has. Learning changes the brain. Feminist critics, however, point out that social explanations are often left out of research reports, giving the impression that any sex/gender differences observed are due to sex, rather than social experience (Eliot 2011; Fine 2013; Bluhm this volume).

Critiquing the Reporting of Results

Kaiser et al. (2009), Schmitz (2011), Meynell (2012), and Fitsch (2014) have brought attention to the way that the brain images are made and how they figure as evidence. All highlight the immense technical expertise it takes to create images and the myriad ways in which the production process is infused with theory and assumptions, including stereotypes. As scientists are trained to read and make brain images, part of that training may be "highly influenced by common-sense notions of gender, sexuality, or ethnicity" (Fitsch 2014: 98). Kaiser et al. demonstrate this by showing how changing the statistical threshold results in changes (switches) in lateralization. For example, changing the significance level from $p < 0.05$ (Bonferroni corrected[2]) to $p < 0.001$ (Bonferroni uncorrected) resulted in changing from left-lateralized activation in Broca's area in both men and women to bilateral activation in men only. This is exactly the sort of space where gendered assumptions can enter into the process of creating brain images and reporting results.

Other criticisms have to do with the way the study is framed. For example, Kaiser et al. (2009) discuss a study in which nine brain areas were examined, but only one showed a sex/gender difference. However, that one difference was put in the abstract. They note: "This minimal dissimilarity is granted space in the abstract of the paper. In this way, the sex/gender variable changes its position from a marginal point within the research setting to a prominent place within the published paper" (55). Given the way that information included in the abstracts is more likely to be cited in the literature, they note, it leads to an increase in the perceived relevance of difference rather than the much greater similarity.

Another framing issue arises when studies introduce the "hardwiring" account of sexual differentiation of the brain (see Bluhm this volume). Dussauge and Kaiser (2012) call this indirect biological determinism or biological determinism by proxy: "where neuroscientific studies do not make explicit biological determinist claims but inscribe themselves in biological-determinist theoretical frames" (138). When the hardwiring paradigm is mentioned in the discussion section of an article, it appears unproblematically as accepted knowledge, even though it has been disputed (Jordan-Young 2010).

Another concern with functional neuroimaging studies is that the hypothesis is not clear. Do women use different brain areas than men to complete the same task (different networks)? Do women use the same areas but to different degrees? Are they using the same areas but the areas are performing different processes in women and men? Bluhm (2013a) asks these questions and points out that some of the methods commonly employed (ROI and within-group) don't actually test between the different hypotheses.

A related concern has to do with performance differences between groups (Kaiser et al. 2009). If you are imaging differences between men and women, and the two groups differ in how well they are completing the task, how do you know if the activation differences are due to "sex" or performance?

Hoffman (2012) draws attention to the ambiguity of the hypotheses and the ability of neuroscience data to be counted as evidence for them. She uses the notion of multiple realizability from the philosophy of mind literature to put a new perspective on the essentialist question. In multiple realizability, the same function (such as solving the math problem 9+4−2) may be accomplished by very different means. It could be the case that there are different female and male brains and that they are associated with different ways of completing a task. For example, women and men may perform equally well on a task but be activating different parts of the brain – the intraparietal sulcus for women and the angular gyrus for men, for example. This shows that a certain type of sex/gender essentialism may not be problematic. Instead of saying that one brain type is doing things the "better" way, it would entail studying male and female brains separately – creating separate male and female brain maps that could not be directly compared, since they're apples and oranges. Such an approach has been suggested by De Vries (2004), which suggests that sex differences in the brain may prevent differences in behavior – a compensation model.

Another problem is that researchers are not standard in their use of the term "sex" or "gender." Kaiser (2012) claims that researchers usually use "sex" when they are interested in reproductive functions and "gender" when dealing with cognition. She argues that

> When using terms such as "sex" and "gender" in the science of the brain, the theoretical and conceptual corpus of knowledge from gender studies remains largely unconsidered. Instead, everyday and general knowledge about women, men, "sex", and "gender" serve as a purported source of know-how.
>
> *(Kaiser 2014: 45–46)*

which results in assumptions, beliefs, and stereotypes being "transported into and assigned to the brain" (46).

I have argued that researchers actually use sex and gender terms interchangeably (Bentley 2015). For example, "gender analysis" is a synonym for "sex analysis" in the quote below:

The results of the *sex analysis* (no baseline differences but greater growth rate in men from teenage years onward) suggest that the corpus callosum in women matures earlier, but differences were weak and our series had nonequal sex distribution (61 women, 29 men). Because the study was not designed for a *gender analysis*, we interpret these results with caution.

(Pujol et al. 1993: 74, emphasis added)

The concern with using sex and gender interchangeably is that it fails to recognize the different possible sources of the difference, whether they be sex-based factors, those related to genes, hormones, and reproduction, or gender-based factors, those that relate to the gendered social environment. Failing to recognize these two factors may contribute to further entrenchment of sex essentialist thinking, especially when the research is framed deterministically (see above and Dussauge and Kaiser 2012) and/or when the study fails to integrate research investigating the social context wherein differences are created (Fine 2013; Bentley et al. 2019a).

Practices for a Feminist Science

In response to the feminist critiques of cognitive neuroscience, and as a further development of feminist neuroscience, feminists are beginning to offer alternate practices and frameworks to overcome the limitations they've been seeing. Kaiser et al. (2009) have pointed out that a sex/gender analysis is easy since it is often automatically collected. Fine and Fidler (2015) draw attention to this as well:

As noted long ago (Maccoby and Jacklin 1974), false-positive errors are exacerbated in the sex/gender differences field to the extent that male/female comparisons are routinely made. The simplicity and "obviousness" of testing for differences between males and females, together with a publication bias towards positive findings, creates good conditions for false-positive findings of difference to be reported, while true-negative findings are not. There is currently no way of knowing whether researchers who do not report sex comparisons have nonetheless tested for them, and it is not known how common such practices are

(1452)

Thus, "sex" is always a possible independent variable to explore. Data-mining practices are starting to be called out in science in general, with concerns about the "replication crisis" in psychology, and the pre-registration of experimental protocols has been offered as a possible solution (Rippon et al. 2014; Bentley et al. 2019b).

A number of scholars have called for studies to report not just differences, but similarities too (Kaiser et al. 2009; Eliot 2011; Schmitz 2011; Rippon et al. 2017; Bryant et al. 2019). Only publishing sex/gender differences leads to the perception that men and women are very different, when it could very well be the case that there are many more similarities than differences (Fausto-Sterling 2000; Hyde 2005). The proliferation of difference publications contributes to a file-drawer problem: findings of no difference (negative findings) are generally considered not interesting to the scientific community, and as such, get filed away in a drawer rather than published: "But such omissions distort the published literature and lead to biased review and meta-analyses" (Eliot 2011: 898).

Another practice-related suggestion is to use different approaches to sex/gender. For example, Jordan-Young and Rumiati (2012) suggest that scientists should ask what it was that we really want to know: "Sex/gender is, for most purposes, at best an imperfect proxy of the variables we actually need to understand" (312). In other words, focusing on "sex" may be obscuring the mechanisms by which neural pathways are built, revised, or maintained. Eliot (2011) suggested using psychological gender role identity rather than "sex," which allows for sex/gender to be a continuous rather than dichotomous variable. Kaiser (2014) and Schellenberg and Kaiser (2018) develop a multiparametric classification scheme. It involves collecting data on behavioral, socialization-related, biological, and cognitive factors and correlating them with behavioral and/or brain data. Kaiser (2014) sees this as

> a very first attempt to segregate the many different aspects and fractions of what, usually, is simplistically called "sex" or "gender" in experimental sessions and what is instead an enormous conglomeration of socialized, behavioral, cognitive, and culturally embedded biomarkers that are then, in a rather imprecise way, called "women" and "men".
>
> *(49–50)*

Collected information may include recalled sex/gender socialization (what sort of activities, toys, or behaviors did your parents encourage/discourage you in), sex/gender identity (belonging to group "women," for example), sex/gender role orientation (how much one relates to gendered stereotypes), sex/gender role behavior (job type, communication style, hobbies, type of relationship), sex/gender expression (gendered appearance), political attitude toward sex/gender issues (sexist beliefs), and culturally embedded biological markers (genes, gonads, genitals, hormones) (Kaiser 2014: 50–52).

More recently, feminists have turned their attention to statistical practices. Bryant et al. (2019) offer several feminist methodological interventions for researchers who work within a binary sex framework "to guarantee that researchers who adopt a classical female-male comparison approach do it in the most informative and useful way." They suggest taking into account effect sizes, which estimate how large the difference is between groups (see also Rippon et al. 2014; Rippon et al. 2017). Similarly, Fine and Fidler (2015) discuss the problem with null hypothesis significance testing and suggest an estimation approach involving confidence intervals, which take into account effect sizes and their associated uncertainty. Bryant et al. also discuss overlap (Hyde 2005; Rippon et al. 2014) and suggest the adoption of an index of similarity to show how much two groups' distributions overlap. They also suggest adopting a randomization control, which involves randomly classifying subjects into two groups to see if significant differences occur. If there is a difference between men and women but no difference between the randomized groups, the sex/gender effect is "real." If there is a difference between men and women and a difference between the randomized groups, then the sex/gender effect is probably a false positive. Another suggestion Bryant et al. offer is permutation testing, a type of nonparametric testing that involves creating a new dataset from randomly sampling the existing dataset that results in a statistical distribution that represents the form of the original dataset: "The subsequent statistical comparison of two groups is test against this null distribution, which permits us to determine whether a true difference exists between the two groups or whether that difference is simply an artefact of the sample's makeup."

Frameworks for a Feminist Science

A number of scholars have offered alternate approaches to neuroscience, cognitive neuroscience, and neuroimaging that range from conservative to radical in scope. Adopting an empiricist strategy (see Borgerson this volume), Maibom and Bluhm (2014) look to resources in science to address sex/gender neuroimaging limitations. They suggest drawing on the situationist approach employed in social psychology, which addresses the fundamental attribution error – the tendency to attribute people's actions to their character. What this would involve, then, is taking into account aspects of the situation that elicit the behavior rather than attributing the behavior to something innate (essentialist). What could be thought of as a stable attribute in women, such as empathy and perspective-taking, may be due to situational factors related to power dynamics, for example, as women are more often in subordinate positions. What's good about this approach is that it highlights the ways that the social environment and experimental environment need to be considered in neuroimaging research.

Another feminist empiricist approach is Rippon et al. (2014). They discuss four principles – overlap, mosaicism, contingency, and entanglement – that come from empirical research that complicates or disputes the sex-essentialist assumption. This new paradigm, which they term the "social context" model, is in opposition to the "essentialist" model that traces all observed sex/gender differences to genes. Overlap draws attention to the considerable similarity in behavioral performance on cognitive tasks between men and women. Mosaicism is the view that nearly all men and women display some stereotypical "masculine" attributes as well as some stereotypical "feminine" attributes. Contingency draws attention to the way that the experimental setup may influence the target behavior ("true" ability may be masked, for example, if stereotype threat is activated in an experimental paradigm). Finally, entanglement highlights the interaction between the biological and social, where both genes and experience together affect the brain. From these key principles, they draw out a number of lessons for experiment design, analysis, and interpretation such as sample size, choice of independent and dependent variables, choice of research models and hypotheses, appropriate statistical methods and corrections for known co-factors or confounds (such as brain/body size), and relying on sex/gender stereotypes to interpret patterns of brain activation and keeping in mind the entanglement of sex and gender factors. Thus, paying attention to these four factors will result in a vastly different approach to and practice of sex/gender neuroimaging. Even though the principles are abstract, Rippon et al. connect the principles to scientific practice, which may be an effective approach for getting scientists on board with a feminist project.

The next set of frameworks is influenced by science and technology studies and continental philosophy. Roy (2012, 2014, 2018) imagines a broader perspective where feminist theory and activism meets neuroscience to construct a new experimental object, paradigm, and set of research questions. She uses Isabelle Stengers' concept of cosmopolitics, which involves creating a new space (ecology) for people from different backgrounds to work together instead of against each other. Her intended audience for this work is the feminist scientist working in the lab: "I would like to envision a new set of practices that will help her to work *with* instead of *against* more traditional forms of scientific practice" (2014: 208). Based upon her experience as a graduate student in a neuroendocrinology lab and as an activist for reproductive justice, she notes how she navigated the different expectations of those communities:

> With time, I learned how to ask and experiment with the questions that related to reproductive health and justice in a way that my fellow scientific practitioners could recognize and see as being relevant, even if these questions were not originally their own.
>
> *(2014: 213)*

Nikoleyczik (2012) is another attempt at a transformative endeavor. Nikoleyczik draws upon the diffractive approach of Karen Barad to imagine productive transdisciplinary work between cognitive neuroscientists and gender and science studies. She proposes a multidimensional framework of sex/gender research that explores the different disciplines' concepts of gender, sex, and gender/sex. She also suggests a prismatic approach where study participants, neuroscientists, and gender and science studies scholars reflect on how the participant and how the neuroscientist are thought about in the context of a research project.

Kraus (2012) offers a radical departure from trying to get multiple interest groups to agree on a research agenda. She suggests a "dissensus framework": "a critical framework centered on the study of conflicts and controversies, including their absence, unsuccessful controversies, etc." (193).

Joel (2011, 2012; Joel et al. 2015) offers a very different feminist empiricist approach to replace the essentialist paradigm: mosaicism. In her view, sex and other factors interact to shape brain structure, resulting in "a multi-morphic, rather than a dimorphic, brain, that is, different individuals will have different combinations of 'male' and 'female' brain characteristics. In this sense brains are neither 'male' nor 'female,' they are 'intersex'" (Joel 2011). Moreover, we all have an "intersex" gender as well, displaying both "masculine" and "feminine" traits (Joel 2012). However, she thinks that accepting that brains do not have a "sex" does not mean that there aren't sex differences in the brain. She states:

> There are sex differences in the brain, and sex is one of several factors that affect the structure of many brain features. It is therefore crucial to study the effects of sex, yet without a priori and implicitly assuming that the effects of sex will be dimorphic and consistent.
>
> *(Joel 2014: 182)*

In order to do such research without the loaded terminology, she suggests using "sex effects" instead of the term "sex differences" and abandoning the term "sexual dimorphism" altogether.

Dussauge and Kaiser (2012) begin to develop a queer perspective for neuroscience, which they believe must reflexively address how neuroscience deals with the categories of sex/gender and sexuality. According to queer theory, as developed by Judith Butler, gender and sexuality are performative and social constructs rather than natural (see also Gupta and Rubin this volume). Thus the sex-essentialist and heteronormative framework is inadequate. To "re-queer" the brain, they advocate for: (1) the development of non-deterministic approaches that investigate the two-way relationship between brain function and brain structure, (2) a performative framework of sex/gender and sexuality to replace the current essentialist framework, (3) the multiplication of gender expressions and sexual desires and the exploration of those differences on their own terms, and (4) the recentering of the interpretation of the study, where neuroscientists self-reflexively ask what the study can actually say to avoid exaggerated extrapolations (Dussauge and Kaiser 2012: 142–143). For Kaiser and Dussauge (2015), a feminist neuroscience investigates the mutual effect of brain material and behavior. From this, they advocate for an "emancipatory" (Rabeharisoa 2003) relationship between activists (feminist and queer) and knowledge creators. The emancipatory model involves neuroscientists working with advocacy groups that will help deliver services to minority groups.

An approach that goes beyond working with advocacy groups and instead works directly with the individuals being researched is exemplified in Einstein (2012). For Einstein, feminist research intends to "give voice to areas of research previously silenced, uncover pockets of ignorance – not just 'knowledge gaps' – turn expectations about the essentialism of biology on its head, and contribute meaningfully to women's lives in all their varieties" (150). She applies feminist principles to the elements of the scientific method – the question, design, methods, and interpretation of results. In studying the neurobiological effects of Female Genital Circumcision/Mutilation/Cutting (FGC), she worked with community members and instituted a community advisory group (CAG) to design the project: "Every aspect of the study design was reviewed by the CAG, from the issue of pain and referred sensation to all of the instruments – qualitative and quantitative" (154). Working with the CAG resulted in revisions of the study instruments and design.

Einstein lists a number of "guideposts" that may work up to a feminist practice of neuroscience, such as exploring the context of the research question, situating the researcher and their prejudices, clarifying the power issues in the paradigm and flattening hierarchies of power, being reflexive, including the "subject" of the research as an active partner, practicing reciprocity, and avoiding overgeneralizing findings (166–167).

Dussage and Kaiser's and Einstein's approaches align with a feminist standpoint framework (see Hundleby this volume). Starting research from the lives of the oppressed, marginalized, or underrepresented has the potential to radically reorient research interests and produce knowledge that helps rather than harms already disadvantaged groups.

Future Directions

As mentioned in the introduction, mainstream psychology, neuroscience, and cognitive neuroscience continue to push back against feminist meddling in science, which keeps feminist neuroscientists and feminist critics of essentialist neuroscience focused on the issues of sex/gender differences and nature/nurture. But feminist neuroscience is a young field and still has opportunity to grow. For example, as feminism continues to confront its historical exclusion of lesbians, women of color, and other marginalized groups, feminist neuroscience should also embrace intersectionality (Jordan-Young 2014; Kuria 2014). As Kuria (2014) has noted, just like the history of "sex" difference research has been used to prove the inferiority of women, research on race has been used to prove the inferiority of certain races.

A strength, however, of the feminist neuroscience and psychology community is its collaborative and interdisciplinary nature. In addition to the rich theoretical and methodological critiques, a number of new feminist theory-driven tools and frameworks have been offered. The challenge to come will be getting research implementing the feminist approaches through peer-review (both in securing grant funding and in publishing research reports) since feminist science conflicts with the traditional notion of science as value-free and nonpolitical.

Related chapters: 6, 7, 14, 23, 24.

Notes

1 See Joel et al. (2016) for a discussion of limits on training computer models to assign sex/gender.
2 A Bonferroni correction is a way to control for false positive findings when making many statistical comparisons.

References

Bentley, V. (2015) Building a Feminist Philosophy of Cognitive Neuroscience (dissertation). http://rave.ohiolink.edu/etdc/view?acc_num=ucin1447691278

Bentley, V., Kleinherenbrink, A., Rippon, G., Schellenberg, D., and Schmitz, S. (2019a) "Improving Practices for Investigating Spatial 'Stuff': Part I: Critical Gender Perspectives on Current Research Practices," *Scholar and Feminist Online* 15(2). http://sfonline.barnard.edu/neurogenderings/improving-practices-for-investigating-spatial-stuff-part-i-critical-gender-perspectives-on-current-research-practices/

Bentley, V., Kleinherenbrink, A., Rippon, G., Schellenberg, D., and Schmitz, S. (2019b) "Improving Practices for Investigating Spatial 'Stuff': Part II: Considerations from Critical NeuroGenderings Perspectives," *Scholar and Feminist Online* 15(2). http://sfonline.barnard.edu/neurogenderings/improving-practices-for-investigating-spatial-stuff-part-ii-considerations-from-critical-neurogenderings-perspectives/

Bluhm, R. (2012) "Beyond Neurosexism: Is It Possible to Defend the Female Brain?," in R. Bluhm, A. J. Jacobson, and H. L. Maibom (Eds.) *Neurofeminism: Issues at the Intersection of Feminist Theory and Cognitive Science*, New York: Palgrave MacMillan, pp. 230–245

Bluhm, R. (2013a) "New Research, Old Problems: Methodological and Ethical Issues in fMRI Research Examining Sex/Gender Differences in Emotion Processing," *Neuroethics* 6(2): 319–330.

Bluhm, R. (2013b) "Self-Fulfilling Prophecies: The Influence of Gender Stereotypes on Functional Neuroimaging Research on Emotion," *Hypatia* 28(4): 870–886.

Bryant, K., Grossi, G., and Kaiser, A. (2019) "Feminist Interventions on the Sex/Gender Question in Neuroimaging Research," *Scholar and Feminist Online* 15(2). http://sfonline.barnard.edu/neurogenderings/feminist-interventions-on-the-sex-gender-question-in-neuroimaging-research/

De Vries, G. J. (2004) "Minireview: Sex Differences in Adult and Developing Brains: Compensation, Compensation, Compensation," *Endocrinology* 145(3): 1063–1068.

Dussauge, I. and Kaiser, A. (2012) "Re-Queering the Brain," in R. Bluhm, A. J. Jacobson, and H. L. Maibom (Eds.) *Neurofeminism: Issues at the Intersection of Feminist Theory and Cognitive Science*, New York: Palgrave MacMillan, pp. 121–144.

Einstein, G. (2012) "Situated Neuroscience: Exploring Biologies of Diversity," in R. Bluhm, A. J. Jacobson, and H. L. Maibom (Eds.) *Neurofeminism: Issues at the Intersection of Feminist Theory and Cognitive Science*, New York: Palgrave MacMillan, pp. 145–174.

Eliot, L. (2011) "The Trouble with Sex Differences," *Neuron* 72: 895–898.

Eliot, L. and Richardson, S. S. (2016) "Sex in Context: Limitations of Animal Studies for Addressing Human Sex/Gender Neurobehavioral Health Disparities," *Journal of Neuroscience* 36(47): 11823–11830.

Fausto-Sterling, A. (2000) *Sexing the Body: Gender Politics and the Construction of Sexuality*, New York: Basic Books.

Fee, E. (1979) "Nineteenth Century Craniology: The Study of the Female Skull," *Bulletin of the History of Medicine* 53: 415–433.

Fine, C. (2010) *Delusions of Gender: How Our Minds, Society, and Neurosexism Create Difference*, New York, London: W. W. Norton & Company.

Fine, C. (2013) "Is There Neurosexism in Functional Neuroimaging Investigations of Sex Differences?," *Neuroethics* 6: 369–409.

Fine, C. and Fidler, F. (2015) "Sex and Power: Why Sex/Gender Neuroscience Should Motivate Statistical Reform," in J. Clausen and N. Levy (Eds.) *Handbook of Neuroethics*, Dordrecht: Spring Science+Business Media, pp. 1447–1462.

Fine, C., Joel, D., and Rippon, G. (2019) "Eight Things You Need to Know about Sex, Gender, Brains, and Behavior: A Guide for Academics, Journalists, Parents, Gender Diversity Advocates, Social Justice Warriors, Tweeters, Facebookers, and Everyone Else," *Scholar and Feminist Online* 15(2). http://sfonline.barnard.edu/neurogenderings/eight-things-you-need-to-know-about-sex-gender-brains-and-behavior-a-guide-for-academics-journalists-parents-gender-diversity-advocates-social-justice-warriors-tweeters-facebookers-and-ever/

Fitsch, H. (2014) "What Goes around Comes around: Visual Knowledge in fMRI and Its Implications for Research Practice," in Schmitz, S. and Höppner, G. (Eds.) *Gendered Neurocultures: Feminist and Queer Perspectives on Current Brain Discourses*, Vienna: Zaglossus, pp. 89–107.

Gungor, N. Z., Duchesne, A., and Bluhm, R. (2019) "A Conversation around the Integration of Sex and Gender When Modeling Aspects of Fear, Anxiety, and PTSD in Animals," *Scholar and Feminist Online* 15(2). http://sfonline.barnard.edu/neurogenderings/a-conversation-around-the-integration-of-sex-and-gender-when-modeling-aspects-of-fear-anxiety-and-ptsd-in-animals/

Hoffman, G. (2012) "What, If Anything, Can Neuroscience Tell Us about Gender Differences?," in R. Bluhm, A. J. Jacobson, and H. L. Maibom (Eds.) *Neurofeminism: Issues at the Intersection of Feminist Theory and Cognitive Science*, New York: Palgrave MacMillan, pp. 30–55.

Hyde, J. S. (2005) "The Gender Similarities Hypothesis," *American Psychologist* 60(6): 581–592.

Joel, D. (2011) "Male or Female? Brains are Intersex," *Frontiers in Integrative Neuroscience* 5: 57.

Joel, D. (2012) "Genetic-Gonadal-Genitals Sex (3G-Sex) and the Misconception of Brain and Gender, or, Why 3G-Males and 3G-Females Have Intersex Brain and Intersex Gender," *Biology of Sex Differences* 3: 27.

Joel, D. (2014) "Sex, Gender, and Brain: A Problem of Conceptualization," in Schmitz, S. and Höppner, G. (Eds.) *Gendered Neurocultures: Feminist and Queer Perspectives on Current Brain Discourses*, Vienna: Zaglossus, pp. 169–186.

Joel, D., Berman, Z., Tavor, I., Wexler, N., Gaber, O., Stein, Y., Shefi, N., Pool, J., Urchs, S., Margulies, D. S., Liem, F., Hänggi, J., Jäncke, L., and Assaf, Y. (2015) "Sex Beyond the Genitalia: The Human Brain Mosaic," *Proceedings of the Natural Academy of Sciences* 112(50): 15468–15473.

Joel, D., Persico, A., Hänggi, J., Pool, J., and Berman, Z. (2016) "Do Brains of Females and Males Belong to Two Distinct Populations?," *Proceedings of the Natural Academy of Sciences* 113(14): E1969–E1970.

Jordan-Young, R. M. (2010) *Brain Storm: The Flaws in the Science of Sex Differences*, Cambridge, MA, London: Harvard University Press.

Jordan-Young, R. M. (2014) "Fragments for the Future: Tensions and New Directions from 'NeuroCultures—Neurogenderings II,'" in Schmitz, S. and Höppner, G. (Eds.) *Gendered Neurocultures: Feminist and Queer Perspectives on Current Brain Discourses*, Vienna: Zaglossus, pp. 373–393.

Jordan-Young, R. and Rumiati, R. I. (2012) "Hardwired for Sexism? Approaches to Sex/Gender in Neuroscience," *Neuroethics* 5: 305–315.

Kaiser, A. (2012) "Re-Conceptualizing "Sex" and "Gender" in the Human Brain," *Zeitschrift für Psychologie* 220(2): 130–136.

Kaiser, A. (2014) "On the (Im)Possibility of a Feminist and Queer Neuroexperiment," in Schmitz, S. and Höppner, G. (Eds.) *Gendered Neurocultures: Feminist and Queer Perspectives on Current Brain Discourses*, Vienna: Zaglossus, pp. 41–66.

Kaiser, A., and Dussauge, I. (2015) "Feminist and Queer Repoliticizations of the Brain," *EspacesTemps.net* https://www.espacestemps.net/en/articles/feminist-and-queer-repoliticizations-of-the-brain/ [accessed on 2/28/19].

Kaiser, A., Haller, S., Schmitz, S., and Nitsch, C. (2009) "On Sex/Gender Related Similarities and Differences in fMRI Language Research," *Brain Research Reviews* 61: 49–59.

Kraus, C. (2012) "Linking Neuroscience, Medicine, Gender and Society through Controversy and Conflict Analysis: A "Dissensus Framework" for Feminist/Queer Brain Science Studies," in R. Bluhm, A. J. Jacobson, and H. L. Maibom (Eds.) *Neurofeminism: Issues at the Intersection of Feminist Theory and Cognitive Science*, New York: Palgrave MacMillan, pp. 193–215.

Kuria, E. N. (2014) "Theorizing Race(ism) while NeuroGendering," in Schmitz, S. and Höppner, G. (Eds.) *Gendered Neurocultures: Feminist and Queer Perspectives on Current Brain Discourses*, Vienna: Zaglossus, pp. 109–123.

Maccoby, E. and Jacklin, C. (1974) *The Psychology of Sex Differences*. Stanford, CA: Stanford University Press.

Maibom, H. and Bluhm, R. (2014) "A Situationist Account of Sex/Gender Differences: Implications for Neuroimaging Research," in Schmitz, S. and Höppner, G. (Eds.) *Gendered Neurocultures: Feminist and Queer Perspectives on Current Brain Discourses*, Vienna: Zaglossus, pp. 127–143.

Meynell, L. (2012) "The Politics of Pictured Reality: Locating the Object from Nowhere in fMRI," in R. Bluhm, A. J. Jacobson, and H. L. Maibom (Eds.) *Neurofeminism: Issues at the Intersection of Feminist Theory and Cognitive Science*, New York: Palgrave MacMillan, pp. 11–29.

Nikoleyczik, K. (2012) "Towards Diffractive Transdisciplinarity: Integrating Gender Knowledge into the Practice of Neuroscientific Research," *Neuroethics* 5: 231–245.

Pujol, J., Vendrell, P., Junqué, C., Martí-Vilalta, J. L, Capdevila, A. (1993) "When Does Human Brain Development End? Evidence of Corpus Callosum Growth up to Adulthood," *Annals of Neurology* 34(1): 71–75.

Rabeharisoa, V. (2003) "The Struggle against Neuromuscular Diseases in France and the Emergence of the "Partnership Model" of Patient Organization," *Social Science & Medicine* 57: 2127–2136.

Rippon, G., Jordan-Young, R., Kaiser, A., and Fine, C. (2014) "Recommendations for Sex/Gender Neuroimaging Research: Key Principles and Implications for Research Design, Analysis, and Interpretation," *Frontiers in Human Neuroscience* 8: 650.

Rippon, G., Jordan-Young, R., Kaiser, A., Joel, D., and Fine, C. (2017) "Journal of Neuroscience Research Policy on Addressing Sex as a Biological Variable: Comments, Clarifications, and Elaborations," *Journal of Neuroscience Research* 95: 1357–1359.

Roy, D. (2012) "Cosmopolitics and the Brain: The Co-Becoming of Practices in Feminism and Neuroscience," in R. Bluhm, A. J. Jacobson, and H. L. Maibom (Eds.) *Neurofeminism: Issues at the Intersection of Feminist Theory and Cognitive Science*, New York: Palgrave MacMillan, pp. 175–192.

Roy, D. (2014) "Developing a New Political Ecology: Neuroscience, Feminism, and the Case of the Estrogen Receptor," in Schmitz, S. and Höppner, G. (Eds.) *Gendered Neurocultures: Feminist and Queer Perspectives on Current Brain Discourses*, Vienna: Zaglossus, pp. 203–219.

Roy, D. (2018) *Molecular Feminisms: Biology, Becomings, and Life in the Lab*. Seattle: University of Washington Press.

Schellenberg, D. and Kaiser, A. (2018) "The Sex/Gender Distinction: Beyond F and M," in C. B. Travis, J. W. White, A. Rutherford, W. S. Williams, S. L. Cook, and K. F. Wyche (Eds.) *APA Handbooks in Psychology Series. APA Handbook of the Psychology of Women: History, Theory, and Battlegrounds*, Washington, DC: American Psychological Association, pp. 165–187. http://dx.doi.org/10.1037/0000059-009

Schmitz, S. (2011) "Sex, Gender, and the Brain – Biological Determinism Versus Socio-Cultural Constructivism," in I. Klinge and C. Wiesemann (Eds.) *Sex and Gender in Biomedicine: Theories, Methodologies, Results*, pp. 57–76.

van Anders, S. (2015) "Beyond Sexual Orientation: Integrating Gender/Sex and Diverse Sexualities via Sexual Configurations Theory," *Archives of Sexual Behavior* 44: 1177–1213. –

26

IMPLEMENTING INTERSECTIONALITY IN PSYCHOLOGY WITH QUANTITATIVE METHODS

Nicole M. Else-Quest and Janet Shibley Hyde

As feminist activists call for greater attention to intersectionality in their social and political movements (e.g. the Women's March, #MeToo, #SayHerName), and scholars grapple with the translation of intersectionality for their diverse methodologies and disciplines (e.g. Cho, Crenshaw, and McCall 2013; Choo and Ferree 2010; Cole 2009; Few-Demo 2014; Spierings 2012), intersectionality has become de rigueur. We wholeheartedly support the call for implementing intersectionality across the social sciences, and also recognize that in the context of calls for more inclusive science and scholarship, calls for intersectionality within the academy may be met with confusion, frustration, or resignation by researchers who doubt the potential, utility, relevance, or suitability of intersectionality for their work. Intersectionality analyzes how multiple social categories are interconnected and constructed by and within power relations, ultimately seeking to empower individuals from marginalized groups and redress power imbalances. As the primary purpose of psychology is to optimize the development and well-being of all people, especially those who are marginalized, intersectionality is, in some ways, a natural fit for psychology.

Epistemological and methodological norms or conventions, however, may make intersectionality seem inaccessible or incompatible, and rigorous intersectional approaches infrequently appear in mainstream psychology journals, which are predominantly characterized by quantitative research. Within psychology, the majority of intersectional approaches are found in valuable research using qualitative methods (e.g. Bowleg, Teti, Malebranche, and Tschann 2013; Calasanti, Pietilä, Ojala, and King 2013; Diamond and Butterworth 2008; Hurtado and Sinha 2008; Settles 2006; Tolhurst et al. 2012). While an emerging body of research implementing intersectionality with quantitative methods exists, the strength and integrity of the intersectional approach vary considerably (Bowleg 2008; Else-Quest and Hyde 2016b). Effectively deploying intersectional approaches in quantitative research throughout psychology will serve to develop both the concept of intersectionality and the field of psychology.

We contend that intersectionality is useful, relevant, and suitable to all feminist scholarship in the sciences, and that it has the potential to transform scientific norms or conventions and even academic disciplines. To that end, in this chapter we synthesize diverse theoretical writings on intersectionality and suggest a handful of methodological techniques or strategies for implementing an intersectional approach. As feminist psychologists, we focus here on the challenge and opportunity of implementing intersectionality in psychology, recognizing that other fields and disciplines may encounter different challenges. Importantly, our distal aim is generative: the techniques and strategies we suggest are a starting place, meant to guide and transform how psychology uses its tools, incorporating feminist and intersectional approaches.

Defining Intersectionality

Writings on intersectionality are diverse in terms of disciplinary conventions and aims, epistemological stances, and perspective in relation to the academy, as well as in how intersectionality itself is defined or construed. Some conceptualize intersectionality as a heuristic or analytic tool (e.g. Collins 2019; Collins and Bilge 2016) or approach (Cole 2009), whereas others understand it as a theory or hypothesis (e.g. Walby, Armstrong, and Strid 2012), or some combination of these (e.g. Hancock 2007). In this chapter we work from the perspective that intersectionality is an analytic approach and critical theory, concerned with both recognizing and analyzing inequalities linked to multiple social categories. As such, it guides the research process but makes no claim to specific testable hypotheses.

Many credit Black feminist theorist and legal scholar Kimberlé Crenshaw (1989, 1991) with having introduced the term *intersectionality* regarding African American women's experience of violence and the US justice system, as well as their marginalization by the feminist and civil rights movements. Crenshaw used the term to describe the simultaneous consideration of race and gender, maintaining that analysis of gender by itself, or of race by itself, typically excludes women of color and that simultaneous consideration of race and gender is necessary for the experiences and voices of women of color to be understood. Nevertheless, Black feminists had described the essence or principle of intersectionality for many years prior (Alexander-Floyd 2012). Hancock (2016) describes the development of "intersectionality-like thought" throughout the nineteenth and twentieth centuries, identifying the diverse origins of intersectionality. For example, in her famous "Ain't I a Woman" speech given extemporaneously outside the Ohio Women's Rights Convention in 1851, Sojourner Truth spoke about the importance of women's rights for all women, not just white women. She detailed how her experience as a Black woman who had been enslaved was distinct on the basis of her gender *and* race within a particular sociopolitical context. Likewise, Anna Julia Cooper (1892) articulated the marginalized and ambiguous status of Black women, noting

> The colored woman of to-day occupies, one may say, a unique position in this country. In a period of itself transitional and unsettled, her status seems one of the least ascertainable and definitive of all the forces which make for our civilization. She is confronted by both a woman question and a race problem, and is as yet an unknown or an unacknowledged factor in both.
> *(Cooper 1892: 134)*

Decades later, Beale (1970) described the "double jeopardy" of being a Black woman and experiencing both racist and sexist marginalization and discrimination. In their Black feminist statement, the Combahee River Collective stated that "the major systems of oppression are interlocking" (Combahee River Collective 1982: 13). Building on the theme of interconnected systems of oppression, Black feminist Patricia Hill Collins described how the experiences of Black women exist within a "matrix of domination characterized by intersecting oppressions" (Collins 2000: 23). And May proposed that intersectional approaches require "matrix" thinking, rather than "single-axis" thinking, and that they are open-ended, dynamic, and "biased toward realizing collective justice" (May 2015: 251). Other examples of Black feminist scholarship invoke consonant themes and imagery regarding the simultaneous membership in multiple social categories and the linked systems of power and inequality (Alexander-Floyd 2012; Berger and Guidroz 2009; Carastathis 2016). Thus, while there is no single gatekeeper or authority on intersectionality, it is unmistakable that the roots of intersectionality are firmly and deeply planted within many decades of Black feminist activism and scholarship.

Reviewing the diverse interdisciplinary scholarship on intersectionality, we discerned commonalities as well as differences. We synthesized three common assumptions constituting the core of intersectionality (Else-Quest and Hyde 2016a). These include recognition that:

1   All individuals are characterized simultaneously by multiple social categories, dimensions, or characteristics, including, for example, gender, race/ethnicity, class, and sexual orientation; these multiple social categories are interconnected or intertwined, such that the experience of each social category is linked to the other categories.
2   Embedded within each of these socially constructed categories is an aspect of inequality or power.
3   Social categories are properties of the individual (i.e. identity) as well as characteristics of the social context inhabited by those individuals (i.e. social structures, institutions, and interpersonal interactions construct the categories and enforce the power inequalities); as such, these categories and their significance or salience may be fluid and dynamic.

While space limitations restrict our discussion to the three common elements, there are also notable areas of divergence within intersectionality scholarship. Perhaps among the most germane to this chapter is McCall's (2005) articulation of three approaches to the empirical application of intersectionality, which diverge on the question of the construal or construction of social categories. One, the *anticategorical* complexity approach assumes that, because categories are socially constructed, segregation and social inequality are inevitable; thus, this approach advocates the deconstruction of categories but still allows for their analysis if it promotes their deconstruction. Two, the *intracategorical* complexity approach, in assuming that categories are ambivalent, emphasizes diversity and heterogeneity within categories. Three, the *intercategorical* complexity approach, like the anticategorical complexity approach, assumes inequality among categories but focuses analysis on the relationships and processes that produce and perpetuate the inequalities. McCall (2005) added that not all empirical intersectional research will clearly fit into each of these approaches, further demonstrating the diversity of perspectives within the intersectionality literature. Along those same lines, these approaches will logically lead to different research questions and designs. For example, within the intercategorical approach, the question of inequalities among categories might be the focus of a research project, which could then be explored with a comparative or between-groups design.

Epistemological Questions: Can Quantitative Psychology Be Intersectional?

Some have argued that intersectionality can only be deployed effectively with qualitative, not quantitative, methods (e.g. Bowleg 2008; Shields 2008). Indeed, much of the psychological research claiming a feminist intersectional approach has been conducted with qualitative methods. While qualitative methods are gaining recognition as essential to psychology as a science, quantitative methods still dominate in mainstream psychological research. If quantitative methods are not included within the toolbox for deploying intersectionality, intersectional feminism will not likely find its way into mainstream psychology, an outcome implying that much psychological research will not benefit from the insights of intersectionality. Our challenge is this: *How can psychologists using quantitative methods implement an intersectional approach in their work, while remaining true to the radical potential of intersectionality as well as rigorous in quantitative technique?*

Inherent in this question are the epistemological foundations of research paradigms and methods. Because epistemologies differ in their specific assumptions about the knower, what is known, and the process of knowing—assumptions which, according to feminism, are systematically interrelated with power and inequality (Harding 1986; Sprague 2016)—some epistemologies may be more compatible with intersectionality than others. As a critical theory, intersectionality assumes that power and inequality are fundamental to the construction of thought, experience, and knowledge.

Traditionally, psychology has adopted a positivist stance, supposing that research aims to create knowledge that is value-neutral, objectively and universally true, and identifiable and quantifiable through rigorous methods (Guba and Lincoln 1994). Feminist psychology challenges these assumptions and more frequently adopts a feminist empiricist stance, maintaining that while sexist bias can interfere with the research process and obfuscate objective truths (Riger 1992), a more careful empiricist approach can eliminate such bias. In light of the positivist stance that the goal of objective and universal truth is achievable and value-neutral, positivism is incompatible with any critical theory (including intersectionality), which is inherently political and emancipatory. Thus, feminist empiricism is a better fit to intersectionality since it seeks to eliminate bias by giving accounts that are more empirically adequate and intersectionality calls for the recognition of the differences that complex social locations make to lived experience. Standpoint theory, working from the position that knowledge is socially constructed, partial, and inextricably linked to power and privilege (Harding 1986; Hare-Mustin and Marecek 1988), has been less frequently used in psychology, but is generally a good epistemological fit with critical theory. In particular, the willingness to engage in reflexivity and acknowledge one's own power and biases are essential to both standpoint theory and critical theory.

Research methods are connected to research epistemologies (Sprague 2016). Often, quantitative methods are erroneously equated with a positivist epistemology and qualitative methods are equated with social constructionist and standpoint epistemologies (Guba and Lincoln 1994; Sprague 2016). Because epistemologies and methods within psychology have not been so clearly and systematically distinguished, such equations are imprecise. Some quantitative researchers have worked from nonpositivist epistemologies, such as standpoint theory, and some qualitative researchers have favored positivism (Sprague 2016). Holding that purposefully choosing research methods not in conflict with one's epistemological stance is important, we maintain that quantitative and qualitative methods can be framed as complementary. While mainstream psychology is dominated by research employing quantitative methods, qualitative methods are frequently marginalized and mixed-methods approaches remain rare. Thus, our purpose here is not to perpetuate the marginalization of qualitative research or to rehash old debates, but rather to promote the dialectic of intersectionality and encourage quantitative researchers to develop, enhance, and enrich their work by deploying an intersectional approach.

What Makes Empirical Research "Intersectional?"

We offer several jumping-off points to facilitate the use and development of intersectionality in quantitative research, acknowledging that our ideas are neither comprehensive nor exhaustive. Within those points, we work from the position that intersectionality is for everybody. To be clear, it is crucial to recognize the origins of intersectionality in Black feminism in order to fully understand what intersectionality is about, but intersectionality is not only about the experiences of women of color. Much intersectionality research focuses on the voices and experiences of women of color and people from other multiply marginalized groups, which is a much-needed correction to the historical trends to focus on men and white people. There is debate about whether expanding intersectionality to study privilege is actually a form of "colonization" of the work (e.g. Alexander-Floyd 2012). We take the position that intersectionality is relevant wherever power and inequality are linked to the multiple social categories that we inhabit. If intersectionality becomes a "content specialization" in women of color (e.g. Hancock 2007), it systematically centers or normalizes the privileged (especially white men), affirms the position of women of color as "Other," overemphasizes differences between groups, and restricts our opportunities to build diverse coalitions and understand mechanisms of privilege. Thus, we must not lose sight of intersectionality's roots in the experiences of women of color *or* the impact that intersectionality can have on the interlocking systems of oppression today. Noting the disciplinary breadth of contemporary intersectionality work, Patricia Hill Collins concluded that "Intersectionality's reach goes beyond the groups who initially advanced its claims through their critical ideas and actions" (2019: 18). Given this generative potential, how do we use intersectionality in empirical research?

For an approach to empirical research to be characterized as intersectional, it must, at a minimum, incorporate each of the three elements we identified in the intersectional literature (Else-Quest and Hyde 2016a). First, intersectional research must attend to the experience and meaning of simultaneously belonging to multiple intertwined social categories. Psychological research has tended to focus on one social category or dimension at a time such as gender *or* race. Within the subfield of the psychology of women and gender, very few articles have explored or analyzed other social categories simultaneously with gender (Eagly et al. 2012). This first assumption, which directly addresses the heterogeneity, diversity, and interdependence of social categories, is necessary but not sufficient for an intersectional approach to empirical research. That is, intersectional research must also incorporate the other two elements.

The second assumption entails examination of power and inequality, explicitly theorizing or analyzing the dimension of power and inequality that characterizes and innervates social categories. McCall (2001) included the explicit role of power and inequality in her definition of intersectionality, maintaining that intersectionality scholarship should be concerned with "multiple, overlapping, conflicting, and changing structures of inequality" (McCall 2001: 14). Fundamental to any feminist perspective is an analysis of power and inequality tied to gender; an intersectional approach extends that analysis to additional social categories with meaning and significance derived from their social construction such as social class, disability, race/ethnicity, immigration, and sexual orientation.

When intersectionality is incorporated into mainstream quantitative psychology today, it is often equated with a statistical interaction (e.g. gender × race/ethnicity × social class interactions in a factorial design analyzed by ANOVA or multiple regression), or with the study of a multiply marginalized group (e.g. American Indian women). These applications of intersectionality include the first element (i.e. simultaneous membership in multiple social categories), but fail to include the second element (i.e. the analysis of power and inequality). Such research is therefore not truly intersectional. Moreover, in naming but not truly analyzing intersectionality, it fails to use intersectionality earnestly or "responsibly" (Moradi and Grzanka 2017).

The analysis of power and inequality connects to the third essential element of intersectional research, which is the attention to social categories as properties of the individual as well as the social context, and consideration of those categories and their significance or salience as potentially fluid and dynamic. In other words, the power that is embedded within social categories is fostered and perpetuated by social contexts *and* experienced by individuals psychologically. In turn, because social categories are properties of the individual as well as of the social context, the meaning and significance or salience of such social categories is not static but fluid and dynamic. In sum, intersectional research must attend to the three common elements of intersectionality theories; we explore specific quantitative methods and techniques for implementing that research next.

Methods and Techniques for Intersectional Approaches

Cho, Crenshaw, and McCall (2013) described two processes crucial to the development of intersectionality. One is a "centrifugal" process, in which intersectionality is adapted by a discipline and the traditional methods within the discipline, and the other is a "centripetal" process, in

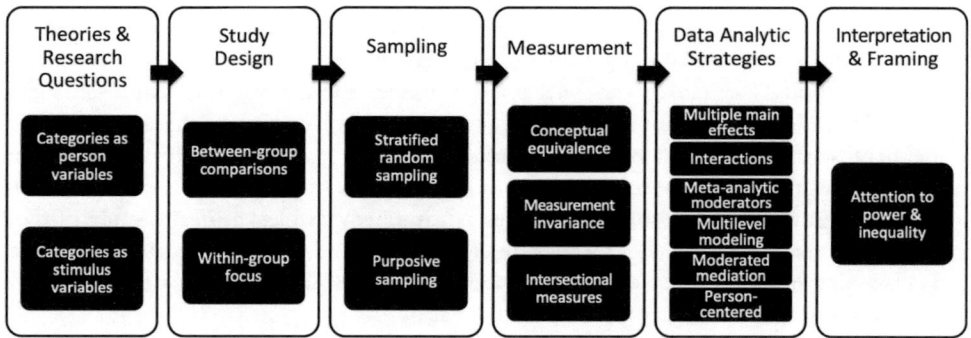

Figure 26.1 Steps of the research process and techniques for implementing an intersectional approach

which a discipline and its methods adapt to intersectionality. Both processes are crucial for the development of intersectionality. Else-Quest and Hyde (2016b) suggested a collection of methods and techniques frequently used within psychology but rarely with an explicitly intersectional approach, which aligns with both processes. We briefly review a handful of those methods and techniques here (see Figure 26.1), and note that it is important that researchers go beyond a specific method to incorporate the complexity and depth of intersectionality as a critical theory and approach, using as many of these techniques as possible and remaining faithful to the social justice aims of intersectionality.

Theories and Research Questions

Theory guides the research process, determining which research questions are asked and, often, leading to testable hypotheses. As a critical theory, intersectionality assumes that power relations construct our thoughts and experiences and seeks ultimately to empower those who are at marginalized intersectional locations and redress power imbalances. Thus, intersectionality is not a theory to be tested or falsified, but instead is a critical theory and analytic approach that guides research aims and reframes or reconceptualizes psychological phenomena. In doing so, intersectionality fosters the development of novel research questions within existing theories and also promotes theory development.

Cole (2009) proposed three questions to guide an intersectional approach at the theory stage of research. One, "Who is included within this category?" points to research questions about within-group heterogeneity and the interdependence of social categories. Two, "What role does inequality play?" re-orients research questions to context and the dimension of power within social categories. Three, "Where are the similarities?" reframes the traditional focus on difference to identify commonalities and build alliances or coalitions. Thus, at the theory stage, researchers may implement an intersectional approach by explicitly starting from these questions, or they may opt to work from theories that are consonant with the core elements of intersectionality such as social dominance theory (Sidanius and Pratto 1999) or the minority stress model (Meyer 2003).

In conceptualizing social categories such as gender, race/ethnicity, and social class, researchers may frame the analysis of social categories in different ways—as person variables or as stimulus variables. That is, gender can be a property of the individual, but it can also be a stimulus to which others respond. Similarly, Bauer (2014) distinguished studying identities and intersectional locations in social and political context from studying processes and policies related to power and inequality. Thus, social categories may be framed as person variables (i.e. as a characteristic of the individual, in context). Framing social categories as person variables does not negate the social construction of the categories, but shifts the focus of analysis to the behavior, characteristics, and experiences of people at particular intersectional locations. For example, the study of identity development or the experience of marginalization or bias among people at a specific intersectional location (e.g. Woods, Buchanan, and Settles 2009) frames social categories as person variables. Yet, the study of identity without analysis of context and power cannot be characterized as intersectional because the analysis of power is essential. In addition, social categories may be framed as stimulus variables that shape or influence how others respond to people. Researchers examining person perception might frame social categories as stimulus variables, as in the study of the perception of Black female leaders (e.g. Livingston et al. 2012). And, social categories may be framed as *both* person and stimulus variables, as in the study of implicit bias by people at one intersectional location toward people at another intersectional location.

Study Design: Within- and Between-Groups

Some intersectional studies have a within-group design or focus, studying people from one specific intersectional location in depth such as a study of gay Asian men. Such a design provides depth and a potentially rich or thick description or characterization of that location. However, a within-groups focus prohibits any comparison with other locations or groups, which is an important aspect of social context.

Other intersectional studies employ between-groups design or comparisons, studying people at different intersectional locations such as a study of Latina lesbian women and Latina straight women. By incorporating multiple intersectional locations, researchers are able to identify similarities as well as differences across those groups but may be less able to identify emergent phenomena that are unique to a given intersectional location. Thus, both of these types of study design offer unique advantages and, ideally, are complementary within a literature.

Sampling: Giving Voice and Group Heterogeneity

A common theme in intersectionality scholarship is that of *giving voice* to people from multiply marginalized groups or locations. Inevitably, this theme points to the attention that some intersectionality research must pay to sampling techniques in order to give voice to those whose voices have historically been excluded. While intersectional research should not be equated with demographically inclusive sampling (May 2015), such sampling may be a critical component of a study's methods. One technique to achieve that goal is stratified random sampling (or quota sampling; Shadish, Cook, and Campbell 2002), which specifies a priori which groups will be studied in the research and then samples to ensure that equal numbers of individuals in each group are included. In addition, purposive (or purposeful) sampling strategies are nonprobability sampling strategies that specify group characteristics and then recruit participants with those characteristics (Shadish et al. 2002; Teddlie and Yu 2007). Both stratified random sampling and purposive sampling strategies are helpful when conducting research with low-frequency groups. Moreover, whenever possible, care must be taken to not impose a particular group identity on research participants and to allow or encourage participants to self-identify their group membership, which is consistent with the theme of giving voice to people from marginalized groups.

Intersectionality also entails attention to the heterogeneity within a given group (e.g. Cole 2009). With regard to sampling, this attention points to the potential to commit the "lumping error," in which samples of heterogeneous groups are treated as homogeneous, in part because they are too small to be divided into appropriate subgroups. For example, a group such as Asian immigrants includes people from diverse countries such as Japan, Korea, India, Vietnam, China, and so on. Such subgroups demonstrate subcultural variations and within-group heterogeneity, with important differences as well as similarities. While some degree of lumping is inevitable, intersectionality researchers must be mindful with their sampling techniques to sample for sufficient heterogeneity to achieve generalizability and to not re-marginalize low-frequency subgroups. Within-group heterogeneity also points to concerns about over- and under-representation within a sample, and the potential for re-marginalization by researchers. In all of these instances, researchers may need to consider innovative recruitment strategies (Williams and Fredrick 2015) or be cautious in their conclusions about representation and generalizability given their sampling techniques. In such instances where there is tension between within-group heterogeneity, statistical power, and generalizability, incorporation of qualitative techniques (i.e. a mixed-methods design) may bolster the strength of a project.

Measurement: Examining Assumptions and Intersectional Phenomena

At the measurement stage of the research process, intersectionality might entail testing the assumptions of conceptual equivalence and measurement invariance, or explicitly attending to intersectionality in self-report measures. Conceptual equivalence (Sue and Sue 2000) is especially relevant to research with between-groups comparisons, examining whether a measure is assessing the same construct across different groups. Conceptual equivalence may often be assumed, but given that many measures have been developed or normed with nonrepresentative samples, this assumption should be tested. Measurement invariance (Meredith1993) is another tenuous assumption often made in research with between-groups comparisons (Corral and Landrine 2010). It presupposes that measures have similar psychometric properties across the different groups being compared. Thus, intersectionality researchers might choose to examine the internal features of a measure (e.g. assessing reliability or internal consistency, patterns of item-total correlations, factor structure, and specific factor loadings) as well as external associations with other variables (e.g. assessing criterion validity and associations with measures of other constructs).

Intersectionality raises the question of experiences or characteristics that are unique to specific intersectional locations such as the gendered racism that Black girls might experience in schools. In these cases, developing specific measures to assess those experiences of intersectional phenomena, which might not be adequately captured by existing measures, is appropriate. In addition, assessing intersectional phenomena represents another technique for "giving voice" to members of historically marginalized groups.

Data-Analytic Strategies

Many traditional data-analytic strategies might effectively be used within an intersectional approach, provided that they are supported by intersectional framing and interpretation. We illustrate these possibilities here with two commonly used techniques to demonstrate ways in which quantitative researchers can apply intersectionality. Other, more complex statistical approaches can also be combined with intersectional approaches, but they are beyond the scope of this chapter (interested readers should see Else-Quest and Hyde 2016b).

While *intersectional effects* involve the identification of unique or novel phenomena at a given intersectional location (e.g. the experience of coming out as lesbian), additive and multiplicative effects involve assessing differences and similarities in group means across multiple intersectional locations. *Additive effects* are a possible finding in quantitative research using intersectional approaches. Note that additive effects are distinct from additive approaches, which consider social categories as entirely independent, distinct, and mutually exclusive and thus are not intersectional (Else-Quest and Hyde 2016a). Moreover, treating two social categories such as race/ethnicity and gender as variables in quantitative data analyses does not imply an additive approach. For example,

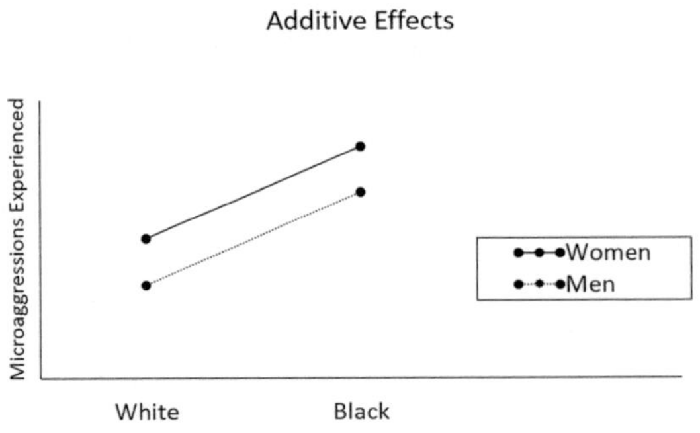

Additive Effects

Figure 26.2 Example of potential additive effects that can be identified with an intersectional approach using quantitative methods

Figure 26.2 depicts hypothetical additive effects, in which the effects of belonging to two marginalized social categories accumulate or accrue. Several techniques (e.g. ANOVA, multiple regression) can be used to test for additive effects, that is, the experience of "double jeopardy" or multiple marginalizations might be quantified by examining multiple statistical main effects. For example, additive effects may be identified when both heterosexist and racist discrimination each (i.e. not interactively) contribute to mental health (e.g. as found by Sandil et al. 2015).

By contrast, the same data-analytic techniques might be used to assess *multiplicative effects* (Else-Quest and Hyde 2016a), which may be reflected in statistical interactions between two or more social categories. Multiplicative effects might occur when two social categories exert effects that contradict one another, as when one category affords privilege and another confers disadvantage. Or, multiplicative effects might occur when membership in one disadvantaged group exacerbates the effects of membership in another disadvantaged group. In sum, traditional analyses of group differences and similarities in means can be used to test hypotheses derived from intersectional research questions. Figure 26.3 depicts two hypothetical sets of results with multiplicative effects.

Meta-analysis is a quantitative method for synthesizing the results of numerous studies on a particular question, e.g. whether there are gender differences in self-esteem (e.g. Hyde 2005). Meta-analysis yields information not only on whether there is a difference, but how large it is. For example, Kling, Hyde, Showers, and Buswell (1999) conducted a meta-analysis of research on gender differences in self-esteem, finding that the widely touted gender difference in self-esteem is small in white samples but non-existent in Black samples. This pattern of results demonstrates the potential errors in "single-axis" thinking and the importance of using an intersectional approach.

Other elements of intersectionality may also be examined with quantitative data-analytic techniques. For example, moderated mediation (Preacher, Rucker, and Hayes 2007) tests whether a mediated effect (e.g. the effect of race/ethnicity on achievement is mediated or explained by social class) is moderated by another variable (e.g. that mediated effect is stronger for boys than for girls). Thus, moderated mediation might be used to test how processes differ across intersectional locations. Alternatively, person-centered methods are a potential set of techniques that can explore intersectional research questions and assumptions about the construction and nature of social categories (e.g. McCall 2005) such as how membership in social categories may be fluid and dynamic. For example, latent transition analysis (Collins and Lanza 2010) might be used with longitudinal data to assess changes or shifts in group membership, as a way to examine social category fluidity explicitly. Such person-centered methods have rarely been employed in an intersectional approach, but represent some of the diverse opportunities that data-analytic techniques present for intersectionality.

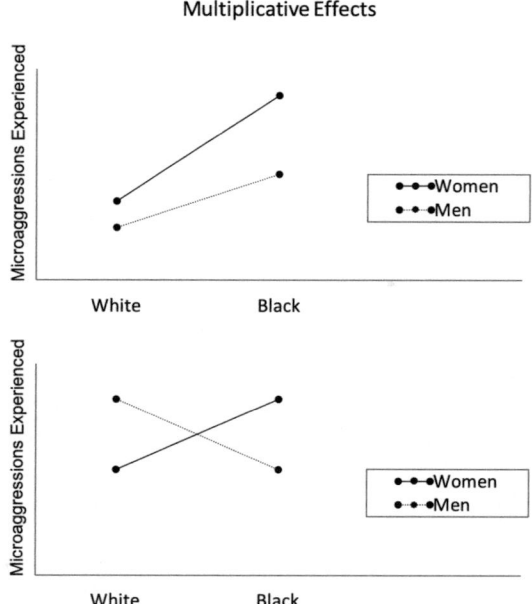

Figure 26.3   Example of potential multiplicative effects that can be identified with an intersectional approach using quantitative methods

Interpretation and Framing

Of course, none of these methods or techniques is truly intersectional without an interpretation and framing that make the role of power and inequality in social categories explicit. Too often, studies of group differences between men and women or of white people and people of color become studies of essentialized differences, reinforcing stereotypes and binary thinking that are antithetical to intersectionality. Thus, of all the techniques we suggest for examining intersectionality, the only essential one is an interpretation of research findings with the knowledge that groups exist within a context in which power and inequality are linked, rooted in, and perpetuated by social categories.

Combining Techniques in an Intersectional Approach: An Example

To illustrate several of the techniques we have described here (and in more detail elsewhere, Else-Quest and Hyde 2016b), consider Williams and Lewis' (2019) quantitative study of Black women's experiences of microaggressions and depressive symptoms. Drawing explicitly on intersectionality as a theory guiding their investigation and research questions, Williams and Lewis examined gender and race as person variables in focusing on the experiences of Black women. The project used a within-groups focus to examine these constructs with a sample of $n = 231$ Black women, who were recruited through purposive techniques. Participants completed surveys explicitly measuring intersectional phenomena. These measures include the Gendered Racial Microaggressions Scale (Lewis and Neville 2015) and a scale of racial identity (Sellers et al. 1997) modified to be specific to Black women. These scales included items such as, "I have received negative comments about my hair when I wear it in a natural hairstyle" and "I am proud to be a Black woman," respectively. To analyze these data and describe how these experiences are connected to coping and depressive symptoms, Williams and Lewis tested a moderated mediation model. In that model,

Williams and Lewis found that Black women's experiences of gendered racial microaggressions were associated with greater experience of depressive symptoms via their coping mechanisms, that is, coping mediated the effect of microaggressions. In addition, gendered racial identity modified this mediation, such that the effects were strongest for women with lower levels of gendered racial identity and weakest for women with higher levels of gendered racial identity. In other words, gendered racial identity, at higher levels, served a protective function in the face of inequitable treatment. In framing the process by which gendered microaggressions have adverse mental health consequences, gendered racial identity can be understood as a psychosocial resource in the context of gendered racism.

Conclusion

In recent years, intersectionality has become a buzzword in the social sciences—frequently invoked but inconsistently undertaken with the critical perspective on social justice in which the approach is rooted. Its widespread and frequent mention is both reasonable and welcome: intersectional approaches can strengthen any research in psychology by incorporating the diversity of human experiences within a rich and complex social context and by promoting the well-being of people from marginalized groups. Yet, while qualitative psychologists have explored and integrated intersectionality to varying degrees, intersectionality is only beginning to find its way into mainstream psychology (Else-Quest and Hyde 2016a, b). With an eye toward shaping the field of psychology to be more effective in its purpose of optimizing the development and well-being of all people, especially those who are marginalized, we have offered some guidance on the quantitative implementation of intersectional approaches. Intersectionality challenges us to revise our research questions and consider diverse perspectives, presenting us with new opportunities for knowledge production. Thus, insofar as intersectional research is rigorous and focused on its original aim of subverting inequality, mainstreaming intersectionality will be beneficial for psychology and ultimately will execute the generative power of intersectionality. We are optimistic that, as more psychologists earnestly undertake an intersectional approach in their research, intersectionality can move beyond buzzword to foster substantive change within the discipline.

Related chapters: 7, 16, 28.

References

Alexander-Floyd, N. G. (2012) "Disappearing Acts: Reclaiming Intersectionality in the Social Sciences in a Post-Black Feminist Era," *Feminist* Formations 24, 1–25.

Bauer, G. R. (2014) "Incorporating Intersectionality Theory in Population Health Research Methodology: Challenges and the Potential to Advance Health Equity," *Social Science & Medicine* 110, 10–17.

Beale, F. (1970) "Double Jeopardy: To Be Black and Female," in Cade, T. (ed.) *The Black Woman: An Anthology* (pp. 90–100), New York, NY: New American Library.

Berger, M. T. and Guidroz, K. (2009) *The Intersectional Approach: Transforming the Academy through Race, Class, & Gender*, Chapel Hill, NC: University of North Carolina Press.

Bowleg, L. (2008) "When Black + Lesbian + Woman ≠ Black Lesbian Woman: The Methodological Challenges of Qualitative and Quantitative Intersectionality Research," *Sex Roles* 59, 312–25.

Bowleg, L., Teti, M., Malebranche, D. J., and Tschann, J. M. (2013) "'It's an Uphill Battle Every Day': Intersectionality, Low-income Black Heterosexual Men, and Implications for HIV Prevention Research and Interventions," *Psychology of Men & Masculinity* 14, 25–34.

Calasanti, T., Pietilä, I., Ojala, H., and King, N. (2013) "Men, Bodily Control, and Health Behaviors: The Importance of Age," *Health Psychology* 32, 15–23.

Carastathis, A. (2016) *Intersectionality: Origins, Contestations, Horizons*, Lincoln, NB: University of Nebraska Press.

Cho, S., Crenshaw, K. W., and McCall, L. (2013) "Toward a Field of Intersectionality Studies: Theory, Applications, and Praxis," *Signs: Journal of Women in Culture and Society* 38, 785–810.

Choo, H. Y. and Ferree, M. M. (2010) "Practicing Intersectionality in Sociological Research: A Critical Analysis of Inclusions, Interactions, and Institutions in the Study of Inequalities," *Sociological Theory* 28, 129–49.

Cole, E. R. (2009) "Intersectionality and Research in Psychology," *American Psychologist* 64, 170–80.

Collins, L. M. and Lanza, S. T. (2010) *Latent Class and Latent Transition Analysis*, Hoboken, NJ: Wiley.

Collins, P. H. (2000) *Black Feminist Thought: Knowledge, Consciousness, and the Politics of Empowerment* (2nd ed.), New York, NY: Routledge.

Collins, P. H. (2019) *Intersectionality as Critical Social Theory*, Durham, NC: Duke University Press.

Collins, P. H. and Bilge, S. (2016) *Intersectionality*, Malden, MA: Polity.

Combahee River Collective (1982) "A Black Feminist Statement," in Hull, G. T., Scott, P. B., and Smith, B. (eds.) *All the Women are White, All the Blacks are Men, But Some of Us are Brave*, Old Westbury, NY: The Feminist Press.

Cooper, A. J. (1892) *A Voice from the South*, Xenia, OH: Aldine.

Corral, I. and Landrine, H. (2010) "Methodological and Statistical Issues in Research with Diverse Samples: The Problem of Measurement Equivalence," in Landrine, H. and Russo, N. F. (eds.) *Handbook of Diversity in Feminist Psychology* (pp. 83–134), New York, NY: Springer.

Crenshaw, K. (1989) "Demarginalizing the Intersection of Race and Sex: A Black Feminist Critique of Antidiscrimination Doctrine, Feminist Theory and Antiracist Politics," *University of Chicago Legal Forum* 1989, 139–67.

Crenshaw, K. (1991) "Mapping the Margins: Intersectionality, Identity Politics, and Violence against Women of Color," *Stanford Law Review* 43, 1241–99.

Diamond, L. M. and Butterworth, M. (2008) "Questioning Gender and Sexual Identity: Dynamic Links over Time," *Sex Roles* 59, 365–76.

Eagly, A. H., Eaton, A., Rose, S. M., Riger, S., and McHugh, M. C. (2012) "Feminism and Psychology: Analysis of a Half-Century of Research on Women and Gender," *American Psychologist* 67, 211–30.

Else-Quest, N. M. and Hyde, J. S. (2016a) "Intersectionality in Quantitative Psychological Research: I. Theoretical and Epistemological Issues," *Psychology of Women Quarterly* 40, 155–70.

Else-Quest, N. M. and Hyde, J. S. (2016b) "Intersectionality in Quantitative Psychological Research: II. Methods and Techniques. *Psychology of Women Quarterly* 40, 319–36.

Few-Demo, A. L. (2014) "Intersectionality as the 'New' Critical Approach in Feminist Family Studies: Evolving Racial/Ethnic Feminisms and Critical Race Theories," *Journal of Family Theory & Review* 6, 169–83.

Guba, E. G. and Lincoln, Y. S. (1994) "Competing Paradigms in Qualitative Research," in Denzin, N. K. and Lincoln, Y. S. (eds.), *Handbook of Qualitative Research*, Thousand Oaks, CA: Sage.

Hancock, A. (2007) "When Multiplication Doesn't Equal Quick Addition: Examining Intersectionality as a Research Paradigm," *Perspectives on Politics* 6, 63–79.

Hancock, A. (2016) *Intersectionality: An Intellectual History*, New York, NY: Oxford.

Harding, S. (1986) *The Science Question in Feminism*, Ithaca, NY: Cornell University Press.

Hare-Mustin, R. T. and Marecek, J. (1988) "The Meaning of Difference: Gender Theory, Postmodernism, and Psychology," *American Psychologist* 43, 455–64.

Hurtado, A. and Sinha, M. (2008) "More than Men: Latino Feminist Masculinities and Intersectionality," *Sex Roles* 59, 337–49.

Hyde, J. S. (2005) "The Gender Similarities Hypothesis," *American Psychologist* 60, 581–92.

Kling, K. C., Hyde, J. S., Showers, C., and Buswell, B. (1999) "Gender Differences in Self-esteem: A Meta-analysis," *Psychological Bulletin* 125, 470–500.

Lewis, J. A. and Neville, H. A. (2015) "Construction and Initial Validation of the Gendered Racial Microaggressions Scale for Black Women," *Journal of Counseling Psychology* 62, 289–302.

Livingston, R. W., Rosette, A. S., and Washington, E. F. (2012) "Can an Agentic Black Woman Get Ahead? The Impact of Race and Interpersonal Dominance on Perceptions of Female Leaders," *Psychological Science* 23, 354–8.

May, V. M. (2015) *Pursuing Intersectionality, Unsettling Dominant Imaginaries*, New York, NY: Routledge.

McCall, L. (2001) *Complex Inequality: Gender, Class, and Race in the New Economy*, New York, NY: Routledge.

McCall, L. (2005) "The Complexity of Intersectionality," *Signs: Journal of Women in Culture and Society* 30, 1771–800.

Meredith, W. (1993) "Measurement Invariance, Factor Analysis and Factorial Invariance," *Psychometrika* 58, 525–43.

Meyer, I. H. (2003) "Prejudice, Social Stress, and Mental Health in Lesbian, Gay, and Bisexual Populations: Conceptual Issues and Research Evidence," *Psychological Bulletin* 129, 674–97.

Moradi, B. and Grzanka, P. (2017) "Using Intersectionality Responsibly: Toward Critical Epistemology, Structural Analysis, and Social Justice Activism," *Journal of Counseling Psychology* 64, 500–13.

Preacher, K. J., Rucker, D. D., and Hayes, A. F. (2007) "Addressing Moderated Mediation Hypotheses: Theory, Methods, and Prescriptions," *Multivariate Behavioral Research* 42, 185–227.

Riger, S. (1992) "Epistemological Debates, Feminist Voices: Science, Social Values, and the Study of Women," *American Psychologist* 47, 730–40.

Sandil, R., Robinson, M., Brewster, M. E., Wong, S., and Geiger, E. (2015) "Negotiating Multiple Marginalizations: Experiences of South Asian LGBQ Individuals," *Cultural Diversity and Ethnic Minority Psychology* 21, 76–88.

Sellers, R. M., Rowley, S. A. J., Chavous, T. M., Shelton, J. N., and Smith, M. A. (1997) "Multidimensional Inventory of Black Identity: A Preliminary Investigation of Reliability and Construct Validity," *Journal of Personality and Social Psychology* 73, 805–15.

Settles, I. H. (2006) "Use of an Intersectional Framework to Understand Black Women's Racial and Gender Identities," *Sex Roles* 54, 589–601.

Shadish, W. R., Cook, T. D., and Campbell, D. T. (2002) *Experimental and Quasi-experimental Designs for Generalized Causal Inference*, Boston, MA: Houghton-Mifflin.

Shields, S. A. (2008) "Gender: An Intersectionality Perspective," *Sex Roles* 59, 301–11.

Sidanius, J. and Pratto, F. (1999) *Social Dominance: An Intergroup Theory of Social Hierarchy and Oppression*, New York, NY: Cambridge University Press.

Spierings, N. (2012) "The Inclusion of Quantitative Techniques and Diversity in the Mainstream of Feminist Research," *European Journal of Women's Studies* 19, 331–47.

Sprague, J. (2016) *Feminist Methodologies for Critical Researchers: Bridging Differences* (2nd ed.). Lanham, MD: Rowman & Littlefield.

Sue, S. and Sue, D. W. (2000) "Conducting Psychological Research with the Asian American/Pacific Islander Population," in Council of National Psychological Associations for the Advancement of Ethnic Minority Interests (Ed.), *Guidelines for Research in Ethnic Minority Communities* (pp. 2–4), Washington, DC: American Psychological Association.

Teddlie, C. and Yu, F. (2007) "Mixed Methods Sampling: A Typology with Examples," *Journal of Mixed Methods Research* 1, 77–100.

Tolhurst, R., Leach, B., Price, J., Robinson, J., Ettore, E., Scott-Samuel, A. … Theobald, S. (2012) "Intersectionality and Gender Mainstreaming in International Health: Using a Feminist Participatory Action Research Process to Analyze Voices and Debates from the Global South and North," *Social Science and Medicine* 74, 1825–32.

Walby, S., Armstrong, J., and Strid, S. (2012) "Intersectionality: Multiple Inequalities in Social Theory," *Sociology* 46, 224–40.

Williams, M. G. and Lewis, J. A. (2019) "Gendered Racial Microaggressions and Depressive Symptoms among Black Women: A Moderated Mediation Model," *Psychology of Women Quarterly* 43, 368–80.

Williams, S. L. and Fredrick, E. G. (2015) "One Size May Not Fit All: The Need for a More Inclusive and Intersectional Psychological Science on Stigma," *Sex Roles* 73, 384–90.

Woods, K. C., Buchanan, N. T., and Settles, I. H. (2009) "Sexual Harassment across the Color line: Experiences and Outcomes of Cross – Versus Intraracial Sexual Harassment among Black Women," *Cultural Diversity and Ethnic Minority Psychology* 15, 67–76.

Suggested Further Reading

Else-Quest, N. M. and Hyde, J. S. (2016a) "Intersectionality in Quantitative Psychological Research: I. Theoretical and Epistemological Issues," *Psychology of Women Quarterly* 40, 155–70. (For a more detailed discussion of the theoretical and epistemological issues of intersectionality in psychology described here.)

Else-Quest, N. M. and Hyde, J. S. (2016b) "Intersectionality in Quantitative Psychological Research: II. Methods & Techniques," *Psychology of Women Quarterly* 40, 319–36. (For a more detailed discussion of the methods and techniques suggested for using quantitative methods within an intersectional approach.)

Social Science

27

FEMINIST ECONOMICS

Drucilla K. Barker and Edith Kuiper

Introduction

Can you define feminist economics for me? What is feminist economics about? Is feminist economics the same as what women economists do? These are some of the questions feminist economists often get asked. They are of course legitimate questions, and although the short answer is simple – doing economics taking women and difference into account – the more useful and extended explanation is considerably more complex. This chapter aims to contribute to that answer by outlining the foundations of feminist economics, briefly discussing the content of the field, and exploring its central epistemological and methodological issues.

Before feminist economics was formally established, feminist scholars in economics applied ostensibly gender neutral economic methods to examine issues that were of particular importance to women. Discussions of the gender wage gap, labor market segregation, and women in development (WID) are cases in point. By the mid-1980s feminism was firmly ensconced, not only in the humanities but also in other social sciences such as sociology and anthropology. Economists – mainly women – began to take a critical look at their own profession and explored various ways to bring feminism to bear on the field of economics. We were an eclectic group and our brands of feminism ran the gamut from liberal to Marxist. It was a heady time, and the epistemological, theoretical, and methodological debates that it sparked are still ongoing. In this chapter we provide the reader with an overview of these debates and how they have evolved over the past three or so decades and articulate both the differences and the commonalities that bind a large subset of economists together under the heading of "feminist economics."

The Foundations

We begin with a brief discussion of the state of the economics profession in the post WWI years. During this time women economists still mainly lacked access to academic positions in economics departments. The Department of Home Economics at the University of Chicago was an exception. It was here that questions about consumption and the economic well-being of families and households were studied, a subject previously ignored by the mostly male mainstream of the profession. In 1923 Hazel Kyrk published *A Theory of Consumption*, and in 1935 Margaret Reid, her PhD student, developed objective and systematic methods for measuring the economic value of household production (Reid 1934). Reid's work on measurement, especially the "third person criterion" – an activity is work if you can pay someone else to do it for you – remains foundational in statistical estimates of the value of household production (UNDP 1995). However, in the years after World War II, interest in the household and families waned, and the home economics department at Chicago was terminated in 1956.

It was during this time that neoclassical economics gained its hegemonic position, particularly in the United States. [1] Although there were other schools of economic thought during this time, American institutionalist (studying the roles of institutions and social norms) and Marxist economics (studying the laws of motion of capitalist society-production, distribution, and accumulation) for example, it was neoclassical economics that became economics unmodified. For reasons that are outside the scope of this chapter, in the United Kingdom and Europe there remained a bit more room for other economic schools, especially Marxism.

Neoclassical economics, also referred to as mainstream economics, was ascendant. It was defined not by its object of study, but rather by its methodology. Its adherents are committed to the belief that rigidly prescribed "scientific" methodology results in unbiased economic science. It assumes that economies are comprised of rational individuals, who maximize their utility subject to the constraints placed upon them by prices, incomes, and time. Formal mathematical models trace the implications of consumers' and firms' behaviors, which simultaneously determine equilibrium prices and quantities (Barker 1999). Although sex may be taken as an explanatory variable in empirical work, there is no ontological difference between women and men in economic theory. Both are rational economic agents. The deductive method and mathematical modeling are at the center of the project as that is where the claim to the scientific status of economics lies.

The ongoing process of professionalization and the increase of mathematical formalism of economics that went with it meant a serious setback for the research on many of the issues that most profoundly affect the lives of women and their families (Kuiper 2008). In addition, the assumption that economies can be described and analyzed in terms of rational economic agents, or informally "economic man," excluded any meaningful discussion of economic interactions motivated by anything other than personal gain. Altruism, for instance, was considered a behavior that occurred only in families, and families lay outside the purview of mainstream economics.

Economics has never been a profession that was particularly welcoming to women. Then, as today, women were small minority and only a limited few made it to associate and even fewer to full professor. To remedy that situation, the Committee on the Status of Women in the Economics Profession (CSWEP) was started in 1971. A standing committee of the American Economic Association (AEA), CSWEP serves professional women economists by monitoring their progress in the field and promoting their careers. As women economists entered the profession they brought their political positions with them and tended to take methodological positions that were in accordance with their positions in the larger feminist community. Liberal feminists used mainly neoclassical approaches (Bergmann 1986; Blau and Ferber 1986; Bergmann 1986), while socialist feminists applied Marxist approaches (Benería and Sen 1981; Elson and Pearson 1981; Folbre 1982; Hartmann 1981). Whether working in the neoclassical or Marxist schools their work emphasized the importance of including gender and difference in economic theory and measurement. It became more and more clear, however, that a feminist lens was necessary in order to reveal and interrogate the masculinist values that are deeply embedded in the concepts of rationality, efficiency, and scarcity that are the methodical core of mainstream economics (Ferber and Nelson 1993).

Establishing a Field

During the1980s and early 1990s, mainstream economics was so firmly entrenched in US academia that it became the standard by which all other schools were judged. Neoclassical economists were able to determine what counted as legitimate economics. Thus, taking a critical stand and conducting economic research outside the mainstream research program could well threaten one's status as "a real economist"; it certainly created a barrier to getting results published in top ranked journals. However, we were not to be deterred. The 1990s saw a coming together of feminist economists worldwide, despite the fact that during that time mentioning the word "feminism" in the economics discipline was like cursing in the church.

Most feminist economists had been working for years in marginalized and isolated positions in their own departments. To be able to work with similarly motivated colleagues worldwide they founded the International Association for Feminist Economics (IAFFE) in 1992. The founding of IAFFE provided them with an epistemological community committed to bringing feminist perspectives on issues particularly germane to women's lives. In 1995 IAFFE launched its journal, *Feminist Economics*, with Diana Strassmann at the helm as its founding editor. The purpose and guiding philosophy of the journal at that time was articulated in the editor's introduction:

> By opening the gates that have for so long protected economics theories from fundamental critique and by subjecting all ideas addressed in the forum to critical scrutiny, *Feminist Economics* will encourage the emergence of a more resilient economics.
>
> *(Strassmann 1995: 1)*

Taking a feminist stand and acknowledging the fundamental role of gender in the economy required uncovering the implicit masculinist values embedded in mainstream concepts, theories, methods, and scientific standards for research. The fundamental rethinking and reorganizing of economics as a science was explored in the anthologies, *Beyond Economic Man* (Ferber and Nelson 1993) and *Out of The Margin* (Kuiper and Sap 1995). These scholars identified gender as the dominant fundamental hierarchical dualism in economics with profound consequences at all levels of the science. This opened the door for further explorations of the importance of including considerations of gender and to a lesser extent, racial differences in economics (Barker 1995; Elson 1991; Feiner 1993; Folbre 1993; Nelson 1995; Pujol 1992; Seiz 1992 Sen and Grown 1987; Strassmann 1995; Williams 1993).

Most of the feminist economic research of this period can be characterized as an example of feminist empiricism (Hankinson Nelson 1990; Harding 1986; Tuana 1992. In this view, sexism and androcentrism in science are biases that can be corrected by adherence to the existing norms of scientific inquiry, while at the same time increasing gender and racial diversity among the academic community. The inclusion of women and people of color is considered necessary to this endeavor as they are the ones most likely to notice the shared andro-and ethnocentric values and implicit assumptions that produce bias both in the context of discovery and in the context of justification.

This period also saw a few brief forays into using qualitative methods such as interviews, archival research, ethnography, and discourse analysis. However, these efforts rarely resulted in publications in *Feminist Economics* unless they also included a healthy dose of either formal modeling or statistical evidence. Mixed methods were encouraged, but articles relying on purely qualitative analyses generally were not. This is not surprising since in order to establish itself as a legitimate field within the larger discipline of economics feminist economists had to establish their scientific credentials. In order to provide a context for a discussion of some of the ongoing epistemological and methodological debates, the next section provides a brief outline of some main subfields where feminist economists made significant contributions.

Labor Economics

Traditionally most feminist economists in the United States and Europe worked in the field of labor economics. (So common in fact that "Are you going into labor?" was a joke commonly circulated among women graduate students.) Dominated by the neoclassical framework of individual optimizing behavior and quantitative research methods during the 1970s and 1980s, emerging feminist economists found a home in labor economics conducting research on wage disparities, occupational segregation, and the impact of motherhood on women's participation in the labor force. (Blau and Ferber 1986; Goldin 1992; Gustafsson 1996). As mentioned earlier, these issues had been previously studied, but not from a feminist perspective. Without an explicit commitment to feminist values, there is strong tendency in this research to explain away the issues by difference in preferences, attitudes toward risk, and other contextual factors between women and men.

This is best illustrated by the influence of the Nobel laureate Gary S. Becker. Becker (1981) extended the core assumptions and methods of neoclassical economics to a theory of the family – previously considered by economists as "black box," a pre-capitalist and a-historical institution. This opened the door for economists to apply mainstream economic theory to research on the relationship between the sexual division of labor in the family and waged and salaried labor. While he may have opened the door, his was not a feminist approach. For example, his analysis of the gender wage gap came to the conclusion that it was the result of women's rational choices to work in lower-paying occupations that required less investment in education and training because these jobs were compatible with their responsibilities as mothers and wives. Becker's claims that men were biologically oriented toward the market and women toward the household invoked the feminist counter argument: that women bear children may be a biological fact; that women bear the sole responsibility for rearing their children is a social construction and can be changed.

Feminist economists took great intellectual pleasure in debunking these and other such arguments and in advancing alternative feminist explanations that questioned why the rearing of children was primarily the responsibility of women. Many turned to more sociological theories and approaches to understanding the complex determinants of disparities in the labor market and including gender and race discrimination. Others used institutionalist approaches that take historical factors into account and investigate institutional characteristics for their impact on economic behavior and decision making (Figart, Mutari and Power 2002). New econometric techniques were developed in order to measure the effects of intersectional identities on wages and occupational segregation (Kim 2007). The gender wage gap remains a central feminist concern. Its significance is now recognized by international institutions such as the World Bank, the International Monetary Fund (IMF), and the World Economic Forum. Nonetheless, feminist perspectives that acknowledge the importance of discrimination, intersectionality, and the recognition of class differences among women are still too often ignored in favor of references to Becker's work.

Domestic Labor and Care Work

The importance of unpaid domestic labor to the functioning of the economy emerged as a concern in own right among feminist economists working in the Marxist tradition. These scholars were located mainly, but not exclusively, in the United Kingdom and Italy. They coined the term "domestic labor" to describe the unpaid housework that was necessary for the support and maintenance of waged labor within capitalism (Dalla Costa and James 1973; Federici 1975; Himmelweit and Mohun 1977). The immediate products of domestic labor such as clean clothes and cooked meals are necessary for the reproduction of the labor force on both a daily and an intergenerational basis. They are crucial to capitalism and central to understanding the oppression of women in capitalist societies.

Other feminist economists, working in the tradition of Hazel Kyrk and Margret Reid, turned their attentions to developing methods for measuring the contribution of women's unpaid household labor to economic growth (Goldschmidt-Clermont 1990; Ironmonger 1996; Waring 1988).

As discussed earlier, measuring domestic labor was premised on the third party criterion: an activity was considered work if you could pay someone else to do it for you. In other words, the activity was separable from the person doing the work. Some types of domestic labor do not fit easily into this definition of work. While who vacuums the floor is irrelevant as long as the job is done well, who cares for the children, the elderly, and the infirm does matter (Himmelweit 1995; Jochimsen 2003). This type of labor is called caring labor. Caring labor is not separable from the person doing it because it is constituted by the relationship between those who give care and those who receive care. The care received by infants, young and school-age children, the ill, the disabled, and the elderly depends upon the quality of the relationship connecting the givers and the receivers of care (Folbre 1995; Folbre and Nelson 2000).

A second question soon arose. Why is care work devalued? Why is it badly paid, if it is paid at all? Folbre (1995) argued that the value of caring labor presented a paradox for feminist economists because the affective nature of care implies that it should be its own reward: however if it does not command an economic return its global supply will be diminished. Folbre and Nelson (2000) resolve this paradox by separating the affective value of care from the activities of care. Caring activities should command an economic reward. Feminist economists also elaborated on the ways that care, broadly understood as applying to both people and the environment, was the foundation for economies that provided for human well-being and were sustainable (Jochimsen 2003; van Staveren 2001).

Other feminist scholars interrogated what they saw as the raced and gendered assumptions in this stream of scholarship (Barker and Feiner 2009; Glenn 1992; Gutiérrez Rodríguez 2010; Hewitson 1999). Without disputing the vital need for care work or its scarcity under global capitalism, these feminist economists reframed the analyses. They pointed out the ways that these discussions naturalized the connection between women and care, masked the racialization implicit in the distinction between the affective and non-affective dimensions of care and failed to address why some groups are entitled to be cared for and cared about, while others are not afforded this privilege.

Globalization has ushered in a significant change in the international division of domestic labor. The increased participation of women in the labor force, aging populations, and low population growth in the global North have increased the demand for domestic workers. Due to structural changes in the global economy more and more poor women, and increasingly men, from the global South migrate to clean the houses and care of the children and elderly in the more affluent parts of the world. The demand for college educated, English speaking Filipinas to care for the children of the elites is the classic example (Parreñas 2015). As these young women migrate, their own children are left behind for many years to be cared for by women further down the ladder of class privilege in the Philippines. Arlie Hochschild (1989) describes this as a "global care chain." These feminist economic insights fly into the face of the neoliberal social and political policies of the United States and Europe that pursue further privatization of care, increased barriers to migration, and decreased state support for the social provision of domestic and care labor (Gutiérrez Rodríguez, 2010).

Development Economics

Feminist analyses of unpaid and non-monetized household labor became an important concern in the field of development economics as well. Briefly, development economics is the field in economics that is concerned with economic growth and modernization in the global South. During the 1960s it was believed that the key to curing poverty in the South was to transform their economies in the image of the North - modern, industrialized, and monetized. However, these plans did not take women into account. Women were typically viewed as "unproductive" housewives despite the important roles they played in agricultural production, urban manufacturing, reproductive labor in the household, and more. Feminist economists pointed out the male bias in this thinking (Elson 1991). The WID framework that emerged stressed the complementarity between women's liberation and the economic goals of efficiency and growth. This framework quickly gave way to the gender and development (GAD) approach, which explored the ways that gender norms are embedded in the structures of economy and connected the gender division of labor in the home to women's position in the paid labor force. Its focus on broader social transformation as the key to women's liberation and empowerment remains salient today.

Just as GAD was gaining currency, however, an enormous structural change occurred in the international economy. The old development model was no longer feasible as the countries of the global South found themselves heavily indebted to the banks of the global North. They turned to the international community for assistance, which was forthcoming in the form of new loans with onerous conditions attached. These conditions were called structural adjustment programs (SAPs) and required changes to the structure of these countries' economies in ways that would result in cost savings, spur growth, and open them to the global market. This was the stated plan anyway.

It took the work of feminist economists to show that these measures did not save money, but rather shifted costs from the monetized sector of the economy into the non-monetized household sector where women were left to cope with higher prices and reductions in social spending. The result was an increase in both their waged and unwaged work (Benería 1995; Benería and Feldman 1992; Elson 1991). In addition, global South countries also had to service (pay the interest plus some portion of the principal) these new loans. They attracted the dollars necessary to do this in two ways: (1) remittances from women's migration to work in transnational domestic labor markets as nannies, maids, and sex workers, and (2) the revenue earned by women working in export production zones sewing clothes, making toys, and assembling electronics (Çatagay, Elson, and Grown 1995). One consequence is today's transnational gendered division of domestic labor (Barker and Kuiper 2014).

Even though SAPs have now been replaced by Poverty Reduction Strategy Papers (PRSP), which require indebted countries to articulate their own strategies for reducing poverty, the reasoning behind this practice has not changed much. Development economics remains a central field of interest for many feminist economists, who employ an expanded set of research strategies, such as gender sensitive budgeting, gender critiques of macroeconomic modeling, time use surveys, and the development of new indices that take gender into account. A large part of this research has been disseminated through policy briefs, reports, and edited volumes. Its importance is also coming to the fore today as many countries in the developed world face austerity measures that have similar gendered effects as SAPs.

History of Economic Thought

Feminist economists have also made important headway in theorizing the history of economic thought. A substantial part of this work is the re-evaluation of women economists' and women economic writers' work and their contribution to economic thinking reaching as far back as the 1800s (Dimand, Dimand and Forget 1995; Barker and Kuiper 2010; Madden, Seiz and Pujol 2004).

Feminism and Anti-feminism in the History of Economic Thought, by Michelle Pujol (1992) laid the basis for a feminist critique of the history of economic thought by showing how sexist values and assumptions structured the course of British political economy in the nineteenth and early twentieth century. Building on Pujol's work Barker (1995), Nelson (1995), and others revealed the implicit assumptions about the masculinist nature of "economic man." Some historians of economics focused on pointing out the way classical economists like Adam Smith, Jeremy Bentham, Jean Baptiste Say, and Nassau Senior wrote about gender and gender relations (see e.g. Dimand and Nyland 2003). Nancy Folbre (2009) published a comprehensive discussion on the conceptualization of "self-interest" throughout the history of economic thought. As the definition of economics is broadened by feminist historians of economics to include women's economic writing, a wider array of texts come under scrutiny (see Kuiper 2014; Madden and Dimand 2019). This work includes interdisciplinary work on women's writing on economic topics and issues in novels, poems, and diaries.

Where the history of economic thought used to be dominated by an internalist historical approach, feminist economists have contributed to a diversification of methods and theoretical frameworks, in which there is ample attention to the impact of the historical context on the development of economic concepts and ideas.

Epistemological and Methodological Reflections

As part of the quest to uncover implicit assumptions and shared masculine values in contemporary mainstream economics, many feminist economists turned their attention to epistemology and methodology. The article by Sandra Harding (1995) in the first issue of *Feminist Economics* was enormously influential at this juncture. Her application of the concept of weak and strong objectivity to economics and critique of the positive/normative distinction set the stage for further discussions and reflections. Julie Nelson in *Feminism, Objectivity and Economics* (1995) elaborated this critical reflection on the development of economic science and indicated the variety of ways sexist values have been impacting the organization of the field, the use of metaphors, theories, concepts, methods, and choice of methodologies.

In a special issue of *History of Political Economy* Janet Seiz (1993) provided an overview of the fundamental questions concerning basic economic concepts addressed by feminist economists, and Nancy Folbre (1993) outlined the then current approaches to feminist economics in terms of "social construction" and "distortion" of scientific inquiry (167). A special issue of *Feminist Economics* provided space for Tony Lawson (1999) to argue that feminist economics could be strengthened by building on a critical realist ontology. This was followed by responses refuting his case on the basis of epistemology (Harding 1999), ethical considerations (Barker 2003), and methodology (Kuiper 2004).

Julie Nelson marked the new millennium by asking,

> Should economics remain defined as rational choice theory – a notion based in a radically Cartesian, anti-body view of the world – feminists will have relatively little to say. I want to change the central question to one of "provisioning" – how we provide for ourselves the means to sustain and enjoy life.
>
> *(Nelson 2000: 1178)*

This reframing of economics would over the years result in a stream of theorizing around economic as provisioning rather than as rational choice.

There were dissenters, of course. Those applying insights of postmodern and the poststructuralist turns in the humanities and other social sciences saw the scientific aspirations of feminist economics in a more problematic light. Gillian Hewitson (1999) was one of the first here. In her monograph, *Feminist Economics*, she points out that by and large feminist economists uncritically adhere to the sex/gender distinction, in which sex is biological and gender is social. This is both politically astute and theoretically problematic. Astute because it provides a familiar terrain upon which to base suggestions and push for changes in the way that economics is done. It is problematic, because it entails the expulsion of the body from the essence of personhood, a philosophical tradition that dates back to Descartes and seventeenth-century French philosophy. By clinging to the mind/body dichotomy feminist economics will, like neoliberal economics, theoretically exclude the specifically female body. If feminist economics is to be truly transformative, its adherents must theorize the production of sexed bodies.

Suzanne Bergeron (2009) and Colin Danby (2007) explore the implicit heteronormative assumptions that haunt feminist economists' scholarship. Bergeron's work reveals the implicit assumptions behind the dominant representations of the household: all adults belong to one of two genders and conform to dominant gender scripts; every adult forms a sexual and reproductive bond with a member of the opposite gender and forms a household with that person; and all households in which care work is performed are understood as being constructed around a heterosexual couple. Even when other household types are considered, single parent and same-sex couples, they are viewed in relationship to the normative paradigm.

Danby (2007) argues that heteronormativity names tacit conceptions about what is socially normal and these conceptions make it possible to think of heterosexuals or homosexuals as essential categories of people. A critique of heteronormativity would make visible a pattern of state repression that makes proper citizens by opposing them to improper ones, a process that simultaneously shapes gender, sexuality, citizenship, and race. If queer theorists are right, the obviousness of heteronormative assumptions to many people is partly a consequence of state efforts to establish and police familial norms. Economics, styling itself as a science of policy, has been reluctant to challenge the state's view of society. Danby makes the point that feminist economics would be greatly enriched taking up this challenge.

The debate on methodological issues over the following ten years furthered by several edited volumes. In *Postcolonialism Meets Economics*, edited by Zein-Elabdin and Charusheela (2004) the authors explore the colonial roots of mainstream and heterodox economics and articulate the ways that economics may start to critically engage with postcolonial themes and issues. *Toward a Feminist Philosophy of Economics*, edited by Barker and Kuiper (2005), provides a space for exploring the contours of the emerging debates between the feminist empiricists and poststructuralist and postcolonial approaches to feminist economics. *Robinson Crusoe's Economic Man: A Construction and Deconstruction*, edited by Grapard and Hewitsen (2011), brings together a group of authors who further deconstruct traditional binary thinking in economics.

The methodological issues that were central in feminist economics at the beginning have been fading in interest over the last decade or so. Recently, Sheba Tejani (2018) investigated three decades of articles in *Feminist Economics*, 1995–2015, and showed that there has been a clear shift toward articles applying quantitative and specifically econometric research methods and secondary datasets, and a decline in theoretical, historical and methodological articles. Based on her data from analysis of 490 articles and reviews, Tejani links this shift of focus in the journal to the internal pressures of the discipline and the self-enforcing process in which the current content of the journal attracts a specific type – mostly econometrically sophisticated feminist articles, which, in turn, results in the selection of associate editors who can handle them.

If we define feminist economics today primarily by what is published in *Feminist Economics*, our assessment is that feminist empiricists "won" the epistemological debate. Scholars working with postpositivist and interdisciplinary approaches have had to find other publication venues such as *Signs*, *Frontiers*, and the *Cambridge Journal of Economics*. This current state of affairs is, we think, due to two factors.

First, the contradictory goals of feminist economics: from the beginning, the main goals of feminist economists were the transformation of the discipline on the one hand and the establishment of validity of feminist economics within the larger profession on the other hand. This need to find acceptance in the larger profession and to increase the number of feminist economists and advance their careers entails compromise. As Barker (2005) argued, feminist economists inherited the power and prestige of economics, a discipline that calls itself the "queen of the social sciences." Poststructuralism and postmodernism, which question the very possibility of science, and postcolonialism, which reveals the racism and domination in Western science, both serve to undermine the scientific aspirations of feminist economists. Postpositivist approaches entail giving up the notion that economics, when grounded in the authority of reason and science, can be separated from power and can work toward a feminist common good. This last point is crucial. Many, perhaps the majority of, feminist economists hold precisely this view, and argue that empiricist methodology is the way toward this goal.

Lourdes Benería, a well-respected and highly influential feminist economist, argued that while postmodern work emphasizing identity, difference, and agency has enriched our understanding of identity politics, postcolonial realities, and the intersections of gender and race it has "run parallel to changes on the material side of life, particularly the resurgence of neoliberalism across countries and to the globalization of markets and of social and cultural life" (Benería 2003: 25). Postmodern work tends to deemphasize the economics and generate an imbalance between the "urgent need to understand economic reality … and the more predominant focus on 'words,' including issues such as difference, subjectivity, and representation" (Benería 2003: 25). It is not that work on these issues is wrong but rather that it needs to be linked to an understanding of the socioeconomic aspects of life. This is the task of feminist economics.

Julie Nelson, whose work has been greatly important in shaping the field, goes even further in her critique. She argues that the lack of deconstructionist or poststructuralist scholarship relative to that in other social sciences is not a drawback because such work creates barriers for scholars not educated in "obscurant literatures/techniques," and "promulgate[s] a bloodless and lifeless view of the world, and fail[s] to take into account lived experience" (Nelson 2000: 1180). Not surprisingly, the contours of this debate reflect what was going on in the larger community of feminist scholars regarding the impact of postmodernism/poststructuralism on feminist theory. Posing this question as an opposition, instead as a source of productive tension and source of theoretical innovation, has pushed feminist economics to the side of feminist empiricism.

The other factor of relevance here in our view is that many, perhaps most, feminist economists have experienced how sexism, discrimination, and masculine values have structured their daily lives and personal experiences as economists. Many in the field have been subject to discrimination and have been engaged in cases brought to committees similar to the Equal Employment Opportunity Commission (EEOC) in the United States to demand a promotion or job they considered themselves (over) qualified for but did not get. Feminist economists understand that the economics discipline is a social institution and are aware of their location within it. Their choice of a theoretical framework may at least be partly based on strategic considerations with respect to the space and opportunity they see for change, their wish to keep open the lines of communications with their colleagues and the field more broadly, and their concerns about continued employment as economists. This could mean that feminist economists have been choosing epistemological and methodological approaches partly based on the power dynamics within the discipline while still pushing the boundaries to make changes happen. Integrating and conceptualizing this activist and power-aware stand of feminist economists and engaging in affiliations with other critical scholars is a direction for a feminist philosophy of economics that may be fruitful for getting beyond the debate between empiricism and postmodernism.

Present State of Feminist Economics

Since the Great Recession of 2008/2009 the context for economics has changed. These events have put neoclassical economics in its place as too ahistorical and too focused on abstract modeling. Historical economic research regained status as the deductive, abstract approach of neoclassical economics has not proven useful in the face of major earth shaking economic problems such as the financial crisis of 2008. The failure of the mainstream has created a demand for new economic thinking in general. This opens a space for pluralism in economics that may engender the transformation of economics that many feminist economists hope for.

Today feminist economics is still fundamentally an empiricist project, although it is likewise eclectic and interdisciplinary in its approach. For researchers in the field, understanding the historical, gendered and racial contexts in which concepts and theories have developed is considered an important part of applying and developing new theories, indicators, indices, and methods. Feminist economists are somewhat freer to make their own choices regarding their approaches to economics, although quantitative methods continue to be privileged over qualitative methods in all the social sciences, not just economics. Many feminist economists have turned their attentions toward policy oriented research on both national and international arenas. For example, Heidi Hartmann founded the highly successful Institute for Women's Policy Research in Washington, DC in order to insert a feminist economics in US policy making. Similarly, Radhika Balakrishnan, Diane Elson, and Heinz (Balakrishnan 2013) were deeply involved in the conversations and debates that led to the adoption of the UN Sustainable Development Goals. Another example is the work that started in 1994 in Australia by Rhonda Sharp and in 1995 in South Africa by Debbie Budlander on Gender Budgeting, and is now supported by the United Nations Development Program (UNDP). It is currently building momentum in both the global South and Europe. Both the UNDP and the European Commission have been working for years now on both gender mainstreaming and bringing gender considerations into the budgeting process on all levels of government policy making. Gender analyses of the impacts of specific policies and the development of improved policy proposals, together with monitoring their progress, require painstaking and detailed work for which new indicators and measurement tools often need to be developed.

While this type of research often does not find its way into academic journals, it constitutes crucial groundwork for transforming economies and improving the conditions of women in countries around the world, thus in many ways fulfilling the aspirations of the feminist empiricists discussed earlier. For example, the continued scholarship and activism by Bina Agarwal starting with her monograph, *A Field of One's Own* (1997), led Indian policy makers to pass the Hindu Succession (Amendment) Act in 2005 which gives all Hindu women, regardless of marital status, equal rights with men in the ownership and inheritance of property. Most recently she has brought together her work on agriculture and food security, women's property rights, and environmental change in a three volume set (Agarwal 2016). She has become one of the most influential economists in India.

As feminist economics increasingly includes voices from the global South and people of color, and builds alliances with feminists in other academic fields and groups outside academia, it clearly strengthens the case of interdisciplinary research. The struggle of feminist economists has for a large part been a fight internal to the discipline; however, the field is now established and recognized. As the economics profession changes we need to make sure that feminist economics is part of these changes.

Related chapters: 8, 26, 28, 29, 33.

Note

1 The term neoclassical is used because it is a synthesis of nineteenth-century classical political economy and twentieth-century marginal analysis.

References

Agarwal, B. (1997) *A Field of One's Own*, Oxford: Oxford University Press.

Agarwal, B. (2016) *Gender Challenges*, 3 vols. Oxford: Oxford University Press.

Barker, D. K. (1995) "Economists, Social Reformers and Prophets: A Feminist Critique of Economic Efficiency," *Feminist Economics* 1 (3), 26–39.

―――― (1999) "Neoclassical Economics: Critique," in O'Hara, P. A. (ed.) *Encyclopedia of Political Economy*, London and New York: Routledge, 793–7.

―――― (2003) "Emancipatory for Whom? A Comment on Tony Lawson and Critical Realism," *Feminist Economics* 9 (1), 103–8.

―――― (2005) "A Seat at the Table: Feminist Economists Negotiate Development," in Barker, D. K. and Kuiper, E. (eds.) *Feminist Economics and the World Bank, 2009–2017*, London and New York: Routledge, 209–207.

Barker, D. K. and Feiner, S. F. (2009) "Affect, Race, and Class: An Interpretive Reading of Caring Labor," *Frontiers, A Journal of Women's Studies* 30 (1), 41–55.

Barker, D. K. and Kuiper, E. (eds.) (2005) *Towards a Feminist Philosophy of Economics*, London and New York: Routledge.

――――and Kuiper (2014) "Gender, Class, and Location in the Global Economy," in Evans, M., Hemmings, C., Henry, M., Johnstone, H., Madhok, S., Plomien, A. and Wearing, S. (eds.) *Handbook of Feminist Theory*, London: Sage Publications, 500–13.

Balakrishnan, R. with Elson, D. and Heintz, J. (2013) "Public Finance, Maximum Available Resources and Human Rights" in Nolan, A., O'Connell, R. and Harvey, C. (eds.) *Human Rights and Public Finance: Budget Analysis and the Advancement of Economic and Social Rights*, London: Hart Publishing, n.p.

Becker, G. S. (1981) *A Treatise on the Family*, Cambridge: Harvard University Press.

Benería, L. (1995) "Towards a Greater Integration of Gender in Economics," *World Development* 23 (11), 1839–950.

―――― (2014) *Gender, Development and Globalization: Economics as if All People Mattered*, New York and London: Routledge.

Benería, L. and Feldman, S. (1992) *Unequal Burden: Economic Crises, Persistent Poverty, and Women's Work*, Boulder: West View Press.

Benería L. and Sen, G. (1981) "Accumulation, Reproduction, and Women's Role in Economic Development: Boserup Revisited," *Signs: Journal of Women in Culture and Society* 7 (2), 279–98.

Bergeron, S. (2009). "Moving Caring Labor off the Straight Path in Development: Some Methodological Considerations," *Frontiers* 39 (1), 55–64.

Bergmann, B. (1986) *The Economic Emergence of Women*, New York: Basic Books.

Blau, F. and Ferber, M. A. (1986) *The Economics of Women, Men, and Work*, Englewood Cliffs: Prentice-Hall Press.

Çatagay, N., Elson, D., and Grown, C. (1995) "Introduction," *World Development* 23 (11), 1827–36.

Dalla Costa, M. and James, S. (1973) *The Power of Women and the Subversion of Community*, Bristol: Falling Wall Press.

Danby, C. (2007) "Political Economy and the Closet: Heteronormativity in Feminist Economics," *Feminist Economics* 13 (2), 29–53.

Dimand, A., Dimand, R. W., and Forget, E. L. (eds.) (1995) *Women of Value. Feminist Essays on the History of Women in Economics*, Aldershot: Edward Elgar.

Dimand, R. and Nyland, C. (2003) *The Status of Women in Classical Economic Thought*, Cheltenham: Edward Elgar.

Elson, D. (1991) "Male Bias in Macroeconomics: The Case of Structural Adjustment," in Elson, D. (ed.) *Male Bias in the Development Process*, Manchester and New York: Manchester University Press, 164–90.

Elson, D., and Pearson, R. (1981). "Nimble Fingers Make Cheap Workers: An Analysis of Women's Employment in Third World Export Manufacturing," *Feminist Review* 7, 87–107.

Federici, S. (2012 [1975]) *Revolution at Point Zero: Housework, Reproduction, and Feminist Struggle*, Brooklyn: Autonomedia, 15–22.

Feiner, S. F. (ed.) (1993) *Race & Gender in the American Economy, Views from across the Spectrum*, Englewood Cliffs, NJ: Prentice Hall.

Ferber, M. A., and Nelson, J. A. (1993) *Beyond Rational Economic Man*, Chicago: University of Chicago Press.

Figart, D. M. Mutari, E., and Power, M. (eds.) (2002) *Living Wages, Equal Wages: Gender and Labour Market Policies in the United States*, London and New York: Routledge.

Folbre, N. (1982). "Exploitation Comes Home: A Critique of the Marxian Theory of Family Labour," *Cambridge Journal of Economics* 6 (4), 317–29.

——— (1993) "How Does She Know? Feminist Theories of Gender Bias in Economics," *History of Political Economy* 25 (1), 167–84.

——— (1995) "Holding Hands at Midnight: The Paradox of Caring Labor," *Feminist Economics* 1 (1), 73–92.

——— (2009) *Greed, Lust, and Gender*, Oxford: Oxford University Press.

Folbre, N. and Nelson, J. A. (2000) "For Love or Money – or Both?," *Journal of Economic Perspectives* 13 (4), 123–40.

Glenn, E. N. (1992) "From Servitude to Service Work: Historical Continuities in the Racial Division of Paid Reproductive Labor," *Signs: Journal of Women, Culture and Society* 18 (1), 1–43.

Goldin, C. (1992) *Understanding the Gender Gap: An Economic History of American Women*, New York: Oxford University Press.

Goldschmidt-Clermont, L. (1990) "Economic Measurement of Non-Market Household Activities," *International Labor Review* 129 (3), 279–99.

Grapard, U. and Hewitson, G. eds. (2011) *Robinson Crusoe's Economic Man*, London and New York: Routledge.

Gustafsson, S. (1996) "Women's Labor Force Transitions in Connection with Childbirth: A Panel Data Comparison between Germany, Sweden and Great Britain," *Journal of Population* 9 (3), 223–46.

Gutiérrez Rodríguez, E. (2010) *Migration, Domestic Work and Affect*, London and New York: Routledge.

Hankinson Nelson, L. (1990) *Who Knows: From Quine to a Feminist Empiricism*, Philadelphia: Temple University Press.

Harding, S. (1986) *The Science Question in Feminism*, Ithaca: Cornell University Press.

——— (1995) "Can Feminist Thought Make Economics More Objective?" *Feminist Economics* 1 (1), 7–32.

——— (1999) "The Case For Strategic Realism: A Response To Lawson," *Feminist Economics* 5 (3), 127–33.

Hartmann, H. (1981) "The Family as the Locus of Gender, Class, and Political Struggle: The Example of Housework," *Signs: Journal of Women in Culture and Society* 6 (3), 366–94.

Hewitson, G. (1999) *Feminist Economics. Interrogating the Masculinity of Rational Economic Man*, Cheltenham: Edward Elgar.

Himmelweit, S. (1995) "The Discovery of 'Unpaid Work': The Social Consequences of the Expansion of 'Work'," *Feminist Economics* 1 (2), 1–19.

——— and Simon, M. (1997) "Domestic Labour and Capital," *Cambridge Journal of Economics* 1 (1), 15–31.

Hochschild, A., with Machung, A. (1989) *The Second Shift: Working Families and the Revolution at Home*, New York: Penguin Books.

Ironmonger, D. (1996) "Counting Outputs: Capital Inputs and Caring Labor, Estimating Gross Household Product," *Feminist Economics* 2 (3), 37–64.

Jochimsen, M. (2003) *Careful Economics: Integrating Caring Activities and Economics Science*, Dordrecht, NL: Kluwer Academic Publishers.

Kim, M. (ed.) (2007) *Race and Economic Opportunity in the 21st Century*, London and New York: Routledge.

Kuiper, E. (2004) "Critical Realism and Feminist Economics: How Well Do they Get Along?," in Lewis, P. A. (ed.) *Transforming Economics: Perspectives on the Critical Realist Project*, London: Routledge, 107–31.

——— (2008) "Feminism in/and Economics," in Davis, J. and Dolfsma, W. (eds.) *The Elgar Handbook of Socio-Economics*, Cheltenham: Edward Elgar, 188–206.

Kuiper, E. (2010) "Introduction," in Drucilla K. Barker and Edith Kuiper (eds.) *Feminist Economics. Critical Concepts in Economics, Vol. I, Early Conversation 1800–1960*, London and New York: Taylor& Francis and Routledge.

Kuiper, E., and Sap, J. C. M. (eds. with Susan F. Feiner, Notburga Ott, and Zafiris Tzannatos) (1995) *Out of the Margin A Feminist Perspective on Economics*, London: Routledge.

Kyrk, H. (1923) *A Theory of Consumption*, Boston and New York: The Riverside Press.

Lawson, T. (1999) "Feminism, Realism and Universalism," *Feminist Economics* 5(2), 25–59.

Madden, K. and Dimand, R. W. (2019) *The Routledge Handbook of the History of Women's Economic Thought*, London and New York: Taylor & Francis and Routledge.

Madden, K. K., Seiz, J. A., and Pujol, M. (2004) *A Bibliography of Female Economic Thought up to 1940*, New York: Routledge.

Nelson, J. A. (1995) *Feminism, Objectivity and Economic*, London and New York: Routledge.

——— (2000) "Feminist Economics at the Millennium: A Personal Perspective," *Signs* 25 (4), 1177–81.

Parreñas, R. (2015) *Servants of Globalization: Migration and Domestic Work*, 2nd. ed., Stanford: Stanford University Press.

Pujol, M. (1992) *Feminism And Anti-Feminism in Early Economic Thought*, Aldershot: Edward Elgar.

Reid, M. G. (1934) *The Economics of Household Production*, New York: J. Wiley & Sons.

Seiz, J. (1992) "Gender and Economic Research," in de Marchi, N. B. (ed.) *Post Popperian Methodology of Economics: Recovering Practice*, New York: Springer Science, 273–326.

——— (1993) "Feminism and the History of Economic Thought," *History of Political Economy* 25 (1), 185–201.

Sen, G. and Grown, C. (1985) *Development, Crisis and Alternative Visions: Third World Women's Perspectives*, New Delhi: Development Alternatives with Women for a New Era (DAWN).

Strassmann, D. (1995) "Creating a Forum for Feminist Economic Inquiry," *Feminist Economics*, 1 (1), 1–5.

Tejani, S. (2018) "What's Feminist about Feminist Economics?," *Journal of Economic Methodology*, DOI: 10.1080/1350178X.2018.1556799.

Tuana, N. (1990) "The Radical Future of Feminist Empiricism," *Hypatia* 7 (1), 110–4.

UNDP (1995) *Human Development Report 1995*, New York: Oxford University Press.

Van Staveren, I. (2001) *The Values of Economics: An Aristotelian Perspective*, London: Routledge.

Waring, M. (1988) *If Women Counted: A New Feminist Economics*, New York: Harper and Row.

Williams, R. (1993) "Race, Deconstruction, and the Emergent Agenda of Feminist Economic Theory," in Ferber, M. A., and Nelson, J. A. (eds) *Beyond Rational Economic Man*, Chicago: University of Chicago Press, 144–43.

Zein-Elabdin, E. O. and Charusheela, S. (eds.) (2004) *Postcolonialism Meets Economics*, New York: Routledge

28

FEMINIST METHODOLOGY IN THE SOCIAL SCIENCES

Sharon Crasnow

Introduction

The twentieth century saw an influx of women into the social sciences. 10% of doctorates were awarded to women in these fields at the beginning of the century whereas by its end slightly more than 40% of social science PhDs were women (Thurgood, Golladay, and Hill 2006).[1] This trend accelerated rapidly during the 1960s, coinciding with second-wave feminism, and resulting in the emergence of explicitly feminist social science – a social science that recognized gender as a social category affecting the distribution of power in ways that had a negative impact on women's lives. Women entering these fields often describe themselves as feeling like outsiders in disciplines where they were relative newcomers. Sometimes they were asking different research questions than those usually investigated by researchers in their disciplines. Feminist economists, for example, noticed that traditional economics did not acknowledge an economic role for unpaid domestic work (done mostly by women) and proposed this as a new area of investigation (see Barker and Kuipers, Chapter 27 of this volume).

Not all women in the social sciences during this period challenged traditional frameworks nor were they all feminists. However, those who saw themselves as feminists viewed social science as potentially valuable for feminism's liberatory and egalitarian goals and sought to produce knowledge that would indeed support those goals. Consequently, feminist approaches in the social sciences elicited concerns about the role of social and political values in science, since feminism involves political and social justice commitments. Such explicit commitments are seemingly in conflict with the value-free ideal of science – the view that science must eschew all political and social values in order to remain objective and thus arrive at a true account of reality. While such an understanding of science has come under scrutiny in recent years, as feminist social science research was emerging the value-free ideal was still the dominant understanding of science (see Chapter 15). Commitment to feminist values thus put feminist researchers at odds with the mainstream work in their disciplines resulting in debates about the nature of their research, including reflection on research methodology. Thus feminist social science of the 70s, 80s, and 90s often challenged the mainstream epistemological and ontological presuppositions of social science disciplines. Debates about what counts as feminist social science methodology are manifestations of these challenges. This chapter explores some aspects of these methodological debates.

Clarifying Terms

It is helpful to begin by explaining some terminology. First, in the context of social science, feminism is not monolithic. There is not agreement among all those who identify as feminists about a precise meaning of the term. Nonetheless, it is possible to identify some shared commitments. Sociologist Joey Sprague describes these commitments in the following way:

> [W]hile feminists are a very heterogeneous group and we disagree on many issues, there are two points on which we have consensus: (1) gender, in interaction with other forms of social relations such as race/ethnicity, class, ability, and nation, is a key organizer of social life; and (2) understanding how things work is not enough – we need to take action to make the social world more equitable (Sprague 2016: 3)

A feminist methodology is one that is shaped by and serves these commitments. It requires adopting approaches and methods that are sensitive to the key structural elements of social life through which power is distributed, most notably gender, although it is important to recognize that most contemporary researchers recognize that gender is interconnected with other social categories that structure power relations such as race, class, ethnicity, religion, sexuality, and ability. Feminist methodology also aims at producing knowledge that can be used to further feminist and other liberatory goals. A recurrent theme when these commitments are addressed is the prescription to start research from the lives of women – the socially situated reality of those lives. It is here that both what matters to those who are oppressed and an understanding of what is involved in that oppression begin. Alison Wylie calls this a commitment "to empower women by recovering the details of their experience and activities" (Wylie 1992: 226).

Analyses of how gender structures society involve recognition of the differential distribution of power that can affect the assumptions made by the researcher located within such matrices of power, access to evidence relevant to those structures, and the ability of those who are marginalized to testify to their experience. Feminist approaches recognize that knowledge is situated – dependent on these and other social and political factors. In addition, the commitment to activism is a commitment to undoing the harms of oppression. Among these harms are various forms of epistemic injustice that occur as a result of women not being included as sources of evidence or being unable to express themselves as a result of being excluded both physically and conceptually from knowledge production (Fricker 2007; Dotson 2011; Medina 2013).

In order to understand what this means it is useful to distinguish method, methodology, and epistemology. Following Sandra Harding, we might think of methods as "techniques for gathering evidence." Interviews, surveys, archival research, creation of data sets, statistical techniques for analyzing data sets, and so on are examples. A methodology, in contrast, is "a theory and analysis of how research should proceed." Epistemology offers accounts of how a particular methodology and methods that might be associated with it produce knowledge (Harding 1987: 2).

Although method, methodology, and epistemology can be conceptually distinguished, in practice they are intertwined. Political scientist Mary Hawkesworth's characterization of the difference between method and methodology indicates one way this is so.

> In contrast to discussions about feminist "methods," which focus on particular tools to collect and analyze specific kinds of data, debates about methodology encompass questions about theories of knowledge, strategies of inquiry, and standards of evidence appropriate to the production of feminist knowledge.
>
> *(Hawkesworth 2006: 4)*

While epistemology can be distinguished from methodology, methodologies are supported by epistemological assumptions and commitments. Put another way, methodology presupposes epistemology.

Sprague acknowledges this stating that

> Each methodology is founded on either explicit or, more often, unexamined assumptions about what knowledge is and how knowing is best accomplished; together, these assumptions constitute a particular epistemology. That is, *a methodology works out the implications of a specific epistemology for how to implement a method.*
>
> *(italics in the original) (Sprague 2016: 5)*

Sociologists Mary Margaret Fonow and Judy Cook point out that methodology and method are related as well.

> Our notion of methodology was, and continues to be, influenced by the philosopher of science Abraham Kaplan, who wrote, "The aim of methodology is to describe and analyze research methods, throwing light on their limitations and resources, clarifying their presuppositions and consequences, relating their potentialities to the twilight zone at the frontiers of knowledge".
>
> *(Fonow and Cook 2005)*

The interplay of epistemology and methodology is prominent in what is perhaps the best known feminist methodological approach – feminist standpoint theory. The phrase "feminist standpoint theory" is ambiguous. It appears as a methodology in feminist social science and is marked by the central idea that research should start from the lives of women – from their "standpoint." But it has also been elaborated as an epistemology by feminist philosophers of science. There it is an account of why it is that such a methodology is able to produce knowledge. Dorothy Smith's remarks in her contribution to Harding's 2004 collection, *The Feminist Standpoint Reader* reflect this ambiguity:

> Feminist standpoint theory, a general class of theory in feminism, was brought into being by Sandra Harding (1986), ... to analyze the merits and problems of feminist theoretical work that sought a radical break with existing disciplines through locating knowledge or inquiry in women's standpoint or in women's experience.
>
> *(Smith 2004b: 263)*

She goes on to say

> I cannot speak here for...others ..., but, for myself, I am very much aware of being engaged with the debates and innovations of the many feminist experiments in sociology that, like mine, were exploring experience as a method of discovering the social from the standpoint of women's experience.
>
> *(Ibid.)*

Smith takes Harding (and other philosophers) to be doing "theoretical work" – epistemology. In contrast, she describes what she and other sociologists were concerned with as questions of research practice. They were experimenting with methods that got at women's experiences and used those experiences as evidence for feminist knowledge production. If we think of epistemology as providing justification of methodology, Harding and others who have offered (epistemological) accounts of standpoint theory are indeed involved in a different project than feminist social scientists like Smith. However, given that methodology depends on epistemological assumptions, it is not always clear that these can be neatly separated.

Feminist standpoint as an epistemology is discussed in more detail in Chapter 7, but it cannot entirely be avoided in discussing methodology. Briefly, the key elements of the epistemology that underpin feminist standpoint methodology are that it takes all knowledge to be socially situated, identifies the potential for epistemic advantage for those situated outside the dominant framework, and describes the fulfillment of that potential as dependent on theorizing how the distribution of power affects knowledge production. This last is an acknowledgment that standpoint is not equivalent to the perspectives of women even though it begins there. Thus a standpoint methodology is justified because it treats social location as highly relevant and takes seriously accounts of experience, recognizing that such accounts stem from a social location that has the potential for epistemic advantage. Finally, it directs researchers to be sensitive to the ways in which dominant knowledge frameworks may be inadequate to capture such experience because of the way the distribution of power affects knowledge production. This last, in part, captures what it is about social location that makes it relevant to knowledge. We will see how this plays out more clearly in the examples discussed in the next section.

While it is not always clear that feminist methodology is explicitly standpoint methodology, the features of the underlying epistemology described here are generally those underpinning feminist research. The primary prescription of standpoint approaches – to start research from the lives of women – appears in many discussions of methodology. Consequently, the examples in the next section can be interpreted as consistent with feminist standpoint methodology.

Before leaving the discussion of standpoint it is important to note a further implication of the directive to start research from the lives of women. If researchers take this directive seriously, the recognition that women do not all share the same lived experiences is unavoidable. Since lives differ as a result of differences in social location the experiences of women are shaped not only by their social location as women but also by other social factors that affect the distribution of power. The work of sociologist Patricia Hill Collins illustrates this. She starts her research from the lives of women in African American communities and describes a Black feminism which develops differently than the middle-class white feminism most closely associated with second-wave feminism. Race, class, ethnicity, sexuality, gender, and disability are all aspects of social location differentially affecting those who live at their intersections. As an advocate of standpoint approaches, it is not at all surprising to see her endorse intersectionality as a crucial tool for social research (Collins and Bilge 2016). Starting from the lives of women reveals not only shared experiences but also the differences in their experiences (Collins 2000/2009).

Feminist Methodology and the Difference It Makes

Methodological choices affect many aspects of research. To understand how this is so consider various stages of inquiry – although "stages" is somewhat misleading as it suggests that they are temporally distinct. Since research is typically iterative, the decisions made at one stage are open to revision as the research proceeds, so while "stages" might be conceptually distinct they may actually occur in any order or even simultaneously. Research questions are developed in response to disciplinary theory and the interests of researchers as they are shaped by their experience. How the objects of inquiry are understood – that is, how they are conceived – influences what questions are investigated and so the connection between questions and concepts is a close one. Research questions may change as the understanding of the objects of inquiry changes in response to changes in theory or as a result of the researcher's interactions with the objects of inquiry. Deciding what data to collect depends in part on what is taken to be significant and significance is dependent on the questions being asked, the understanding of the objects of inquiry, and the interests and goals of the researchers, which may change as the research proceeds. The generation and sampling of data, the analysis of the data, including the choice of technique for analysis, and decisions about when to end the collection and analysis of data are all aspects of research as well. Methodological choices make a difference throughout the research process and thus ultimately to the conclusions that are drawn from the research.

In her work on methodology, sociologist Marjorie DeVault (1999) addresses the connection between methodology and the various stages of research explicitly with examples from her own work and those of other feminist social scientists. She frames the central dilemma of the feminist researcher in the following way:

> The dilemma for the feminist scholar, always, is to find ways of working within some disciplinary tradition while aiming at an intellectual revolution that will transform that tradition (Stacey and Thorne 1985). In order to transform sociology…we need to move toward new methods for writing about women's lives and activities without leaving sociology altogether. But the routine procedures of the discipline pull us insistently toward conventional understanding that distort women's experiences (Smith 1987, 1989).
>
> *(DeVault 1999: 59)*

She goes on to give examples of the various ways in which feminist scholars tackle this dilemma. At the stage of developing research questions – which she calls "constructing topics" – she considers her own research which "examines household routines for planning, cooking, and serving meals" (DeVault 1999: 63). She first thought of this as a study of housework but she found that how the concept was dealt with in the literature did not capture all that she wanted to explore, primarily because housework was understood as a type of work. What DeVault wanted to get at was how the various tasks involved in this sort of caring activity organized the daily lives of those who did it and the ways the various tasks involved were thought of as meaningful. She came to focus on what she first thought of as "providing food" and then ultimately as "feeding the family" as a specific example.[2] When described in this way, she found that she was able to ask questions of her informants that revealed the organizational nature of the daily activities in ways that conceiving of the preparation of food as work traditionally understood did not. The difficulty was that the sort of activity she was investigating did not fit into the work/leisure dichotomy – a dichotomy understood in her discipline as a standard way of classifying human activities. Further reflecting on the process that led her to alter her understanding of her topic she writes, "This particular insufficiency of language is an example of a more general problem, a more pervasive lack of fit between women's experiences and the forms of thought available for understanding experience" (DeVault 1991: 5). DeVault needed to find a way to ask her informants about their experiences that allowed them to describe what she was trying to investigate. The open-ended nature of her description of the topic did this. She found the informants able to talk easily about their daily tasks, and interestingly, they were sometimes concerned that they were not describing what she was interested in – that what they were talking about could not be what a sociologist wanted to find out. Their preconceptions of social science knowledge based on the dominant conceptual framework did not seem appropriate to their experience.

DeVault describes a similar reconceiving of topic in the research of criminologist Elizabeth Stanko who, in order to study what might be thought of as self-defensive steps that women take to avoid assault, asked her informants about "the things we do to keep safe" (Stanko 1997). In both cases the researchers were interested in the broader structural causes and effects surrounding women's daily routines. In addition, both were conscious that the behaviors surrounding these activities are often routinized and normalized so that they are invisible both to those who engage in the behaviors and to researchers operating within traditional research frameworks. This is because the ideas of both researchers and those researched are shaped by the dominant culture. For DeVault's research, the way that caring work permeates the entire life of those who do it obliterates the traditional disciplinary distinction between work and leisure. For Stanko, while the things women do to keep safe are really forms of self-defense in some broad sense, they do not fit into the traditional understandings of that concept.

Altering the understanding of the objects of inquiry – changing the concepts used to describe the features of the world that are of interest to the researcher and relevant to the lives of their informants – affects the framing of the questions put to the informants. The feminist methodology operating here involves having an awareness that standard understandings may not be appropriate for gathering the data relevant to the project. In doing so, it challenges the ontological presuppositions underlying mainstream research. Consequently one feature of feminist methodology – one way in which feminists approach research – is that they come to the research with an openness to the possibility that the concepts in their disciplinary toolbox may not fit the experiences of those they are researching.[3] Sociologist Dorothy Smith refers to this awareness as a sensitivity to "lines of fault" – points where women's experiences are in conflict with the culture or ideology of the society in which she lives (Smith 1987: 49).

Methodological issues are also pertinent for data production, collection, and interpretation. DeVault's account of changes in her interview practices illustrates this. Traditional interview transcribing practice called for eliminating or smoothing over hesitations and verbal missteps – the "ums" and "you knows" in the responses of those interviewed. But as DeVault transcribed her interviews she came to realize that such features of ordinary conversation often marked topics that needed to be returned to or reflected emotional states that were relevant to the research questions she was investigating (DeVault 1999: 78). Specifically, she noted that these hesitations often occurred when her interviewees were searching for some way to describe what they were thinking or doing because the usual categories of description – those of the dominant voices in society – did not quite capture what was that they wanted to say. Consequently, she came to reject some aspects of the transcribing practices she had been taught. DeVault, "starting from the lives of women," was attentive to the emotional valences of her informants. What was usually discarded from the interview became evidence that her interviewees were struggling with finding appropriate descriptions of their experiences. DeVault's understanding of what was significant – what phenomena were data – was altered. The epistemological presuppositions about what counts as evidence that come with standard practice are thus open to question when a feminist framework is adopted.

The methods researchers use in these examples are primarily qualitative. As noted in the previous section, methodology does have something to say about method, but it is not straightforward. We cannot read method directly from methodology and it would be a mischaracterization to treat feminist methodology as prescribing that research be qualitative. The next section considers the question of whether, and if so how, feminist methodology constrains choices of method.

Methods: Qualitative vs. Quantitative?

In some of their earliest forms methodological debates revolved around the question of whether feminist methodology required a rejection of traditional methods of social science disciplines – methods that were increasingly quantitative in many of the social sciences. Some feminists argued that quantitative methods were inadequate to address feminist interests and goals and that qualitative methods – ethnography, participant observation, and case studies for example – were better suited to feminist research goals. Others disagreed, embracing the power of quantitative research and the prevailing norms of their disciplines. The remarks from Fonow and Cook cited in section

2 suggest that feminist thinking about methodological questions should be more nuanced than such a framing would suggest. A more accurate characterization describes feminist methodology as calling for reflection about the choice of method – how and why it is being used in any particular research project – rather than a rejection of any particular technique for eliciting evidence. Feminists were not alone in questioning the growing dominance of the quantitative approach, as Sprague's overview of the current state of play regarding methodology indicates:

> Contemporary critiques of mainstream knowledge have fed a kind of methodological schizophrenia in the social sciences. On one side are a legion of committed practitioners of quantitative methods who, aided by the rush of technological developments, are pursuing ever-increasing levels of technical precision, mostly untouched by the swirl of doubt about the validity of their product. On the other side, many critical researchers are rejecting quantitative methods because of their skepticism about assumptions of objectivity, impartiality, and control. Instead, they are relying on qualitative methods, believing that these have more potential for avoiding some of the major pitfalls of the past.
>
> *(Sprague 2016: 29)*

To understand what is at stake it is helpful to consider the difference in the sort of evidence each type of method produces.

Quantitative – statistical – methods are well-suited for studying average effects in populations. They examine differences in sample populations in order to draw conclusions about the characteristics of the populations that have been sampled. There are questions relevant to feminist goals that are appropriate to answer through this sort of evidence. If one is interested in investigating a wage-gap in the salaries of women in comparison to men there are appropriate quantitative tools for doing so. However, many of the questions that feminist social scientists are concerned with are more directly related to the ways that gender makes a difference in the experiences of individuals. Qualitative methods are good at getting at the differences *within* populations but do not provide strong support for conclusions *about* the characteristics of populations.

Different methods produce evidence suited to answer different types of research questions and so the dominance of these methods in the social sciences privileges some research questions over others (questions about populations). But there are indeed questions about populations that feminist researchers may want answers to. Statistical (quantitative) methods in the social sciences have often been used to good effect in support of feminist goals as illustrated by the extensive research documenting the wage-gap mentioned earlier. Londa Schiebinger offers another example from primatology:

> In the 1970s Jeanne Altmann drew attention to representation sampling methods in which all individuals, not just the dominant and powerful, were observed or equal periods of time. (Primatologists had previously used "opportunistic sampling," merely recording whatever captured their attention). Representative sampling required that primatologists evaluate the importance of events by recording their frequency and duration. Commonplace events such as eating, grooming, and lolling thus claimed their place next to the high drama of combat and sexual encounters, allowing for a more nuanced and egalitarian vision of primate society.
>
> *(Schiebinger 1999: 7)*

The work of Susan Greenhaigh and Jiali Li (1995) on "missing" girls in China using demographic techniques similar to those used by Amartya Sen on missing women globally (1990) provides yet another case in which quantitative methods are able to serve feminist ends.

Examples like these count against identifying feminist methodology with qualitative methods. Nonetheless, there are questions that arise during research that cannot be answered quantitatively.

The conceptual shifts that were described in the previous section depend on attention to the experiences of individuals. Recognizing these strengths of qualitative methods has led some feminist social scientists to be more supportive of qualitative research than other (non-feminist) researchers in their disciplines. As the social sciences have become increasingly impressed with the power of quantitative and formal methods and so dominated by them, the research questions that might be better addressed through qualitative methods have been marginalized. Feminists focusing on issues better suited to qualitative research have consequently been marginalized as well.

Quantitative methods have sometimes been thought to be more "scientific." There are several reasons for this. The first is the idea that physics – a field in which quantitative methods dominant – is taken as paradigmatic of science. Such an understanding suggests that in order to be better as a science, any discipline should aim for the clarity and precision found in that field. While this is a somewhat naïve view, it does still have some resonance in the social sciences. However, the idea that physics should provide a model for all of the sciences is open to question. There are reasons to think that the physical world and the social world have differences that would count against such a comparison, but even more compelling are arguments that the complexity that we find in biology is much more akin to the social world than is physics (see for example Mitchell 2009).

A second and somewhat related reason that quantitative methods are thought more scientific is the idea that they lead to greater objectivity. This idea seems to be based on the belief that numbers are not affected by the subjectivity that sometimes seems problematic in the social sciences. Thus turning to numerical representations of the features of interest is a way of eliminating problematic subjectivity (Porter 1995). While this idea has had broad appeal, when examined it too is revealed to be overly simple. It is only plausible when the sources of numerical values that play a role in their social sciences are not considered.

There are a variety of ways in which numbers are affected by factors that might be considered subjective. Whenever we count, we make judgments about what things qualify as the sort of thing we are counting. For the social sciences, those judgments need to be based on a clear understanding (a precise definition) of what we are counting. Conceptualization proceeds measurement (see Chapter 29; see also Cartwright, Bradburn, and Fuller 2017). Economists who did not conceptualize work as inclusive of unpaid domestic work did not include such work in their quantitative research. Their judgment of how to conceptualize work was disputed by feminist economists.

Science does not provide a complete description of the world but rather an account of the features of the world that are significant for those investigating and in relation to particular goals of investigation. The data upon which quantitative work is based should not be understood as self-evident. Decisions about what to count – how and what to measure – rely on theory and the concepts related through theory. In other words, all research, including quantitative research, involves determinations about what aspects of the (social) world matter – what is significant. As Smith puts it, "From the point of view of "women's place" the values assigned to different aspects of the world are changed. Some come into prominence while other standard sociological enterprises diminish" (Smith 2004a: 21).

Suppose, for example, one wished to study, as Stanko did, how often in the day women engage in behaviors that are self-defensive. The sorts of things that Stanko thought were appropriate to consider depend on how the researcher conceives of self-defense and so which behaviors to count. Stanko starts with the lives of the women in which routinized behavior such as avoiding dark streets at night, choosing a time to go to the laundromat, choosing what to wear all are seen as forms of self-defense when the researcher elicits them through conceptualizing them as "things we do to stay safe".

Another example is the work of Pamela Paxton (2000) who challenges the dominant understanding of democracy through which the "three waves" of democracy are identified (Huntington 1991). Paxton points out that the understanding of suffrage used as one of the key elements of democracy is "universal male suffrage" and that if one understands suffrage as inclusive of all adults there appear to be two rather than three waves. How to identify what counts matters.

Research by Julia Brines provides a final example of how concepts matter. Brines (1994) offers a reevaluation of the standard account given for the unequal distribution of housework in households with male and female partners. The economic model that Brines challenges is a dependency model. What needs to be accounted for is that in heterosexual marriages, women typically do more housework than men do, even when both partners are employed outside of the home. The standard explanation given for this phenomenon is that women typically earn lower wages than their husbands and hence are economically dependent on them. As a result they have less bargaining power when it comes to housework. They do more housework because they lose when negotiating as a result of their economic dependence. The underlying model is a game theoretic model that treats marriage as an economic bargain. The account fails to explain an anomaly, however. It seems that in households where women earn more than men, men do even *less* housework. The standard explanation predicts that given the increased bargaining power of women in such circumstances, men should be doing more of the housework, not less.

Brines argues that the explanation based on the dependency model is unable to accommodate this anomaly because it ignores the way gender structures power relations in society (and consequently in marriages), focusing solely on economic dependence. In so doing the dependency account fails to recognize other factors about the social significance of gender that are relevant to bargaining. Brines points out that the social institution of marriage is not only an economic institution but also provides a venue in which husbands and wives perform gender. Wage earning is conceived of as a masculine performance. Housework is traditionally feminine. When women perform masculinity through higher wage earning men compensate by performing more masculinity at home (less housework). On this account, the way gender structures relationships plays out in marriage so that it affects the bargain. Brines does not reject the dependency model per se but rather offers a revision that incorporates gender relations. She modifies the model so that it incorporates the power dynamic produced by the gendered structure of social life. The bargaining is not solely economic – the loss of (men's) economic bargaining power calls for a renegotiation and compensation of power as exemplified through gender role. Understanding such behavior calls for attention to the way that gender structures social relations. This, in turn, calls for a sensitivity to the difference in experience depending on one's social location in a world where power relations are structured along many lines, only one of which is economic.

In each of these examples, shifts in understanding are brought about through feminist methodology. Stanko starts from the lived experience of her informants. Paxton simply notices the way gender was not taken to be a relevant factor in democracy studies. Brines is conscious of the importance of gender as a social phenomenon and the power dynamics gender identity generates. Their work reflects the dual commitments that Sprague notes feminist methodology shares: the acknowledgment of gender as a key organizer of social life and the desire to address the injustice that results from the inequitable distribution of power resulting from that social organization.

While it would be inaccurate to characterize feminist methodology as qualitative, it is true that feminists have often been among the advocates for the potential benefits of qualitative research. Interviews, participant observation, ethnography, and case studies are research methods that pay attention to details of context, pick out factors that reveal differences in social location, and hence are likely to be more sensitive to the way that gender matters. Nonetheless, as previously noted, there are many circumstances in which the sorts of information that quantitative research reveals are relevant for feminist goals. Feminist social science has always employed a plurality of methods and more recently, feminist researchers have been at the forefront of those advocating for "mixed methods" research in the social science. Mixed methods research involves combining qualitative and quantitative methods within the same research project. Brines's research is an example. She employs formal models (game theory), statistical information (how much work is done, on average), and qualitative evidence gained from understanding the meaning that women and men attribute to the tasks they carry out. She engages in mixed methods research, recognizing that different types of approaches may be useful for understanding different aspects of the phenomenon.

Feminist methodology is consistent with a variety of methods; however, it does imply something about how these methods should be chosen, specifically in relation to which methods are best suited for particular projects. Consequently, feminist methodology is self-reflective and pragmatic. While quantitative methods are powerful tools for discovering trends in populations and correlations between variables exhibited in those populations, feminist methodology often leads to reexamination of the concepts through which the social world is understood. Often this reexamination is needed because the ways in which gender structures social spaces have not been incorporated into the data collection or analysis. Qualitative methods are better suited to the local and specific features of the social variations from the presumed norm or dominant framework. Sprague describes this as the power of qualitative methods to challenge inequalities that are apparent from "the downside of the social hierarchies" – from the lives of those who are the subjects of research (Sprague 2005: 114).

Feminist Methodology as Pluralist

While the previous section argues that a variety of methods can be useful for the production of knowledge that supports feminist liberatory goals, nonetheless it is possible that methods can carry value commitments antithetical to feminist research in some contexts. As explained in the last section, quantitative methods in the social sciences are aimed at providing information about average effects in large populations. Such information can be useful in establishing that social injustice has occurred but when disciplinary norms treat methods that produce average effects evidence as the only acceptable research – for publication, for tenure, for grants – difficulties can arise. Privileging knowledge of average effects can lead to the disappearance of relevant differences within populations. This can result in a de facto de-valuing of minority populations and individuals. Such an approach threatens to have a disproportionate impact on those who are not "average" in the requisite sense. This is not a problem if knowledge of average effects is adequate to the problem under investigation. If we want to know *that* there is a causal relationship but do not need to know precisely what the mechanism that produces that relationship is then the methods that produce average effects evidence may be adequate. Nor is it a problem if the policy decisions need to be based on average effects. But sensitivity to these issues requires attention to the choice of method, its adequacy to goals, and the values that will be promoted in such policy decisions.

The adept use of quantitative methods may ameliorate these worries to some degree. If populations are recognized as heterogeneous then there is motivation to engage in further research on subpopulations. This move addresses some aspects of the problem of erasure of difference, but which factors determine relevant subpopulations will depend on paying attention to the lived experience of those who are studied and so this points again to the need for pluralism regarding methods. Feminist attention to intersectionality and the ways power differentially structures society through gender, race, ability, sexuality, socioeconomic class, ethnicity, and other salient features of individuals raise a further worry. It is not entirely clear how a subpopulation strategy can address social justice issues intertwined in this way. There has been some work in this direction. For example, Bright, Malinsky, and Thompson (2015) offer one approach to examining intersectional causality. But intersectional effects are not to be understood as additive – once again counting against reliance on exclusively quantitative approaches. Finally, there is the problem that further partitioning of the reference class may also decrease the sample size and thus undermine the power of the method. In any case, these worries further support the feminist intuition that methods must be chosen consciously and with attention to the aims of research.

The dominance of quantitative methods can have another dampening effect on research, particularly in disciplines where recognition and advancement require demonstrating one's skill with such methods. Economics provides an example. Sheba Tejani examines the methods used in articles published in *Feminist Economics* from 1995 through 2015 and finds that the majority use econometric (quantitative) methods. Qualitative methods such as interviews and mixed methods approaches constitute only a small proportion of the published work in this journal. Tejani suggests that "institutional barriers to methodological plurality, professional pressures, the background and training of researchers, and a bias towards empiricism in feminist thought might account for some of these shifts" (Tejani 2018: 3). In other words, the normative standards in economics influence the way that all research in that field is carried out – including feminist research. Tejani's worry suggests that disciplinary pressures can push against the pluralism many feminist methodologists advocate for.[4]

In summary, feminist methodology sometimes reveals that standard disciplinary conceptions of the objects of inquiry do not get at what is significant relative to questions that matter for addressing social justice. Furthermore, while no particular method is uniquely feminist, there are important differences in the types of evidence that each produces. Some methods are suitable for studying large populations but not for revealing what we may need to know about individuals or smaller populations. The appropriateness of methods depends on a variety of factors – among them, the aims of research. Finally, feminist social scientists have among their aims liberatory goals. In other words, their social justice goals and values shape their research. Methodology is thus connected to ontological, epistemological, and value commitments. Each of these requires conscious consideration when research methodology is chosen. The pluralism, pragmatism, and critical approach of feminist research motivate this sort of reflective awareness of the interrelationship among all aspects of knowledge production.

Related chapters: 7, 26, 29.

Notes

1 This trend continued in this century with the women PhDs in the social sciences at nearly 49% as of 2016 (National Center for Science and Engineering Statistics/National Science Foundation https://ncses.nsf.gov/pubs/nsf19304/data).
2 DeVault's *Feeding the Family: The Social Organization of Caring as Gendered Work* (1991) is the result of this research.
3 Arguably this is the case for all research; however, because the lives of women were not traditionally the target of social research the point is particularly relevant in the context of *feminist* social science.
4 Tejani's work is also discussed in Barker and Kuiper (Chapter 27).

References

Bright, L. K. Malinsky, D. and Thompson, M. (2015) "Causally Interpreting Intersectionality Theory," *Philosophy of Science* 83(1), 60–81.

Brines, J. (1994) "Economic Dependency and the Division of Labor," *American Journal of Sociology* 100(5), 652–688.

Cartwright, N., Bradburn, N., and Fuller, J. (2017) "A Theory of Measurement," in McClimans, L. (ed.) *Measurement in Medicine: Philosophical Essays on Assessment and Evaluation*, London: Rowman & Littlefield International, Ltd, 73–88.

Collins, P. H. (2000/2009) *Black Feminist Thought*, New York: Routledge.

——— and Bilge, S. (2016) *Intersectionality*, Cambridge, MA: Polity Press.

DeVault, M. L. (1991) *Feeding the Family: The Social Organization of Caring as Gendered Work*, Chicago: University of Chicago Press.

————. (1999) *Liberating Method: Feminism and Social Research*. Philadelphia: Temple University Press.

Dotson, K. (2011) "Tracking Epistemic Violence, Tracking Practices of Silencing," *Hypatia: A Journal of Feminist Philosophy* 26(1), 236–257.

Fonow, M. M. and Cook, J.A. (2005) "Feminist Methodology: New Applications in the Academy and Public Policy," *Signs: Journal of Women in Culture and Society* 30(4), 2211–2236.

Fricker, M. (2007) *Epistemic Injustice: Power and the Ethics of Knowing*, Oxford: Oxford University Press.

Greenhaigh, S. and Li, J. (1995) "Engendering Reproductive Policy and Practice in Peasant China: For a Feminist Demography of Reproduction," *Signs* 20(3), 601–641.

Harding, S. (1986) *The Science Question in Feminism*, Ithaca: Cornell University Press.

————. (1987) "Is There a Feminist Method?" in Harding, S. (ed.) *Feminism and Methodology*, Bloomington: Indiana University Press, 1–14.

Hawkesworth, M. (2006) *Feminist Inquiry: From Political Conviction to Methodological Innovation*, New Brunswick, NJ: Rutgers University Press.

Huntington, S. (1991) *The Third Wave: Democraticization in the Late Twentieth Century*, Norman: University of Oklahoma Press.

Medina, J. (2013) *The Epistemology of Resistance: Gender and Racial Oppression, Epistemic Injustice, and Resistant Imaginations*, Oxford: Oxford University Press.

Mitchell, S. (2009) *Unsimple Truths: Science, Complexity, and Policy*, Chicago: University of Chicago Press.

National Center for Science and Engineering Statistics (National Science Foundation) (2016) https://ncses.nsf.gov/pubs/nsf19304/data.

Paxton, P. (2000) "Women's Suffrage and the Measurement of Democracy: Problems of Operationalization," *Studies in Comparative International Development* 43, 92–111.

Porter, T. M. (1995) *Trust in Numbers: The Pursuit of Objectivity in Science and Public Life*, Princeton, NJ: Princeton University Press.

Schiebinger, L. (1999) *Has Feminism Changed Science?* Cambridge, MA: Harvard University Press.

Sen, A. (1990) "More Than 100 Million Women Are Missing," *The New York Review of Books* 37(20), 61–66.

Smith, D. E. (1987) *The Everyday World as Problematic*, Boston, MA: Northeastern University Press.

————. (1989) "Sociological Theory: Methods of Writing Patriarchy," in Wallace, R. A. (ed.) *Feminism and Sociological Theory*, Newbury Park, CA: Sage, 34–64.

————. (2004a) "Women's Perspective as Radical Critique of Sociology," in Harding, S. (ed.) *The Feminist Standpoint Theory Reader*, New York: Routledge, 21–33.

————. (2004b) "Comment on Hekman's 'Truth and Method: Feminist Standpoint Theory Revisited,'" in Harding, S. (ed.) *The Feminist Standpoint Theory Reader*, New York: Routledge, 263–289.

Sprague, J. (2005) *Feminist Methodologies for Critical Researchers: Bridging Differences*, Lanham, MD: AltaMira Press.

————. (2016) *Feminist Methodologies for Critical Researchers: Bridging Differences*, Lanham, MD: Rowman & Littlefield Publishers.

Stacey, J. and Thorne, B. (1985) "The Missing Feminist Revolution in Sociology," *Social Problems* 32(4), 301–316.

Stanko, E. A. (1997). "Safety Talk: Conceptualizing Women's Risk as a 'Technology of the Soul,'" *Theoretical Criminology* 1, 479–499.

Tejani, S, (2018) "What's Feminist about Feminist Economics?" *Journal of Economic Methodology* 26(2), 99–117.

Thurgood, L. Golladay, M. J. and Hill, S. T. (2006) *National Science Foundation, Division of Science Resources Statistics, U.S. Doctorates in the 20th Century*, NSF 06–319.

Wylie, A. (1992) "Reasoning About Ourselves: Feminist Methodology in the Social Sciences," in Harvey, E. and Okruhlik, K. (eds.) *Women and Reason*, Ann Arbor: University of Michigan Press, 611–624.

Suggestions for Further Reading

Collins, P. H. (1986) "Learning from the Outsider within: The Sociological Significance of Black Feminist Thought," *Social Problems* 33 (Special Theory Issue): S14–S32. (An overview of Collins approach to standpoint theory and Black feminist thought.)

Harding, S. (1991) *Whose Science? Whose Knowledge? Thinking from Women's Lives*, Ithaca: Cornell University Press. (An elaboration of standpoint theory and Harding's notion of "strong objectivity.")

Hesse-Biber, S. (ed.) (2011) *Handbook of Feminist Research* (Second Edition), New York: Sage Publications. (A guide to doing feminist research.)

Intemann, K. (2010) "Twenty-Five Years of Feminist Empiricism and Standpoint Theory: Where Are We Now?," *Hypatia: A Journal of Feminist Philosophy* 25(4), 778–796. (An argument that feminist empiricism and feminist standpoint theory have merged in many ways.)

Ramazanoglu, C. with Holland, J. (2002) *Feminist Methodology: Challenges and Choices*, Thousand Oaks: Sage Publications. (A guide to feminist methodology.)

Reinharz, S. (1992) *Feminist Methods in Social Research*, New York: Oxford University Press. (A classic of feminist methodology.)

29

FEMINIST APPROACHES TO CONCEPTS AND CONCEPTUALIZATION

Toward Better Science and Policy

Amy G. Mazur

Concept formation lies at the heart of all social science endeavors. It is impossible to conduct work without using concepts. It is impossible to even conceptualize a topic, as the terms suggest, without putting a label on it. Concepts are integral to every argument for they address the most basic question of social science research; what are we talking about.

(Gerring 2012: 112)

Introduction

Alongside the fundamental role of concepts highlighted in the opening quote from Gerring's 2012 non-feminist social science methodology textbook stands the crucial feminist project of concept formation and development in the social sciences, with the following major goals:

- To challenge gender-biased, gender-blind and gender-neutral concepts used in social science research.
- To "gender" existing concepts or create new concepts to take on board systems of power and domination in their full "intersectional[1]" complexity.
- To use those concepts in empirical analysis that generates theories, understanding and empirical renderings of what is actually happening through the core principles of reliability and validity, in other words to promote good science.
- And, in doing better science, to suggest and, for some feminist analysts, to pursue paths of action that can address persistent, deep-seated and multi-form gender-based inequities located in the public and private spheres, including within their own fields of study.

The goal of this chapter is to map out feminist approaches to concepts in the social sciences, which have emerged over the past 30 years and to discuss how these efforts have affected knowledge production and theory-building in feminist and non-feminist studies, with a particular emphasis on the internationalized field of gender and politics within the purview of the discipline of political science. This focus on gender and politics is not as narrow as it seems. Feminist political studies are interdisciplinary often eschewing the typical conventional boundaries in the social sciences (Siim 2004; Celis et al. 2013). For example, one of the first systematic feminist treatments of concepts in gender and politics, *Contested Concepts in Gender and Social Politics*, was conducted by sociologists, historians, policy scholars and political scientists (Hobson et al. 2002). With several active scholarly associations and numerous international journals, gender and politics research engages over 1,000 scholars across the globe who publish and work in this area.[2] Moreover, the very heart of the feminist enterprise in political science has been to broaden what has been seen as the quite narrow notion of politics in conventional political science to cover the public sphere and the complex intersectional systems of power and domination located in the private sphere where informal biases, institutions and norms operate to block gender equality.

Increasingly, many feminist scholars who are " ... normatively interested in gender struggles and [attentive] to social epistemologies (Ackerly and True 2018: 263)" differentiate between feminist and non-feminist analysis. On the one hand, feminist scholars seek to bring gender meaningfully into scientific endeavors to promote more gender just societies. On the other hand, non-feminist analysts tend to ignore gender, in its full complexity, as a significant analytical object and the issue of gender justice in politics or the discipline itself. There is a sexual division of labor between these two approaches – feminist scholars are for the most part women, with an increasing number being self-identified as trans gender – and non-feminist scholars are predominantly men. This chapter shows the ways in which feminist analysts have critiqued, challenged and sought to change how non-feminist work operates and the persistent divide between feminist and non-feminist studies. Despite over 40 years of feminist efforts to "gender" political science, "malestream," political science remains remarkably unaffected by the advances in the highly productive and growing field of politics and gender, even when feminist scholarship directly engages with that work.

The rest of this chapter first discusses why concepts and conceptualization are so important to non-feminist and feminist studies. Next, a deeper dive is made into feminist approaches to concepts through the approach of "gender concepts in analysis." The chapter then turns to the disappointing and limited reception of feminist conceptualization by non-feminist political science, including the 2012 textbook cited earlier, which makes no reference to the growing feminist work on concepts and methodologies covered in this chapter. The conclusion discusses implications and next steps in light of the work on feminist conceptualization and the persistent lack of engagement of most non-feminist political studies for political science more generally.

Why Do Concepts Matter? Non-feminist Analysis

As Gerring clearly states earlier, conceptualization is at the center of social science research and understanding. Naming phenomena, identifying how they operate in the real world and then measuring those phenomena through the development of indicators and collecting data according to those indicators are part and parcel of any social science study. Many argue, particularly from a more qualitative perspective, that the use of concepts that are both valid – that accurately measure the phenomenon under study, and reliable – that can be consistently applied across a wide range of cases – is a part of doing good social science that produces theories that are replicable and transparent; thus, good conceptualization is a "fundamental link between concept, research, data collection and theory-building (Goertz and Mazur 2008b: 20)."

The importance of concepts and conceptualization in scientific research was placed on the scholarly agenda in the early 1990s in US political science in the context of the debate between quantitative and qualitative approaches, that is, whether good research should examine phenomena in large samples with statistical analysis to identify trends and patterns and the "effects of causes" or in smaller samples or single cases to identify causal mechanisms or the "causes of effects" (Goertz and Mahoney 2012). The highly controversial book by King, Keohane and Verba (1994) asserted that small "n" qualitative research should be more like large "n" quantitative research because it was not good social science. Political scientists who were more open to the advantages of small 'n' research responded vocally (Brady and Collier 2010). They argued that qualitative research was legitimate, scientific and not just descriptive and had contributions to the social science process that were quite different and even better than quantitative approaches. This group of scholars pursued efforts to bridge the "qualitative- quantitative" divide between what Goertz and Mahoney describe as "two cultures" (Goertz and Mahoney 2012) in the 2000s. From this dialog came a movement that was institutionalized within the American Political Science Association, in a new division to promote qualitative approaches and research that took a multi method approach that combined the tools of statistical analysis with case studies, qualitative comparative analysis and other qualitative research tools (https://www.maxwell.syr.edu/moynihan/cqrm/qmmr/).

The call for better conceptualization was a major part of this methodological movement. Scholars argued that more attention to developing better nominal and operational definitions that could be used in quantitative and qualitative analysis would assure higher levels of validity and reliability (Goertz and Mahoney 2012; Goertz 2020). They criticized researchers who only used statistical analysis of over-simplifying key "variables" and using proxy measures instead of acknowledging the empirical complexity of most social science concepts. For example, economic well-being was far more complicated than simple increases in GDP per capita. The core concept of democracy and its various indicators and measurements were hotly debated with calls for more accurate and less ethnocentric systems of measurement, to reflect a more fluid notion of democracy – a "democracy with adjectives" (Collier and Levitsky 1997). Early on, renowned comparativist, Giovanni Sartori asserted that it was necessary to develop better concepts that were able to "travel" across national boundaries without "stretching" their meaning to conduct good comparative politics research (1970). Collier and Mahoney (1993) further developed the idea of concept traveling by arguing for the use of "radial" and "family resemblance" concepts rather than a classical approach to concepts that were hierarchically structured. While this effort to promote good conceptualization came out of the "tale of two cultures" (Goertz and Mahoney 2012) in the United States, it has since traveled to an international level. For example, there is an official research group on Concepts and Methods in the International Political Science Association. Goertz dedicated a book to showing how concept formation and definition should be pursued in the interest of better research (2020). Ten guidelines for good conceptualization, published in a book on gender and politics concepts, summarized Goertz's treatise in ten steps or "guidelines (Goertz and Mazur 2008b: 15)." Thus, a clear road map was laid out for good conceptualization and applied to the discussion of gender and politics concepts by top gender and politics scholars – nine concepts were treated in the edited volume (see Table 29.1).

Why Do Concepts Matter? Feminist Analysis

The place of concepts and conceptualization is arguably more crucial in feminist studies than the pursuit of good conceptualization within American political science. Indeed, the critical approach to concepts was an operating premise and a central part of the identity of feminist scholars working in the social sciences regardless of their specific epistemological or methodological approach. The argument is that conventional and long-established concepts have completely ignored or even purposefully omitted gender, systems of domination and intersectional considerations. Feminist analysts of all stripes, therefore, had to first tackle this gender-blind, gender-neutral and/or gender-biased nature of analytical concepts before even beginning their studies. While there has been, and still is, a diversity of feminist approaches, feminist scholars agree that concepts are a central tool of analysis.[3] Discussions of feminist methodology in political analysis generally assert that concepts are an integral part of feminist approaches as well. "Feminist methodology provides an approach to identify important questions, clarifying concepts and engaging in research practice (Ackerly and True 2018: 260)." As the coeditors of the 2013 Oxford University Press Handbook on Gender and Politics (henceforth OUP Handbook) point out, one of the three aims of the new field is to "… challenge existing political science in terms of its concepts, subject matter and methods (Celis et al. 2013: 3)." There is a separate section on concepts and all 30 chapters in the volume begin with a critique of how the concept which they are covering has been dealt with by non-feminist analysis. Similarly, the editors of the new *European Journal of Politics and Gender*, in their introductory essay to the first double issue of the journal, make a call for "… robust discussion of the conceptualization and operationalization of central concepts in politics and gender research" (Aherns et al.: 9). Eight out of the 14 articles place conceptual critiques and reconceptualization at the center of their analyses.

Unlike the US-based effort to focus on conceptualization in non-feminist studies, the feminist approaches to concepts and conceptualization were internationalized from its beginnings, arguably quite distinct from American gender and politics of research which has been highly influenced by American behavioralism's lack of interest in deep conceptualization (Tripp and Hughes 2018). Berrebi-Hoffmann, Giraud, Renard and Wobbe's study (all European scholars) of the use of work as a category in German and French public policy discussions from a historical perspective asserts the process of categorization itself as an international process "…the space of categorization in the field of gender equality at work seems to be a transnational one (2019: 28)." European gender and politics research is more interdisciplinary and influenced by constructivism, with links to sociology, and the interpretivist turn in research where underlying meanings and power constructs are placed front and center. Linking theories to action through praxis is far more a part of European approaches to feminist studies as well (Kantola and Lombardo 2018). As Siim explains, one of the three elements for "cross cultural and cross-national" research on gender in Europe has been the "contested and constructed nature of concepts" (2004: 99).

To be sure, feminist treatments of concepts are quite international and incorporate many different approaches, but as many analysts have shown about feminist studies more broadly speaking there is a problem with ethnocentrism and the dominance of the "Global North (Parpart et al. 2002)" or "the West," essentially the wealthier and post industrial countries in Western Europe, North America, Australia and New Zealand. In the early days of feminist studies in the 1980s, much deliberation and discussion was limited to the Anglo-American English speaking scholarly world, thus leading to the critiques of ethnocentrism and colonialism. Today, while the lingua franca remains English for the purposes of communication, the community of gender and politics scholars has become increasingly international. Conceptualization and concepts are not only the province of English-language publications, for example Jenson and Lépinard propose a new conceptual typology for gender policy analysis in a French political science journal (2009). Concepts about the global South or that are used to make global comparisons, such as development, have also been shown by feminist analysts to be quite biased with a normative agenda, based on colonial notions of showing the North should be an aspirational model for the South (Sen and Grown 1987; Staudt 2008). While some comparative studies have made efforts to be more sensitive to cultural contexts and use concepts that travel (McBride and Mazur 2010; Spierings 2015; Htun and Weldon 2018), the field of gender and politics today is still grappling with the continued underrepresentation of scholars from and issues about the global South. As Medie and Kang (2018), two postcolonial feminist scholars from the global South state

> Although there have been changes in Western-produced scholarship, these international inequalities remain largely understudied… reflect[ing] global power imbalances in knowledge production as the perspectives of critical feminist scholars [from the global South] have not been fully incorporated unto the sub-discipline.
>
> *(Medie and Kang 2018: 49)*

The methodological diversity and international nature of politics and gender research underpin the three streams in which feminist scholars have worked on concepts: **contested concepts,**

concepts in analysis and the **political use of concepts**. Feminist analysts who examine the "contested nature" of concepts assume that social science concepts and their meaning are a product of context and power struggles, in other words tend toward interpretivist or constructivist approaches (Hobson et al. 2002; Siim 2004; Kantola and Lombardo 2018). From that, it is necessary to first understand how a given concept has been used by non-feminist analysts –"mapping out the conceptual landscape of social politics that follows the markings of gender, which alters the choice of key concepts" (Hobson et al. 2002) – and then to show how bringing in considerations of gender and power can change the conventional concept. In Hobson et al. (2002) the established concepts under the microscope were citizenship, care, social exclusion, contractualization, commodification, representation, agency and empowerment, and social capital. As the editors of this path-breaking volume state "Our central question is whether and to what degree gender has destabilized or altered these concepts" (Hobson et al. 2002: 2).

A part of treating concepts in the context of contestation is that there are necessarily disagreements between feminist analysts, reflecting the diversity of approaches about the core meaning of crucial concepts like gender, feminism and gender equality. Moreover, there is a debate about how much disagreement exists on concepts. Feminist scholars with more constructivist tendencies argue that "There is no agreement in feminist theory about key concepts like emancipation, feminism, and gender equality, and the nature of women's interests and who has the power to define them" (Siim 2004: 98). Some feminist analysts have argued that to provide a single operational definition of a concept contributes to that concept's "reification," that is making an absolute that does not reflect the highly intersectional and unequal nature of gender relations.

Feminist analysts who are closer to the empirical tradition of analysis and fall within the **concepts in analysis** stream argue that it is necessary to provide an operational definition of a key concept, within the concept of a given piece of research, in order to move a given project's agenda forward (e.g. McBride and Mazur 2010). It is not that the terrain of conceptualization is neutral, but that it is important to move beyond identifying disagreements and challenging established concepts to be able to conduct research and to develop empirically based theory. Mazur (2002) in her study of feminist policy in post industrial democracies, for example, developed the following operational definition of feminism, often identified as one of the most "contested" concepts in gender and politics analysis (e.g. Siim 2004; Dhamoon 2013) from a review of over 300 pieces of work on feminist policy formation. She asserts that the two-pronged operational definition she forwards for her comparative study reflects a consensus in the research community that studies feminist policy in a comparative perspective.

- The improvement of women's rights, status or situation to be in line with men's (however rights, status and situation are culturally defined within a given context).
- The reduction/elimination of gender-based hierarchies or patriarchy.

The Research Network on Gender Politics and the State (RNGS), a 40 member group that conducted a 17 country study of the impact of women's policy agencies in Western post industrial democracies, took this same approach to concepts in analysis. The scholars from all 17 countries in the study, from quite different backgrounds, worked together and developed operational definitions of key concepts, including feminism, to produce valid and reliable findings about the impact of women's policy agencies.[4]

The third stream of feminist approaches to concepts is the **political use of concepts**. Here, analysts examine how political and social actors use key concepts and how politics is a struggle over the meaning of concepts. Analysis tends to examine how policy actors use conceptual categories over time and thus often involves historians. As Berrebi-Hoffmann et al. suggest, "The role of social actors in the formation and imposition of categories is at the centre of our analysis"

(Berrebi-Hoffmann et al. 2019: 24). A large multi-national cross-national project, MAGEEQ, developed a method to examine the political use of concepts – "critical frames analysis" and showed how the concept of gender equality is "stretched" and "bent" to serve the needs of the dominant power group in public policy making (Lombardo et al. 2009). This focus on the political construction of concepts used by actors is also associated with the tradition of looking at how social conditions, like gender inequities, become defined into public problems with a focus on the framing of public issues and the struggle over political meaning. Inspired by E. E. Schattschneider's observation "that the mobilization of bias is the supreme instrument of power (1960)," numerous feminist analysts have studied problem definition often in relation to agenda-setting in the policy process (e.g. McBride and Parry 2011). Bacchi (1999) formalized this focus on "the construction of policy problems" and has been identified with the "what's the problem approach" where analysts need to identify what is the political construction of policy problems and solutions.

The Scientific Promise of Feminist Concepts in Gender and Politics Analysis: A Deeper Dive

The ultimate aim for scholars who use concepts in analysis is to do better science through providing more accurate definitions of concepts that measure what is actually happening in the real world when one looks at that world in terms of gender relations and intersectional inequities and discrimination. Since feminist researchers began to turn their attention to conducting systematic analysis, developing precise and meaningful definitions of crucial concepts was a necessary first step. As Goertz and Mazur assert in their edited volume on gender and politics concepts, feminist approaches to concepts hold great promise for doing better and more meaningful science.

> The goal of gendering has been to improve the explanatory power of empirical theory-building that uses core concepts as well as the very process of concept formation itself. Political scientists who gender concepts assert that research, methodology and theory building that ignores gender as a complex analytical concept is not good science.
>
> *(Goertz and Mazur, 2008a: 6)*

The notion of gendering social science analysis is by no means, however, a new one. As early as 1998, Joni Lovenduski, a British feminist analyst provided a road map for the rationale, process and expected outcomes for introducing gender seriously into political analysis. Twenty years later, she pointed out in a keynote presentation at the European Conference on Politics and Gender in Uppsala, Sweden in 2017 that this gendering project by necessity was an example of "slow science," thus a careful yet time consuming process to be pursued over the long-haul.

It is important to note that feminist analysts seldom use static definitions of these core concepts; rather they are constantly being updated and reconfigured to reflect a rapidly changing society.

This caveat applies most certainly to the concept that is at the core of any feminist analysis – gender. With sociologist Joan Acker (1990) and historian Joan Scott leading the way, the notion of gender has gone from being considered a synonym for sex – biologically defined initially in terms of men and women – to gender as a "useful category of analysis (Scott 1986)." The US-based journal, *Politics and Gender,* dedicated a portion of its inaugural issue in 2005 to essays on "the concept of gender." Mary Hawkesworth's chapter in the OUP Handbook shows how sex, gender and sexuality went from "a naturalized presumption to analytical categories" as well (Hawkesworth 2013: 31). With work by DiStefano (1983), Connell (1995) and others focusing on masculinities, gender increasingly moved away from just considering women, but rather assessing the construction of men's and women's identity in asymmetric relation to one another. Today, gender (men and women) is no longer socially constructed along lines of biological sex understood dichotomously as male and female, in part because biological sex is now understood as a gray zone concept aligned along a spectrum rather than in what is seen as an accurate binary notion.[5] The challenge for feminist scholars today is to develop operational concepts with indicators that account for this full intersectional complexity of gender. For example, gender equality can no longer be seen in terms of equality between men and women; it must incorporate vectors of inequality based on sexual orientation, race, culture, age, class, disability without losing its core notion that male domination and heteronormative hierarchies of power are a part of the structural biases of power that discriminate against individual women or individuals who are recognized and/or identify as women.

Table 29.1 presents representative work on established concepts that have been gendered or "adjectivized" and new concepts that have been developed to better understand and study gender-specific phenomena; it is by no means an exhaustive inventory.[6] Goertz and Mazur (2008a) invited gender and politics scholars to apply their formal guidelines to good conceptualization in mapping out nine of the concepts presented in Table 29.1, indicated by an ★. The chapters, therefore, provide critical reviews of the definition and use of each concept and make suggestions on how to better operationalize and use each concept in empirical comparative analysis, including calls, made by Sartori (1970) and others, to make concepts "travel" so as to have more reliable and valid findings and theories An example of each category of concept is presented in more detail below.

Gendering Democracy – Paxton (2008) parsimoniously, yet explosively, adds the requirement of women's suffrage to the operational definition of democracy. Although women's suffrage was implied by the procedural definitions of democracy she examines, in the operationalization of the concept, only manhood suffrage was used. Indeed, many students of democracy argued that women's suffrage was irrelevant in the era of manhood suffrage in Western democracies given it was not on the political agenda (Paxton 2008: 54). Changing the transition dates to democracy from when adult men received the right to vote to when adult women were able to *fully* benefit from voting rights has significant descriptive impacts. Paxton examines four different studies that establish a measure of democracy based on universal manhood suffrage. In those, the transition

Table 29.1 Scholarly Treatment of Core Concepts in Politics and Gender Analysis

Gendered Concepts	
★Democracy	Paxton (2008)
★Representation	Celis and Lovenduski (2018)
★Development	Sen and Grown (1987); Staudt (2008)
★Welfare State	Ciccia and Sainsbury (2018)
★ Governance	Waylen (2008)
★Ideology	Duerst-Lahti (2008)
Empowerment	Parpart, Rai and Staudt (2002)
Institutions	Krook and MacKay (2011); Chappell (2006)
Adjectivized Concepts	
Women's Interests	Jónasdóttir (1988)
★Women/ Feminist Movements	Beckwith (2013); McBride and Mazur (2008)
Gender Transformation	Engeli and Mazur (2018)
New Concepts	
★Intersectionality	Mügge and Montoya (2018)
★ State Feminism	Mazur and McBride (2008)
★Resistance	Verloo (2018); Celis and Lovenduski (2018)

dates for countries are on average 14–38 years later when universal female suffrage is included. Similarly, in another commonly used measure of democracy, Polity's "high democracies," the average difference between when men received suffrage and women were enfranchised was 40 years (ibid.: 57). In two Swiss cantons women did not receive the vote until 1990; universal male suffrage was established in 1848.

Using women's suffrage as the transition point also transforms our understanding of the process democracy's emergence. Paxton shows how Huntington's propositions about the different waves of democracy are incorrect. Including real and full suffrage to women essentially puts many of the 68 countries in later "waves" and shows that in more than two discrete "waves" of democratization between 1893 and 1958, with two "reversals" along the way, there was actually one "… long continuous democratization period from 1893 to 1958, with only one war-related reversal" (Paxton 2008: 65). Finally, making women's enfranchisement a requirement for democratic status raises a new set of explanatory factors, many of which are gender-specific. Among the most salient drivers that need to be assessed for democratization, Paxton argues we should now add (international) feminist and women's movements, individual feminist activists and women's policy agencies, as well as general public opinion rather than only elite opinion (Paxton 2008: 66–67).

Adjectivizing Transformation – Engeli and Mazur (2018) through the Gender Equality Policy in Practice Approach take on the highly difficult task of not just mapping out how gender equality policies are implemented effectively but also developing a conceptual measure to assess their impact in terms of gender transformation.[7] Bearing in mind the complexity of the construct, the new concept is based on the extent to which a given gender equality policy resulted in either the tangible step toward the transformation of gender relations in society or toward the sole equalization of opportunities without challenging the core of the gendered and heteronormative hierarchies of power. Obviously, gender transformation is a long-term process that can also be fueled by the cumulative impact of incremental changes, including equalizing ones. Progression in transformation is very often to be assessed in light of the context where it takes place and the institutional and sectoral legacies in effect (Krizsán and Roggeband 2018). The process of gender transformation is not linear either. Reversal and backlash happen and are probably more likely when policies aim at radical change rather than incremental adjustments (Verloo 2018). Finally, the process is very often asymmetric as well. Gender policies across Western Europe tend to fare better for some categories of women, i.e. white middle-class women who can benefit from a number of privileges. Engeli and Mazur (2018: 121–123) develop four possible levels of gender change for assessing gender equality policies in practice: neutral, rowback, accommodation and transformation.

A New Concept: Resistance – "Resistance" or "opposition" has been developed recently, used and explored by many feminist scholars in the context of the continued persistence of gender bias and norms that serve as formidable obstacles to gender equality and the rise of neoliberal and "illiberal" opposition to gender equality coming from right-wing and extreme right movements (Verloo 2018). This notion has been discussed in terms of backlash as well. Celis and Lovenduski (2018) assert that in the arena of women's representation, the lack of significant progress is due to the persistent resistance and opposition to gender equality and to the achievement of real and meaningful equality between men and women in political offices; feminists must struggle and fight against what can often be quite latent and hidden resistance. Neoliberal government agendas have been identified as an important impediment to gender equality through the drastic downsizing of social policies and state programs that are often necessary to pursue gender equality. More recently, the rise of populist extreme-right governments and actors throughout the Western world, "illiberal democracies," has supported policy proposals and ideas that are blatantly antifeminist and homophobic in an unprecedented manner, at least in the post World War II period. Thus, the concept of resistance is an excellent example of how feminist conceptualization is highly responsive to the rapidly changing political context around us. Moreover, the development and operationalization of this concept are closely connected to promoting action that can counteract these growing efforts to blatantly block well-established principles of social justice and equality and civil rights.

Integration between Feminist and Non-feminist Work?
Up Against the "Glass Wall"

While there has been a strong and clear effort of feminist analysts to dialog and build from established political science concepts and research in the interest of addressing the inherent biases and silences of existing research, little "malestream" work in political science takes into consideration the plethora and wealth of feminist scholarship on similar or adjacent topics. As Mazur's analysis shows in the area of comparative public policy,

> ... the potential of newer feminist scholarship to improve knowledge and theory building in the "mainstream" is at best mitigated, if not all together thwarted by a glass wall, even when feminist scholars seek to directly contribute to research and theory-building that is not explicitly feminist.
>
> *(Mazur 2012: 533)*

This observation that there is a "profound gender blindness" (Celis et al. 2018) is echoed in more recent reviews of feminist work in contemporary political science as well (e.g. Celis et al. 2013; Ackerly and True 2018; Aherns et al. 2018). Unfortunately, we are still far away from the "contextual" and "gender sensitive" European political science that Siim had hoped for in 2004. The bottom line is that while feminist approaches to concepts may have "much to offer" and many feminist analyses have contributed "to building bridges between different approaches in political science" (Kantola and Lombardo 2017: 2), the uptake of gender and politics scholarship in the non-feminist world of political science has been incremental and piecemeal at best and more often than not ignored. The Gerring methodology textbook cited in the beginning of this chapter is an example. Even with its central emphasis on conceptualization and the close ties of John Gerring to feminist analysts in the context of qualitative methods movement, he makes no mention of any feminist work on methodology. This echoes Ackerly and True's (2018: 266) findings of gender silences and blindness across their review of methodology textbooks in the social sciences. Similarly, Pamela Paxton's analysis of what happens to theorizing about democracy when we add the cut-off of women's suffrage was initially rejected by several different top-level Comparative Politics journals. The Goertz and Mazur book on concepts and gender, which was aimed at a non-feminist audience uninformed about feminist work on concepts, was cited significantly less often – 585 times on Google scholar, when compared to Goertz's 2005 book on social science concepts more generally – 1,588 times.

To be sure, there are exceptions to this rule. Gary Goertz first became interested in the feminist conceptualization project after looking at the example of work on gendering welfare states in the 1990s, which had put into question and gendered Esping-Andersons "Three World's of Welfare Capitalism" typology which had completely ignored any gender component (Ciccia and Sainsbury 2018). A 2000 review of welfare state research by Paul Pierson in *Comparative Political Studies*, a top-ranked political science journal, included the new feminist scholarship on gendering welfare state in his review.

The industry of datasets and measures for democracy – Polity, Freedom House, used by many comparativists who follow a quantitative approach to theory-building in comparative politics, despite the existence of a plethora of gender equality indices – has recently seen some signs of change. The World Values Survey added a battery of questions on gender equality in the past 10 years, but unfortunately these do not seem to be informed by the sophisticated feminist scholarship on gender and politics. The four questions on gender equality in the survey seem to reinforce ethnocentric Western notions about women's rights and gender equality, which are quite outmoded and inappropriate in the context of today's postcolonial work by scholars from the global South. The V-Dem dataset, a comprehensive cross-national and longitudinal data project, on an impressive range of democratic attributes at a high level of fine grained detail now includes a good number of gender/sex-specific indicators on social exclusion, equality and a separate index on women's empowerment (https://www.v-dem.net/). This in large part has come with Pamela Paxton's participation on the leadership team.

There has been little integration between the burgeoning large-scale comparative feminist policy projects, with numerous top-level publications and large-scale scientific grants and malestream work in adjacent areas,[8] although some inroads have been made in recent years. See, for instance, Lieberman's use of the international RNGS study as an example of good practice in mixed methods analysis (2009). The subfield of international relations in general has been particularly resistant, according to Prügl and Tickner (2018: 18), where feminist analysts are still in the stage of challenging "…seemingly unproblematic concepts and framings" in non-feminist work in this area.

Conclusion: Next Steps

Whether one should regard the state of feminist approaches to conceptualization in terms of a glass half full or empty in terms of intersecting with and influencing non-feminist research remains to be seen. What is certain is that feminist conceptualization is a permanent and important part of the scientific enterprise in the social sciences. Building from the epistemological and methodological diversity of feminist studies more generally speaking, a rich and multi-form scholarship on concepts has emerged that informs in a significant manner thinking, understanding, theorizing and in some cases action on gender-related issues. Indeed, the level of publications in books and refereed journal articles and the new feminist political science journals as well as significant levels of scientific funding for projects that place conceptualization at their center indicate the health and institutionalization of feminist work that places conceptualization at its core.

Of course, given the iterative and critical nature of the feminist scholarly enterprise itself, there is room for improvement and new directions. As Atchison (2018) and others show, there is a link between the continuing low status of women in the discipline in terms of positions and salary to the marginalization of feminist approaches, research findings and conceptualizations. The inclusion of the gender dimension in the V-Dem data project with the arrival of Pam Paxton in the leadership team shows the importance of having gender experts brought into non-feminist projects in a meaningful way. Therefore, a part of the feminist agenda in this area should be an effort to promote the status of women in political science including women of color, women from the global South as well as non binary, trans and gay individuals. Ongoing efforts made by the Politics and Gender Standing Group of ECPR through its semi-annual conference on gender ECPG and its new journal EJPG to create a big tent for the full variety of work on gender that goes beyond the social construction of men's and women's identities, to apply the full complexity of gender in terms of the contemporary reality of sexual identity and to include formerly excluded scholar groups need to be continued and followed by other feminist academic groups as well. More attention should be given to operationalizing the new complex concepts to incorporate intersectional considerations so that studies can be more valid and reflect the complexity of ongoing power structures.

Concepts should be developed to travel across different methodological approaches that include qualitative, interpretive and quantitative methodologies through mixed methods research (Tripp and Hughes 2018). To make this happen more successfully, an awareness needs to be developed that conceptualization needs to be adapted for a variety of methodological contexts. The new feminist work on experimental methods is very promising, but current concepts need to be adapted to this experimental setting (Waylen 2018). Moreover, given the slow progress, often with significant reversals and even backlash, of gender equality in the real world, more attention should be placed on praxis, that is, how to get stakeholder, practitioners and policy makers to take into account and be held accountable for the findings and recommendations of feminist studies. More effort needs to be devoted to building bridges between the feminist scientific and policy practitioner world as well.[9] Thus, in the final analysis feminist approaches to concepts and conceptualization not only have the potential to make social science more scientific but also to apply that science to solving the wicked and thorny problems around gender justice and inequality that still confront the world.

Related chapters: 14, 27, 28.

Notes

1 Intersectionality is a relatively new and core concept in feminist studies that asserts inequities, discrimination and gender equality are based on far more than socially constructed differences between men and women. They include many different and, sometimes, intertwined forms of domination and discrimination that have cumulative and divergent affects depending on the ascriptive identity of the individual or group – including biological sex, which is no longer considered dichotomous, sexual identity, ethnicity, culture, national origins, race, age, disability and class, to name the more prominent vectors of inequality. For a recent review of intersectionality scholarship see for example Mügge and Montoya (2018) and Chapter 30 in this volume.

2 The European Conference on Gender and Politics affiliated with the Standing Group on Gender and Politics of the European Consortium of Political Research (https://www.ecpg.eu/) has been held on a biannual basis since 2009 and now hosts over 700 feminist researchers. The standing group is one of the most active subgroups of the ECPR and just recently launched a new journal, the *European Journal of Political and Gender*. Similarly, the Women and Politics section of APSA maintains a large membership – 800 plus – and has since 2005 published its own journal as well, *Politics and Gender*. Other feminist journals that dialog with gender and politics research include *Journal of Women, Politics and Policy, International Journal of Feminist Politics* and *Social Politics*.

3 For a discussion of the major feminist approaches in political analysis see Kantola and Lombardo (2017). In this volume, see the section on theoretical frameworks to see the diversity of feminist approaches across all sciences.

4 For more details on this process, see the discussion of conceptualization in McBride and Mazur (2010) and chapters in the edited volume by Goertz and Mazur (2008a) on state feminism and women's policy agencies (Mazur and McBride 2008) and women's and feminist movements (McBride and Mazur 2008).

5 For more on sex and gender as analytical concepts see Chapter 22 in this volume.

6 Goertz and Mazur (2008b) identify these three different ways that gender and politics scholars have used to pursue good conceptualization.

7 For more on the GEPP network and project go to http://www.csbppl.com/gepp/.

8 Including GEPP, there have been ten major international gender and politics research projects on post industrial democracies alone in the past 15 years (Mazur 2012). Htun and Weldon's study (2018) is another well-funded international project that covers gender and politics issues in all regions of the world.

9 See section III of this volume for more on the intersection between science and policy.

References

Acker, J. (1990) "Hierarchies, Jobs, and Bodies: A Theory of Gendered Organizations," *Gender & Society* 4, 139–158.

Ackerly, B. and True, J. (2018) "With or Without Feminism? Research Gender and Politics in the 21st Century," *European Journal of Politics and Gender* 1(1), 259–279.

Aherns, P., Celis, K., Childs, S., Engeli, I., Evans, E. and Mügge, L. (2018) "Politics and Gender: Rocking Political Science and Creating New Horizons," *European Journal of Politics and Gender* 1(1), 3–16.

Atchison, A. L. (2018) "Towards the Good Profession: Improving the Status of Women in Political Science," *European Journal of Politics and Gender* 1(2), 279–298.

Bacchi, C. L. (1999) *Women, Policy and Politics. The Construction of Policy Problems*. Thousand Oaks, CA: Sage.

Beckwith, K. (2013) "The Comparative Study of Women's Movements," in Waylen, G., Celis, K., Kantola, J. and Weldon, S. L. (eds.) *The OUP Handbook of Gender and Politics*. New York: Oxford University Press, 411–436.

Berrebi-Hoffmann, I., Giraud, O., Renard, L. and Wobbe, T. (eds.) (2019) *Categories in Context: Gender and Work in France and Germany, 1900-Present*. New York and Oxford: Berghahn Books.

Brady, H. E. and Collier, D. (eds.) (2010) *Rethinking Social Inquiry: Diverse Tools, Shared Standards*. Boulder: Rowman & Littlefield.

Celis, K. and Lovenduski, J. (2018) "Power Struggles; Gender Equality and Representation," *European Journal of Politics and Gender* 1(1), 149–165.

Celis, K., Kantola, J., Waylen, G. and Weldon, S. L. (2013) "Introduction. Gender and Politics: A Gendered Discipline," in Waylen, Celis, Kantola and Weldon (eds.) *The OUP Handbook of Gender and Politics*. New York: Oxford University Press, 1–25.

Chappell, L. (2006) "Comparing political institutions: Revealing the gendered 'logic of appropriateness,'" *Politics & Gender* 2(2), 223–234.

Ciccia, R. and Sainsbury, D. (2018) "Gendering Welfare State Analysis: Tensions between Care and Paid Work," *European Journal of Politics and Gender* 1(1), 93–110.

Collier, D. and Mahoney, J. E. (1993) "Conceptual Stretching' Revisited: Adapting Categories in Comparative Analysis," *American Political Science Review* 87(40), 845–855.

Collier, D. and Levitsky, S. (1997) "Democracy with Adjectives: Conceptual Innovation in Comparative Research," *World Politics* 49, 430–451.

Connell, R.W. (1995) *Masculinities*. Berkeley: University of California Press.

Dhamoon, R. K. (2013) "Feminisms," in Waylen, Celis, Kantola and Weldon (eds.) *The OUP Handbook of Gender and Politics*. New York: Oxford University Press, 88–110.

DiStefano, C. (1983) "Masculinity as Ideology in Political Theory: Hobbesian Man Reconsidered," *Women's Studies International Forum* 6, 633–644.

Duerst-Lahti, G. (2008) "Gender Ideology: Masculinism and Feminalism," in Goertz, G. and Mazur, A. G. (eds.) *Politics, Gender and Concepts: Theory and Methodology*. Cambridge: Cambridge University Press, 159–192.

Engeli, I. and Mazur, A. G. (2018) "Taking Implementation Seriously in Assessing Success: The Politics of Gender Policy in Practice," *European Journal of Gender and Politics* 1(1), 11–29.

Gerring, J. (2012) *Social Science Methodology: A Unified Framework*. Cambridge: Cambridge University Press.

Goertz, G. (2020) *Social Science Concepts: A User's Guide*. Princeton, NJ: Princeton University Press.

Goertz, G. & Mahoney, J. (2012) *A Tale of Two Cultures: Qualitative and Quantitative Research in Social Science*. Princeton and Oxford: Princeton University Press.

Goertz, G. and Mazur, A. G. (eds.) (2008a) *Politics, Gender and Concepts*. Cambridge: Cambridge University Press.

Goertz, G. and Mazur, A. G. (2008b) "Mapping Gender and Politics Concepts: Ten Guidelines." in Goertz, G. and Mazur, A. G. (eds.) *Politics, Gender and Concepts: Theory and Methodology*. Cambridge: Cambridge University Press, 14–43.

Hawkesworth, M. (2013) "Sex Gender, and Sexuality: From Naturalized Presumption to Analytical Categories," in Waylen, Celis, Kantola and Weldon (eds.) *The OUP Handbook of Gender and Politics*. New York: Oxford University Press, 31–56.

Hobson, B., Lewis, J. and Siim, J. (2002) *Contested Concepts in Gender and Social Politics*. Cheltenham: Edward Elgar.

Htun, M. and Weldon, S. L. (2018) *The Logic of Gender Justice: State Action and Women's Rights around the World*. Cambridge: Cambridge University Press.

Jenson, J. and Lépinard, E. (2009) "Penser le Genre en Science Politique: Vers une Typologie des Usages du Concept," *Revue Française de Science Politique* 2(59), 183–201.

Jónasdóttir, A. G. (1988) "On the Concept of Interests, Women's Interests and the Limitation of Interest Theory," in Jones, K. B. and Jónasdóttir, A. G. (eds.) *The Political Interests of Gender*. London: Sage Publications, 123–148.

Kantola, J. and Lombardo, E. (2018) *Gender and Political Analysis*. London: Palgrave Macmillan.

King, G., Keohane, R. and Verba, S. (1994) *Designing Social Inquiry: Scientific Inference in Qualitative Research*. Princeton, NJ: Princeton University Press.

Krizsán, A. and Roggeband, C. (2018) *The Gender Politics of Domestic Violence*. New York: Routledge.

Krook, M. L. and MacKay, F. (eds.) (2011). *Gender, Politics and Institutions: Towards a Feminist Institutionalism*. Houndmills: Palgrave Macmillan.

Lieberman, E. (2009) "Bridging the Qualitative-Quantitative Divide: Best Practice in the Development of Historically Oriented Replication Databases," *Annual Review of Political Science* 13, 37–59.

Lombardo, E., Meier, P. and Verloo, M. (eds.) (2009) *The Discursive Politics of Gender Equality: Stretching, Bending and Policymaking.* London: Taylor & Francis.

Mazur, A. G. (2002) *Theorizing Feminist Policy*, Oxford: Oxford University Press.

Mazur, A. G. (2012) "A Feminist Integrative and Empirical Approach in Political Science: Breaking Down the Glass Wall," in Kincaid, H. (ed.) *The Oxford University Press Handbook on Philosophy of Social Sciences.* Oxford: Oxford University Press, 533–558.

Mazur, A. G. and McBride, D. (2008) "State Feminism," in Goertz, G. and Mazur, A. (eds.) *Politics, Gender and Concepts: Theory and Methodology.* Cambridge: Cambridge University Press, 244–269.

McBride, D. E. and Mazur, A. G. (2008) "Women's Movements, Feminism and Feminist Movements," in Goertz, G. and Mazur, A. (eds.) *Politics, Gender and Concepts: Theory and Methodology.* Cambridge: Cambridge University Press, 219–243.

McBride, D. E. and Mazur, A. G. (2010) *The Politics of State Feminism: Innovation in Comparative Research.* With participation of Joni Lovenduski, Joyce Outshoorn, Birgit Sauer and Marila Guadagnini. Philadelphia, PA: Temple University Press.

McBride, D. E. and Parry, J. (2011) *Women's Rights in the USA: Policy Debates and Gender Roles.* New York City: Routledge.

Medie, P. A. and Kang, A. J. (2018) "Power, Knowledge and the Politics of Gender in the Global South," *European Journal of Politics and Gender* 1(1), 37–54.

Mügge. L. and Montoya, C. (2018) "Intersectionality and the Politics of Knowledge Production," *European Journal of Politics and Gender* 1(1), 17–36.

Parpart, J. L., Rai, S. M. and Staudt, K. (2002) *Rethinking Empowerment: Gender and Development in a Global/ Local World.* London and New York: Routledge.

Paxton, P. (2008) "Gendering Democracy," in Goertz, G. and Mazur, A. (eds.) *Politics, Gender and Concepts: Theory and Methodology.* Cambridge: Cambridge University Press, 47–70.

Pierson, P. (2000) "The Three Worlds of Welfare State Research," *Comparative Political Studies* 33(67), 791–821.

Prügl, E. and Tickner, A. (2018) "Feminist International Relations: Some Research Agendas for a World in Transition," *European Journal of Politics and Gender* 1(2), 75–92.

Sartori, G. (1970) "Concept Misformation in Comparative Politics," *American Political Science Review* 64, 1033–1053.

Schattschneider, E.E. (1960) *The Semisovereign People: A Realist's View of Democracy in America.* New York: Holt, Rinehart & Winston.

Scott, J. (1986) "Gender a Useful Category of Historical Analysis," *American Historical Review* 91, 1053–1075.

Sen, G. and Grown, C. (1987) *Development, Crises, and Alternative Visions: Third World Women's Perspectives.* New York: Monthly Review Press.

Siim, B. (2004) "Towards a Contextual and Gender Sensitive European Political Science?," *European Political Science* 3(2), 97–101.

Spierings, N. (2015) *Women's Employment in Muslim Countries: Patterns of Diversity.* London: Palgrave Macmillan.

Staudt, K. (2008) "Gendering Development," in Goertz, G. and Mazur, A. (eds.) *Politics, Gender and Concepts: Theory and Methodology.* Cambridge: Cambridge University Press, 136–157.

Tripp, A. and Hughes, M. (2018) "Methods, Methodologies and Epistemologies in the Study of Gender and Politics," *European Journal of Politics and Gender* 1(2), 241–258.

Verloo, M. (ed.) (2018) *Varieties of Opposition to Gender Equality in Europe.* New York: Routledge.

Waylen, G. (2008). "Gendering Governance." in Goertz, G. and Mazur, A. (eds.) *Politics, Gender and Concepts: Theory and Methodology.* Cambridge: Cambridge University Press, 114–155.

Waylen, G. (2018) "Nudges for Gender Equality? What Can Behaviour Change Offer Gender and Politics," *European Journal of Politics and Gender* 1(1–2), 1671–84.

Further Reading

Waylen, G., Celis, K., Kantola, J. and Weldon, S. L. (eds.) (2013) *The OUP Handbook of Gender and Politics.* New York: Oxford University Press.

Goertz, G. and Mazur, A.G., (eds.) (2008) *Politics, Gender and Concepts.* Cambridge: Cambridge University Press.

Special Inaugural Double Issue. European Journal of Politics and Gender 1(1–2), 2018.

The Physical Sciences

30

WHAT IS IT LIKE TO BE A WOMAN IN PHILOSOPHY OF PHYSICS?

Laura Ruetsche

Prolegomenon

My title is a special case of a more general question that is also the name of a blog: "what is it like to be a woman in philosophy?" (https://beingawomaninphilosophy.wordpress.com/) The blog chronicles episodes, ranging from being talked over in seminar to being pawed over in office hours (and beyond), in which women in philosophy are diminished, delegitimated, devalued, and even assaulted. So, what is it like to be a woman in philosophy *of physics*? Some answers would fit right into the real blog.

- 1994. I submit a paper, my first ever, to *Foundations of Physics Letters*. Standard practice is to send papers to two referees. Mine is sent to six.
- 2008. I am giving a talk at a foundations of physics conference whose audience is over-whelmingly male. I am using an overhead projector and covering the part of the transparency I'm not talking about yet with a sheet of paper. This technique is widespread. But from the audience, a prominent physicist grouses: "Don't do a strip tease." From another part of the audience, a prominent philosopher of physics interjects: "But I want Laura to do a strip tease." In a stirring confirmation of the existence of stereotype threat, I go on to deliver the worst professional presentation of my career. (And that's saying something.)
- 2015. I present a talk sharing the paper's title at Pittsburgh's Center for Philosophy of Science. Reports reach me that a few days before my talk, two senior male philosophers of physics associated with the Center are cracking wise about its title. Their example of what it's like: "You get invited to all the conferences."

But some vignettes illustrating what it's like to be a woman in philosophy of physics operate in a register not well-documented on the real blog.

- Early 90s. Although there are other women in my graduate student cohort and in neighboring ones, I am the only one working in philosophy of physics. (According to Paxton's (2015) analysis of data from *PhilPapers Surveys*, the representation of women with *AOS: phil of physical sciences* is 5%, against a base rate of 18% in the survey sample. It's the AOS (on Paxton's list) with the lowest representation of women. I can't even find data about the representation of women of color in philosophy of physics.) Other women graduate students have projects that more directly engage feminist themes and approaches. It is suggested that my choice of research agenda exposes me as a "traitor to my gender."

- 1992. Looking into the nature of my treachery to my gender, I take a feminist philosophy course from Tamara Horowitz. We read Sandra Harding's *The Science Question in Feminism*, which considers the claim that "A woman scientist is a contradiction in terms" (1986: 59). There is a nearby-feeling possible world in which *I* am a woman scientist. Yet I don't feel like a victim of false consciousness, nor do I expect that my counterpart in the nearby-seeming possible world would feel like a victim of false consciousness. (But then again, we wouldn't.)
- 2013. I attend the inaugural Diversity in Philosophy conference in Dayton, Ohio. More than once, and not for the first time, I hear the sentiment expressed that some kinds of philosophy are the problem, or part of it. The example inevitably given: philosophy of physics.

But is philosophy of physics the problem? I've most often encountered the claim that it is in informal contexts, delivered as a jocose-bitter aside rather than developed *ex cathedra*. This leaves uncertainty about what is being claimed. It could be an empirical generalization about the *people* who are philosophers of physics and how they tend to differ from the people who are (say) feminist philosophers. So understood, the claim posits that, as an empirical matter of fact, philosophers of physics are more likely to engage in problematic behavior than feminist philosophers are. A diagnostic variation on this is: departments where feminist philosophy is a strong and respected presence are departments with healthy climates, but not *mutatis mutandis* for philosophy of physics. The diagnostic variation likens feminist philosophers to canaries in a coal mine: if you find yourself in a department where you can't see any, or where there used to be some but there aren't any anymore, you better wonder whether the environment is toxic.

Here's another way to understand the claims that philosophy of physics is the problem: there's some incompatibility between "technical" work in philosophy and philosophical feminism, between equation mongering in the service of articulating firm foundations for physics and doing the ameliorative and critical work characteristic of feminism. I'll call this the *incompatibility thesis*. As vague (and I hope implausible[1]) as it is, it will be my focus here. One reason for the focus is that the incompatibility thesis feels more philosophical than the other options I've considered. Whereas those address social dynamics, the incompatibility thesis addresses conceptual dynamics. It addresses the question of whether some modes of philosophical engagement *radically preclude* other modes of philosophical engagement. In the philosophy of causation, regularity theorists and necessitarians are at odds. But either takes the other to appear on a shared philosophical landscape. By contrast, according to the incompatibility thesis, technical philosophy operates in a landscape so inhospitable to feminist philosophy that feminist philosophical approaches can't find expression there. I'll also discuss a neighboring thesis: the *hostility thesis*, which holds that those accustomed to navigating technical philosophical landscapes are *in virtue of their habitation to those landscapes* too dubious of feminist philosophical theses to give them a hearing before dismissing them.

A second reason I want to focus on these theses is, even if they lack official supporters, they exert a pull. Informal asides, allegations of gender treachery, the question I get (but usually only in bars) about how *I*, given my documented interests in such obviously compelling topics as quantum measurement, could be interested in philosophical feminisms—all of these are evidence that the pull is operating, which is unfortunate. The theses are not merely wrong but also pernicious. They obscure important truths and important commonalities.

Later in this essay, I describe a technical vehicle in which a feminist thesis I take to be both interesting and plausible might travel. This is meant to suggest, against the incompatibility thesis, that feminist views *can* find expression and even support in technical landscapes. Later still, I offer an account of my own technical work on the foundations of quantum field theory (QFT), emphasizing continuities between aspects of that work and philosophical feminisms. This is meant to suggest, against the hostility thesis, that operating in technical landscapes might cultivate openness to key components of feminist approaches. Before undertaking the heavy lifting of countering the incompatibility and hostility theses, I'll conduct a warmup exercise and issue some (very) belated terminological disclaimers.

Warmup Exercise: Intersectionality

I am a woman. I am a philosopher of physics. Understanding each of those predicaments individually does not eventuate in an understanding of what it is like to be a woman philosopher of physics. It may be that *qua* woman in philosophy I fall under suspicion of having submitted a manuscript so deviously bad that it requires triple the usual complement of referees to detect its weaknesses. And it may be that *qua* philosopher of physics I fall under suspicion of contributing to philosophy's gender problem. But neither categorization on its own or in any obvious mechanical combination generates my suspected status as a traitor to my gender. This is a (relatively low-stakes!) example of *intersectionality*.

A higher-stakes example is DeGraffenreid v. General Motors Assembly Line (discussed in Crenshaw 1989, which coins the term "intersectionality"). In a wave of "last hired first fired" layoffs prompted by the crash of the American auto industry during the early 70s, all but one Black woman working at a GM plant in St. Louis loses her job. Five Black women bring a class action suit against the company. The suit contends that historical patterns of discrimination are why Black women were hired last, and that the "last hired first fired" policy perpetuates those patterns. In a 1975 decision, the court dismisses the suit on the grounds that the plaintiffs had established neither that GM discriminated against women nor that GM discriminated against Blacks during the period in question[2]— as though discrimination against Black women per se could only exist in the form of a mechanical concatenation of discrimination against Blacks and discrimination against women. If you think something has gone wrong here, you think intersectionality matters.

One of the technical notions important to my work is the notion of an *algebra*. There's nothing tricky about it. An algebra is just a collection of things, along with an operation that, given any two things in the collection, determines a third thing in the collection that is the (unique) combination of the first two things. One way—not the most complete or profound way, but a way nonetheless—to express why intersectionality matters is: *oppressions don't form an algebra*. Getting a grip on gender oppression and getting a separate grip on racial oppression (supposing either such monumental achievement were possible) wouldn't suffice to issue in a grip on the varieties of oppression that develop when racial and gender oppression interact. This claim—that oppressions (identities, predicaments, categories [particularly of interactive kinds (Hacking 1999)]) don't form an algebra—strikes me as a very natural way to articulate something important.

Reflecting on the observation that *oppressions don't form an algebra* is a good warmup exercise for later discussions of possible alliances and continuities between technical and feminist philosophy. By illustrating that technical tools can be used to articulate feminist theses, the italicized observation suggests that technologies themselves aren't necessarily the problem. And even if articulating a thesis in a technical language in no way deepens or illuminates the thesis, it could have the rhetorical effect of getting some interlocutors to recognize the thesis for the first time, because it's presented in their vernacular. As a matter of practical politics, that's a way to advance a thesis. And technologies could have not just rhetorical/political but also conceptual value. For it could be that there are feminist contributions that require apt technical tools to fully realize. It could be, that is, that certain technologies stand to certain advances in feminism in something like the way differential geometry stands to general relativity: they afford a language for framing understandings not readily articulable without that language. This is a possibility distinct from, and more provocative than, the possibility that feminist theses couched in a technical vernacular might encounter wider uptake by audiences fluent in that vernacular.

To say that oppressions don't form an algebra is to draw attention to a phenomenon worth understanding, and such that we need to move outside the framework of algebras to understand it. This underscores the fact that *frameworks—formal, social, analytic—have histories. They have uses and they have purposes.* A framework, like the framework of algebras, carries along with it no specification of the scope of its appropriate application, and no guarantee that that scope is limitless. A characteristically philosophical task is to reflect on not only the capacity of frameworks to promote their purposes but also the legitimacy of the purposes themselves. Whether the framework is a gender hierarchy or a QFT, this is true. Something (beyond the reflective attitude itself) that can unify such reflection across disparate domains is a commitment to the legitimacy, for the reflective project, of certain sorts of analytical resources.

Terminology

I'm asking whether *technical* philosophy is incompatible with, or an incubator of hostile attitudes toward, *feminist* philosophy. It's high time that I say something about what I take the words just italicized to mean.

Start with "technical." An important disclaimer: it's the wrong word. All philosophy is technical. It's just that the technical vocabulary of some parts of philosophy ("cause," "woman") sounds a lot more like natural language than the technical vocabulary of other parts of philosophy ("algebra," "singularity"). Here, and with apologies, I'll use "technical" to mark a popularly recognized distinction between the kind of philosophy whose practitioners are liable to use LaTeX and the kind whose practitioners are liable to use Word, or the kind in which equations appear and the kind in which words like "intersectionality" appear. Technical philosophy makes conspicuous (and sometimes gratuitous) use of logical or mathematical apparatus. (As in: "Let R be a Type III von Neumann algebra.") As specious as I take the technical/non-technical distinction to be, I also take it to be a (faulty!) faultline in a powerful cultural divide.

A further disclaimer: "feminist" is also the wrong word for present purposes. Here I'll be treating feminist approaches as members of a family of approaches united less by content than by a commitment to the legitimacy of a particular sort of analytical resource—a commitment often accompanied by the hope that real good can be done by tapping resources of that sort. I'll sketch the family by contrasting it with other ways to approach philosophy.

There are perennial philosophical questions—the natures of knowledge and justification, the makeup of the universe and our selves, the possibility and contours of moral agency, …—and there are the ephemeral embodied enquirers who occupy historically contingent social locations and who raise those questions. These inquirers might wonder: to what extent, if any, should consideration of contingent histories, and the social locations they condition, inform inquiry into philosophical questions? This is a question of philosophical method.

A possible answer is: not at all. Here is John Earman on what it takes to understand scientific knowledge:

> To my mind, the most interesting aspect [of science] is the epistemic one. I insist (in my Bayesian mode) that this aspect be explained in Bayesian terms. This implies that all valid rules of scientific inference must be derived from the probability axioms and the rule of conditionalization.
>
> *(Earman 1992: 204)*

Earman expresses a methodological commitment to what I'll call *pristine* philosophy. Pristine philosophical accounts are unadulterated by the intrusion of "insufficiently general" considerations such as accidents, brute contingencies, and the like. Pristine philosophy has a number of attractions. One is the aesthetic of disciplining oneself to confront the deepest of questions with the sparest of resources. Another, brought home to the perennial question of epistemology, is the not unreasonable thought that when Socrates asked what knowledge was, he asked the same question we are asking today or that an alien intelligence might be raising in some distant galaxy. One might expect or hope that an adequate answer to such a question would be grounded in considerations that abstract away from circumstances that aren't common to all inquirers who might raise the question. Relying only on "probability axioms and the rule of conditionalization" is one way to get a suitably pristine answer to this perennial question.

Adherents to the methodological ideal of pristine philosophy abjure appeal to accidents of human psychology, sociology, history, geography, and so on. Subjecting the epistemology of science to the ideal of pristine philosophy issues in the judgment that scientific knowledge can happen in contexts where gender doesn't, and so ought to be understood without bringing gender into it.

A sound way to market philosophy as widely relevant is to observe that it licenses and promotes methodological self-reflection—a practice all inquirers, no matter what their disciplinary allegiance, may be well advised to consider. Philosophy faces the imperative to reflect upon its own methods. One form this reflection might take is to ask whether pristine resources are adequate to the tasks philosophy takes on. As I understand the family of approaches which includes philosophical feminisms, they share a commitment to the proposition that "insufficiently general" commodities, such as historical contingency and social context, are variables that should inform and would improve our inquiries into philosophical questions, including some evidently perennial ones. Foregrounding the insufficiently general resources of gender and its social conditions, philosophical feminisms are members of the family, but not its entirety. Gender is only one of many, many potentially salient varieties of analytic category commended by philosophy methodologically committed to contextual resources.

Although a methodological commitment to the philosophical relevance of gender as a contextual resource is an important component of philosophical feminism, it is not its entirety. One can adopt the methodological commitment without sharing features essential to a fully realized philosophical feminism as many understand it. These features might include a motivating concern with gender oppression and gender inequality, and the hope that philosophical accounts might help us combat them. Focusing here on the methodological commitment, I won't take a stand on what *else* it might take to fully realize philosophical feminism.

Now take one of my favorite questions: the nature of empirical/scientific justification. Given how I want to understand "feminism," feminist epistemology of science is epistemology of science that contends that gender/social location/the histories and structures that constitute social locations matter to the epistemic dimension of scientific inquiry. One way to do feminist epistemology of science is to argue that an adequate understanding of scientific justification requires taking social locations and their dynamics into account. The next section develops an example where technical devices convey and support such an argument.

Against Incompatibility

Here is a caricature, inspired by Wylie 1997's discussion of Gero 1983, and meant to make a variety of possible epistemic shortcomings vivid. Consider the Bison-Mammoth construct. It is the 1970s. Edgeware analysis—using techniques including microscopy to study the physical morphology of archeological relics with a view toward identifying their likely uses—is a feminized subdiscipline of archeology. It is also low-status—low enough status that the mainstream is slow to take up evidence from edgeware analysis. Instead, the mainstream is busy articulating the Man-the-Hunter paradigm, theorizing about bison and mammoth hunting practices and the social, technological, linguistic, and cognitive adaptations they required, as well as collecting evidence (for instance from kill sites) informing these theories. A felt puzzle arises for the Man-the-Hunter paradigm: when the big game Man Hunted and relied upon (according to the paradigm) for His subsistence went extinct, how did Man avoid following suit? The (ignored) testimony of edgeware analysis: Many of Man-the-Hunter's tools served foraging purposes; ergo Man-the-Hunter wasn't in fact subsisting solely on bisons and mammoths. The puzzle dissolves, and the paradigms that frame it are revealed to be incomplete.

This is the story of a community that persisted in embracing a theory longer than it should have. It persisted even after evidence challenging the theory surfaced, as it were offstage from where the tastemakers were theorizing. *Even if* every member of the taste-making community diligently updated their credences by conditionalizing via Bayes' rule on the evidence at hand, they were engaging in *epistemic bad behavior*—the behavior of a community striated by status in such a way that its hierarchies induce hierarchies of credibility that retard or prevent relevant evidence from informing belief dynamics. This diagnosis of epistemic bad behavior is a normative epistemological claim that qualifies as a feminist thesis because it attributes *epistemic* significance to *contingent social structures*, ones that happen often to be indexed by race, gender, and so on. If status hierarchies carry epistemic costs, then we can't understand everything we ought to understand about empirical justification if we ignore social structures and the contingencies they condition.

Now, imagine harnessing technical apparatus (Bayesian confirmation theory or formal learning theory or social network theory) in the service of the thesis that status hierarchies carry epistemic costs. Such technical work could generate a result supporting the feminist epistemological thesis inspired by Wylie's tale of the Bison-Mammoth construct. I call such a result a *Wylie Theorem*.

The literature already contains results lying in the direction of a Wylie theorem.

Hegselmann and Krause (2006) offer one model. A community of inquirers is organized into subcommunities D_i. These can be subdisciplines or labs or demographics. Status is operationalized by a *respect (or uptake) matrix* W_{ij} of weight one, where each matrix element w_{ab} lies between 0 and 1 inclusive, and reflects subcommunity D_a's propensity to respond to considerations that move subcommunity D_b. When $w_{ab} = 1$, subcommunity D_a adopts D_b's beliefs wholesale; $w_{ab} = 0$ means D_a flat-out ignores considerations D_b deems salient. The respect matrix is not required to be symmetric. So for instance if $0 < w_{ab} \neq w_{ba} = 0$, D_a is responsive to D_b's considerations, but D_b declines to return the favor. Asymmetries and patterns of entries in the respect matrix can model hierarchies of credibility keyed to status and markers of status.

Now consider inquiry into some quantity x whose true value is x_T; let x_i represent what subcommunity D_i believes x's value to be. Hegselmann and Krause introduce a master equation describing a discrete-time belief dynamics:

$$x_i(t+1) = \alpha_i x_T + (1 - \alpha_i)\Sigma_i W_{ij} x_j(t)$$

The left hand side is what subcommunity D_i believes x's value to be at time step $(t+1)$. The right hand side expresses this as resulting from two forces, one (captured by the first term on the right hand side) emanating from the truth, and the other (captured by the second term on the right hand side) emanating from a social process involving the beliefs of other subcommunities, and hence the considerations that move those beliefs.

Regarding the first term, $\alpha_i x_T$, on the right hand side: taking a value between 0 and 1 inclusive, α_i gauges subcommunity D_i's "truth-sensitivity." The higher α_i's value, the more strongly the truth attracts the subcommunity's belief. A limiting case, corresponding to a subcommunity driven ineluctably and immediately to the truth, is $\alpha_i = 1$. No matter what its prior beliefs, such an extremely truth-sensitive subcommunity is delivered after one cycle of belief revision to the belief that x takes its true value x_T. Another limiting case is $\alpha_i = 0$, which corresponds to a subcommunity whose own considerations fail utterly to move it toward the truth.

As crude as the model is, attributing different subcommunities different truth-sensitivities captures the important feature of inquiry, that different subcommunities latch on and respond to salient evidence at different rates. The crude model enables us to trace the impact of that feature on community-wide belief dynamics without developing detailed accounts (which would be substantive epistemological work interesting in its own right) of what makes evidence salient or what constitutes latching on to its salience.

The second term on the right hand side of the master equation, $(1 - \alpha_i)\sum_i W_{ij} x_j(t)$, conveys the impact, on subcommunity D_i's belief at time step $(t+1)$ about x's value, of the beliefs at time step t of other subcommunities about that value. The expression $\sum_i W_{ij} x_j(t)$ represents a weighted (by respect) average, over subcommunities, of their beliefs about x's value. The more one subcommunity respects another, the more its beliefs move in response to their beliefs, and the closer this weighted average will be to their beliefs. The coefficients α_i and $(1 - \alpha_i)$ establish a tradeoff between how strongly the truth moves a subcommunity directly and that subcommunity's susceptibility to belief revision by means of social processes. A subcommunity with 0 truth sensitivity can nevertheless be guided to the truth if its beliefs are sufficiently moved by other communities whose beliefs are headed in the direction of the truth.

With respect to the belief dynamics they model, Hegselmann and Krause prove a proto-Wylie theorem they call the "Leader of the Pack" theorem. A subcommunity D_i is *truth-sensitive* if and only if $\alpha_i > 0$. The "Leader of the Pack" theorem is: (even a single) truth-sensitive subcommunity will drive community-wide consensus to the truth if there's a "chain of respect"—think of this as a sequence of intervening subcommunities such that each subcommunity in the sequence has non-zero respect for the next subcommunity in the sequence—connecting every subcommunity to every other.

This result is Wylie theoremesque because it harnesses a technology to characterize status-indexed credibility hierarchies and their potential epistemic cost. Such hierarchies can institute *silencing*, understood in terms of the formal model as follows: one subcommunity is silenced with respect to another if and only if the respect matrix contains no chain of respect leading from the first subcommunity to the second. (Notice that the model says nothing at all about *how* or *why* chains of respect might go awry. Dotson (2011) and Davis (2018) develop conceptual tools for better understanding this.) The Leader of the Pack theorem identifies a condition under which any subcommunity's truth sensitivity leads the whole community to the truth, whatever it may be. When any subcommunity is silenced, this condition fails. And if—as in the Bison-Mammoth construct tale—a high status community silences a low status community that has latched onto truth-revelatory considerations, those consideration will percolate more slowly to other (less truth-sensitive) subcommunities, even ones connected by chains of respect to the low status subcommunity, because those other subcommunities will be following the high status community's (misleading) epistemic lead. (See Hegselmann and Krause 2006, §4, for examples illustrating this pattern.)

As preliminary as these results are, they suggest that more sophisticated Wylie theorems could well secure support for nuanced and interesting feminist philosophical theses. This is not to say that feminist epistemology will or should be re-expressed as a collection of Wylie theorems. For one thing, it isn't obvious that the full range of considerations feminist approaches seek to unearth can be crammed in to any of the technical apparatus available off the shelf right now. That apparatus tends to focus on credence revision, but credence revision is hardly the only activity of epistemological interest: there is also hypothesis formation, question identification, research protocol, and so on. When it comes to the epistemic costs of ignoring, downplaying, or misplacing the testimony and perspective of those with salient sensitivities, these arenas are probably even more important than the arena of belief revision. Thus a probable limitation of technical frameworks for modeling belief revision is that a lot of epistemology, including a lot of feminist epistemology, could fall through their cracks. Nevertheless, the Wylie theorem project suggests that, limited though they may be, technical frameworks could enable the statement and delivery of some feminist epistemological morals. And that's enough to establish that the incompatibility thesis, the thesis technologies are so conceptually inhospitable to feminisms as to preclude expressing or establishing feminist theses by technical means, fails.

Against Hostility?

The last section rebutted the incompatibility thesis by illustrating how technical resources could be harnessed to express and evaluate some feminist philosophical theses. This section addresses the concern that the rebuttal amounts to an idle gesture at an unoccupied region of conceptual space. According to the hostility thesis, working in technical landscapes predisposes practitioners of technical philosophy to be hostile—unwilling to entertain—feminist approaches. I undertake here to counter the hostility thesis by exhibiting continuities between some recent work in the purportedly technical field of philosophy of QFT and philosophical feminisms.

The recent technical work is mine. The overt topic of my 2011 book is QFT and other quantum theories of systems with infinitely many degrees of freedom, which I gather under the heading of QM_∞. The book argues that QM_∞ presents important philosophical puzzles that do not arise in connection with simpler quantum theories. These include puzzles about how and why a physical theory comes to be equipped with *content*, in the rough sense of a picture of how, according to the theory, the world is put together. QM_∞ is not straightforward. Articulating and addressing the puzzles it poses require a certain amount of technical apparatus, including functional analysis and operator algebra theory.

The project I've just described hardly sounds feminist. Still, I take it to share a commitment central to feminism as I understand it. This is a commitment to the possibility that philosophical accounts crafted in accordance with pristine methodological ideals might fail to be adequate to the purposes we should have for them—and thus the possibility that philosophically adequate accounts need to draw on resources broader than the resources admitted by those ideals. These broader resources include messy contingent contextual ones. Gender is a species of this genus. Although other species—initial and boundary conditions, the shape of projects in aspirational physics, explanatory and modeling needs—roam the book, to make a case for the philosophical relevance of the genus is to dismantle one source of resistance to philosophical feminism. That source is the perennialist's conviction that the sorts of considerations feminist work brings to bear just aren't philosophically fruitful or even philosophically admissible.

The book's argument for the relevance of messy contingent contextual considerations sets out from the mathematical core of the physical theories in question. The key difference between simpler quantum theories and theories of QM_∞ is that the latter, but not the former, fall outside the scope of the Stone-von Neumann theorem. This is a mathematical result, but its significance can be explicated without too much jargon. Theories of classical physics are different in structure from theories of quantum physics. Take the simplest case of a single particle of mass m confined to a line. The classical theory assigns the particle a state by equipping it with precise values for its position on the line and its momentum along the line. All of the particle's other properties are determined by its position and momentum—for instance, the particle's kinetic energy is its momentum squared, divided by twice its mass. Thus, given the particle's classical state, we can predict with certainty the values of all its other physical properties. The laws and symmetries of the theory are expressed by how these properties are inter-related, inter-relations which are captured by a gadget called the Poisson bracket.

By contrast, the quantum theory of our particle attributes it a state which is a vector in a vector space, and associates position, momentum, and other properties with gadgets called operators on that vector space. Typically, the state vector does not fix the values of these properties but instead offers a probability distribution over possible values. Given a pair of quantum properties, there is usually a tradeoff in the informativeness of the probability distributions the state vector defines over their possible values: the more accurately the state vector predicts the value of one property, the less accurately it predicts the value of the other. A gadget called the commutator bracket sets the terms of this tradeoff, and also structures the collection of quantum properties in a way that expresses the quantum theory's laws and symmetries.

As different as quantum and classical theories are, they are also similar. At their hearts lie a structuring of physical magnitudes afforded in the classical case by the Poisson bracket and in the quantum case by the commutator bracket. This inspires a recipe for generating, from a classical theory, a quantum theory that is its quantization. To follow this Hamiltonian quantization recipe, start with the Poisson bracket between the classical position and momentum magnitudes, and try to find a vector space on which act a pair of operators satisfying a commutator bracket that mirrors the classical Poisson bracket. What you are looking for is a vector space *representation of the canonical commutation relations* (or CCRs) defining the quantum theory you seek. Once you find a representation of the CCRs, you're off to the races: identifying the operators furnishing your representation as quantum mechanical position and momentum magnitudes, use those operators to generate a panoply of other quantum magnitudes standing to one another in functional and nomic relationships; having thus assembled your collection of quantum magnitudes, define a family of quantum states as those that assign well-behaved probabilities to possible values of those magnitudes.

Recipes are only as good as their results are consistent. About this Hamiltonian quantization recipe, we might worry: is it possible to follow it starting from the *same* classical theory and obtain *different* quantum theories? The Stone-von Neumann theorem assures us that it is not—provided that the classical theory we start from concerns systems with finitely many degrees of freedom (finitely many particles moving in finitely many dimensions, say). No matter how different a pair of representations of the CCRs quantizing such a theory might seem, those representations will always prove to be notational variants on one another. They'll agree about what's physically possible, as well as about what structures of properties physical possibilities instantiate. If a classical theory is suitably finite, its quantization is essentially unique. (This is a slight oversimplification. For complications irrelevant to the questions posed in this essay, see Ruetsche 2011: Chapters 2–3.)

Classical field theories aren't suitably finite. The systems they address are fields, specified (in the simplest case) by assigning a number (the field's strength) to each point of space. Because there are infinitely many points of space, a field enjoys infinitely many degrees of freedom. We can still follow the Hamiltonian quantization recipe to quantize a classical field theory. The result is a QFT—but not a unique one. We have moved outside the scope of the Stone-von Neumann theorem, and there are (shockingly!) infinitely many apparently physically distinct ways to construct quantizations of a given classical field theory. Different quantizations can differ on such physically basic questions as whether there are particles at all, and if there are, whether it's possible to have only finitely many of them. In the case of a theory of ordinary QM, we at least know what vector space structure that theory has. (We just don't know how to make sense of it!) In the case of a QFT, there are infinitely many rival vector space structures, keyed to infinitely many distinct representations of the CCRs constituting the theory, that seem equally qualified to serve as that QFT. This circumstance calls for some reflection—about what quantum theories are, about what criteria of identity they obey, about what it really takes to be a quantum state or a quantum property—as well as about how to frame and adjudicate answers to questions such as the foregoing.

Two broad strategies of response to the non-uniqueness suggest themselves immediately. The *privileging* strategy is to identify the theory with a unique physically significant vector space representation of the CCRs, and consign rival representations to the dustbin of mathematical artifacts. Ascending a level of abstraction, the *abstraction* strategy identifies the theory with features all representations of the CCRs share—thereby consigning features parochial to particular representations to the dustbin of physically superfluous structure. Much of Ruetsche (2011) is devoted to examining uses to which theories of QM_∞ are put, in the hopes that a winning interpretive strategy, a strategy that makes the most sense of the most uses, will emerge. I think that what makes the book interesting (if anything does) is that these hopes are dashed. Theories of QM_∞ are used in many contexts—particle physics, cosmology, black hole thermodynamics, solid state physics, homely statistical physics—, and with many aims—to model, explain, predict, and serve as launching pads for the development of future physics. An interpretive strategy that secures one aim in one context may frustrate another aim in another—or even in the same—context.

The privileging strategy has worked capitally for standard particle physics, which privileges a representation by requiring obedience to the symmetries of a particularly simple spacetime (Minkowski spacetime); the representation privileged anchors a fundamental particle notion. Still, there are aspects of standard particle physics—for instance, the "soft photons" involved in certain scattering experiments—that can't be modeled in the privileged representation, but can be modeled by discarded representations. And some explanatory agendas involving particles exceed the confines of a single privileged representation: accounts of cosmological particle creation, for instance, appeal to different (and rival) representations, privileged at different epochs in the history of the cosmos. What's more, QM_∞ abounds in other explanatory agendas—symmetry breaking, ferromagnetism, superconductivity, the dynamics of an expanding universe—that invest a variety of representations with physical significance. These explanations would be hamstrung by the privileging strategy. The abstraction strategy lends aid and comfort to some of these agendas. But not all of them. For instance, among the "surplus" properties the abstraction strategy consigns to physical irrelevance are the order parameters that distinguish between the distinct phases in a phase transition, as well as the properties that enable us to makes sense of the dynamics of mean field models. There are worthwhile physical projects promoted by each strategy, worthwhile projects frustrated by each strategy, and worthwhile physical projects frustrated by both strategies.

A winning strategy for interpreting QM_∞ has failed to emerge. Does it follow that we don't understand QM_∞? On the contrary, or so I would contend. Noticing the failure—noticing that equipping a theory of QM_∞ with constitutive CCRs leaves open a host of interpretive questions, questions which can be and in practice are answered in different ways in different contexts of aim and application— *is* understanding QM_∞.

The book concludes with a suggestion and a guess. The suggestion is that the sort of semantic indecision that enables QM_∞ to admit a variety of interpretations is an underappreciated scientific virtue: a resource of constrained adaptability that enables QM_∞ to compete in the scientific jungle red in tooth and claw. It's a resource that equips QM_∞ to meet the demands, many and varied, a living scientific theory faces. The guess—and it is really a guess, one that can be falsified by the future of science—is that because semantic indecision is a scientific resource, successor theories will share with QM_∞ the feature that no single interpretation emerges as the best. Semantic indecision isn't a passing frailty of present science but a critical strength of science as humans practice it.

As distant as the foundations of QM_∞ may seem from the natural habitat of philosophical feminism, I take the book to support the theme that messy contingent contextual considerations matter. Their mattering can be made visible through inquiry that begins with questions ("What does QFT say the world is like?") and with methods (representation theory for abstract algebras) that many believe to be insulated, even detached, from broader contextual considerations. I am *not* suggesting that work on the interpretation of QM_∞ affords an understanding of, much less a means to combat, gender oppression. That is, I am not suggesting that it is fully realized feminist work. But I am claiming that from technical work on the interpretation of QM_∞ there can emerge a methodological commitment essential to philosophical feminism: the commitment to the philosophical significance of context, including in arenas where the philosophical tradition has sought to operate without appeal to contextual considerations.

The species of this genus of commitment distinctive of feminism uses the analytical resource of gender. And the reader will no doubt have noticed that I haven't named *gender* as one of the contextual contingencies that matter to the content of QM_∞! The truth is that I don't know how to make that case. (One way might be to contend that both privileging and abstraction are somehow inherently gendered—only I'm not moved by such wholesale suspicions, and indeed I surmise that suspicions of those sorts lend credibility to the incompatibility and hostility theses!) But I do know that to make it at all requires acknowledging that context matters.

Technical work in philosophy of physics can challenge what I've called the ideal of pristine interpretation. This is the assumption that everything that's physically possible is physically possible in the same way. The assumption has the methodological consequence that interpreting QM_∞ is the pristine business of bringing metaphysical scruples and technical/mathematical sophistication to bear on the theory's laws. One can imagine related ideals: the ideals of pristine epistemology (perhaps expressed by Earman on his Bayesian days), or pristine moral theory (perhaps expressed by the proposal that justice be understood in terms of agreements rational agents would reach behind veils of ignorance). It's a non-trivial philosophical claim that an interpretation (epistemology, moral theory) conforming to the pristine ideal is adequate. If such ideals *do* underwrite adequate accounts, feminist philosophical positions—positions informed by and attentive to unpristine considerations including contingent accidents like gender and its structure— are at best optional and at worst false. One way to be hostile to feminist theses is to treat as unquestionable the assumption that these ideals underwrite adequate accounts.

Being a philosopher of QM_∞, one enlightened enough to take seriously the suggestion that the ideal of pristine interpretation hampers attempts to make sense of physical theories, can be good practice for taking feminist philosophical theses seriously.

Acknowledgments

For feedback on earlier versions, I am grateful to audiences at UCLA and the University of Pittsburgh; to Jingyi Wu, Rehan Alhas, and other members of a reading group at the Munich Center for Mathematical Philosophy; to Doreen Fraser and the FemPhys Club at the University of Waterloo; and to the editors of this volume.

Related chapters: 5, 7, 31.

Notes

1 My discussion later of "Wylie Theorems" offers a toy example of technical work that advances feminist theses. Real examples of such work abound, including Bright, Malinsky, and Thompson (2016), which harnesses the apparatus of causal Bayes nets to express empirical theses about intersectionality; O'Connor and Bruner (2019), which harnesses evolutionary game theory to understand how unbalanced demographics lead to patterns of collaboration and recognition that disadvantage minorities; Hancox-Li and Hancox-Li (ms), which uses the notion of a Reimann surface to articulate significant features of a duly ameliorative concept of gender. Despite these and other examples, in-the-field suspicion of incompatibility persists.

2 An oversimplification—the court punted on the question of racial discrimination on the grounds that the question was the subject of another, ongoing, case that the plaintiffs could join.

References

Bright, L.K., Malinsky, D. and Thompson, M. (2016) "Causally Interpreting Intersectionality Theory," *Philosophy of Science*, *83*(1), pp.60–81.

Crenshaw, K. (1989) "Demarginalizing the Intersection of Race and Sex: A Black Feminist Critique of Antidiscrimination Doctrine, Feminist Theory and Antiracist Politics," *University of Chicago Legal Forum*, p.139.

Davis, E. (2018) "On Epistemic Appropriation," *Ethics*, *128*(4), pp.702–727.

Dotson, K. (2011) "Tracking Epistemic Violence, Tracking Practices of Silencing," *Hypatia*, *26*(2), pp.236–257.

Earman, J. (1992) *Bayes or Bust? A Critical Examination of Bayesian Confirmation Theory*, Cambridge: MIT Press.

Gero, J.M. (1983) "Gender Bias in Archaeology: A Cross-cultural Perspective," *The Socio-politics of Archaeology*, ed. Gero, David M. Lacy, and Michael L. Blakey (Research Reports, 23) (Amherst: Dept. Anthropology, Univ. Massachusetts), pp. 51–7.

Hacking, I. (1999) *The Social Construction of What?*, Cambridge: Harvard University Press.

Hancox-Li, L. and Hancox-Li, S. (ms), "Gender is a Riemann Surface."

Harding, S.G. (1986) *The Science Question in Feminism*, Ithaca: Cornell University Press.

Hegselmann, R. and Krause, U. (2006) "Truth and Cognitive Division of Labor: First Steps Towards a Computer Aided Social Epistemology," *Journal of Artificial Societies and Social Simulation*, *9*(3), p.10.

O'Connor, C. and Bruner, J. (2019) "Dynamics and Diversity in Epistemic Communities," *Erkenntnis*, *84*(1), pp. 101–119.

Paxton, M. (2015) "Potential Sources of Gender Disparities in Philosophy: An Examination of Philosophical Subfields and Their Relation to Gender," www.apaonline.org/resource/group/bf785b0d-eb59-41f8-9436-1c9c26f50f8e/Paxton_ Gender_Disparities_i.pdf

Ruetsche, L. (2011) *Interpreting Quantum Theories: The Art of the Possible*, Oxford: Oxford University Press.

Wylie, A. (1997) "The Engendering of Archaeology Refiguring Feminist Science Studies," *Osiris*, *12*, pp.80–99.

31
INCLUSIVITY IN ENGINEERING EDUCATION

Donna Riley

Introduction: Framing the Conversation

Diversity, inclusion, and equity in engineering have lagged behind other STEM fields in terms of observable outcomes. In the United States, the National Science Board (2007: 5) observed that federal investments had "not led to systematic change in perceptions and retention of engineers." By conventional measures, the proportion of women, minorities, and people with disabilities graduating with bachelor's degrees in engineering has either not increased, or not kept up with changing population demographics (National Science Board 2018). It would appear that systems of inequality are durably resistant to significant investment of time and effort across a range of strategies from student support, to research, to institutional transformation.

While certain aspects of diversity climates have changed over time, reports of hostile climates pushing women out of engineering persist (AAUW 2010; Fouad 2014). The number of white females on engineering faculties has increased, but the progress for underrepresented minority women and men is slower (National Science Board 2018). We have barely begun to track other aspects of diversity such as sexual orientation, gender identity, first generation college status, or income. Categories of difference are still predominantly viewed individually with little understanding of intersectional realities, or of the multitude of differences in experience contained within category labels like Latinx or LGBTQ+.

It may be that change leaders have acted upon incorrect analysis, or that our change models have precluded the kinds of collective action true systemic change demands. A search of the engineering education literature indicates how few researchers name root causes of inequity operating in engineering education systems such as colonialism, racism, sexism, heterosexism, ableism, and economic inequality stemming from unfettered capitalism and neoliberal education policies. Thus, it is unclear whether or how strategies to address diversity, inclusion, or equity in engineering education and practice might effectively address these root causes.

Diversity in engineering is conventionally conceived as a collection of people ("participation," "representation," "celebration") rather than resistance to power and privilege. More recently "equity and inclusion" have been added to indicate a desire to also focus on issues such as climate or student success outcomes. Still, progress is counted primarily by demographics and to a lesser extent by changes in climate, which resist quantification and are thus epistemically vulnerable to de-prioritization. A great deal of literature in engineering education is focused on adapting individuals to hostile and inequitable settings by enhancing their self-efficacy, or, more recently, grit (Cosgrove and Blaisdell 1996; Mamaril, Usher, Li, Economy, and Kennedy 2016; Kirn and

Benson 2018). While the impetus to assist individuals faced with immediate survival needs in engineering is understandable, until the focus is on changing the settings themselves, it is not surprising that we do not see lasting transformation. Ultimately, the broadened focus on diversity, equity, and inclusion may also be insufficient if power and systemic forms of oppression are not taken into account.

The most common frame for discussing diversity in engineering education continues to be the recalcitrant pipeline metaphor, which persists despite some three decades of critique from feminist scholars. Alice Pawley and Jordana Hoegh (2011) summarize these arguments as follows: pipelines are unidirectional, orderly, and rigid, with only one entry point, masking the lived realities of women in engineering and marking different timing or paths as "leaks"; it is unclear what counts as "in" or "out" of the pipeline; leaks are assumed to be random failures, not systematic occurrences; it implies that the goal is to patch leaks, not redesign the system; the metaphor justifies a time lag in organizational change; and finally, it does not account for privilege and power operating in the system starting with K-12, including biases built into systems of meritocracy.

Susan Walden, Deborah Trytten, Randa Shehab, and Cindy Foor (2018) critique the label "underrepresented minorities" for focusing on proportional representation as the thing to be remedied, rather than inclusion, proposing instead the label "excluded identities," shifting the focus from minoritized individuals (who may not, as in the case of women, actually number less than half the population) to structures of power and forces of exclusion. Considering relationships of power and privilege might move the conversation into a justice frame. In so doing, questions of agency arise. Who has authority to define justice? Who benefits, and who bears the costs of these definitions and the actions ultimately indicated?

In order to understand and address diversity, equity, and inclusion at the level of root causes, we must consider not only ethical questions but also epistemic and pedagogical questions identifying how we can expand engineering ways of knowing and learning, and ontological questions characterizing what it would mean for engineering to *be* inclusive. Throughout this chapter I draw lessons from feminist ontology, epistemology, pedagogy, and ethics in order to create more diverse, inclusive, and equitable approaches to engineering education and practice.

Being Inclusive: Ontologies of Inclusion

Engineering ontology, for the most part, is rooted in presumptions that reality comprises only fixed biological, physical, and chemical phenomena, precluding recognition of phenomena such as gender, race, disability, or sexual orientation as socially constructed (Haslanger 1995). At the same time, engineers are creating and inhabiting virtual worlds in which gender and other identities are in fact constructed, and having to confront social ontologies (Searle 1995). Feminist studies of engineering academics and practitioners have empirically documented how engineering is gendered. Alice Pawley (2012), for example, analyzes engineering faculty narratives to demarcate gendered boundaries of the field, which place certain actors at the center, others at the margins, and others outside of engineering, broadly to the benefit of male actors in masculine spaces. Pawley's observations are bolstered by the historical analysis of Amy Sue Bix (2002) illustrating how home economics became carved out as a separate course of study for women deemed "not engineering," despite obvious parallels as female students applied principles of chemical, mechanical, or electrical engineering, with the same preparation in math and physics, to household appliances used by women, rather than to industrial equipment.

Cindey Foor and Susan Walden (2009) demonstrate through ethnographic interviews how Industrial Engineering has come to occupy a marginal, feminized space within engineering, offering a limited flexibility and inclusivity with regard to gender norms. Wendy Faulkner's workplace ethnographies (2000, 2007, 2009a, b) detail how engineering practice encodes gender through a mutually reinforcing relationship between narrowly technical engineering identities and hegemonic masculinities that emphasize hands-on orientations on the one hand, while women are more readily identified with the broader sociotechnical and people-oriented reality of engineering work. These workplace gender dynamics set up conditions in which women must continually prove that they are both "real engineers" and "real women" (Faulkner 2007).

Other accounts of gendered workplace norms come from Paulina Borsook (1996), who describes the sexism of Silicon Valley tech culture through her career as a journalist for *Wired*. Prominent females like Esther Dyson were scrutinized based on whom she was dating or what she wore, a standard not applied to male tech leaders. When Borsook spoke to her editor about this sexism, he remarked that he saw Dyson as neither male nor female; in other words, she was invisible as a woman to him. Other scholars (e.g. Seymour 1995) have observed this androgyny or invisibility of gender as a strategy some female engineers consciously take on to avoid negative or unwanted attention from male peers.

Hacker (1989) employed authoethnography to describe a rigorous weed-out culture in a deeply masculinized and militarized engineering curriculum. She observed students physically disciplining their bodies in deprivation of food, sleep, and even pleasure, directed toward success on high stakes tasks such as exams or problem sets.

Karen Tonso (1996) uses discourse analysis to explore gender norms as articulated in an engineering design class via sexual innuendo, profanity, and violent metaphors, communicating a presumption of male professional ability linked to male sexual desirability. Similarly, Akpanudo, Huff, Williams, and Godwin (2017) identify hegemonic masculine social norms operating in engineering, characterized by "emotional regulation, the exertion of significant dominance over others, and the desire to win."

Some of the descriptions rendered by Tonso and Hacker in particular note incidents of violence embedded in engineering; indeed, sexual and gender-based violence are prevalent across STEM disciplines (Clancy, Nelson, Rutherford, and Hinde, 2014). The dominant expected response is to ignore, grin, and bear this violence to develop a trait once characterized as persistence, but now rebranded or "improved" as grit (Slaton, Cech, and Riley 2019).

While engineering is a cis-masculine profession, it is simultaneously also a white (Slaton 2010), straight (Cech and Waidzunas 2011), able-bodied (Svyantek 2016), professional class (Smith and Lucena 2015) activity. Theorists of intersectionality (Anzaldúa 1987; Crenshaw 1989; Hill Collins 1998) have shown how these effects are not merely additive, requiring focused study of intersectional experience. Some of the work referenced earlier explicitly considers this multiplicity, while for others it is implicit, or even rendered invisible as whiteness, straightness, able-bodiedness, and middle-class existence are the presumed normative state of being (Hammonds 1994). Engineering education research is beginning to narrate intersectional experiences, particularly for women of color, but also considering intersectionality in the lived reality of LGTBQ students (e.g. Camacho and Lord 2013; Leyva 2016; Revelo, Mejia and Villanueva 2017; Cech and Rothwell 2018). Some have raised concerns with ontological slippage that has emerged in some treatments of intersectionality (Carbin and Edenheim, 2013); a construct emerging from structuralist Black Feminist Thought has found application in poststructuralist as well as liberal feminist settings, at times without a careful unpacking of ontological presumptions and without challenging the tensions inherent in operationalizing intersectionality in both a structural sense and at an individual/agential level. Slaton and Pawley (2018: 147) also note this problematic, stating that "we are a long way from being sure that these two analytic objects – individual and collective experiences of identity – even represent two distinct ontological categories." They propose using queer theory to embrace these incompatibilities with critical reflexivity, allowing ontology to inform epistemology, acknowledging that, "if it does not question our sense of what counts as progress, or fails to treat privileged researchers' ideas of progress as themselves continuously productive of privilege, research is unlikely to produce social change" (148).

Having reviewed critiques of gender applied to engineering, let us turn now to how we might apply feminist ontologies that challenge gender (and other) binaries (Butler 1990; Haslanger and Ásta 2017) to generate an ontologically feminist engineering.

Wendy Faulkner (2007) envisions a world in which the sociotechnical reality of engineering is acknowledged and accepted by engineers, dispensing with the current state of affairs in which elevation of a masculinist, purely technical, "real engineering" strains gender relations in the engineering workplace. Pat Treusch (2017) and Laura Forlano (2017), in two separate pieces in a special issue of *Catalyst* on re-making feminist science, posit new human-machine relations that enact (and ontologically engineer) feminist futures. Treusch (2017) seeks to account for affective labor and co-production between humans and machines in the work of robotics development, queering capitalist notions of automation into a process that reflectively and consciously constructs human-machine relations. Forlano (2017) critically reflects on her own use of smart technologies to manage diabetes as an intimate human-machine relationship, from which a lived experience of feminist data practice emerges, contributing to new understandings of crip time specifically and ontologies of difference more broadly. These works inhabit new ways of *being* engineering.

Expanding the engineering imaginary can help us conjure new ways that engineering can be inclusive. Pawley (2017) posits a reality in which diversity is the normative state of existence in engineering education, while Slaton et al. (2019) suggest that sometimes the best action in the interest of inclusion is *not* to engineer. Inclusive engineering is not only an ontological project but must also simultaneously entail knowing and doing. Thus, we turn now to the potentialities of inclusive epistemologies in engineering, and to developing a more inclusive engineering ethics and inclusive pedagogical practices.

Knowing Inclusion: Inclusive Epistemologies

Engineering's epistemologies are traditionally based narrowly on that of the physical sciences, and thus are subject to critiques offered by feminist philosophers of science. Central to these critiques is science's (or engineering's) presumed objectivity, characterized by reductionism and abstraction, logical positivism, and a "view from nowhere" that evacuates emotion and normative values and cleanly separates knowers from their subjects (Keller 1985; Longino; Haraway 1991; Harding 1991). However, feminist ontologies would require that some realities are socially constructed, and not accounting for this possibility causes scientists to mistakenly naturalize gendered or raced power structures (e.g. Fausto-Sterling 1986; Proctor 2006). Harding (2006: 125) questions the universalism of masculinist epistemology in science for its epistemic hubris, observing that "the ideal of one true science obscures the fact that any system of knowledge will generate systematic patterns of ignorance as well as of knowledge." These patterns, Harding notes, privilege male perspectives and interests, as well as white perspectives and interests, and colonial interests of the global North.

Within engineering education, Alicia Waller (2006) has noted that research projects in engineering disciplines are predominately rooted in positivist assumptions about knowledge, and seeks to make these epistemic assumptions visible to colleagues. Montfort, Brown, and Shinew (2014), in a study of civil engineering professors' personal epistemologies, identify ontologies that affirm the existence of objective reality, the truth of which can be ascertained through empirical observation, even as some affordances were made for the complexity of the world.

Montfort et al. (2014) additionally identified engineering's instrumentalist leanings in civil engineering professors' personal epistemologies, valuing knowledge for its truth as well as its practical usefulness. Slaton (2012: 7) notes that such instrumentality systematically displaces attention from the actors involved: those who bring about engineering (engineers and their patrons and clients) and those who live with its benefits (material comfort and well-being, profit, lucrative and secure employment, etc.) and costs (to health, safety, national and personal security, labor conditions, etc.).

This raises the question of epistemic justice (Fricker 2007): who has the authority to know and be believed, and the unequal distribution of knowledge and expertise in society. In the engineering setting, Bauschpies, Douglas, Holbrook, Lambrinidou, and Lewis (2018) and Riley and Lambrinidou (2015) further a notion of epistemic humility for engineers rooted in listening. Gwen Ottinger (2017) explores the hermeneutic injustices experienced by communities located near energy facilities struggling to gain the knowledge resources necessary to collect, analyze, and interpret air quality data. She explores the promises and limitations of using storytelling as an alternative resource for meaning making related to air quality. These narratives provide a more holistic accounting of harms to health and the disrespect encountered by communities than methods afforded by strictly scientific frameworks. Likewise, there is much more work to be done to decolonize engineering knowledge and epistemology, and recognize indigenous knowledge and ways of knowing (Leydens, Morgan, and Lucena 2017; Hughes, Prpic, Goldfinch, and Kennedy 2018; Kutay and Leigh 2018; Turner, Burt, and Mann 2018).

Inclusive epistemologies recognize relations of power embedded in knowledge processes. Amy Slaton's (2010) history of racial inequality in American engineering education reveals the close relationship between conceptualizations of engineering's disciplinary rigor and ideological commitments to meritocracy. She shows how the two work together to create standards that, while appearing objective or value-neutral, in fact have served to limit access to the profession for African Americans in particular. Even where educational administrators were well-meaning (or convinced by workforce needs) and sought to improve access for African Americans, a commitment to "high standards" and concerns about "diminished rigor" led to the perpetuation of underrepresentation of African Americans in the field. Building on Slaton's critique, Riley (2017) in her send-up of rigor in engineering education research seeks to address head-on (pun intended) the phallic implications of rigor, and its functions as an epistemic gatekeeper and disciplinarian.

Erin Cech (2013) shows how the technical–social dualism in engineering forms the basis of an ideology of depoliticization that renders concerns about equity or justice as lying beyond the field's scope. She further argues that meritocratic thinking naturalizes inequities as the logical result of a rewards system assumed to operate on a level playing field, privileging only those who deserve it due to talent or effort. These inequities, then, need no correction, stymieing efforts toward diversity in engineering or broader engagement of the profession with social issues.

The knowledge gaps produced by these epistemic commitments suggest that epistemologies of ignorance (Tuana 2006) hold promise for explaining the recalcitrance of engineering to recognize the social as part of its purview, to acknowledge the political interests inherent in the profession and its applications, or to deal directly with inequity in engineering education and the workplace. Riley (2019) has employed epistemologies of ignorance to explore the denial of gender-based violence in STEM and the systematic exclusion of sexual violence as an explanatory factor in women's underrepresentation in engineering.

Acknowledging situated knowledges shaped by the positionality of the knower (Haraway 1988) is one avenue for approaching inclusive engineering epistemologies. Keller's (1985) "dynamic objectivity" eschews value neutrality and detachment, centering one's relationship to the subject of study, while Harding's (1993) "strong objectivity" calls for a reflexivity that is transparent about the social relations of knowledge: its biases, assumptions, and interests. While representations will be necessarily partial truths, by triangulating among multiple representations one might build a stronger, more valid representation than any single representation alone.

Helen Longino (1990) also uses multiple accounts to arrive at a feminist rendering of objectivity, but her process is one of democratic discussion that does not rely on standpoint epistemology. Through a cooperative process of public participation, criticism and evaluation, and response to criticisms, more objective knowledge may result. This is predicated on all participants having equal epistemic authority. The presumption of equal epistemic authority is not easily achieved, as recognized by Patti Lather (2007) in *Getting Lost*. Among other explorations, she considers the loss of authority and expertise as an epistemic method that seeks to reckon honestly with the power/ knowledge relations inherent in research. The practice of diffraction (Haraway, 1992; Barad, 2007) and Fortun and Bernstein's (1998) reality both seek to represent subjective understandings along with their twists, turns, gaps, slippages, and unknowings.

Doing Inclusion: Inclusive Ethics

Riley (2013) draws on the work of feminist ethicists to present possibilities for feminist approaches to engineering ethics, beginning with Margaret Urban Walker's (1989) thematic analysis of failures of masculinist ethics. As with feminist ontology and epistemology, feminist ethics finds fault in an abstract, decontextualized, and depersonalized positionality in masculinist ethics, arguing that feminist ethics must be relational, and contexualized through the use of narrative. As with scientific knowledge, feminists caution against universals that set aside people and their relationships.

As with feminist approaches to scientific knowledge, relational approaches get messy, and Walker notes that this results in the acknowledgment of "moral remainders" – ethical decision-making is imperfect and there are inevitable consequences to relationships. In this way, communication and iteration become crucial elements of ethical action. Feminist ethicists have thus developed approaches rooted in care (Tronto, 1987; Warren 2000) and relational re-conceptualizations of justice (Young 1990; Jaggar 2009).

Feminist ethical practices constructed as engineering processes of iterative design (Whitbeck 1998; Pantazidou and Nair 1999) can open a space for re-imagining, re-thinking, re-making, and re-doing engineering (Riley and Longmaid 2013). Engineering and ethics are both normative activities, ways of seeking the world as it could be. Exploring the following questions posed by feminist ethicists can, in turn, generate new forms of inclusive engineering ethics.

Who is a moral agent? Who has power to act, on whose behalf? Questions of power, agency, and identity help us imagine an engineering accessible to marginalized communities, where problems are identified and defined not by disconnected experts or prevailing interests but by those whose lives are most impacted (Narayan and Harding 1998). How can attention to power and agency in design produce different engineering processes, opportunities, and results? What structures facilitate or inhibit agency, and how can these be better designed? Caroline Whitbeck (1995) transformed engineering ethics case studies in order to emphasize students as engaged moral agents rather than passive judges. Bauschpies et al. (2018) re-imagine an engaged, community-led engineering that dismantles "the flaw of the awe" – the unjust power relations and community disempowerment that extends from an un-listening wielding of engineering expertise.

What new knowledge emerges from lived experience? As feminist ethics has emerged from women's lives where intersectional specificities matter and resist universalization as both unattainable and undesirable (Walker 1989), an inclusive engineering will attend to specific persons rather than abstractions. To shape a relational practice of engineering, *what narratives can we construct that provide new contexts for engineering? What new forms of communication come into being and serve as new methods and methodologies of engineering?* Riley and Lambrinidou (2015) engaged in a thought experiment in which engineering's ethics canons were arranged as if people mattered. Using Social Work's ethical commitments as a guide, the new ethics canon centered on helping others, structurally addressing injustice, honoring all people, centering human relationships, and seeking peace.

How can alternative epistemologies of feminist ethics construct un-reasonable or empathic engineering? Feminist ethics claims a role for affect and emotion, ways of knowing that are outside or beyond objectivity. This begs a question in response from traditional engineers: can the bridge still stand up, or the plane stay in the sky (if we would still engineer bridges or planes), without universals and absolutes, without depersonalization and abstraction? Feminist ethics deals with the complexities of imperfect solutions, with moral remainders and associated relational fallout. How would using feminist philosophers' alternate constructions of validity or robustness (e.g. Haraway, Harding, Lather) change engineering? Slaton et al. (2019) posit a queering of engineering's desires, bending away from gritty capitalist strivings and embracing the strange.

Who benefits, and who loses (and who decides)? Who cares for and about whom? Feminist ethics has been a generative site for doing ethics differently, offering both ethics of care and feminist conceptions of justice as alternatives to masculinist ethics. How does attention to these central questions reshape the content and boundaries of engineering knowledge and practice? These questions have been well explored by engineering colleagues, with engineering design processes reconceived along lines of care and empathy, with worked and lived examples (Pantazidou and Nair 1999; Hess, Beever, Strobel, and Brightman 2017; Nair and Bulleit 2019).

Similarly, the Engineering, Social Justice, and Peace community has sought to envision new realities for engineering in a justice frame (e.g. Catalano 2006; Riley 2008; Baillie and Catalano 2009; Leydens and Lucena 2017; Siller and Johnson 2018), responding to previous critiques identifying engineering's cultures of depoliticization and meritocracy (Cech 2013), or the field's pervasive militaristic foundations (Nieusma and Blue 2012).

Learning Inclusion: Inclusive Pedagogies

Enacting inclusion in the classroom requires consideration of ontology, epistemology, and ethics. It begins with the recognition that we are already teaching diversity, inclusion, and equity in engineering classrooms – and most typically we are teaching a null curriculum in which the absence of diversity is tacitly permitted, sending the message that not only is diversity unwelcome but also that engineering classrooms and the engineering discipline are not safe spaces to even raise the issue for discussion.

It is past time for diversity, equity, and inclusion to be an integral part of engineering pedagogy and curriculum. And yet, engineering is still in a place where diversity education is not valued. Worse than dismissing it as a "soft skill," like communication, teaming, or ethics, diversity and inclusion literally have no place in engineering curricula today. Finding commitment to make room in the curriculum for the kinds of sustained experiences that transform hearts and minds is a long way off yet.

From a pragmatic standpoint, one would make any and every case for this transformation: the business case, national competitiveness, changing demographics, legal requirements, ethics, accreditation, university mission, basic human decency, and social justice (Riley 2018). For decades, pragmatic feminist scholars have sought to create readily adoptable and easily assimilated strategies for inclusive engineering teaching and learning (e.g. Rosser 1997; Schiebinger 2008). While these are places to start, the transformation required is much more fundamental.

Pedagogically, engineering operates very much in the "banking model" critiqued by Freire (1970). Curricularly, it does not recognize it has a canon, let alone being open to a critique of it as privileging Western and male civilization (hooks 1994; Riley 2003). Engineering is ahistorical; teaching engineering faculty to reckon with past injustices of, say, colonialism, and engineering's role in that, is both necessary and a tall order. But these conversations about power and privilege are essential if engineering education is ever to become inclusive. Significant transformative experiences with faculty would be necessary to create the circumstances under which comprehensive diversity education could be undertaken for all students. Merely working diversity education in on top of or alongside engineering knowledge, or implementing it using traditional banking model pedagogies will not work.

Still, there is a movement to bring liberative (e.g. critical, feminist, anti-racist, decolonizing, queer, crip, culturally sustaining) pedagogies into engineering. While there are many nuances and differences within this family of pedagogies, in engineering education they currently enjoy mutual support toward common goals: deliberately cultivating a learning environment where all students are treated equitably, have equal access to learning, and feel welcome, valued, and supported in their learning; attending to student identities and seeking to change the ways systemic inequities shape dynamics in teaching-learning spaces; and incorporating a critical perspective on power relations, encouraging reflective action and holistic selves.

For example, Riley, Pawley, Tucker, and Catalano (2009) provide some seeds of ideas of how engineering education might need to *be* in order to prepare engineers to enact new relational realities in the classroom. Kehdinga Formunyam (2017) has begun a conversation envisioning a decolonized engineering education curriculum in South Africa. Oglala Lakota College has instilled the Lakota value of non-abandonment of students in its engineering education practice (LaGarry, Tinant, Higa, and Sandoval, 2014). Hoople, Mejia, Chen, and Lord (2018) are implementing culturally sustaining pedagogies in a transdisciplinary course on energy engineering. Villanueva and colleagues (2018) are identifying the characteristics of engineering's hidden curriculum in order to design effective educational interventions supporting diversity, inclusion, and equity.

Just as inclusive pedagogy and curriculum in engineering weave together considerations of ontology, epistemology, and ethics, so too do objections to this work. On an epistemic level, we must counter presumptions of engineering's objectivity by asking "whose engineering are we teaching?" and pointing out the ways in which engineering has supported regimes of inequality. We might remind colleagues that our teaching is already political, as is our silence. Pragmatic objections to the "crowded curriculum" drive at our ethics: what do we value? Is diversity, equity, and inclusion important enough for engineering as a discipline and a profession? And finally, ontologically, we come up against the perception that this just isn't what engineering is. This is especially difficult to counter when colleagues are unable to see the ways in which these ontologies are replicated through engineering's power/knowledge structures.

Conclusion

This chapter has explored what might be entailed in inclusive engineering education, reviewing critiques and possible transformations in engineering's ways of being, knowing, doing, and learning. Among these, it is perhaps the ontological questions that prove most challenging: can there ever even *be* a feminist, anti-racist, decolonizing engineering? When engineering begins to adopt feminist ways of being, knowing, doing, or learning, it often ceases to be recognized *as* engineering. For there to be an inclusive engineering education, engineering must be (or become) the thing it cannot be. Resolving this contradiction is ultimately a relational project, taking care and caring for others in co-constructing new forms of knowledge, practice, learning, and existence.

Related chapters: 5, 7, 15, 16, 18, 19, 30, 33.

References

Akpanudo, U. M., Huff, J. L., Williams, J.K., and Godwin, A. (2017) "Hidden in Plain Sight: Masculine Social Norms in Engineering Education," *IEEE Frontiers in Education Conference (FIE)*, Indianapolis, IN, pp. 1–5.

American Association of University Women (AAUW) (2010) *Why So Few: Women in Science, Technology, Engineering, and Mathematics*, Washington, DC: AAUW. https://www.aauw.org/research/why-so-few/

Anzaldúa, G. (1987) *Borderlands/La Frontera: The New Mestiza*, San Francisco, CA: Aunt Lute Books.

Baillie, C. and Catalano, G. D. (2009) *Engineering and Society: Working Toward Social Justice*, San Rafael, CA: Morgan and Claypool.

Barad, K. (2007) *Meeting the Universe Halfway: Quantum Physics and the Entanglement of Matter and Meaning*, Durham: Duke University Press.

Bauschpies, W., Douglas, E. P., Holbrook, J. B., Lambrinidou, Y., and Lewis, E. Y. (2018) "Reimagining Ethics Education for Peace Engineering," WEEF-GEDC, Albuquerque, NM. https://weef-gedc2018. org/wp-content/uploads/2018/11/47_Reimagining-Ethics-Education-for-Peace-Engineering.pdf.

Bix, A. (2002) "Equipped for Life: Gendered Technical Training and Consumerism in Home Economics, 1920–1980," *Technology and Culture 43*, pp. 728–54.

Borsook, P. (1996) "Memoirs of a Token," in Cherney, L. and Weise, E. R. (eds.) *Wired_Women: Gender and New Realities in Cyberspace*, Seattle, WA: Seal Press, pp. 24–41.

Butler, J. P. (1990) *Gender Trouble: Feminism and the Subversion of Identity*, New York: Routledge.

Camacho, M. M. and Lord, S. M. (2013) *The Borderlands of Education: Latinas in Engineering*, Boulder, CO: Lexington Books.

Carbin, M. and Edenheim, S. (2013) "The Intersectional Turn in Feminist Theory: A Dream of a Common Language?" *European Journal of Women's Studies 20* (3), pp. 233–48.

Catalano, G. D. (2006) *Engineering Ethics: Peace, Justice, and the Earth*, San Rafael, CA: Morgan and Claypool.

Cech, E. A. (2013) "The (Mis)Framing of Social Justice: Why Meritocracy and Depoliticization Hinder Engineers' Ability to Think about Social Injustices," in Lucena, J. (ed.) *Engineering Education for Social Justice: Critical Explorations and Opportunities*, New York: Springer, pp. 67–84.

Cech, E. A., and Rothwell, W. (2018) "LGBTQ Inequality in Engineering Education," *Journal of Engineering Education 107*(4), pp. 583–610.

Cech, E. A. and Waidzunas, T. J. (2011) "Navigating the Heteronormativity of Engineering: the Experiences of Lesbian, Gay, and Bisexual Students," *Engineering Studies 3*, pp. 1, 1–24. doi: 10.1080/19378629.2010.545065.

Clancy, K. B. H., Nelson, R. G., Rutherford, J. N., Hinde, K. (2014) "Survey of Academic Field Experiences (SAFE): Trainees Report Harassment and Assault," *PLoS ONE 9*(7): e102172. https://doi. org/10.1371/journal.pone.0102172.

Cosgrove, C. R., and Blaisdell, S. L. (1996) "A Theoretical Basis for Recruitment and Retention Interventions for Women in Engineering," American Society for Engineering Education Annual Conference, Washington, DC. https://peer.asee.org/6348.

Crenshaw, K. (1989) "Demarginalizing the Intersection of Race and Sex: A Black Feminist Critique of Antidiscrimination Doctrine, Feminist Theory, and Antiracist Politics," *University of Chicago Legal Forum 1989*, pp. 139–68.

Faulkner, W. (2000) "Dualisms, Hierarchies and Gender in Engineering," *Social Studies of Science 30*, pp. 759–92.

Faulkner, W. (2007) "'Nuts and Bolts and People' Gender-Troubled Engineering Identities," *Social Studies of Science 37*, pp. 331–56.

Faulkner, W. (2009a) "Doing Gender in Engineering Workplace Cultures. I. Observations from the Field," *Engineering Studies 1*, pp. 3–18.

Faulkner, W. (2009b) "Doing Gender in Engineering Workplace Cultures. II. Gender In/authenticity and the in/visibility Paradox," *Engineering Studies 1*, pp. 169–89.

Fausto-Sterling, A. (1986) *Myths of Gender: Biological Theories about Women and Men*, New York: Basic Books.

Foor, C. E., and Walden. S.E. (2009) "'Imaginary Engineering' or 'Re-imagined Engineering': Negotiating Gendered Identities in the Borderland of a College of Engineering," *NWSA Journal, 21*, pp. 41–64.

Forlano, L. (2017) "Data Rituals in Intimate Infrastructures: Crip Time and the Disabled Cyborg Body as an Epistemic Site of Feminist Science," *Catalyst: Feminism, Theory, Technoscience 3*(2), pp. 1–28. https:// catalystjournal.org/index.php/catalyst/article/view/28843/pdf_13.

Formunyam, K. G. (2017) "Decolonising Teaching and Learning in Engineering Education in a South African University," *International Journal of Applied Engineering Research 12* (23), pp. 13349–58.

Fortun, M. and Bernstein, H.J. (1998) *Muddling Through: Pursuing Science and Truths in the 21st Century*, Washington, DC: Counterpoint.

Fouad, N. A. (2014) "Leaning In, but Getting Pushed Back (and Out)," American Psychological Association Annual Convention, Washington, DC. https://www.apa.org/news/press/releases/2014/08/pushed-back.pdf

Freire, P. (1970) *Pedagogy of the Oppressed*, New York: Herder & Herder.

Fricker, M. (2007) *Epistemic Injustice*, Oxford: Oxford University Press.

Hacker, S. (1989) *Pleasure, Power & Technology: Some Tales of Gender, Engineering, and the Cooperative Workplace*, Boston: Unwin Hyman.

Hammonds, E. (1994) "*Black* (W)Holes and the Geometry of Black Female Sexuality. *differences: A Journal of Feminist Cultural Studies 6*, pp. 126–45.

Haraway, D. (1988) "Situated Knowledges: The Science Question in Feminism and the Privilege of Partial Perspective," *Feminist Studies 14*, pp. 575–99.

Haraway, D. (1991) *Simians, Cyborgs, and Women: The Reinvention of Nature*, New York: Routledge.

Haraway, D. (1992) "The Promises of Monsters: A Regenerative Politics for Inappropriate/d Others," in Grossberg, L. C., Nelson, C. and Treichler, P. A. (eds.) *Cultural Studies*, New York: Routledge, pp. 295–337.

Harding, S. (1991) *Whose Science? Whose Knowledge?*, Ithaca: Cornell University Press.

Harding, S. (1993) "Rethinking Standpoint Epistemology: 'What is Strong Objectivity?'," in Alcoff, L. and Potter, E. (eds.) *Feminist Epistemologies*, New York: Routledge, pp. 49–82.

Harding, S. (2006) *Science and Social Inequality: Feminist and Postcolonial Issues*, Champaign: The University of Illinois Press.

Haslanger, S. (1995) "Ontology and Social Construction," *Philosophical Topics 23*(2), pp. 95–125.

Haslanger, S. and Ásta (2017) "Feminist Metaphysics," in Zalta, E. N. (ed.) *The Stanford Encyclopedia of Philosophy* (Fall 2017 Edition). The Metaphysics Research Lab, Center for the Study of Language and Information, Stanford University, Stanford, CA. https://plato.stanford.edu/archives/fall2017/entries/feminism-metaphysics/.

Hess, J. L., Beever, J., Strobel, J., & Brightman, A. O. (2017) "Empathic Perspective-taking and Ethical Decision-making in Engineering Ethics Education," in Michelfelder, D., Newberry, B. and Zhu, Q. (eds.) *Philosophy and Engineering: Exploring Boundaries, Expanding Connections*, Dordrecht: Springer, pp. 163–79.

Hill Collins, P. (1998) "The Social Construction of Black Feminist Thought," in Myers, K. A., Risman, B. J. and Anderson, C. D. (eds.), *Feminist Foundations: Towards Transforming Sociology*, Thousand Oaks, CA: Sage Publications, pp. 371–96.

hooks, b. (1994) *Teaching to Transgress: Education as the Practice of Freedom*, New York: Routledge.

Hoople, G. D., Mejia, J. A., Chen, D. A., Lord, S. M. (2018) *Reimagining Energy: Deconstructing Traditional Engineering Silos Using Culturally Sustaining Pedagogies*, ASEE Annual Conference, Salt Lake City, UT. https://peer.asee.org/30929.

Hughes, M., Prpic, J. K., Goldfinch, T., and Kennedy, J. (2018) "He Awa Whiria: Weaving Indigenous and Western Perspectives and Creating Inclusion in Australasian Engineering Education" Australasian Association for Engineering Education Annual Conference, Hamilton, NZ.

Jaggar, A. (2009) "The Philosophical Challenges of Global Gender Justice," *Philosophical Topics 37*(2), pp. 1–15.

Keller, E. (1985) *Reflections on Gender and Science*, New Haven: Yale University Press.

Kirn, A. and Benson, L. (2018) "Engineering Students' Perceptions of Problem Solving and Their Future," *Journal of Engineering Education 107*, pp. 87–112. doi:10.1002/jee.20190.

Kutay, C. and Leigh, E. (2018) *Validating Storytelling for Indigenous Knowledge Teaching*, Hamilton, NZ: Australasian Association for Engineering Education Annual Conference.

LaGarry, H. E., Tinant, C. J., Higa, A., and Sandoval, D. (2014) "Undergraduate research and place-based constructivism at Oglala Lakota College, Pine Ridge Reservation, South Dakota," *Geological Society of America Abstracts with Programs 46*(4):7.

Lather, Patti. (2007) *Getting Lost: Feminist Efforts Toward a Double(d) Science*, Albany: State University of New York Press.

Leyva, L. A. (2016) "An Intersectional Analysis of Latin@ College Women's Counter-stories in Mathematics," *Journal of Urban Mathematics Education 9*(2), pp. 81–121.

Leydens, J. A. and Lucena, J. C. (2017) *Engineering Justice: Transforming Engineering Education and Practice*, New York: Wiley IEEE Press.

Leydens, J. A., Morgan, T. K. K. B., and Lucena, J. C. (2017) "Mechanisms by Which Indigenous Students Achieved a Sense of Belonging and Identity in Engineering Education," ASEE Annual Conference, Columbus, OH. https://peer.asee.org/28661.

Longino, H. E. (1990) *Science as Social Knowledge: Values and Objectivity in Scientific Inquiry*, Princeton, NJ: Princeton University Press.

Mamaril, N. A., Usher, E. L., Li, C. R., Economy, D. R. and Kennedy, M. S. (2016) "Measuring Undergraduate Students' Engineering Self-Efficacy: A Validation Study," *Journal of Engineering Education, 105*, pp. 366–95. doi:10.1002/jee.20121.

Montfort, D., Brown, S., and Shinew, D. (2014) "The Personal Epistemologies of Civil Engineering Faculty," *Journal of Engineering Education 103*(3), pp. 388–416.

Nair, I. and Bulleit, W. M. (2019) "Pragmatism and Care in Engineering Ethics," *Science and Engineering Ethics*. https://doi.org/10.1007/s11948-018-0080-y

Narayan, U. and Harding, S. (1998) "Border Crossings: Multicultural and Postcolonial Feminist Challenges to Philosophy (Part I)," *Hypatia 13*(2), pp. 1–6.

National Science Board. (2007) *Moving Forward to Improve Engineering Education*, Washington, DC: National Science Foundation. https://www.nsf.gov/pubs/2007/nsb07122/nsb07122_2.pdf.

National Science Board. (2018) *Science and Engineering Indicators 2018*, Washington, DC: National Science Foundation. https://www.nsf.gov/statistics/2018/nsb20181/report.

Nieusma, D. and Blue, E. (2012) "Engineering and War," *International Journal of Engineering, Social Justice, and Peace 1*(1), pp. 50–62.

Ottinger, G. (2017) "Making Sense of Citizen Science: Stories as a Hermeneutic Resource," *Energy Research & Social Science 31*, pp. 41–9.

Pantazidou, M. and Nair, I. (1999) "Ethic of Care: Guiding Principles for Engineering Teaching and Practice," *Journal of Engineering Education 88*, pp. 205–12.

Pawley, A. L. (2012) "What Counts as 'Engineering'?: Towards a Redefinition," in Baillie, C., A Pawley, A. L., and Riley, D. (eds.) *Engineering and Social Justice: In the University and Beyond*, West Lafayette: Purdue University Press, pp. 59–85.

Pawley, A. L. (2017) "Shifting the "Default": The Case for Making Diversity the Expected Condition for Engineering Education and Making Whiteness and Maleness Visible," *Journal of Engineering Education 106*, pp. 531–33. doi:10.1002/jee.20181.

Pawley, A. L. and Hoegh, J. (2011) "Exploding Pipelines: Mythological Metaphors Structuring Diversity-oriented Engineering Education Research Agendas," ASEE Annual Conference, Vancouver, Canada.

Proctor, R. (2006) *Racial Hygiene: Medicine under the Nazis*, 2nd ed. Cambridge, MA: Harvard University Press.

Revelo, R. A., Mejia, J. A., and Villanueva, I. (2017) "Who are We? Beyond Monolithic Perspectives of Latinxs in Engineering," ASEE Annual Conference, Columbus, OH. https://peer.asee.org/29125.

Riley, D. (2003) "Employing Liberative Pedagogies in Engineering Education," *Journal of Women and Minorities in Science and Engineering 9*, pp. 137–58. https://doi.org/10.1615/jwomenminorscieneng.v9.i2.20.

Riley, D. (2008) *Engineering and Social Justice*, San Rafael, CA: Morgan and Claypool.

Riley, D. (2013) "Hidden in Plain View: Feminists Doing Engineering Ethics, Engineers Doing Feminist Ethics," *Science and Engineering Ethics 19*(1), pp. 189–206.

Riley, D. (with Emily Rider-Longmaid) (2013) "How Feminist Ethics Brings Pluralism to Engineering: The Prison-Industrial Complex and Drone Warfare," Feminist Ethics and Social Theory Conference, Tempe, AZ.

Riley, D. (2017) "Rigor/Us: Building Boundaries and Disciplining Diversity with Standards of Merit," *Engineering Studies 9*(3), pp. 249-265.

Riley, D. (2018) "Refuse, Refute, Resist: Alt-Right Attacks on Engineering and STEM Education Diversity Scholarship," CoNECD – The Collaborative Network for Engineering and Computing Diversity Conference, Crystal City, Virginia. https://peer.asee.org/29572.

Riley, D. (2019) "Pipelines, Persistence, and Perfidy: Institutional Unknowing and Betrayal Trauma in Engineering," *Feminist Formations 31* (1), pp. 1–19.

Riley, D., and Lambrinidou, Y. (2015) *Canons against Cannons? Social Justice and the Engineering Ethics Imaginary*, ASEE Annual Conference, Seattle, Washington, DC. 10.18260/p.23661

Riley, D., Pawley, A. L., Tucker, J. and Catalano, G. D. (2009) "Feminisms in Engineering Education: Transformative Possibilities," *National Women's Studies Association Journal 21*(2), pp. 21–40.

Rosser, S.V. (1997) *Re-engineering Female Friendly Science*, New York: Teacher's College Press.

Schiebinger, L. (2008) *Gendered Innovations in Science and Engineering*, Palo Alto, CA: Stanford University Press.

Searle, J. (1995) *The Construction of Social Reality*, Cambridge, MA: MIT Press.

Seymour, E. (1995) "The Loss of Women from Science, Mathematics, and Engineering Undergraduate Majors: An Explanatory Account," *Science Education 79*(4), pp. 437–73.

Siller, T. J. and Johnson, G. (2018) *Just Technology: The Quest for Cultural, Economic, Environmental, and Technical Sustainability*, San Rafael, CA: Morgan and Claypool.

Slaton, A. E. (2010) *Race, Rigor, and Selectivity in U.S. Engineering: The History of an Occupational Color Line*, Cambridge, MA: Harvard University Press.

Slaton, A. E. (2012) "The Tyranny of Outcomes: The Social Origins and Impacts of Educational Standards in American Engineering," American Society for Engineering Education Annual Conference, Austin, TX.

Slaton, A. E., Cech, E. A. and Riley, D. M. (2019) "Yearning, Personhood, and the Gritty Ontologies of American Engineering Educaton," in Fifield, S. and Letts, W. (eds.) *STEM of Desire: Queer Theories in Science Education*, Amsterdam: Brill, pp. 319–40.

Slaton, A. E. and Pawley, A. L. (2018) "The Power and Politics of Engineering Education Research Design: Saving the 'Small N'," *Engineering Studies* 10(2–3), pp. 133–57, DOI: 10.1080/19378629.2018.1550785.

Smith, J. M., & Lucena, J. C. (2015) *Making the Funds of Knowledge of Low Income, First Generation (LIFG) Students Visible and Relevant to Engineering Education*, ASEE Annual Conference & Exposition, Seattle, Washington, DC. 10.18260/p.24464.

Svyantek, M.V. (2016) "Missing from the Classroom: Current Representations of Disability in Engineering Education," ASEE Annual Conference & Exposition, New Orleans, LA. 10.18260/p.25728.

Tonso, K. L. (1996) "The Impact of Cultural Norms on Women," *Journal of Engineering Education*, *85*, pp. 217–25.

Treusch, P. (2017) "The Art of Failure in Robotics: Queering the (Un)making of Success and Failure in the Companion-Robot Laboratory," *Catalyst: Feminism, Theory, Technoscience*, *3*(2), pp. 1–27. https://catalyst-journal.org/index.php/catalyst/article/view/28846/pdf_16.

Tronto, J. C. (1987) "Beyond Gender Difference to a Theory of Care," *Signs 12*(4), pp. 644–63.

Tuana, N. (2006) "The *Speculum of Ignorance*: The Women's Health Movement and Epistemologies of *Ignorance*," *Hypatia 21*(3), pp. 1–19.

Turner, J., Burt, A. and Mann, L. (2018) "Connection to Country – Building a Connection to Indigenous Knowledges in Engineering Education," Australasian Association for Engineering Education Annual Conference, Hamilton, NZ.

Villanueva, I., Gelles, L. A., Di Stefano, M., Smith, B., Tull, R. G., Lord, S. M., Benson, L., Hunt, A. T., Riley, D. M., Ryan, G. W. (2018) "What Does Hidden Curriculum in Engineering Look Like and How Can It Be Explored?" ASEE Annual Conference & Exposition, Salt Lake City, Utah. https://peer.asee.org/31234.

Walden, S. E., Trytten, D. A., Shehab, R. L., and Foor, C. E. (2018) *Critiquing the "Underrepresented Minorities" Label*, CoNECD – The Collaborative Network for Engineering and Computing Diversity Conference, Crystal City, Virginia. https://peer.asee.org/29524.

Walker, M. U. (1989) "Moral Understandings: Alternative 'Epistemology' for a Feminist Ethic," *Hypatia 4*(2), pp. 15–28.

Waller, A. A. (2006) "Special Session – Fish is Fish: Learning to See the Sea We Swim in: Theoretical Frameworks for Education Research," Frontiers in Education Conference Proceedings, San Diego, CA.

Warren, K. (2000) *Ecofeminist Philosophy: A Western Perspective on What It Is and Why It Matters*, Lanham, MD: Rowman & Littlefield.

Whitbeck, C. (1995) "Teaching Ethics to Scientists and Engineers: Moral Agents and Moral Problems," *Science and Engineering Ethics 1*(3), pp. 299–308.

Whitbeck, C. (1998) *Ethics in Engineering Practice and Research*, New York: Cambridge University Press.

Young, I. M. (1990) *Justice and the Politics of Difference*, Princeton, NJ: Princeton University Press.

Public Policy

32

RETHINKING DEBATES ABOUT PEDIATRIC VACCINE SAFETY

A Feminist View

Maya J. Goldenberg

Introduction

This chapter frames the problem of vaccine hesitancy as a problem of public mistrust of scientific institutions and rejects the common thinking that vaccine hesitancy persists because members of the public misunderstand the science or endorse science denialism and/or anti-intellectualism. This rethinking of the supposed war on science and expertise, and how the war allegedly animates pediatric vaccine hesitancy and refusal, is informed by key themes from feminist philosophy of science about the nature and aims of science as well as unequal power relationships between experts and non-experts. These considerations resituate the problem in poor scientific governance rather than the moral and epistemic failings of the public. Vaccine outreach tends to focus on education interventions, but attention and resources need to be additionally directed toward building and maintaining public trust. This chapter will conclude with recommendations for how to build this trust.

The common framing of public controversies over scientific claims, such as the safety and efficacy of vaccines, is in combative terms. The combatants are: science vs nonsense, reason vs ignorance, expertise vs non-expertise. We frequently hear that science, rationality, and expertise, the rightful vanguards of public health and well-being, are losing the war. Furthermore, the public[1] is to blame. Despite continued science communications efforts to expose pseudoscientific claims and counter misinformation, members of the public seem all too ready to question, challenge, and ignore scientific advice.

The stakes are high in the so-called "war on science" (Otto 2016). Public rejection of scientific claims and scientific experts is viewed as deeply dangerous and threatening to democracy. The influential Bodmer Report on Public Understanding of Science (London Royal Society 1985) sounded the alarm in 1985, offering the somber assessment that:

> the nature and extent of public understanding of science and technology are limited by a general lack of scientific literacy and are inadequate for the majority of the population to play an active informed role in an advanced industrial democracy.

This concern is still expressed today, albeit the language and tone are more emotive. See, for example, the anxious titles of a fast-growing genre of popular science books dedicated to the woeful fallout of science denialism: *Unscientific America: How Scientific Illiteracy Threatens Our Future* (Mooney and Kirshenbaum 2009), *Denialism: How Irrational Thinking Harms the Planet and Threatens Our Lives* (Specter 2010), *Reality Check: How Science Deniers Threaten Our Future* (Prothero

2008), *Deadly Choices: How the Anti-Vaccine Movement Threatens Us All* (Offit 2011), and many more. It is this common linking of science denial and existential threat that led science communicator Neil deGrasse Tyson to decree that "science illiteracy is a tragedy of our time" (Davey 2015; in response, see Goldenberg 2015) and made it politically expedient for progressive politicians to declare their allegiances to science—for instance, Hillary Clinton told convention delegates "I believe in science" in her acceptance speech for the US Democratic Party's presidential candidate in 2016 (Lehmann 2016).[2] Similarly, in his 2017 inaugural agency-wide address as the new director of Center for Disease Control and Prevention (CDC), Robert Redfield Jr. became "overcome with emotion" as he promised to defend science in his new role. He mentioned vaccine refusal as a particular challenge (Sun 2018).

Yet there is good reason to challenge this dominant narrative that there is a culture war on science and expertise. The challenge draws from themes that are well-grounded in feminist thinking about science regarding the nature of science and its entangled relationship with society. Whereas feminist science studies has often been accused of attacking science, ignoring expertise, or threatening the objectivity of science, this feminist reframing of vaccine hesitancy and disavowing of the war on science will refute this misconception of feminist contributions to science and science studies. It follows feminist science studies themes of recognizing and interrogating the values and power dynamics that influence policy-relevant science with an eye toward improving scientific practices in ways that enhance the objectivity and/or the credibility of science.

The analysis begins with feminist social epistemologist Helen Longino's influential character-ization of science as *social practice*, a conceptual framework that highlights the process by which multiple stakeholders, standing in relationship to each other, co-constitute objective scientific knowledge (Longino 1990). Social understandings of science have highlighted the importance of trust and credibility in the successful operations of scientific institutions—both within re-search communities and in relation to the public (Hardwig 1991; Rolin 2002; Goldenberg 2016; Grasswick 2017). Expert knowledge is thereby legitimated as expert, or refused as such, amidst relations of unequal power and epistemic dependency. These considerations will be brought to-gether to defend the novel thesis that public resistance to scientific claims (about vaccines and other hot-button issues like GMOs, climate change, and water fluoridation) stems from a crisis of trust rather than a war on science, specifically poor public trust in scientific institutions rather than poor public understanding and appreciation of science. Poor trust, not scientific illiteracy or anti-intellectualism among the general public, leads some people to reject the strong scientific consensus on vaccines. Thinking about vaccine hesitancy as a symptom of public mistrust also redraws the lines of responsibility. Vaccine hesitancy is a problem of scientific governance rather than a problem of the public.

The Nature of Science

The divisive "us versus them" discourse of the so-called war on science is grounded in a mis-understanding of the nature of science and how it operates in society. Science, its embattled defenders assume, cuts through partisan politics and rationalizes democratic choice by informing and directing the populace. Scientific experts are indispensable to the flourishing of the polis in a constrained advisory role, as a division of cognitive and deliberative labor is thought to balance the democratic need for both informed policy and political legitimation through public partici-pation (Kappel 2014). Democracy needs science to ensure social stability, and public resistance to scientific claims destabilizes the social order. The lines of responsibility are equally clear in the war on science framework: insofar as it is the public challenging the science, the problem lies squarely with the public, while science and its institutions require little or no scrutiny. At most, one hears the tepid criticism that scientists could be better communicators. Whether it's a war on evidence, or a war on experts, public resistance to scientific claims is similarly envisioned as a high-stakes battle between established knowledge and destructive ignorance.

The difficulty with this framing as a war on science and expertise is that science, even sound science, does not work that way—cutting through politics and confusion to produce optimal policies and optimal social benefit. One can in principle agree to the scientific superiority of the majority view on vaccine safety and efficacy without supporting the presumed corollary claims that the best policies follow from the best science, and that the public are the problem. Against the "linear model" of science-to-policy (Pielke Jr. 2007), science and society are far more entangled. Underlying this relationship, like all relationships, is the matter of trust. Public acceptance or rejection of scientific claims pivots on the perceived trustworthiness of the scientific bodies underwriting those claims. Feminist social epistemologies of science offer insight into the role of trust in science.

Trust, Trust within Science, Trust in Science

Trust is heavily theorized across multiple disciplines, including moral theory, philosophy of science, and social theory, with particular attention provided by feminist theorists in these domains due to their focus on the relational aspects of morality, knowledge production, and social structures, especially relationships involving imbalances of power between participants. By highlighting how knowledge and justification occur at the community level (Nelson 1993), and thereby rely on relationships of interdependence, relations of trust are understood to have epistemic significance. The concept of trust is generally taken to mean having confidence in someone or something. Discussions about trust in the context of science usually refer to *epistemic trust*; "to invest *epistemic trust* in someone is to trust her in her capacity as a provider of information" (Wilholt 2013). Wilholt's helpful theorizing of epistemic trust adopts Baier's (1986) distinction between "mere reliance" and "trust"; epistemic trust is more than mere reliance insofar as it is the kind of dependence that makes the trusting person dependent on the trustee's good will.

Trusting others is risky, but it is also unavoidable. This tension makes trust ripe for ethical analysis. When we find ourselves in situations where we lack adequate information to know for ourselves—and this happens often!—we must trust others. Knowing the risk that this trust may be broken requires that the trustor partake in what has been described as "leaps of faith" in the sociology literature (Lewis and Weigert 1985; Mollering 2006).[3]

This "leap" refers to the necessary bridging of an information gap in situations of risk. Epistemically dependent people will fill any perceived knowledge gap with "a kind of suspension or bracketing-off of uncertainties" (Brownlie and Howson 2005; see also Luhmann 1979). The confidence with which the trustee "leaps" is captured in the expectation of the trusted expert's goodwill (or perhaps the less demanding expectation of the expert's moral integrity is enough (McLeod 2002))[4]; there is the additional expectation that the expert will be properly motivated to act (Jones 1996). Confidence hinges on the moral character, not just the epistemic credentials, of the expert. Hardwig (1985) argued that the moral legitimacy of the expert source is a prerequisite for rational epistemic dependence.

There has been qualitative research into parental decision-making regarding vaccines that has highlighted multiple "leaps of faith" taken in the face of incomplete knowledge and anxiety over future unknowns (Brownlie and Howson 2005). In Brownlie and Howson's (2005) interviews with British parents facing childhood immunization decisions in the early 2000s (the height of the media frenzy over the alleged link between the MMR vaccine and autism), some parents were able to continue, but others suspended, the previous familiar routineness of childhood immunizations. The hesitators, moreover, bracketed off the uncertainties unevenly, sometimes choosing to vaccinate one child but not another, or to choose some vaccines and skip others. The choice depended on the health history of that particular child and the timing of media reports. The dynamism of the trusting leap or refusal was made by parents carefully gathering warrants to trust or distrust the assurances made by public health agencies and medical providers that vaccines had been and continued to be safe, effective, and necessary (Goldenberg 2016).

Parents' trusting "leaps" are taken or denied based on vaccine advice from relations of familiarity such as peers, family members, and health professionals, and also with regard to perceptions of the trustworthiness of scientific bodies or institutions. Trust is thereby a means of social cohesion (Misztal 1996) through affective commitments (Jones 1996).

What is important for thinking about vaccine acceptance, hesitancy, and refusal is that it is *not* the growing mountain of data that are convincing parents to vaccinate their children, but a willing "leap" in favor of the scientific consensus.[5] Similarly, vaccine hesitators and refusers situate themselves in different spheres of familiarity that disqualify the majority view on vaccines. These calculations, these moves toward in-group belonging, are not well explained by risk assessment accounts of trust. Instead Baier (1991) seems correct to describe trust as cognitive, affective, and conative.[6]

Feminist science studies and STS scholarship strongly advance the position that trust is endemic to science—it supports knowledge creation, including the management of dissent, as well as consensus building. Yet trust *within* science communities operates invisibly: so opaque are these lines of trust that science is commonly thought to be rigorous by being *wary* of trust. Rather than listen to authority, scientific rationality allegedly demands that one examines the evidence for oneself. This popular understanding of science made STS's alternative thesis radical: trust relations make scientific progress possible. Because there is too much to know for oneself, to test all background assumptions, and to check all features of a colleague's work in large collaborative projects, trust enables much of the knowledge that we call science. In short, trust operates behind the scenes within science in establishing and legitimating knowledge as true and universal (that is, true for everyone) (Shapin 1995).

Moving from science communities to science-public relationships, epistemic dependency and trust in experts are more visible. Members of the public rely on scientific knowledge to inform everyday choices and practices. Lacking the time and skill to check each claim for ourselves, non-experts routinely look to experts for advice. When the channels of knowledge transfer, translation, and mobilization work well, this move from expert advice to non-expert action can go smoothly. Public resistance to science occurs because relations between experts and non-experts are not so secure.

Members of the public are arguably well advised to defer to scientists with relevant expertise (Hardwig 1985). Collins and Evans (2008) explain that those scientists' "interactional expertise," the expertise that arises from being part of the relevant knowledge community, makes it likely that their judgments will be better than our own. Thus, the public benefits from well-placed trust.[7] The challenge is knowing when that trust is well placed. The risk of harm remains, but non-experts can work to reduce that risk by assessing expert advice. Where non-experts are presumably unqualified to assess the content of scientific claims, we can evaluate the character of the scientific expert or the integrity of the institutions they represent. Members of the public will follow expert advice *if* those experts are trusted to be both epistemically and morally responsible.[8] The rationality of following expert advice hinges on trust and credibility: experts must be trustworthy and non-experts must recognize them as such.[9] And so, relations of trust mediate successful exchanges between scientific institutions and the public.

Because knowledge is produced in communities, disagreement between members can be expected. Science even encourages it: dissent and disagreement are seen as signs of healthy epistemic enterprise. Trust therefore does not preclude disagreement, but it can help manage it. Disagreement does not need to undercut trust either. The avenues for managing dissent and disagreement in science follow from a generally accepted democratic orientation toward truth seeking and consensus building: one that is public and accountable. Social epistemologists view these mechanisms favorably and even make recommendations to *improve* the democratic tenor of science, for example by increasing diversity in scientific communities to make dissent and criticism more robust (Keller 1985; Longino 1990, 2002; Kitcher 2011). They also recommend limiting spurious dissent that is meant to be obstructionist rather than knowledge seeking (see Solomon's three conditions for normatively appropriate dissent (2002) and De Melo Martin and Intemann's (2018) useful distinctions between helpful and harmful forms of dissent[10]). It is with these communicative practices in place that robust scientific consensus can arise on some issues, while points of disagreement can still respectfully endure without rupturing community cohesion.

The contrast between debates within scientific communities and public controversies over science is striking. In the public arena, where science-public controversy takes place, we do not find comparable shared rules for the management of disagreement and for consensus building. There are no shared frames of reference; instead we have facts, "alternative facts" (Blake 2017), and disagreements over which side wields legitimate science while the other side is "junk science."[11] Conflicts of interest are often present on both sides of the disagreement. These disputes over facts have been popularly characterized as "post-truth"—Oxford Dictionary's 2016 word of the year.[12,13] But "post truth" wrongly suggests that there was a prior time when truth was unequivocally stable. Instead, facts have always been negotiated and contested by those who see and experience the world differentially. Case studies in the history of science offered by Poovey (1998) and Shapin and Schaffer (1985) detail how the modern scientific fact came to be established—that is, how the criteria for "matters of fact" (as Shapin and Schaffer called them) were historically negotiated as a feature of the European Enlightenment. Acknowledging the social history of facts should not be misunderstood as downgrading the epistemic status of facts,[14] as there is no contradiction in saying that facts are true *and* that they are embedded in social mechanisms that establish their epistemic authority as claims that demand universal uptake. The current problem of post-truth should signal that it is these mechanisms that need attention. Re-establishing the common ground for constituting matters of fact requires trusting relationships among differently situated people rather than a powerful reinstatement of science and expertise.

It can be a hard realization that science cannot singularly guide us to good policy and right action, neutralize political partisanship, and rationalize democratic decision-making. It is shocking to many vaccine advocates that the scientific consensus on vaccines does not settle public concern. Instead the consensus gets positioned as one side of a debate, where scientific experts must jockey for legitimacy against seemingly disreputable opponents proclaiming to have science (and moral credibility) on *their* side.

We should appreciate the surprise. After all, the consensus functions to settle debate, not invite it, by representing the majority view of those most suited to pronounce on the issue. Consensus claims can also serve a public function: to educate the public on issues and promote appropriate response, whether personally or politically. The failure to achieve these aims is no doubt frustrating. Doesn't the consensus deserve more deference? The consensus claim, when done well, is the best approximation of scientific truth; it is produced by the best of science's truth-seeking procedures and practices. The universal applicability of the findings rests in the methods of consensus building. For the public to suggest impartiality is to reject an elaborate set of epistemic, methodological, and institutional mechanisms meant to ensure reliable knowledge and public benefit from that knowledge. Science isn't something you are supposed to "believe" in or be against. To say otherwise is to say that science depends on trust. But it does.

It is important to consider that much of what members of the public know about vaccines pivots on epistemic trust. Almassi (2012) has drawn the same conclusion regarding our beliefs on climate change. Tied into the consensus statement is a claim to the epistemic and moral legitimacy of its authors and their institutions. Vaccine hesitators and more strident vaccine refusers reject those claims of legitimacy.

So what is the appropriate response when the consensus does not fulfill its function of engendering public trust? Here is what is happening now: vaccine hesitators and refusers are ridiculed for raising concerns. Against the democratic tenor of science, science journalists write articles like "Why You Have No Business Challenging Scientific Experts" (Mooney 2014) to convey sincere disgust over the current state of affairs. Why, vaccine supporters ask incredulously, would one take the word of a media-savvy celebrity mom who attributed her knowledge of autism to the "University of Google" over *many* expert scientists? (van Heuvel 2013).

Consensus claims are expert-generated directives for epistemically dependent outsiders; yet the mechanisms used to ensure the trustworthiness of that information—the negotiation of conflicting views in academic conference settings and in expert journals, replication of findings, peer review, and so on—are internal to the scientific community and are therefore largely shielded from public view. Thus the final step in the expert-lay exchange, where (if all goes well) the public accepts the scientific consensus view, requires some degree of a trusting "leap of faith" that the scientific experts have done their due diligence and reported responsibly. The trust requirement places the outsider in a vulnerable position, and there is no sympathy for that predicament. The public are then implored to "*trust* science": trust in a process whose trustworthiness lies in it being shielded from public opinion or other non-expert contributions (cf. (Scheman 2001). Without an eye on, or participation in, the innermost practices of scientific knowledge and consensus building, with various threats of sanction for *not* accepting the findings, the public is instructed to trust. Some are not willing to do so.

When parents make vaccine decisions, the trusting leaps or refusals are surely influenced by the misinformation peddled on the internet (CBC Radio 2017), but those dubious claims only gain traction because they fit with a broader narrative of perilous health care. Informed news consumers are well aware of problems in health research and practice. The replication crisis, the weaknesses of the peer review system, disease mongering (Payer 1992; Moynihan and Henry 2006), and law suits against pharmaceutical companies are part of health consumers' background knowledge. They draw on these narratives when they evaluate new information about vaccine risks.

As I write this manuscript, mainstream news media are widely reporting on how deceptive sales tactics employed by a pharmaceutical company allegedly led to the current opioid crisis (Martin 2019). Just two months earlier, the International Consortium of Investigative Journalists (ICIJ) published "The Implant Files," the results of a global investigation undertaken by 250 journalists in 36 countries into lax regulation of medical devices (Shiel 2018); those devices have caused devastating harm to patients and many deaths worldwide (ICIJ 2018).[15] While these alarming issues do not tie directly to vaccines, consumers do not know whether the next scandal *will* make that connection. Faced with unknowns and uncertainty, trust and mistrust tend to travel (Grasswick 2018).

Some might object that following the consensus view is not about trust. It's a numbers game: if you follow the majority view, you are more likely to get the right advice. But the number in the numbers game is only given weight if the majority represents a convergence from multiple sources. Those numbers would not add any epistemic weight to the claim if they represented crowds of scientists slavishly repeating the same dogma. The numbers would not be directive in knowing what to believe (Goldman 2001). This caveat reveals that trust is *not* avoided by following the numbers.[16] Again, that is because those of us standing outside of the community of experts do not have a clear view of how impartially the consensus was negotiated. At best we have a partial view gained by our varying immersion in the effort to read the document, check the supporting literature, analyze media reports, and make freedom of information requests. Most of us will not dive deep, due to lack of time, skill, and energy. Instead we fill the gaps in with trusting leaps or distrusting refusal.

Both vaccine advocates and critics can claim to meet Goldman's stipulations for whether or not to follow expert advice. Vaccine advocates can make a case for going by the numbers: they can point to the robustness of the weight of evidence coming from a diversity of research teams from different countries supporting the consensus that vaccines are safe, effective, and that there is no correlation to autism or mercury poisoning.

Vaccine critics interpret the weight of evidence and the consensus differently. With strong ties to industry, the "scientific consensus" does *not* meet the requirement of plural and independent sources. If all vaccine science communicators are complicit with vaccine manufacturers, or silenced into submission by their medical boards and health authorities, then health care consumers ought not to be swayed by the numbers. The point here is that it is poor trust that keeps some of the public from accepting the consensus view, that is, from reasonably "going by the numbers." The trust issue remains even in the presence of a strong consensus.

The current climate of parental decision-making on vaccines is difficult. Many parents say they don't know which side to believe. It is due to poor trust that the institutions tasked with protecting the public good are not able to carry out their mandate by offering the definitive voice of reason. The consensus does not fulfill its public function of guiding parents regarding childhood vaccines.

Poor Public Trust: Implications for Medico-Scientific Institutions

There is general agreement that the public need science, but the point being made here is that science needs the public too. The fulfillment of many institutional mandates hinges on positive public relations. Science strives to create universally applicable knowledge, and this knowledge is universal only insofar as it is accepted by a variety stakeholders. This places a demand on scientific communities to earn and maintain the trust of the public.

Research institutes and agencies rely on stable relations with the outside, at minimum to ensure access to public research funds and to enjoy little interference with their work. When that minimal level of public trust is in place, science can operate smoothly.

In policy-relevant science—research motivated by practical goals like furthering human, animal, and environmental welfare—there are more elaborate ties to the public. These practical goals require scientific claims to be accepted by stakeholders outside of its specialized epistemic communities (Scheman 2001; Wilholt 2009; Whyte and Crease 2010). Policy-relevant science can only provide those public benefits if its institutions are regarded as trustworthy by members of the public.

Public health science, for example, can only improve population health if the general public largely accept and follow its recommendations. Health recommendations and consensus statements bank on the public's trust in these institutions' conscientious and honest efforts to inform and protect. Earning and maintaining the public trust are crucial for fulfilling public health mandates. Offering the best science and the most carefully considered action-directives is not enough. The science must be trustworthy but also trusted by all public health stakeholders. Persistent vaccine hesitancy indicates institutional failure to engender public trust. This warrants self-reflection about institutional trust-building practices.

Conclusions and How to Build Trust

The evidence most of the public accept on vaccines turns crucially on epistemic trust. It is poor trust in the expert sources that gives rise to vaccine hesitancy and refusal. Consensus claims will not convince anyone if the source(s) is not perceived as trustworthy. To confront public resistance of scientific claims, what if we focused on building that trust rather than educating the misinformed public or puzzling over their moral and epistemic failings?

Doing this does not discount that public health agencies have the science on their sides. It *does* mean that we have to recognize that the best science is not enough. This is not a war with the public or a war over science. I have offered a different picture of science in relation to the public than science as the firm anchor mooring liberal democratic political organization. Science should still be understood to hold firm ground (i.e. this is not a case for evidentiary relativism), but the idea that the *evidence speaks*, or dictates right policy, is a fiction. All evidence is subject to interpretation, and political and policy decision-making requires numerous extra-scientific considerations. The language of "evidence based" is misleading in that respect. Scientific evidence operates within a constellation of social influences that guide personal decision-making and policy formation. Good trust relations between scientific institutions and the public ensure that science stands prominently within social policy frameworks. The current tendency to criticize and publicly shame the skeptical public for failing to appreciate the primacy of scientific reasoning and the authority of experts does not address the problem of public mistrust of scientific institutions. If anything, it exacerbates the mistrust by entrenching a polarizing us vs them mentality.

Trust is built and maintained in relationships that are respectful, open, and honest (Peters, Covello, and McCallum 1997). Primary care providers need the time to respond patiently and non-judgmentally to parents' questions, and to build on shared goals like ensuring children's health and safety. Listening to parents' concerns will lead to more effective responses. For instance, the many parents who think vaccines are generally safe but may not be safe for their *own* child will not have their fears allayed when well-meaning health care providers point to the latest epidemiological study demonstrating vaccines to be safe at the population level (Goldenberg 2016). Patients also want honest information, which may require admitting to gaps in the research, for instance, regarding what causes serious adverse events. Admitting to uncertainty does not undermine trust, as patients look for providers who have their interests at heart, and communicate honestly, more than they look for unequivocal scientific pronouncements (Larson, Cooper, Eskola, Katz, and Ratzan 2011). Public health organizations can also remedy perceptions of dishonest advice by addressing seeming contradictions in public health messaging to parents. With the advancement of precision medicine comes the promise of no more "one size fits all" medicine in treatment and prevention (Health 2.0 2018). Parents see it as justified to ask why this isn't the same for vaccines. Also, breastfeeding promotion valorizes the natural and conveys the message that immunity is conferred to the child through breast milk. It is hardly surprising to hear parents disparaging vaccines as "unnatural" and unnecessary when coupled with prolonged breastfeeding strategies (Dubé et al. 2015; Reich 2016).

The best way to counter mistrust is to find and remedy its sources. Industry influence on health care is ubiquitous and harmful to public trust. This makes reconfiguring industry ties to health care both stubbornly difficulty and urgently necessary. Empirical research shows lower public trust in scientists and physicians perceived to suffer from financial conflicts of interests or loss of independence (Hargreaves et al. 2013). Feminist analyses have further shown commercialization and an inattentiveness to the needs and interests of certain groups to be damaging to public perception of science and health care (Intemann and De Melo-Martin 2014; Jukola 2019). Current mild remedies such as disclosure statements and sunshine lists are not enough to ensure the levels of public confidence needed to stave persistent vaccine hesitancy. For those who think curtailed industry ties and longer appointment times to talk to patients are impossible demands on health care systems, consider that so too is the public health burden of vaccine hesitancy and refusal.

Related chapters: 9, 12, 15, 17, 19, 33.

Notes

1 Communications studies scholars prefer the term "the publics" instead of "the public" or "public sphere" in order to disabuse the notion that there is a unified body of lay people that interact with expert science in the same way. Instead there are a plurality of non-expert modes of engaging with science. I support this pluralist interpretation but still employ the singular "public" that is familiar to philosophy audiences.
2 Clinton had likely taken cue from Barak Obama, who had promised in his 2009 inaugural speech as 44th President of the United States to "restore science to its rightful place" (Cohlan 2009).
3 This terminology can be traced back to classic contributions by Georg Simmel, and is still being used contemporaneously by Mollering (2006) and others.

4 The expectation of moral integrity is less demanding because there is no requirement that the trustee have positive feelings toward the trustor in particular. This seems better suited for trust in the collective enterprises of science or in any institution, rather than trust in a particular individual. Baier's (1986) and Jones's (1996, 1999) focus on goodwill might be because they, as moral theorists, focused on interpersonal relationships rather than relations with institutions.

5 The scientific consensus is the collective judgment or opinion of a community of scientists. While "medical consensus conferences" are no longer widely used to establish collective expert opinion, consensus claims can still be inferred from other aggregating practices used in health research such as position statements from professional groups, and meta-analyses and systematic reviews of the relevant evidence. Regarding childhood vaccines, a consensus can be inferred from the wide endorsement of vaccines by numerous professional scientific, medical, and health bodies, including the American Academy of Pediatrics, Centers for Disease Control and Prevention, National Institutions of Health, National Academy of Sciences, and World Health Organization. .

6 This "trilogy of mind" was well represented in the cognitive psychology literature at the time of Baier's moral theory research into trust. See Hilgard (1980).

7 Luhmann (1979) adds that we benefit more over time as less vigilance is required.

8 The epistemic and moral qualities that we look for in a trustworthy expert have been detailed by Hardwig (1994), Elliott (2010), De Melo Martin and Intemann (2018), and others.

9 Kristina Rolin (2002) made the important distinction between trustworthiness and credibility (the perception of trustworthiness by others) in an important feminist corrective to Hardwig on epistemic trust.

10 De Melo Martin and Intemann (2018) still insist that because it can be extremely difficult to distinguish between normatively appropriate dissent and normatively inappropriate dissent, a more productive route is focusing on the social conditions that lead to unhelpful forms of dissent such as lack of trust in scientific institutions and their personnel.

11 Used as the contrast class to "sound science," "junk science" refers to scientific data, research, or analysis considered to be inaccurate or fraudulent. The term also points to research being driven by ideological motives, whether political, financial, or anything unscientific. Charges of "junk science" are often made in political and legal contexts where facts and scientific results have a great amount of weight in making a determination. See, for example, Brandt (2007) and chapter 1 of Douglas (2009) for a philosophical analysis of sound vs. junk science.

12 Post-truth is defined as "relating to or denoting circumstances in which objective facts are less influential in shaping public opinion than appeals to emotion and personal belief." Discussion of post-truth is usually in the context of "post-truth politics," where lies and deception are used to promote political agendas—notably the Brexit "leave" campaign and the election of Trump as president of the United States, both in 2016. The term "post-truth" therefore signals two unsavory issues: the diminished epistemic status of truth and factuality, and the moral issue of speaking truthfully and lying. This two part-distinction was offered by Lynch (2017).

13 By virtue of being a new and popular word, the meaning of the word "post-truth" is still evolving. Furthermore, because the word has the demanding task of summing up an in-the-moment cultural preoccupation, it runs into other new terminology similarly tasked such as Collins English Dictionary's 2017 word of the year, "fake news," as well as "post-fact," "anti-expertise," and "anti-intellectualism." These latter terms are also still evolving in their meaning and use.

14 Social constructivism is often misread as relativizing factuality and truth by pointing to the active and human construction of our strongly held facts, doctrines, and theories. This charge led to the heated Science Wars of the 90s (Segerstrale 2000) and underpins the more recent disciplinary dispute within STS over the discipline's alleged responsibility for post-truth (Collins, Evans, and Weinel 2017; Lynch 2017; Sismondo 2017).

15 In the United States alone, 10 years of injury reports to US regulators potentially link more than 1.7 million injuries and nearly 83,000 deaths to medical devices (ICIJ 2018).

16 For more extensive argumentation on why the mere majority agreement is not sufficient for trust or an epistemically reliable consensus, see Beatty (2006), Miller (2013), and Intemann (2017).

References

Almassi, B. (2012). Climate Change, Epistemic Trust, and Expert Trustworthiness. *Ethics & the Environment,* *17*(2), 29–49.

Baier, A. (1986). Trust and Anti-Trust. *Ethics, 96*(2), 231–260.

Baier, A. (1991). Trust and Its Vulnerabilities. In *Tanner Lectures on Human Values*. Salt Lake City: University of Utah Press, 109–136.

Beatty, J. (2006). Masking Disagreement among Experts. *Episteme, 1–2*, 52–67.

Blake, A. (2017, January 22). Kellyanne Conway Says Donald Trump's Team has "alternative facts." Which Pretty Much Says It All. Retrieved January 9, 2019, from Washington Post: https://www.washingtonpost.com/news/the-fix/wp/2017/01/22/kellyanne-conway-says-donald-trumps-team-has-alternate-facts-which-pretty-much-says-it-all/?utm_term=.07a5a753f2c3

Brandt, A. M. (2007). *The Cigarette Century: The Rise, Fall, and Deadly Persistence of the Product that Defined America*. New York: Basic Books.

Brownlie, J., & Howson, A. (2005). "Leaps of Faith" and MMR: An Empirical Study of Trust. *Sociology, 39*(2), 221–239.

CBC Radio. (2017). Just Five Minutes on the Internet Can Sow Seeds of Doubt about Vaccines. Retrieved January 24, 2019, from CBC Radio: https://www.cbc.ca/radio/whitecoat/an-outbreak-of-doubt-1.4373395/just-five-minutes-on-the-internet-can-sow-seeds-of-doubt-about-vaccines-1.4374871.

Cohlan, A. (2009, January 20). *Obama to Restore Science to Its Rightful Place*. Retrieved January 24, 2019, from New Scientist: https://www.newscientist.com/article/dn16452-obama-to-restore-science-to-its-rightful-place/

Collins, H., & Evans, R. (2008). *Rethinking Expertise*. Chicago: University of Chicago Press.

Collins, H., Evans, R., & Weinel, M. (2017). STS as Science or Politics? *Social Studies of Science, 47*(4), 580–586.

Davey, M. (2015, August 3). Neil deGrasse Tyson on Q&A Calls Scientific Illiteracy a Tragedy of Our Time. The Guardian.

De Melo Martin, I., & Intemann, K. (2018). *The Fight Against Doubt*. Oxford: Oxford University Press.

Douglas, H. (2009). *Science, Policy, and the Value-Free Ideal*. Pittsburgh: University of Pittsburgh Press.

Dubé, E., Vivion, M., Sauvageau, C., Gagneur, A., Gagnon, R., & Guay, M. (2015). "Nature Does Things Well, Why Should We Interfere?" Among Mothers. *Qualitative Health Research, 26*(3), 411–425.

Elliott, K. E. (2010). *Is a Little Pollution Good for You? Incorporating Societal Values Into Environmental Research*. Oxford: Oxford University Press.

Goldenberg, M. J. (2015, September 3). *Scientific Illiteracy Is Not the Tragedy of Our Time*. Retrieved from Impact Ethics: https://impactethics.ca/2015/09/03/scientific-illiteracy-is-not-the-tragedy-of-our-times/

Goldenberg, M. J. (2016). Public Misunderstanding of Science? Reframing the Problem of Vaccine Hesitancy. *Perspectives on Science, 24*(6), 552–581.

Goldman, A. I. (2001). Experts: Which One Should You Trust? *Philosophy and Phenomenological Research, 63*(1), 85–110.

Grasswick, H. (2017). Trust and Testimony in Feminist Epistemology of Science. In S. J. Anne Gary (Ed.), *Routledge Companion to Feminist Philosophy* (pp. 256–267). New York: Routledge.

Grasswick, H. (2018, November 4). Pluralizing the Relationships of Public Trust in Science. *Philosophy of Science Association Biennial Meeting*. Seattle. Retrieved from https://psa2018.philsci.org/en/74-program/program-schedule/program/93/trust-and-the-feminist-politics-of-science

Hardwig, J. (1985). Epistemic Dependence. *Journal of Philosophy, 82*, 335–349.

Hardwig, J. (1991). The Role of Trust in Knowledge. *Journal of Philosophy, 88*, 693–708.

Hardwig, J. (1994). Toward an Ethics of Expertise. In D. Wueste (Ed.), *Professional Ethics and Social Responsibility*. Lantham: Rowman and Littlefield, 83–101.

Hargreaves, I., et al. (2013). *Towards a Better Map: Science, the Public, and the Media*. London: Economic and Social Research Council.

Health 2.0. (2018). *One Size Fits All Is Out, Precision Health Is In*. Retrieved January 19, 2019, from Health 2.0: https://health2con.com/news-iteam/8992-2-2/

Heuvel, K. v. (2013, July 22). *Jenny McCarthy's Vaccination Fear-Mongering and the Cult of False Equivalence*. Retrieved January 18, 2019, from The Nation: https://www.thenation.com/article/jenny-mccarthys-vaccination-fear-mongering-and-cult-false-equivalence/

Hilgard, E. R. (1980). The Trilogy of Mind: Cognition, Affection, and Conation. *Journal of the History of Behavioural Sciences, 16*(2), 107–117.

ICIJ. (2018, November 25). *Medical Devices Harm Patients Worldwide as Governments Fail On Safety*. Retrieved January 28, 2019, from International Consortium of Investigative Journalists: https://www.icij.org/investigations/implant-files/medical-devices-harm-patients-worldwide-as-governments-fail-on-safety/

Intemann, K. (2017). Who Needs a Consensus Anyway? Addressing Manufactured Doubt and Increasing Public Trust in Climate Science. *Public Affairs Quarterly, 31*(3), 189–208.

Intemann, K., & De Melo-Martin, I. (2014). Addressing Problems in Profit-driven Research: How Can Feminist Conceptions of Objectivity Help? *European Journal of Philosophy of Science, 4*(2), 135–151.

Jones, K. (1996). Trust as an Affective Attitude. *Ethics, 107*(October 1996), 4–25.

Jones, K. (1999). Second-Hand Moral Knowledge. *Journal of Philosophy, 96*(2), 55–78.

Jukola, S. (2019). Commercial Interests, Agenda Setting, and the Epistemic Trustworthiness of Nutrition Science. *Synthese.* https://doi.org/10.1007/s11229-019-02228-3.

Kappel, K. (2014,). *The Proper Role of Science in Liberal Democracy.* Retrieved January 15, 2019, from un-published conference paper: https://www.academia.edu/7017103/The_Proper_Role_of_Science_in_Liberal_Democracy

Keller, E. F. (1985). *Reflections on Gender and Science.* New Haven, CT: Yale University Press.

Kitcher, P. (2011). *Science in a Democratic Society.* Amherst, NY: Prometheus Books.

Larson, H., Cooper, L. Z., Eskola, J., Katz, S. L., & Ratzan, S. (2011). Addressing the Vaccine Confidence Gap. *Lancet, 378*, 536–535.

Lehmann, E. (2016, July 29). *Hillary Clinton Declares, "I Believe in Science."* Retrieved from Scientific Ameri-can: https://www.scientificamerican.com/article/hillary-clinton-declares-i-believe-in-science/

Lewis, J. D., & Weigert, A. (1985). Trust as a Social Reality. *Social Forces, 63*(4), 967–85.

London Royal Society. (1985). *The Public Understanding of Science: Report of a Working Party.* London: London Royal Society.

Longino, H. (1990). *Science as Social Knowledge: Values and Objectivity in Scientific Inquiry.* Princeton, NJ: Princeton University Press.

Longino, H. (2002). *The Fate of Knowledge.* Princeton, NJ: Princeton University Press.

Luhmann, N. (1979). *Trust and Power.* Toronto: Wiley.

Lynch, M. (2017). STS, Symmetry, and Post-Truth. *Social Studies of Science, 47*(4), 593–599.

Martin, L. (2019, January 25). *Lawsuit Alleges Purdue Pharma "Caused Much Of The Opioid Epidemic."* Re-trieved January 25, 2019, from CBS Boston: https://boston.cbslocal.com/2019/01/25/oxycontin-purdue-pharma-sackler-family-opioid-lawsuit-hearing-healey/.

McLeod, C. (2002). *Self-Trust and Reproductive Autonomy.* Cambridge, MA: MIT Press.

Miller, B. (2013). When is Consensus Knowledge Based? Distinguishing Shared Knowledge from Mere Agreement. *Synthese, 190*(7), 1296–1316.

Misztal, B. A. (1996). *Trust in Modern Societies: The Search for the Bases of Social Order.* Cambridge: Polity Press.

Mollering, G. (2006). *Trust, Reason, Routine, Reflexivity.* Amsterdam: Elsevier.

Mooney, C. (2014, May 30). *This Is Why You Have No Business Challenging Scientific Experts.* Retrieved from Mother Jones: https://www.motherjones.com/environment/2014/05/harry-collins-inquiring-minds-science-studies-saves-scientific-expertise/

Mooney, C., & Kirshenbaum, S. (2009). *Unscientific America: How Scientific Illiteracy Threatens our Future.* New York: Basic Books.

Moynihan, R., & Henry, D. (2006, April). The Fight against Disease Mongering: Generating Knowledge for Action. *PLoS Medicine, 3*(4), 3191.

Nelson, L. H. (1993). Epistemological Communities. In L. Alcoff & E. e. Potter (Eds.), *Feminist Epistemologies* (pp. 121–159). London: Routledge.

Offit, P. (2011). *Deadly Choices: How the Anti-vaccine Movement Threatens Us All.* New York: Basic Books.

Otto, S. L. (2016). *The War on Science: Who's Waging It, Why It Matters, What We Can Do about It.* Minne-apolis: Milkweed Editions.

Payer, L. (1992). *DiseaseMongers: How Doctors, Drug Companies, and Insurers Are Making You Feel Sick.* New York: John WIley & Sons.

Peters, R., Covello, V., & McCallum, D. (1997). The Determinants of Trust and Credibility in Environ-mental Risk Communication: An Empirical Study. *Risk Analysis, 17*, 43–54.

Pielke Jr., R. S. (2007). *The Honest Broker: Making Sense of Science in Policy and Politics.* Cambridge: Cambridge University Press.

Poovey, M. (1998). *A History of the Modern Fact: Problems of Knowledge in the Sciences of Wealth and Society.* Chicago, IL: University of Chicago.

Prothero, D. R. (2008). *Reality Check: How Science Deniers Threaten Our Future.* Bloomington: Indiana Uni-versity Press.

Reich, J. A. (2016). Of Natural Bodies and Antibodies: Parents' Vaccine Refusal and the Dichotomies of Natural and Artificial. *Social Science and Medicine, 157*, 103–110.

Rolin, K. (2002). Gender and Trust in Science. *Hypatia, 17*, 95–120.

Scheman, N. (2001). Epistemology Resuscitated: Objectivity as Trustworthiness. In N. M. Tuana (Ed.), *Engendering Rationalities,* (pp. 23–52). Buffalo, NY: Suny Press.

Segerstrale, U. (2000). Science and Science Studies: Enemies of Allies?. In U. Segerstrale (Ed.), *Beyond the Science Wars: The Missing Discourse* (pp. 1–28). Albany, NY: SUNY.

Shapin, S. (1995). Trust, Honesty, and Authority in Science. In R. Bulger, E. Bobby, & H. Fineberg (Eds.), *Society's Choices: Social and Ethical Decision Making in Biomedicine* (pp. 388–408). Washington, DC: National Academy Press.

Shapin, S., & Schaffer, S. (1985). *Leviathan and the Air Pump: Hobbes, Boyle and the Experimental Life*. Princeton, NJ: Princeton University Press.

Shiel, F. (2018, November 25). *About the Implant Files Ivestigation*. Retrieved January 28, 2019, from International Consortium of Investigative Journalism: https://www.icij.org/investigations/implant-files/about-the-implant-files-investigation/

Sismondo, S. (2017). Post-truth? *Social Studies of Science, 47*(1), 3–6.

Specter, M. (2010). *Denialism: How Irrational Thinking Harms the Planet and Threatens our Lives*. London: Penguin.

Sun, L. J. (2018, March 29). *In Emotional Speech, CDC's New Leader Vows to Uphold Science*. Retrieved from Washington Post: https://www.washingtonpost.com/news/to-your-health/wp/2018/03/29/in-emotional-speech-cdcs-new-leader-vows-to-uphold-science/?utm_term=.4eefdf5ea842

Whyte, K. P., & Crease, R. (2010). Trust, Expertise and the Philosophy of Science. *Synthese, 177*, 411–425.

Wilholt, T. (2009). Bias and Values in Scientific Research. *Studies in History and Philosophy of Science Part A, 40*(1), 91–101.

Wilholt, T. (2013). Epistemic Trust in Science. *British Journal of Philosophy of Science, 62*(2), 233–253.

THE HARD SELL OF GENETICALLY ENGINEERED (GE) MOSQUITOES WITH GENE DRIVES AS THE SOLUTION TO MALARIA

Ethical, Political, Epistemic, and Epidemiological Issues in Global Health Governance

Zahra Meghani

Introduction

Researchers from a variety of disciplines have argued that scientific activity is shaped by political or economic considerations in a multitude of ways (e.g. Haraway 1988; Longino 1990; Nelson 1990; Ong 2006; TallBear 2013). This chapter contributes to that body of scholarship by analyzing the 'hard sell' of genetically engineered (GE) mosquitoes with gene drives as the solution to the high incidence of mosquito-borne diseases in socio-economically marginalized communities in parts of the global South. Specifically, it analyzes two papers by scientists with ties to Target Malaria, a research organization developing GE mosquitoes with gene drives. In the interest of a contextualized reading of those scientific articles, this chapter examines the normative commitments of the Bill and Melinda Gates Foundation, a non-governmental organization (NGO) that is Target Malaria's key backer. The Gates Foundation is a powerful actor on the global health governance stage.

The analysis developed here is influenced by the work that Kim TallBear and Aihwa Ong have done to de-mythologize genetic research. Drawing on her expertise in science and technology studies, and Native American and Indigenous Studies, TallBear (2013) has challenged the notion of value-neutral genetic research on 'race'. She has analyzed the political interests and corporate concerns that birth and influence the scientific projects that attempt to *find* 'Native American DNA' and which are served by such projects. Anthropologist Ong (2016) has traced the range of values (including corporate, nationalist, and regionalist interests) that motivate and shape genetic research in Singapore that divides up and 'constructs' populations in Asia as (genetically) distinct kinds. Pharmaceutical companies, for instance, stand to reap considerable benefits from that research as it creates targeted markets for their products.

A defining characteristic of the two scientific papers analyzed in this chapter is that they 'biologize' (or 'naturalize') the significant prevalence of mosquito-borne diseases in certain poorer regions of the South. They do not appropriately acknowledge the structural, systemic factors that are *partially* responsible for the public health problem that the GE mosquitoes are intended to biosolve. The biologization of the public health problem, in conjunction with the effort to pose GE mosquitoes with gene drives as the solution (or as a necessary part of any solution) to mosquito-borne diseases, qualifies as a hard sell of the bio-technology.

Public health experts and social scientists have argued that the disproportionate incidence of infectious diseases (including mosquito-borne diseases) in certain socio-economically marginalized regions of the South is not a purely biological phenomenon (e.g. Allotey et al. (2010), Hausmann-Muela & Eckl (2015), and Birn et al. (2017)). Rather there are structural, systemic factors at work too. However, some powerful non-state actors remain committed to a biomedical model of disease that construes the significant prevalence of infectious diseases in certain poorer regions of the South as a primarily (if not purely) biological phenomenon that can *only* be effectively and comprehensively addressed by means of medical or scientific-technological solution (Cueto 2013; Birn et al. 2017). The point here is that the hard sell of a high-tech patented technology (in this case, GE mosquitoes with gene drives) as the solution to the substantial incidence of an infectious disease among the poor in parts of the global South is not a unique phenomenon. However, as it is a new development, it merits scrutiny.

Hard sells of patented medical or scientific-technological solutions for public health problems that primarily impact the poor tend to have epistemic, epidemiological, and gendered ethico-political costs. A scientific account of the causal factors for the disproportionately high occurrence of an infectious disease among marginalized groups is epistemically flawed if it fails to acknowledge key political and economic determinants of the public health problem. Moreover, such scientific narratives may undermine efforts to address the structural, systemic factors that are partially responsible for the substantial incidence of *multiple* infectious diseases in certain socio-economically marginalized global South regions. That would have particular negative implications for persons who are afflicted with those diseases and the women and girls that are responsible for providing care to sick family members. In virtually all societies, in low-income households, unpaid care work is performed disproportionately by female family members (Rathgeber & Vlassoff 1993; Sepúlveda Carmona 2013).

The chapter begins with a brief description of GE mosquitoes with gene drives. Some of the key ethico-political, ecological, and epidemiological questions entailed by the use of the modified mosquitoes are also delineated. After that, a contextual analysis of the substantial prevalence of infectious diseases in socio-economically marginalized regions of the global South is provided. It is followed by a brief account of the concepts and theoretical frameworks used to evaluate scientific articles that biologize the public health problem and advocate the use of GE mosquitoes with gene drives as the solution to it. Then, two papers are analyzed that 'construct' GE mosquitoes with gene drives as *the* (or as a necessary part of the) solution. Their hard push for a patented bio-technology as the solution to malaria is traced to the normative commitments of the Gates Foundation, a private philanthropic organization. It is argued that the NGO's support of a high risk, high-tech, market-based intervention for malaria is rooted in its neoliberal orientation.

Gene Drives and Crucial Ethico-Political, Ecological, and Public Health Issues

The term "genetically engineered" (GE) denotes entities that have been modified by modern biotechnology methods (such as recombinant DNA technology or by means of gene editing techniques (e.g. ZFNs, TALENS, and CRISPR)) *as well as* organisms that have inherited the alterations. Recently developed gene editing techniques are presented in mainstream media as more precise than the older methods for affecting genetic change, but whether they are in fact as precise as they are purported to be remains a question (see, for instance, Ono et al. 2019; Sirinathsinghji 2019, Norris et al. 2020).

GE organisms with gene drives have intentionally altered heritable traits that are supposed to be passed on to their progeny at a percentage higher than the one described by Mendel's law of inheritance (Burt 2003; Champer et al. 2016; Hammond & Galizi 2017; Harvey-Samuel et al. 2017). The 2016 US-based National Academies of Sciences, Engineering, and Medicine (NASEM) report on gene drives identified mosquito strains that are vectors of diseases as prime candidates for this form of genetic modification (NASEM 2016). Proponents of the use of mosquitoes with gene drives envision using modified versions of various strains of mosquitoes as public health measures to eradicate or reduce the incidence of diseases such as malaria, dengue, and Zika. The possibility of designing gene drives for mosquitoes has captured the attention of non-state and state actors. The Gates Foundation gave US$75 million to Imperial College which houses the Target Malaria initiative and Tata Trust of Mumbai awarded University of California at San Diego with US$70 million for research and development of gene drives that could be used to eliminate mosquitoes that cause malaria (Regalado 2016a; Robbins & Fikes 2016). Gene drive research is also being funded by the US Defense Advanced Research Projects Agency in excess of US $65 million (Callaway 2017).

Different kinds of gene drives are under development. Gene drives for mosquitoes can be classified as "population suppression gene drives" or "population replacement gene drives."[1] The former type of modification is intended to reduce (or eliminate) the population of future generations of the wild-type mosquito strains that transmit parasites or viruses to humans. For instance, researchers are devising strains of GE mosquitoes that pass on a lethal trait to their offspring. The GE insect is expected to mate with its wild-type counterpart in the target area. The GE offspring (i.e., the progeny with the heritable lethal trait) are not expected to survive. Over time, with periodic additional releases of the GE mosquitoes with the heritable lethal trait (that is passed on at a rate greater than the one described by Mendel's law), all other things being equal, the expectation is that the future population of the wildtype mosquito will be either smaller than it might have been otherwise or non-existent. (Generations of wildtype mosquitoes that would have been born, grown to adulthood, and reproduced would not do so because the mating between the GE mosquito and its wildtype counterpart would not produce viable offspring.)

However, the use of GE mosquitoes (with intentionally altered heritable traits) to suppress the population of its wildtype mosquito is not necessarily an unproblematic solution. The population crash of a strain of mosquito that transmits malaria could result in its ecosystem niche being taken over by a different strain of mosquito that also transmits malaria parasites or is a vector for another disease (Wilke et al. 2018; Shaw & Catteruccia 2019). Consider that an estimated 40 species of mosquitoes transmit malaria, while West Nile virus has been found in approximately 60 strains of mosquitoes and both the *Aedes aegypti L.* and the *Aedes albopictus* strains of mosquitoes transmit the dengue virus, Zika virus, yellow fever virus, and chikungunya virus (Wilke et al. 2018). There is also the possibility that the target organism may mutate and develop resistance to the genetic modification. In addition, the impact of the modification on hybrids and other species in the ecosystem could be substantial given the entangled existence of species. So, while an approach that focuses on a particular strain of mosquito with the aim of suppressing its population (by changing its feature(s) so that its offspring are not viable) may appear to be a targeted solution, but, arguably, depending on the mosquito strain, the parasite or virus, the particulars of the population at risk, and the ecological context, it may be akin to playing a very expensive and ultimately ineffective game of whack-a-mole.[2]

There are plans to develop GE mosquitoes with population replacement gene drives too. The intent is to use them to affect change in the wild-type of the targeted mosquito strain (by modifying or eliminating existing traits, or introducing new ones). If GE mosquitoes with a germline modification that knocks-out a pest trait (for instance, the ability to pass pathogens to humans) are designed and they mate with their wild-type counterpart, the resulting progeny would not inherit the 'pest' trait. With this type of gene drive too, the genetic modification would be inherited at a rate greater than the one described by Mendel's law, with a population level effect that is manifested over multiple generations.

Gene drives can also be classified as self-sustaining or self-limiting. Lab experiments and mathematical modeling suggest that if GE mosquitoes with self-sustaining (population replacement) gene drives are released in sufficient numbers, the introduced change may spread in the wild-type population such that over time, all other things being equal, the entire population may be affected.[3] Even the (uncaged) field trial release of GE mosquitoes with self-sustaining (population replacement) gene drives may mean that the introduced trait could spread uncontrollably, resulting in unanticipated and unintended widespread harmful consequences (Harvey-Samuel et al. 2017).

Gene drive researchers have argued that the risk of unanticipated, unintended irreversible ecological harm from GE mosquitoes with self-sustaining gene drives could be 'managed' if mosquitoes with *self-limiting* gene drives are used instead (see, for instance, Gould et al. 2008; Rasgon 2009; Harvey-Samuel et al. 2017). The heritable genetic modification 'carried' by engineered insects with a self-limiting gene drive would only pass on to a certain number of future generations. It would peter out over time, unless there were additional periodic future releases of the GE mosquito. Developers of GE mosquitoes contend that the need for additional periodic releases entailed by the use of GE mosquitoes with self-limiting gene drives can be a means of controlling the presence and spread of the introduced trait and preventing irreversible harm. For instance, Harvey-Samuel et al. (2017: 1685) claim that

> [a] benefit of this limited persistence is that the deployment of self-limiting systems is inherently reversible: if unintended consequences were to arise from a release programme, the target population could be allowed to return to its unmodified state simply through the cessation of releases.

However, the question must be asked if the optimism about the ability of researchers to control the effects of the release and reverse unintended harms is warranted. Ecosystems are complicated, complex dynamic systems with emergent traits so that it may be difficult to predict and precisely control how they may react to change, such as the introduction of GE mosquitoes with self-limiting (or self-sustaining) gene drives. The result of the change may be irreversible.[4] There could be much at risk whether GE mosquitoes with self-limiting or self-sustaining gene drives are used.[5]

To bring into focus other significant ethical, political, and public health issues, it is useful to consider the case of the OX513A GE *Aedes aegypti* mosquito (that has been modified to have a new heritable feature that is intended to have a population suppression effect on its wildtype counterpart). The OX513A GE mosquito was developed by Oxitec Ltd. As of January 2016, 150 million OX513A GE mosquitoes had been released in field trials (Oxitec 2016: 9). The GE mosquito is a patented bio-product that is marketed as a public health measure; it is intended to reduce the spread of mosquito-borne diseases. Since 2009 the GE mosquito has been field-trialed or used in Brazil, the Cayman Islands, Malaysia, and Panama. It has a heritable synthetic genetic sequence that makes the laboratory modified organism and approximately 95% of its progeny dependent on tetracycline for their survival. The 2016 US FDA's Environmental (Risk) Assessment report on the proposed (uncaged) field trial of the OX513A GE mosquito notes that in areas where tetracycline is not present in sufficient quantity, the heritable tetracycline dependency trait would limit the spread of the GE insects and their GE progeny (that are the result of them mating with their wildtype counterpart and have inherited the introduced trait), whilst resulting in the suppression of the population of its wildtype counterpart in those locations:

> Biological containment would be afforded by the introduction of the conditional lethality trait[6] into the OX513A *Ae. aegypti* line, where on mating with the local females of the same species, >95% of the progeny will not survive to functional adulthood in the absence of tetracycline (Harris et al. 2011), leading to the overall reduction in the population of *Ae. aegypti* at a given site.
>
> *(FDA 2016: 17–18)*

The field trial protocol is to periodically release male OX513A GE mosquitoes in high enough numbers such that they out-compete their male wild-type counterpart to mate with their female wild-type, resulting in offspring who, by and large, inherit the tetracycline dependency trait and thus do not survive to adulthood (in the absence of tetracycline).[7] As a result, over multiple generations, the population of the wild-type of the *Aedes aegypti* mosquito in the target area is expected to be smaller than it would have been otherwise, all other things being equal. (Generations of mosquitoes that would have been born and reproduced do not come into existence.) Presumably, resulting in lower incidence of transmission of dengue and Zika in humans.

The 2016 NASEM report on gene drives places OX513A GE mosquito (that has a population suppression effect) on a continuum with GE mosquitoes (under development) that have gene drives that rely on the newly developed genetic editing techniques. The report recommends that the risk assessment of GE mosquitoes with gene drives (developed using one of the new genetic editing techniques) should be modeled on the risk evaluation protocol used for the OX513A GE mosquito (NASEM 2016: 103). Taking its cue from the NASEM report, this chapter uses the case of the OX513A GE mosquito to identify some of the key ethico-political issues with the use of GE mosquitoes with gene drives that are under development.[8]

The cost of licensing the patented bio-product requires consideration. For regulatory purposes, in the United States, the OX513A GE mosquito has been classified by its developer as a bio-pesticide and not as a public health measure that reduces the incidence of mosquito-borne diseases in humans.[9] The World Health Organization (WHO) has encouraged the collection of data about the cost of the public health intervention "during the evaluation of new vector control products, particularly during phase IV studies" (2017: 7). Nations with limited resources need accurate information about the price of new interventions (such as GE mosquitoes with germline modification including those designed using gene editing) relative to that of the standard ones to make policy decisions that are effective in terms of public health, whilst being cost effective too.

The Oxitec OX513A GE mosquito has been used in Piracicaba, Brazil. The city of 391,449 was charged approximately US$1.1 million for use of the bio-product for a two-year period at the rate of US$10 per person in the target area (Servick 2016). This is a significant price tag for low-income and even middle-income countries given the need for periodic re-releases every year and even seasonally (Meghani & Boëte 2018). The profit potential of the OX513A GE mosquito (and presumably other such modified insects) appears to be considerable. In 2015, Intrexon, a US biotechnology company, purchased Oxitec, a British company, for US$160 million (Nickel & Gillam 2015). Investors may consider mosquitoes with gene drives to have a significant profit potential as they are patented bio-products.

In 2017, the WHO Expert Advisory Group on vector control recommended that studies should be conducted that track the efficacy and effectiveness of different types of new vector control measures (such as GE mosquitoes) in reducing the incidence of vector-borne diseases. The WHO defines "efficacy" as "an intervention measured when it is implemented under ideal, highly controlled circumstances … (during) phase III studies" (2017a: iv). In contrast, "effectiveness" is evaluated during Phase IV studies to determine the "degree of benefit of an intervention measured when it is delivered and used operationally under routine, 'real-world' conditions" (WHO 2017a: iv). The WHO's Vector Control Advisory expert group stipulated the following standards for new vector control measures (such as GE mosquitoes):

> For any new vector control tools in new product classes, WHO requires evidence from at least two well conducted, randomized controlled trials with *epidemiological* outcomes and follow up over at least two transmission seasons. With limited funds available for disease control, Member States are required to implement the most effective intervention for their local

context. *Epidemiological* trials should therefore be conducted in different entomological and epidemiological settings in order to verify the public health value of the new product class or product variation. Two trials is the minimum number needed to assess generalizability (my italics).

(WHO 2017a: 2)

The OX513A GE mosquito's developer contends that its bio-product is meant for use with other methods for reducing the incidence of mosquito-borne diseases. So, it may be complicated to conduct trials to ascertain the effectiveness of the GE insect as a public health measure. With respect to GE mosquitoes with (self-limiting or self-sustaining gene edited) gene drives, evidence would be needed of their effectiveness in decreasing the prevalence of diseases in humans over time, relative to other known-to-be effective measures.

The two trials criterion of the WHO's Vector Control Advisory expert group requires unpacking. Specifically, governments considering permitting the controlled randomized trials of GE mosquitoes to get realistic data about their effectiveness in reducing the incidence of disease would have to consider the following issues:

i   The trials would entail the *open (uncaged) environmental release of GE mosquitoes with intentionally altered heritable traits*: the risks would not be limited to persons in the trial area (the point here is that as the GE mosquitoes (and their GE offspring) are mobile, fairly small organisms, the possibility that they may go (or be inadvertently transported) beyond the trial site area cannot be dismissed, so in an important regard, it is a misnomer to call these uncaged field research experiments 'controlled' trials).

ii   If female GE mosquitoes are inadvertently released, they and their female offspring could be vectors of diseases, *and* male GE mosquitoes (like their male wild-type counterpart) could function as indirect vectors of diseases insofar as they have the capacity to produce germ cells that could result in female mosquitoes that could be disease vectors. (This worry would be relevant whether the GE mosquitoes are intended to have a population suppression or population replacement effect because, among other things, of the possibility of the genetic modification having an error rate.)

iii The randomized 'controlled' trials (aimed at getting real world data about the effectiveness of the bio-technology) would expose infants, young children, pregnant women, breastfeeding women, persons with disabilities, immunocompromised persons, and other vulnerable populations in the trial area (and possibly, beyond) to known and unknown risks in contexts where mosquito-borne and other infectious diseases (attributable to different kinds of pathogens) are endemic.

iv   The risks might be compounded if the local populace is suffering from malnutrition or undernutrition and is not assured of preventative and therapeutic health care.

Governments considering subjecting populations to (the above) risks from the experimental research about the 'real world' effectiveness of GE mosquitoes in reducing the incidence of mosquito-borne diseases would have to take into account that there are already cost-effective and known-to-be effective methods of controlling those infectious diseases, which include structural, systemic changes (more on this below).[10] Moreover, any state contemplating permitting trials of the new bio-technology would have to contend with the fact that complexes of local, national, and international socio-political-economic-ecological factors are partial determinants of emerging and re-emerging vector-borne diseases *and* their substantial prevalence in socio-economically marginalized regions. (See, for instance, the discussion below of the re-emergence of malaria in parts of Venezuela.) Governments would also have to recognize that given the variability between ecosystems and given that ecosystems have emergent traits, the results from the trials of a GE organism in a specific ecosystem may not hold for other ecosystems. In addition, the countries contemplating permitting epidemiological trials of GE mosquitoes have an ethico-political obligation not to move forward with the biotechnology unless those who would or could be affected give their *free, informed consent* (Meghani & Boëte 2018; and especially, Meghani 2019). As the population that could be impacted could extend beyond national borders, the autonomy of various countries would be at stake as would the right to self-determination of indigenous peoples who could be affected (Meghani 2019). This, of course, is not an exhaustive list of the issues that countries considering 'controlled' (uncaged) research trials of GE mosquitoes would have to address.

There is another serious ethico-political issue at stake. If the treatment by US and Canadian courts of cases of genetic drift of GE seed is considered the relevant precedent, then even nations that did not sign-up for the GE mosquitoes with germline modification might find themselves beholden to pay for the patented GE animals if they encroach on their territories (see Meghani (2019) for a detailed discussion of this matter). The penalty for unlicensed use may be onerous for low-income and middle-income countries.

Below, context is provided for this chapter's critique of scientific articles that naturalize the high incidence of mosquito-borne diseases in certain marginalized parts of sub Saharan Africa. Specifically, the role of structural, systemic factors in creating and sustaining the public health problem is outlined.

A Contextual Examination of the Substantial Prevalence of Infectious Diseases in Socio-Economically Marginalized Regions

The recent re-emergence of malaria as a public health problem in poorer regions of Venezuela is illuminating.[11] While the South American country was the first to eradicate malaria in 1961, during the 2000–2015 period there was a 365% increase in reported cases of the illness. In 2016, over 240,000 cases of malaria were officially documented and the numbers were worse in 2017, with a 68% increase in the cumulative number of cases compared to 2016 by October 2017 (WHO 2017a: 53; Grillet et al. 2018). This rise in the number of malaria cases is attributable to a complex of structural, systemic factors, including a generalized failure of the nation's health care system that can be traced to economic and political mismanagement (Grillet et al. 2018). The increase in malaria is concentrated, with 43% of the reported cases in 2016 in the municipality of Sifontes in areas where gold is mined illegally. Mining in Sifontes has entailed erosion of the rainforest and the practice of "creating pools of stagnant water to extract gold" (Pardo 2014). The mosquitoes flourish in the muddy standing waters in the humid climate. The miners live in sheds constructed of black plastic bags with roofs of sheets of zinc, next to the areas where they mine for gold (Pardo 2014). The lack of government oversight of those mining ventures and the want of housing facilities (with standard utilities), coupled with limited state funds for purchasing malaria commodities (such as insecticides, medication, mosquito nets, and diagnostic supplies) and (infection) surveillance efforts, have contributed to the public health crisis. Other contributory factors include the decision of infected persons to not continue with the treatment once they feel better (Pardo 2014) as well as, presumably, the impracticality of the advice to miners (at risk for mosquito bites) to wear long sleeved shirts and pants, given that they work outdoors in hot, humid conditions. The re-emergence of malaria is partially attributable to an assemblage of political, economic, legal and cultural norms, institutions, laws, policies, and practices that interact in complicated ways.

The WHO has noted that poorer communities with significant prevalence of mosquito-borne diseases experience high incidence of other diseases, which are traceable to some of the same structural, systemic factors that are partially responsible for the high rates of mosquito-borne diseases in the affected populations (WHO 2008, 2017b; also see Allotey et *al.* 2005; Manderson et al. 2009; Garchitorena et al. 2017). It matters whether neighborhoods have safe water, sanitation facilities, and waste treatment processes, and whether the residents have access to sufficient nutrition, and medical and preventative care. The ability of populations to avoid an illness or heal from it is not always merely a question of them receiving the appropriate therapeutic or preventative medical measure. For instance, according to the 2017 World Malaria Report, malnutrition (a feature of poverty and inequality) impairs the biological ability of children and pregnant women to fight malaria. Structural, systemic factors play a role in determining the vulnerability of populations to mosquito and other vector-borne diseases and their capacity to fight the infections and recover from them (WHO 2017b).[12]

However, there is a trend among some global health governance actors to biologize the high incidence of infectious diseases among the poor of low-income nations, whilst underplaying or ignoring the role of socio-political and economic factors. Allotey *et al.* (2005) note that "[p]ublic health efforts to control communicable (read: infectious) diseases continue to have a predilection towards biomedical models which discount the systematic long-term evaluation of social and environmental interventions in disease control" (p.13). The biomedical model of disease with a narrow focus on pathogens or disease vectors (like mosquitoes) seems to lend itself to and, in turn, motivate efforts to devise technical solutions that ignore the relevant socio-political and economic causal factors (see also Allotey et al. (2010)).

Given that the disproportionate prevalence of malaria in certain global South regions is not a purely biological phenomenon that has no socio-political or economic components, Target Malaria's proposal to use GE mosquitoes with gene drives to eradicate malaria (and other mosquito-borne diseases) should not be accepted unquestioningly. The next section delineates the methodology used to scrutinize two scientific articles by key developers of GE mosquitoes that aggressively push for the use of the insects with synthetic gene drive.

A Methodology for Analyzing the Hard Sell of GE Mosquitoes with Gene Drives

This chapter analyzes as *scientific narratives* two scientific papers that hard sell GE mosquitoes with gene drives as the (or as a necessary part of the) solution to the high incidence of mosquito-borne diseases in certain poorer regions of the global South. The notion of narratives (including scientific narratives) invoked here is that of accounts that describe and organize events into coherent structures to help humans make sense of the world (Cronon 1992). 'Stories' or accounts of the world—whether they are constructed by laypersons or scientists—are always situated. This chapter too is situated: it is from a standpoint that recognizes that power inequities between groups and countries play a significant role in creating and maintaining public health problems that disproportionately impact marginalized populations.

Narratives aim to affect particular purposes; neither the descriptions nor the structuring of the events is value-neutral (Cronon 1992). Elaborating on the value-ladenness of narratives, Cronon notes that the creation of 'stories' invariably involves the use of "a rhetorical razor that defines included and excluded, relevant and irrelevant, empowered and disempowered" (Cronon 1992: 1349). Narratives constitute exercise of power as "they sanction certain voices and views while silencing others" (Cronon 1992: 1349).

Narratives (including scientific narratives) may be judged on the basis of the interests that they represent or undermine. Scientific papers that biologize the high incidence of public health problems that primarily affect socio-economically marginalized communities may play a role in securing political and financial support for the research as well as the development and use of high-tech solutions to them. Given the respect (including trust) generally afforded to science (Douglas 2009) *and* given that such papers minimize (or utterly mask) the role of structural, systemic factors in determining the disproportionate disease burden on marginalized populations, such scientific narratives in effect legitimize the decision of state actors to avoid addressing the inequitable power relations between groups (at the local, national, and international levels) that play a critical role in determining the substantial prevalence of infectious diseases in certain socio-economically marginalized communities. Such scientific narratives are morally charged in another way too. They may undermine efforts to change inequitable power relations between groups as part of the solution to reduce the disease burden shouldered by marginalized populations.

A defining feature of such scientific narratives is that they reclassify problems that are at least partially caused by *structural, systemic factors* as the type of problems that are 'natural' and which require scientific-technical solutions. The term "structural factors"[13] denotes dynamic, entangled *assemblages* of political, economic, legal and cultural norms, institutions, laws, policies, and practices that interact in complicated ways to produce, shape, or maintain certain phenomena such as public health crises that disproportionately affect the poor. The term "assemblage" (Ong 2016) is used to indicate that both the factors and the relationships between them have an element of contingency. In other words, structural factors generate systems that create, configure, and sustain them. However, this is not to deny that specific agents should be held morally responsible for public health crises that primarily affect the poor.

The naturalization of (socio-political and economic) problems tends to be a two-step process. First, the role of structural, systemic factors in creating, shaping, or sustaining the public health problems (that disproportionately affect the marginalized) is minimized (or disappeared). The second step involves the conversion of that public health problem into a problem of nature (i.e. biology), specifically a problem that is apolitical and best amenable to scientific-technical interventions. Those re-tellings relieve the dominant classes of the responsibility of addressing the ethico-political problem of inequitable power relations among groups that accounts for the disproportionate incidence of the disease among the socio-economically marginalized.

There is another dimension to this re-casting and naturalization of political-economic problems. The science that is charged with solving these problems is construed as 'objective' science, i.e., a (mythical) science that is untainted by any normative considerations and whose motives should never be questioned. While researchers from various disciplines have argued in painstaking detail that scientific theorizing and practice is shaped by a variety of values in a multitude of ways, the conception of science as a purely epistemic activity continues to flourish, providing it with (some degree of) immunity from normative scrutiny. This chapter calls into question the immunity of a subset of scientific narratives that biologize the high prevalence of malaria in poorer communities in some parts of the global South and advocate the use of GE mosquitoes with gene drives as key solution to it.

The Hard Sell of GE Mosquitos with Gene Drives as the Solution to the High Incidence of Malaria in Parts of Sub Saharan Africa

This section examines two papers by scientists with ties to Target Malaria that push GE mosquitoes with gene drives as the solution to the substantial prevalence of mosquito-borne diseases in socio-economically marginalized regions of the global South. A defining feature of these papers is that they biologize that public health problem. While it goes beyond the scope of this chapter to track the prevalence of such biologizing narratives, the aim here is to show that it does happen and is problematic. (Other chapters have addressed this similar tendency of biologizing narratives in evolutionary psychology (Meynell, this volume) and literature that discusses disability (Tremain, this volume).) It must be re-iterated that there is a history of certain global health governance actors biologizing public health problems that primarily impact the poor (Birn et al. 2017).

Target Malaria is a research consortium that has been and, presumably, continues to be funded primarily by the Gates Foundation. It plans to release GE mosquitoes with gene drives in Burkina

Faso, Mali, and Uganda to eradicate malaria. The Foundation's approach to addressing public health problems that disproportionately impact the poor of low-income countries[14] is mirrored by Target Malaria's plan for a narrowly focused, high-tech market-based approach (that entails considerable uncertainty and risks) to malaria intervention in the three sub Saharan countries.

Scientists with significant ties to Target Malaria (and thereby, presumably, the Gates Foundation) have been presenting GE mosquitoes with gene drives as key solution to malaria. Research and review papers in well-regarded science journals can be instrumental in convincing the larger scientific community as well as investors and governments to consider a certain approach to a problem as a good one or, in this case, as *the* solution and generate support for its use.

A 2018 article by the two scientific leaders of Target Malaria, Austin Burt and Andrea Crisanti, presents the high incidence of vector-borne diseases (such as malaria, dengue, etc.) in poorer regions in parts of the global South as a phenomenon that must be addressed by chemical means and by targeting biological processes, ideally, by use of GE organisms with gene drives (including GE mosquitoes with gene drives). (Target Malaria is Burt's brainchild and Crisanti is the Principal Investigator for the initiative). Burt and Crisanti write,

> Even now, more than 700,000 people die every year from vector-borne diseases, and there is an additional heavy burden of nonlethal morbidity. Much of disease control is, ultimately, chemical, with efficacy largely determined by the degree to which production and delivery can be targeted and affordable. Vaccination can be among the most cost-effective of all health interventions because it uses the adaptive immune system to generate and deliver the active agents. *The promise of genetic approaches and gene drive in particular is again to use biological processes*—mating, meiosis, transcription, translation, etc. *for targeted, cost-effective delivery of appropriate chemicals* (e.g., nucleases or antimicrobial peptides) *that will substantially reduce disease transmission.* Indeed, *gene drive may take efficiency a step further, with a single release* (perhaps with periodic "booster" releases) *giving area-wide, population level control.* Important steps have been made toward realizing this potential, though there remains much more to do (my italics).
>
> *(2018: 344)*

The first step of this two-step argument categorizes chemical interventions as among the most effective solutions for vector-borne diseases (including mosquito-borne diseases) because they focus on biological processes. Then, genetic modification, and especially gene drive, is presented as a method that also targets biological processes. In other words, the argument deftly aligns the genetic approaches, including gene drives, with chemical ones that it identifies as most effective. Then, modified organisms with gene drives are presented as possibly the most efficient way to reduce infection transmission from mosquitoes and other disease vectors to humans.

To make the case for mosquitoes with gene drives, some scientists with ties to Target Malaria have asserted a strong uni-directional causal connection between mosquito-borne diseases and poverty. In a 2017 article in *Pathogens and Global Health*, Hammond and Galizi (both at Imperial College which houses Target Malaria) write,

> Mosquitoes are considered to be the most dangerous animals on earth. Through the transmission of deadly pathogens, they are believed to cause more than 700,000 deaths each year and are considered a major underlying cause of poverty in developing countries.
>
> *(412)*

The relationship between mosquito-borne diseases and poverty is not that simple. It is complicated (Garchitorena et al. 2017; WHO 2017b). Populations are rendered vulnerable to mosquito *and* other vector-borne diseases in the context of inequitable (socio-political and economic) power relations, which translate into circumstances of poverty. Lack of sanitation facilities in poorer neighborhoods means that there is open sewage in the waterways and canals. Solid waste blocks the flow of water, with the contaminated water serving as a breeding ground for disease vectors. Homes that are open to the elements (for instance, lack screened windows and doors) do not protect the residents from vectors of disease. In the absence of garbage disposal services in poorer neighborhoods, residents have no way of disposing of containers that collect rainwater, which may then function as breeding grounds for mosquitoes. The lack of reliable, safe water sources in poorer neighborhoods means that households have to store water in containers, which can act as breeding sites for mosquitoes. Inadequate nutrition[15] and a dearth of preventative and therapeutic health care are determinants of the prevalence of the incidence of mosquito-borne diseases too. For the person who is sick with a vector-borne illness (or some other ailment), the harm from the loss of wages is amplified if the society she lives in does not have much or any social protection for those who are poor, unwell, or disabled. The ripple effect of the harm extends to her family and community members. So, the relationship between poverty and mosquito-borne diseases (and other vector-borne diseases) is complex.

Arguing for the use of GE mosquitoes with gene drives, Hammond and Galizi (2017) appear to imply that the development of the infrastructure (which, presumably, includes safe living and work environments (i.e. homes that are not open to the elements, safe water supply, sanitation facilities, waste disposal measures), a comprehensive primary health care system as part of a national health system, roads, etc.) that is needed to provide standard mosquito control measures and treatments in poorer and remote regions of the world is an onerous task that should be avoided:

> Gene drives-based strategies have a number of advantages that make them particularly suited for malaria control, most notably their self-sustaining nature. For instance, these approaches can offer a solution for malarial zones that are difficult to reach with conventional interventions as they do not require continuous or mass releases (of, presumably, GE mosquitoes with self-limiting gene drives). *This (approach) would also negate the requirement for substantial infrastructure that may be unfeasible in the vast rural areas of Africa where the burden of malaria is most severe.* Because gene drives spread by mating and rely upon highly specific target site recognition, the technology can potentially be restricted to single species or perhaps even local populations. Together, *these features depict this technology as the most targeted and cost-effective intervention currently under development with the potential to benefit entire communities irrespective of their socio-economic conditions.* Nevertheless, gene drives should not be seen as a silver bullet but rather, a complementary intervention that can be used alongside other malaria control strategies such as bed-nets, antimalarial drugs and vaccines.
>
> *(Hammond and Galizi 2017: 420)*

There is an incompatibility between the expression of the egalitarian concern (of reaching across class lines using mosquitoes with gene drives) *and* the implication that providing basic necessities of life to those who live in remote, poorer parts of the world and who are affected or afflicted by vector-borne diseases is a burden to be side-stepped. Also, the question must be asked from *whose* perspective is it unfeasible to develop such infrastructure? Surely not from the perspective of those who live in such regions and experience those diseases. Regarding the development of infrastructure in rural, remote, and poor parts of the world, the 2013 report by Magdalena Sepúlveda Carmona, the United Nations Special Rapporteur on extreme poverty and human rights, states that:

> the time burden of unpaid care work for women living in poverty can be significantly alleviated if there is adequate *infrastructure* in place in their communities—particularly through reduced time spent on travel to workplaces or markets (or health care facilities), meal preparation,

water collection and fuel collection. *The availability, access to, and use of, critical infrastructure must therefore be significantly improved, prioritizing disadvantaged areas such as remote rural communities and informal settlements, explicitly seeking to provide better access for these communities to work and services.*

The construction of new roads, affordable public transport, low-cost electricity, solar and water energy for domestic purposes, and water and sanitation infrastructure are particularly crucial in this regard. In addition, States should increase construction of health care facilities and schools in underserved areas … (my italics)

(23–24)

There is some evidence that key managerial personnel of Target Malaria and the Gates Foundation have been pushing GE mosquitoes with gene drives as the solution to malaria in mainstream news articles. Such articles are likely to shape the public's perception of scientific endeavors, so what is or is not said in them and by whom matters. In an interview for a 2018 article for ABC News,[16] the stakeholder engagement manager of Target Malaria justified her employer's plan to use GE mosquitoes with gene drives in (parts of) sub Saharan Africa on the following grounds:

"Some time ago, we found that efforts against malaria in Africa had hit a plateau," said Delphine Thizy, the stakeholder engagement manager of Target Malaria. "Even though the number of mosquito nets and insecticides and sprays being used was increasing, the number of cases of infection had stopped decreasing." She continued, "The problem was that the local mosquitoes had adapted—they had grown resistant to the chemicals and also learned to bite in times and places when people were not under the net".

(Biswas 2018)

It is important to contextualize this statement. It is illuminating to consider a 2017 study that evaluated the importance of malaria risk factors for children in 16 sub Saharan African countries. It assessed the risk at the country, community, household, and individual level (Mfueni Bikundi & Coppieters 2017). At the country level, Madagascar, Senegal, Angola, Benin, Tanzania, Rwanda, and Burundi were categorized as low malaria risk, while Uganda, Malawi, Liberia, Mozambique, Ivory Coast, Burkina Faso, Democratic Republic of Congo, Nigeria, and Mali were classified as high malaria prevalence. At the household level (in gendered societies), the most significant risk factors were the low wealth status of the household, absence of electricity, and low education level of mothers (and presumably, other females who provide direct care to very young children). The latter phenomenon is a significant risk factor for the incidence of malaria in very young children *because* in poorer communities where girls and women are provided with little or no education and they do virtually all of the care work for their families and where community health workers are not present, mothers may not know about malaria prevention or when they should seek medical care for their young children who are sick with what may be malaria. The risk is compounded in contexts where there is limited or no public transportation to medical facilities.[17] Walking on unpaved roads or trails carrying a sick young child for miles to the nearest medical care facility is no easy matter, especially if a mother has other young children to look after and she has to manage a household and work to support herself and her family. In addition, it may be difficult for mothers (and other family caregivers) to follow complex instructions entailed by combination therapies for malaria if they are not literate (Schäfer 2014: 101) or if medical personnel do not speak the local (indigenous) language of the mother (or other family caregivers) (Schäfer 2014: 65).

It is also useful to consider a recent *PLoS One* article that reported the results of a systemic review and meta-analysis of research on malaria in sub Saharan Africa. Degarege et al. (2019) noted the climatic and environmental conditions conducive to the survival of mosquitoes as well as insecticide resistance development in the mosquito species that transmit malaria as responsible factors in the high incidence of the disease. Like Mfueni Bikundi and Coppieters (2017), they identified a variety of socio-economic determinants of the disease. Residents that did not have homes that were protected from the elements and disease vectors were vulnerable as were populations that did not have the means to purchase insecticides or treated bed nets (in areas where those things are not available for free). The high prevalence of the mosquito-borne diseases was also traceable to treatment not being available for free to populations that could not afford to buy medication or pay for other medical costs associated with malaria.[18] Moreover, Degarege et al. (2019) recognized the connection between lack of education and low levels of knowledge about malaria prevention and treatment (see above analysis of this factor). Thus, they recommended policy changes that would improve the socio-economic circumstances of affected and at-risk population to reduce their risk of malaria in particular sub Saharan African countries.

The high incidence of malaria in some parts of sub Saharan Africa is attributable to a complex of biological and social factors. The claim that the prevalence of the disease is the same in all of Africa is inaccurate as is the assertion that the high incidence of the disease is purely (or primarily) a biological phenomenon. If such claims are made in mainstream newspaper interviews or in the pages of scientific journals by those that are considered epistemic authorities and who have ties to powerful organizations with a well-developed international brand as a global health governance actors, they may, for instance, undermine efforts to address the underlying socio-political and economic determinants of the high incidence of malaria *and* other diseases of poverty in marginalized communities. That would have a particular impact on those who experience those illnesses, and in gendered societies, it would also have significant implications for the girls and women (from socio-economically marginalized neighborhoods) who are responsible for providing care for their unwell family and community members.

Some key employees of Target Malaria and the Gates Foundation have implied (or asserted) that nations where malaria is endemic among the poor will not make the needed socio-political and economic changes. They seem to believe that their organization must take charge and solve the problem by deploying GE mosquitoes with gene drives. A 2016 *MIT Technology Review* article on gene drives for mosquitoes makes that clear:

> "Malaria is a problem of poverty, of instability and lack of political will," says Andrea Crisanti, the Italian parasitologist and genetic engineer who developed the (GE) insects at Imperial College (and is the principle investigator of the Target Malaria initiative). "We are asking the drive to do what we can't do politically or economically."
>
> *(Regalado 2016b)*

That *MIT Technology Review* article also notes that "The Gates Foundation has said it no longer believes that malaria can be wiped out without a gene drive. 'You can't walk around with a bed net on you all the time. That's not going to eliminate malaria,' says (Fil) Randazzo (deputy director of the Foundation). With a gene drive, 'human behavior change is not required'" (Regalado 2016b).

As Target Malaria's focus is three sub Saharan African nations, the pessimism about change appears to be about populations in those regions. The views expressed by these key (scientific and managerial) personnel of Target Malaria and the Gates Foundation in scientific papers and mainstream news articles, arguably, could be understood as influenced by and advancing the Foundation's goals and approaches to public health problems of poverty in regions of the global South. In the interest of a deeper understanding of the Foundation's push for a market-based, high-tech approach (that entails substantial uncertainty and risks), it is useful to examine its normative commitments and its complicated relationship with the WHO.

The Gates Foundation and the WHO: Values, Risks, and Patented High-tech Technologies

The UN global health agency has recognized that structural, systemic factors play an important role in the spread of vector-borne diseases, including mosquito-borne diseases (WHO 2017b). The WHO is committed to the idea that the "enjoyment of the highest attainable standard of health (is) … one of the fundamental rights of every human being without distinction of race, religion, political belief, economic or social condition" (WHO 1995). Therefore, a leadership priority of the WHO is "advancing universal health coverage: enabling countries to sustain or expand access to all needed health services and financial protection, and promoting universal health coverage as a unifying concept in global health" (2017c: 27). The WHO has also expressed an unequivocal commitment to "[a]ddressing the social, economic and environmental determinants of health as a means to promote health outcomes and reduce health inequities within and between countries" (2017c: 27).

The WHO, in effect, advocates radical change that would raise living standards and end poverty as part of a comprehensive solution to the high incidence of mosquito and other vector-borne diseases among the poor. Implementing the global public health agency's proposal for structural, systemic change would mean that, among other things, nations would need to re-configure (political, economic, and legal) power relations between populations so that they are equitable (WHO 2008; Ottersen et al. 2014). Egalitarian power relations between groups are the key to substantive (rather than just formal) equality (Schwartzman 2006). Equitable power relationships among groups would, among other things, mean decent living standards for all groups, and comprehensive primary health care for everyone. That would reduce the disproportionate disease burden on the poor (Manderson et al. 2009).

This account of the WHO's stance does not mean to deny that the organization has been deeply influenced by neoliberal actors (Brown et al. 2006; Storeng 2014). The agency has made significant compromises to its historical mission because of dwindling financial support from its member nations. It has entered into collaborations with private donors, including NGOs who are interested in specific approaches or goals. The Gates Foundation is one such actor.

The degree to which the Gates Foundation is substantively responsive to the WHO's message of eliminating health disparities among and within nations by eradicating poverty is an open question.[19] The Foundation has expressed a desire to collaborate with governments of the global South to fill in the gaps in health care needs that they are unable to meet (Youde 2013). However, it does not believe that it has a responsibility to establish or maintain comprehensive primary care health systems that meet the health care needs of the poor at no cost (Chen 2006: 663). The Foundation has poured funds into vertical, targeted, market-based, high-tech interventions for diseases, such as malaria and tuberculosis, that have a substantial prevalence in certain parts of low- and middle-low-income countries (McNeil 2008). Patented GE mosquitoes with gene drives qualify as that type of intervention.

The Foundation's worldview can be heard in Bill Gates's keynote speech at the May 2005, 58th World Health Assembly where WHO member countries establish policy and vote on key issues:

> Some point to the better health in the developed world and say that we can only improve health when we eliminate poverty. And eliminating poverty is an important goal. But the world didn't have to eliminate poverty in order to eliminate smallpox—and we don't have to eliminate poverty before we reduce malaria. We do need to produce and deliver a vaccine.
>
> *(Gates 2005, quoted in Birn 2014:11)*

Historically, the WHO and the Gates Foundation have had different attitudes toward what qualifies as acceptable levels of risks from preventative or therapeutic interventions. The latter has displayed a much higher tolerance for the possibility of harm from preventative measures or therapies to vulnerable populations than the WHO. The global health agency's lower risk tolerance could be attributed to it being responsible to its member nations. The Gates Foundation has no such constraint on its actions. Its high tolerance of risk may account for its support of GE mosquitoes with gene drives even though, among other things, there is uncertainty about their epidemiological effectiveness and environmental risks.

It is illuminating to consider a case of conflict between the WHO and the Gates Foundation about acceptable risk from a preventative intervention for infants of poorer families in global South regions that have a substantial prevalence of malaria. Intermittent preventive treatment for infants (IPTi) entails giving infants doses of an older anti-malaria drug, Fansidar, at 2 months, 3 months, and 9 months. While early studies showed that it decreased malaria cases by about 25 percent, each dose provided protection for only a month (McNeil 2008). As Fansidar is a sulfa drug that may cause a rare but fatal reaction and as the incidence of Fansidar-resistant malaria was growing, WHO scientists had reservations about its use as a preventative public health measure for infants. McNeil (2008) notes the Gates Foundation and Gates supported scientists aggressively disagreed with the WHO scientists. The Foundation asked the US-based Institute of Medicine (IOM) to convene an expert committee to evaluate the evidence about IPTi with sulfadoxine-pyrimethamine (IPTi-SP). The IOM accepted the Foundation's request and the expert committee it marshaled concluded that "IPTi-SP is a valuable strategy for decreasing morbidity from malaria infections among infants who are at high risk because they reside in malaria-endemic areas in sub-Saharan Africa" (IOM 2008: 1). In 2010, the WHO recommended IPTi-SP's use through the Expanded Program on Immunization. It is not obvious why the WHO disregarded its own experts' opinion and accepted the IOM's recommendation.[20]

While in a crucial sense the Gates Foundation's and the WHO's fundamental values (including stance on acceptable level of risks for preventative and therapeutic measures) have been at odds, the former is the third largest donor to the global health agency, so it has considerable clout in global health policy formulation (Levich 2015: 717–718). During 2014–2017, the Gates Foundation gave funds in excess of US$1 billion to the WHO (Seitz & Martens 2017).

However, compared to the collective contribution of high-income nations, the Gates Foundation's expenditure on global health is relatively small. In 2012, the United Kingdom gave US$1.3 billion, and the United States gave US$7 billion for global health, while the Gates Foundation spent US$899 million (McGoey 2015: 154). Also, consider that during the 2004–2008 period, the US President's Emergency Plan for AIDS Relief spent US$18.1 billion on global HIV/AIDS programs, utterly outdistancing the Gates Foundation's total spending record on global health to date (McGoey 2015:154).

Given those figures, McGoey has asked "... why is Bill Gates a regular presence at summits such as the G20 in 2011? Why does the WHO consult with the Gates Foundation on major policy decisions?" (2015: 154). Such questions about the legitimacy of the Gates Foundation as a global health governance actor are worth asking because its priorities and agendas continue to be at odds with the expert opinion of public health researchers and practitioners and the concerns of affected nations (Sridhar, quoted in Bowman 2012).[21] For instance, the Gates Foundation has advocated for polio eradication even though public health experts and nations in the global South consider spending scarce resources on elimination efforts to be a misuse of monies and personnel given that the disease is mostly under control with a relatively limited number of cases per year in very few countries (McGoey 2015: 155).

Ruth Levine, a former Program Director of Global Development and Population at the William and Flora Hewlett Foundation, has identified the 'cost' of the domination of the global health arena by the Gates Foundation and other such well-endowed private philanthropic entities. It means a one-track approach to public problems based on opaque decisions, without appropriate regard for the trade-offs (Levine 2015). In light of Levine's analysis, the following question must be asked: what would it mean if the Gates Foundation throws its full weight in support of the use of GE mosquitoes with gene drives as the solution to (or as a necessary part of any effort to address) the high incidence of mosquito-borne diseases in the global South? For the affected nations, presumably, there would be the substantial cost of the patented biotechnology as well as ecological risks and uncertainties. Most importantly, there may be risks to vulnerable populations (such as infants, young children, pregnant women, breastfeeding women, persons with disabilities or compromised immune systems) that are affected by the use of GE mosquitoes.

McGoey traces the outsize influence of the Gates Foundation on global health policy to a "strong public profile, significant marketing and publicity expenditure" (2015: 154). The public relations campaigns ensure substantial positive media coverage of the private actor's global health agenda. It may explain why many global health organizations are eager to collaborate with the Foundation, "including bilateral donors, which collectively contribute ten times more resources to global health each year than does the BMGF (the Gates Foundation) itself" (Birn 2014: 10).[22]

The Gates Foundation's clout as a global health governance actor could also be partially attributed to it piggybacking on the efforts of the World Bank and the International Monetary Fund to undermine efforts in the global South to create or maintain publicly funded health care systems for the poor. The international financial institutions' neoliberal policies do not permit nations that are their debtors to create or maintain state-funded comprehensive primary care systems for their socio-economically marginalized populations that was envisioned in the 1978 Alma-Ata Declaration.[23] That places at significant risk the well-being and even survival of those who are unable to purchase health care services and goods. To meet the needs of their vulnerable populations, debtor nations are compelled to rely on non-state actors, including private philanthropic organizations. As a result, organizations like the Gates Foundation have acquired the power to decide which health care need they will provide for, by what means, to what extent, and the conditions under which they will provide assistance to those countries. Given the Gates Foundation's clout on the global health governance stage, its aggressive push (via Target Malaria) for the use of GE mosquitoes with gene drives is not something to be taken lightly. The two scientific articles analyzed here could be read as part of that hard sell of the biotechnology.

The complicated relationship, including value and agenda conflict, between the WHO and the Gates Foundation should be understood as shaped by the latter's neoliberal orientation. It could also be used to make sense of its support of the Target Malaria initiative that aims to use a patented, high risk, high-tech approach to eradicate malaria. The Gates Foundation's work has been labeled "philanthrocapitalism" by global health governance scholars.[24] However, it would be more appropriately categorized as "neoliberal-philanthropism." There are significant similarities between philanthrocapitalism and neoliberal-philanthropism, but the differences between them are marked. Prior to the neoliberal shift of the United States, the Rockefeller Foundation was the embodiment of private philanthropy based on principles of capitalism. Birn (2017) describes it as a model of philanthropy that "… infuses business principles into philanthropy (proffering handsome investment returns), essentially justifies wealth accumulation on the backs of ordinary people." It played a defining role in shaping the public health agenda during the nineteenth century, advancing projects that served its corporate goals and US interests. For example, Birn (2014: 4–5) examines the Rockefeller Foundation's interest in controlling the incidence of yellow fever among French and Caribbean workers so that the construction of the Panama Canal could be completed. However, the Rockefeller Foundation did not attempt to undermine the public health system of nations or shift control of health provisioning into the hands of the private sector, including business-modeled NGOs (Birn 2014: 16).

In contrast to philanthrocapitalism (aimed at health care), neoliberal-philanthropism (focused on health care) aims to give control over the provisioning of health care services and goods to corporations and (neoliberal) NGOs so that they get to decide which health care needs they will meet and how they will address them. That agenda is rooted in the neoliberal conviction that the democratic state is an inefficient and unfair allocator of goods and services and thus it must defer to the wisdom of the market. Proponents of neoliberalism believe that the market (unlike the democratic state) fairly rewards hard working, enterprising individuals who are rational risk-takers; it does not coddle those who lack initiative and a work ethic by providing them with social protections (Bonanno 1998, p.233: 227; Harvey 2005). From a neoliberal perspective,[25] arguably, health care and basic living amenities (such as safe water supplies, sanitation facilities, waste treatment processes, and regular garbage disposal services) that are funded by the democratic state for socio-economically marginalized populations are a form of undeserved social protection. That may explain why neoliberal-philanthropic organizations tend to prefer market-based, high-tech technologies as solutions to public health problems (that disproportionately affect the poor), rather than projects that entail ensuring safe water, sanitation facilities, funding regular garbage disposal services, comprehensive primary healthcare as part of state funded national health systems in socio-economically marginalized regions of the global South, *at no cost to those communities and nations.*

A characteristic of (neoliberal) philanthropic organizations is their commitment to "social entrepreneurship," i.e. generate profits (for yourself or others) as you do good and earn accolades (McGoey 2015). Ideally, the corporate benefactors or partners of the neoliberal-philanthropic organization would be the ones who benefit in some way.[26] The Gates Foundation's support of patent rights for pharmaceuticals is well-known and has been subject to serious criticism (Birn 2014; McGoey 2015: 198–202). The Foundation has attempted to deflect those criticisms by divesting itself of direct pharmaceutical holdings (Birn 2014: 15). However, Warren Buffett is one of the three trustees of the Foundation (Bill and Melinda Gates are the other two). The Gates Foundation has substantial investment in Buffett's Berkshire Hathaway, which has holdings in Johnson & Johnson, Sanofi Aventis, and other pharmaceutical companies (Birn 2014: 15). As discussed earlier, GE mosquitoes with gene drives are a *patented* public health measure.

In 2013, the Gates Foundation in conjunction with JP Morgan Chase created the Global Health Investment Fund (GHIF). In 2014, the Fund's website stated that its goal was to help "bring about significant improvements in the treatment and prevention of disease, and in family planning, and the reduction of maternal and child mortality, *along with the prospect of a net financial return for investors*" (GHIF 2014, quoted in Erikson 2015: S311 (Erikson's italics)). There were 11 initial investors in the Fund, including the Gates Foundation, JP Morgan Chase, the World Bank, Merck, Pfizer Foundation, and GlaxoSmithKline (Erikson 2015: S311).

According to the current (i.e. 13 October 2019) GHIF website, the Fund finances the development of:

> drugs, vaccines, diagnostics and *other interventions against diseases that disproportionately burden low- and middle-income countries. GHIF supports late-stage innovations for public health challenges such as malaria*, pre-eclampsia, *cholera*, HIV *and river blindness, with an emphasis on infectious diseases* and maternal/infant health issues *that cause significant morbidity and mortality in resource-limited settings ... Products with 'dual market' potential are of greatest interest; i.e., those that will have a clear impact on public health in developing countries but also have value in high-income countries* (my italics).
>
> *(GHIF 2016)*

GE mosquitoes with gene drives are a market-based, technological solution for a public health problem that disproportionately affects (some socio-economically marginalized communities in certain) low- and middle-income countries.[27] Moreover, high-income countries that are fearful of the spread of mosquito-borne diseases, such as dengue and Zika, could be future markets for the patented, high-tech intervention. As a product with "dual market potential," GE mosquitoes may be of the "greatest interest" to the Gates Foundation and other investors in the GHIF. The hard sell of GE mosquitoes with gene drives by the Foundation (and Target Malaria) as the solution to the high incidence of mosquito-borne diseases in sub Saharan African countries could be read, arguably, as an effort to position it as key solution for any and all countries that are worried about the spread of mosquito-borne diseases within their borders given globalization, global warming, environmental degradation (including de-forestation), increasing concentration of socio-political-economic power in the hands of a few, and inadequate, fragmented, and underfunded public health care systems.

Conclusion

This chapter analyzed the hard sell of GE mosquitos with gene drives as the solution to the high incidence of mosquito-borne diseases in parts of the global South. The hard sell overlooks (or underplays) the structural, systemic factors that are partially responsible for the public health problem that disproportionately affects the poor. The epistemic credence (including the assumption of neutrality) generally afforded to scientists (and especially scientists with ties to an organization that has an internationally recognized brand as a key global health governance actor) means that scientific papers that aggressively push GE mosquitoes with gene drives as *the* solution could play a role in convincing the publics, governments, health governance actors, funders, and the larger scientific community of the rightness of their approach. Given the Gates Foundation's relationship to the WHO and the World Bank, the possibility of the hard sell (of GE mosquitoes with gene drives) getting uptake in policy circles needs to be taken seriously.

The aggressive sell of GE mosquitoes with gene drives could undermine efforts to bring about political and economic changes that would improve the general life conditions of the populations in the global South afflicted or affected by vector-borne diseases. Among other things, that is significant epidemiologically and ethico-politically because the structural, systemic factors that are partially responsible for the high incidence of mosquito-borne diseases also contribute to the substantial prevalence of other infectious diseases in those regions.

This chapter's intersectional analysis is attentive to the gender-class-race-nationality significance of the hard sell of patented GE mosquitoes with gene drives (under conditions of neoliberalism that spans across borders). It should be read as signaling the need for research programs that track the pervasiveness of biologizing scientific narratives about mosquito-borne (*and* other infectious) diseases that disproportionately affect the poor in certain regions of the South. Moreover, this chapter should be considered an argument for research projects that map the influence of such biologizing scientific narratives on researchers (who study those public health problems), funders, policymakers, and regulatory authorities. (For instance, do such scientific narratives impact the decision of regulatory bodies to greenlight high-tech solutions to public health problems that primarily affect the socio-economically marginalized?) Last, but not least, it should be viewed as making the case for ethics analyses *and* empirical research projects that trace the impact of such scientific narratives on the lives of the populations afflicted or affected by those diseases, including the girls and women from socio-economically marginalized communities responsible for the care of sick family and community members.

Acknowledgment

The author would like to thank the editors for their thoughtful comments on the drafts of this chapter and the multiple time extensions.

Related chapters: 8, 9, 11, 15, 22.

Notes

1 See Macias et al. (2017) for a detailed account of the development of human made (i.e. synthetic) gene drives modeled on gene drives found in nature. For a description of gene drives found in nature see Burt and Trivers (2009).

2 For a detailed account of the complexity of the problem, see Wilke et al. (2018).

3 See Alphey (2016) for discussion of studies demonstrating this possibility.

4 To appreciate the complexity of ecosystems as systems with emergent traits and the challenge of predicting how they might act, it is useful to consider the recent case of mass die-off of the critically endangered saiga antelope in the Kazakhstan steppe. In Spring 2015, in the span of a week, 200,000 saiga antelopes (approximately 60% of the world's population) died. Researchers identified the responsible complex of factors as an increase in temperature to 37C and an increase in humidity above 80%, which stimulated the bacterium *Pasteurella multocida* that usually exists harmlessly in the tonsils of "some, if not all, of the antelopes … to pass into the bloodstream causing haemorrhagic septicaemia, or blood poisoning" in the antelopes (Derbyshire 2018). In other words, the bacterium that was usually harmless to the saiga antelopes became a devastating pathogen (i.e., an emergent trait) under a specific set of circumstances.

5 For an overview of possible risks from the environmental release of GE organisms with gene drives, see Sirinathsinghji (2019: 4–5).

6 The tetracycline dependency trait is termed a conditioned lethality trait because it is only lethal to the insect (or larvae) that has the synthetic genetic sequence *if* tetracycline in sufficient quantity is not available to the organism with the modification.

7 Female GE mosquitoes are not intentionally released because like their wild-type counterpart they can be disease vectors.

8 Some of the same issues, presumably, may hold for patented mosquitoes modified with *wolbachia* bacteria that are meant to reduce the transmission of dengue and which are regulated as a type of biopesticide by the US Environmental Protection Agency (EPA).

9 In the United States, regulatory jurisdiction is dependent on how a developer classifies its product. If a bioproduct is classified as a pesticide (rather than a means for preventing or reducing the incidence of mosquito-borne diseases in humans), it is regulated by the EPA and not the FDA. If a regulatory agency decides that it does not have authority over a certain type of product, it will send a letter to the developer and ask it to consult with the appropriate agency.

10 See Schäfer 2014, for instance, for a detailed account of the national and international socio-political and economic factors responsible for the failure of standard therapies in the context of Burkina Faso.

11 While Venezuela is classified as a middle-income country, its circumstances have changed drastically in recent years.

12 For a comprehensive account of the social determinants of health and disease, see Kelly and Doohan (2012).

13 This notion of structural, systemic factors draws from Young's (1990) analysis of the causal elements of oppression.

14 See Youde (2013).

15 See, for instance, Burki (2013).

16 The news article was about Target Malaria importing 5,000 genetically modified mosquito eggs to its laboratory in Burkina Faso in preparation for the eventual environmental release of GE mosquitoes with gene drives (that are created using gene editing) in that country.

17 See Schäfer's (2014) carefully detailed analysis of the social determinants of the high incidence of malaria in parts of Burkina Faso.

18 See Schäfer (2014) for an account of the combination of biological and social factors responsible for development of resistance to mono-therapies for malaria and the high prevalence of the disease in the context of Burkina Faso.

19 See Kickbusch and Szabo (2014) on the complexities of governance in the global public health domain.

20 See McGoey (2015) (especially, pages 194–197) for other cases of conflict between the WHO and the Gates Foundation.

21 Also see Harman's analysis of the substantive lack of legitimacy of the Gates Foundation as a global health governance actor. Moreover, see Harman (2016) and Clarke (2019) on proposals for reforming the Gates Foundation and other such private philanthropic organizations so that they are accountable to and responsive to the constituencies they seek to help.

22 Also see Youde (2013) on the ability of the Foundation to attract private and public organizations to support its agenda and priorities.

23 See, for instance, Schrecker (2016) and Sridhar et al. (2017). For a history of the development of the Alma-Ata Declaration, see Cueto (2004). For an account of the subversion of the goals of the Alma-Ata Declaration by the World Bank's neoliberal policies and programs, see Colgan (2002).

24 See, for instance, Youde (2013) and Birn (2014).

25 There are many forms of neoliberalism (see, Cahill et al. 2018).

26 Arguably, in this regard, neoliberal-philanthropic entities stand in contrast to philanthropic-capitalist organizations. For instance, historically, the Rockefeller Foundation aimed to improve the reputation of Rockefeller and his business enterprises and make amends for the harm that he inflicted on his employees (McGoey 2015).

27 It is worth noting that cholera and river blindness are both diseases of poverty. According to the US Centers for Disease Control and Prevention (2018), cholera epidemics are attributable to water or food contaminated with the feces of a person infected by the cholera bacterium and "the disease can spread rapidly in areas with inadequate treatment of sewage and drinking water" (Centers for Disease Control and Prevention undated). River blindness also known as onchocerciasis is another vector-borne disease that is classified as a neglected tropical disease that disproportionately compromises marginalized populations in poorer regions of the global South (WHO 2020).

References

Allotey, P., Gyapong, M., & UNICEF. (2005). "The gender agenda in the control of tropical diseases: a review of current evidence" (No. TDR/STR/SEB/ST/05.1). Geneva: World Health Organization, available at: https://apps.who.int/iris/bitstream/handle/10665/69067/TDR_STR_SEB_ST_05.1.pdf

Allotey, P., Reidpath, D.D., & Pokhrel, S. (2010) "Social Sciences Research in Neglected Tropical Diseases 1: The Ongoing Neglect in the Neglected Tropical Diseases," *Health Research Policy and Systems*, 8(1), p. 32. https://link.springer.com/article/10.1186/1478–4505-8–32.

Alphey, L. (2016) "Can CRISPR-Cas9 Gene Drives Curb Malaria?," *Nature Biotechnology*, 34 (2), p. 149.

Birn, A.E. (2014) "Philanthrocapitalism, Past and Present: The Rockefeller Foundation, the Gates Foundation, and the Setting(s) of the International/Global Health Agenda," *Hypothesis*, 12(1), p. e8.

Birn, A.E. (2017) "Philanthropy: The Politics of Giving," *Nature*, 544(7648), p. 31.

Birn, A.E., Pillay, Y., & Holtz, T.H. (2017) *Textbook of Global Health*, New York: Oxford University Press.

Biswas, J. (2018). "Scientists will release genetically-engineered mosquitoes in Africa to fight

Malaria," ABC News, September 14, available at: https://www.abc15.com/news/national/scientists-will-release-genetically-engineered-mosquitoes-in-africa-to-fight-malaria

Bonanno, A. (1998). "Liberal democracy in the global era: Implications for the agro-food sector," *Agriculture and Human Values*, 15(3), 223–242.

Bowman, A. (2012) "The Flip Side to Bill Gates' Charity Billions," *New Internationalist*, April, available at: http://newint.org/features/2012/04/01/bill-gatescharitable-giving-ethics

Brown, T.M., Cueto, M., & Fee, E. (2006) "The World Health Organization and the Transition from 'International' to 'Global' Public Health," *American Journal of Public Health*, 96(1), pp. 62–72.

Burki, T.K. (2013) "Malaria and Malnutrition: Niger's Twin Crises," *The Lancet*, 382(9892), pp. 587–588.

Burt, A. (2003). "Site-Specific Selfish Genes as Tools for the Control and Genetic Engineering of Natural Populations," *Proceedings of the Royal Society of London. Series B: Biological Sciences*, 270(1518), 921–928.

Burt, A., & Crisanti, A. (2018) "Gene Drive: Evolved and Synthetic," *ACS Chemical Biology*, 13(2), pp. 343–346.

Burt, A., & Trivers, R. (2009) *Genes in Conflict: The Biology of Selfish Genetic Elements*, Boston: Harvard University Press.

Cahill, D., Cooper, M., Konings, M., & Primrose, D. (2018) *The SAGE Handbook of Neoliberalism*, Thousand Oaks: Sage.

Callaway, E. (2017) "US Defence Agencies Grapple with Gene Drives," *Nature News*, 547(7664), p. 388.

Centers for Disease Control and Prevention. (2018). "Cholera—Vibrio Cholerae Infection, General Information, Frequently Asked Questions: How Does a Person Get Cholera?," available at: https://www.cdc.gov/cholera/general/index.html

Champer, J., Buchman, A., & Akbari, OS. (2016) "Cheating Evolution: Engineering Gene Drives to Manipulate the Fate of Wild Populations," *Nature Reviews Genetics*, 17(3), pp. 146–159.

Chen, I. (2006) "Thinking Big about Global Health," *Cell*, 124(4), pp. 661–663.

Clarke, G. (2019) "The New Global Governors: Globalization, Civil Society, and the Rise of Private Philanthropic Foundations," *Journal of Civil Society*, 15(3), pp. 197–213.

Colgan, A.L. (2002) *Hazardous to health: The World Bank and IMF in Africa*, Africa Action Position Paper, April.

Cronon, W. (1992) "A Place for Stories: Nature, History, and Narrative," *The Journal of American History*, 78(4), pp. 1347–1376.

Cueto, M. (2004) "The Origins of Primary Health Care and Selective Primary Health Care," *American Journal of Public Health*, 94, pp. 1864–1874.

Cueto, M. (2013) "A Return to the Magic Bullet? Malaria and Global Health in the Twenty First Century," in Biehl, J. and Petryna, A. (eds.), *When People Come First: Critical Studies in Global Health*, Princeton: Princeton University Press, pp. 30–53.

Degarege, A., Fennie, K., Degarege, D., et al. (2019) "Improving Socioeconomic Status May Reduce the Burden of Malaria in Sub Saharan Africa: A Systematic Review and Meta-analysis." *PLoS One*, 14(1), p. e0211205.

Derbyshire, D. (2018) "The Terrifying Phenomenon that Is Pushing Species towards Extinction," *The Guardian*, available at: https://www.theguardian.com/environment/2018/feb/25/mass-mortality-events-animal-conservation-climate-change

Douglas, H. (2009) *Science, Policy, and the Value-free Ideal*, Pittsburgh: University of Pittsburgh Press.

Erikson, S.L. (2015) "Secrets from Whom? Following the Money in Global Health Finance," *Current Anthropology*, 56(S12), pp. S306–S316.

Garchitorena, A., Sokolow, S.H., Roche, B., et al. (2017) "Disease Ecology, Health and the Environment: A Framework to Account for Ecological and Socio-economic Drivers in the Control of Neglected Tropical Diseases," *Philosophical Transactions of the Royal Society B: Biological Sciences*, 372(1722), p. 20160128.

Global Health Investment Fund (2014) "An Audience with Bill Gates," available at: http://ghif.com/ (accessed August 31, 2014 (by SL Erikson)).

Global Health Investment Fund (2016) "About," available at: http://www.ghif.com/

Gould, F., Huang, Y., Legros, M. et al. (2008) "A Killer–Rescue System for Self-limiting Gene Drive of Anti-pathogen Constructs," *Proceedings of the Royal Society B: Biological Sciences*, 275(1653), pp.2823–2829.

Grillet, M.E., Villegas, L., Oletta, J.F., et al. (2018) "Malaria in Venezuela Requires Response," *Science* 359(6375), pp. 528–528.

Hammond, A.M., & Galizi, R. (2017) "Gene Drives to Fight Malaria: Current State and Future Directions," *Pathogens and Global Health*, 111(8), pp. 412–423.

Haraway, D. (1988) "Situated Knowledges: The Science Question in Feminism and the Privilege of Partial Perspective." *Feminist studies*, 14(3), 575–599.

Harman, S. (2016) "The Bill and Melinda Gates Foundation and Legitimacy in Global Health Governance," *Global Governance*, 22, p. 349.

Harris, A.F., Nimmo, D., McKemey, A.R., Kelly, N., Scaife, S., Donnelly, C.A., Beech, C., Petrie, W.D., & Alphey, L. (2011) "Field Performance of Engineered Male Mosquitoes," *Nature Biotechnology*, 29, pp. 1034–1037.

Harvey, D. (2007). *A Brief History of Neoliberalism*. New York: Oxford University Press.

Harvey-Samuel, T., Ant, T., & Alphey, L. (2017) "Towards the Genetic Control of Invasive Species," *Biological Invasions*, 19(6), pp. 1683–1703.

Hausmann-Muela, S., & Eckl, J. (2015) "Re-imagining Malaria–A Platform for Reflections to Widen Horizons in Malaria Control," *Malaria Journal*, 14(1), p. 180.

Institute of Medicine (IOM) (2008) *Committee on the Perspectives on the Role of Intermittent Preventive Treatment for Malaria in Infants, Assessment of the Role of Intermittent Preventive Treatment for Malaria in Infants: Letter Report*, Washington, DC., available at: http://www.nap.edu/catalog.php?record_id=12180

Kelly, M.P., & Doohan, E. (2012) "The Social Determinants of Health," in Merson, M. H., Black, R.E., & Mills, A. J. (eds.), *Global Health: Diseases, Programs, Systems, and Policies*, 3rd ed., Burlington: Jones & Bartlett Learning, pp. 75–114.

Kickbusch, I., & Szabo, MMC. (2014) "A New Governance Space for Health," *Global Health Action*, 7(1), p. 23507.

Levich, J. (2015) The Gates Foundation, Ebola, and Global Health Imperialism," *American Journal of Economics and Sociology*, 74(4), pp. 704–742.

Levine, R. E. (2015). "Power in Global Health Agenda-Setting: The Role of Private Funding: Comment on 'Knowledge, Moral Claims and the Exercise of Power in Global Health,'" *International Journal of Health Policy and Management*, 4(5), p. 315.

Longino, H. (1990) *Science as Social Knowledge: Values and Objectivity in Scientific Inquiry*, Princeton: Princeton University Press.

Macias, V.M., Ohm, J.R., & Rasgon, J.L. (2017) "Gene Drive for Mosquito Control: Where Did It Come from and Where Are We Headed?" *International Journal for Environ Research and Public Health*, 14(9), p. 1006.

Manderson, L., Aagaard-Hansen, J., Allotey, P., et al. (2009) "Social Research on Neglected Diseases of Poverty: Continuing and Emerging Themes," *PLoS Neglected Tropical Diseases*, 3(2), e332.

McGoey, L. (2015) *No Such Thing as a Free Gift: The Gates Foundation and the Price of Philanthropy*, Brooklyn: Verso Books.

McNeil, D. (2008) "Gates Foundation's Influence Criticized," *New York Times*, February 16 available at: https://www.nytimes.com/2008/02/16/science/16malaria.html

Meghani, Z. (2019) "Autonomy of Nations and Indigenous Peoples and the Environmental Release of Genetically Engineered Animals with Gene Drives," *Global Policy*, 10(4), pp. 554–568.

Meghani, Z., & Boëte, C. (2018) "Genetically Engineered Mosquitoes, Zika and Other Arboviruses, Community Engagement, Costs, and Patents: Ethical Issues," *PLoS Neglected Tropical Diseases*, 12(7), p. e0006501. https://doi.org/10.1371/journal.pntd.0006501.

Mfueni Bikundi, E., & Coppieters, Y. (2017) "Importance of Risk Factors Associated with Malaria for Sub-Saharan African Children," *International Journal of Environmental Health Research*, 27(5), pp. 394–408.

National Academy of Sciences, Engineering, and Medicine (NASEM) (2016) *Gene Drives on the Horizon: Advancing Science, Navigating Uncertainty, and Aligning Research with Public Values*, Washington, DC: National Academy Press.

Nelson, L. (2010). *Who knows: From Quine to a Feminist Empiricism*. Philadelphia: Temple University Press.

Nickel, R. & Gillam, C. (2015). "Market Turbulence or Not, North American Investors Plow into Farm Tech. Technology News," *Reuters*. September 15th.

Norris, A.L., Lee, S.S., Greenlees, K.J., et al. (2020) "Template Plasmid Integration in Germline Genome-edited Cattle," *Nature Biotechnology*, 38(2), pp. 163–164.

Ong, A. (2006). *Neoliberalism as Exception: Mutations in Citizenship and Sovereignty*. London: Duke University Press.

Ong, A. (2016) *Fungible Life: Experiment in the Asian City of Life*, Durham: Duke University Press.

Ono, R., Yashuhiko, Y., Aisaki, K., et al. (2019) "Exosome-mediated Horizontal Gene Transfer Occurs in Double-strand Break Repair During Genome Editing," *Communications Biology*, 2, p. 57.

Ottersen, O.P., Dasgupta, J., Blouin, C., et al. (2014) "The Political Origins of Health Inequity: Prospects for Change," *The Lancet*, 383(9917), pp. 630–667.

Oxitec (2016) "Oxitec's Vector Control Solution: A Paradigm Shift in Mosquito Control," available at: https://www.dshs.state.tx.us › docs › OxitecsVectorControlSolution

Pardo, D. (2014) "The Malaria Mines of Venezuela," *BBC Mundo*, available at: http://www.bbc.com/news/health-28689066.

Rasgon, J.L. (2009) "Multi-locus Assortment (MLA) for Transgene Dispersal and Elimination in Mosquito Populations," *PLoS One*, 4, p. e5833.

Rathgeber, E.M., & Vlassoff, C. (1993) "Gender and Tropical Diseases: A New Research Focus," *Social Science & Medicine*, 37(4), pp. 513–520.

Regalado, A. (2016a) "Bill Gates Doubles His Bet on Wiping Out Mosquitoes with Gene Editing," *MIT Technology Review*, available at: https://www.technologyreview.com/s/602304/bill-gates-doubles-his-bet-on-wiping-out-mosquitoes-with-gene-editing/

Regalado, A. (2016b). "The extinction invention," *MIT Technology Review*, available at: https://www.technologyreview.com/s/601213/the-extinction-invention/

Robbins, G., & Fikes, B.J. (2016) "India's Tata Gives UCSD $70M in Hot Area of Genetics," *The San Diego Union Tribune*, available at: http://www.sandiegouniontribune.com/news/science/sd-me-tata-gift-20161018-story.html

Schäfer, F. (2014) *Diagnosis and Therapy of Malaria Under the Conditions of a Developing Country-the example of Burkina Faso* (dissertation), available at: https://opus.bibliothek.uni-wuerzburg.de/opus4-wuerzburg/frontdoor/deliver/index/docId/10286/file/Schaefer_Frauke_Malaria.pdf

Schrecker, T. (2016) "Neoliberalism and Health: The Linkages and the Dangers," *Sociology Compass*, 10(10), pp. 952–971.

Schwartzman, L.H. (2006) *Challenging Liberalism: Feminism as Political Critique*, College Park: Penn State Press.

Seitz, K., & Martens, J. (2017) "Philanthrolateralism: Private Funding and Corporate Influence in the United Nations," *Global Policy*, 8, pp. 46–50.

Sepúlveda Carmona, M. (2013) "Promotion and Protection of Human Rights: Human Rights Situations and Reports of Special Rapporteurs and Representatives," in *Report of the Special Rapporteur on Extreme Poverty and Human Rights*, United Nations, General Assembly, Sixty-eighth session, Item 69 (c) of the provisional agenda, available at SSRN 2534341.

Servick, K. (2016) "Brazil Will Release Billions of Lab-grown Mosquitoes to Combat Infectious Disease: Will it work?" *Science Magazine*, October 13, doi:10.1126/science.aal0253

Shaw, W.R., & Catteruccia, F. (2019) "Vector Biology Meets Disease Control: Using Basic Research to Fight Vector-borne Diseases," *Nature Microbiology*, 4(1), pp. 20–34.

Sirinathsinghji, E. (2019) "Transferring the Laboratory to the Wild: An Emerging Era of Environmental Genetic Engineering," *Biosafety Briefing*, ThirdWorldNet, November.

Storeng, K.T. (2014) "The GAVI Alliance and the 'Gates approach' to Health System Strengthening," *Global Public Health*, 9(8), pp. 865–879.

TallBear, K. (2013) *Native American DNA: Tribal Belonging and the False Promise of Genetic Science*, St. Paul: University of Minnesota Press.

US Food and Drug Administration (FDA) (2016) *Environmental Assessment for Investigational Use of Aedes aegypti OX513A*, August 5.

Wilke, A. B., Beier, J. C., & Benelli, G. (2018). "Transgenic Mosquitoes–Fact or Fiction?," *Trends in Parasitology*, 34(6), 456–465.

World Health Organization (1995) *Constitution of the World Health Organization*. Geneva, available at: https://apps.who.int/iris/bitstream/handle/10665/121457/em_rc42_cwho_en.pdf

World Health Organization (2008). *Commission on Social Determinants of Health. Closing the Gap in a Generation: Health Equity Through Action on the Social Determinants of Health. Final report of the Commission on Social Determinants of Health.* Geneva.

World Health Organization (2017a) *Design of Epidemiological Trials for Vector Control Products: Report of a WHO Expert Advisory Group*, Geneva, 24–25 April.

World Health Organization (2017b) *Fourth WHO Report on Neglected Tropical Diseases: Integrating Neglected Tropical Diseases Into Global Health and Development*, Geneva.

World Health Organization (2017c) *Twelfth General Programme of Work: Not Merely the Absence of Disease (2014–2019)*, Geneva.

World Health Organization (2020) "Prevention, Control and Elimination of Onchocerciasis: Onchocerciasis," available at: https://www.who.int/onchocerciasis/control/en/

Youde, J. (2013) "The Rockefeller and Gates Foundations in Global Health Governance," *Global Society*, 27(2), pp. 139–158.

Young, I.M. (1990). *Justice and the Politics of Difference*. Princeton: Princeton University Press.

INDEX